ADVANCES IN BOTANICAL RESEARCH

Series Editors

Jean-Pierre Jacquot
Professeur, Membre de L'Institut Universitaire de France, Unité Mixte de Recherche INRA, UHP 1136 "Interaction Arbres Microorganismes", Université de Lorraine, Faculté des Sciences, Vandoeuvre, France

Pierre Gadal
Professor honoraire, Université Paris-Sud XI, Institut Biologie des Plantes, Orsay, France

ADVANCES IN BOTANICAL RESEARCH

Genome Evolution of Photosynthetic Bacteria

VOLUME SIXTY SIX

Advances in BOTANICAL RESEARCH

Genome Evolution of Photosynthetic Bacteria

Volume Editor

J. THOMAS BEATTY

Department of Microbiology and Immunology,
University of British Columbia,
Vancouver, BC, Canada

AMSTERDAM • BOSTON • HEIDELBERG • LONDON
NEW YORK • OXFORD • PARIS • SAN DIEGO
SAN FRANCISCO • SINGAPORE • SYDNEY • TOKYO

Academic Press is an imprint of Elsevier

Academic Press is an imprint of Elsevier
525 B Street, Suite 1900, San Diego, CA 92101-4495, USA
225 Wyman Street, Waltham, MA 02451, USA
32 Jamestown Road, London NW1 7BY, UK
The Boulevard, Langford Lane, Kidlington, Oxford, OX5 1GB, UK
Radarweg 29, PO Box 211, 1000 AE Amsterdam, The Netherlands

First edition 2013

ISBN: 978-0-12-397923-0

For information on all Academic Press publications
visit our Web site at store.elsevier.com

ISSN: 0065-2296

Printed and bound in USA
13 14 15 10 9 8 7 6 5 4 3 2 1

CONTENTS

PREFACE

The best evidence indicates that biological photosynthesis arose on Earth more than 3.5 × 109 years before the present, and that the earliest photosynthetic system was anoxygenic, meaning that the photochemical process did not split water (Blankenship, 2010). The bacterial subjects of the chapters in this book perform anoxygenic photochemistries, catalysed by evolutionarily related pigment–protein complexes called reaction centres. Although the most rigorous definition of the word 'photosynthesis' means harvesting of light energy to synthesise organic carbon from carbon dioxide, this word is commonly used to generally describe chlorophyll-based processes that drive electron transfer reactions in living cells. The term 'phototrophy' is sometimes favoured as a general descriptor of the harvesting of light energy, to include organisms such as the aerobic anoxygenic phototrophic bacteria that synthesise little organic carbon from carbon dioxide. However, there are also phototrophs that contain rhodopsin-related pigments, not chlorophyll, and use light energy to drive proton transport pathways directly. Finally, the term 'photosynthetic bacteria' has been used for more than half a century to describe anoxygenic phototrophs that use chlorophylls to harvest light, sometimes referred to as purple or green (phototrophic or photosynthetic) bacteria. These considerations led to the compromise represented by the title of this book, *Genome Evolution of Photosynthetic Bacteria*. Because of the variety of terms used in the literature, contributors were free to wander between the photosynthetic/phototrophic terminology, and so the reader should keep this flexibility in mind when reading the chapters in this volume.

The contributions to this book review data ranging from fossil evidence through genomes of classes or physiological groups of existing organisms, to studies of particular types of proteins, metabolic processes, and cellular responses to environmental cues. Some details may appear to be in conflict because different approaches or interpretations may lead to different conclusions, but such differences are thought to reflect the vibrant, active nature of evolutionary research.

The first chapter (Mulkidjanian & Galperin, 2013) uses a broad brush to start before biological photosynthesis, by reviewing the evidence for abiogenic photosynthesis, and moving on to use gene sequences to suggest a geochemical context for the evolution of photosynthesis genes in six key bacterial phyla. In Chapter 2, Gupta (2013) uses his original gene insert/deletion (indel) analysis of sequences to arrive at plausible origins of photosynthesis genes.

The Heliobacterial genome is evaluated by Sattley and Swingley (2013) in Chapter 3, and Bryant and Liu (2013) provide a wide-ranging coverage of members of the phylum Chlorobi in Chapter 4, and delve deeply into evolutionary scenarios relating to the origins of pigment biosynthesis and reaction centre genes.

The so-called purple bacteria are the focus of the remaining chapters. Nagashima and Nagashima (2013) compare photosynthesis gene clusters and evaluate the likelihood of horizontal gene transfer in Chapter 5. In Chapter 6, Gomelsky and Zeilstra-Ryalls (2013) describe changes in the genome-wide transcriptome of *Rhodobacter sphaeroides* in response to changes in the concentration of molecular oxygen and light intensity. The evolutionary history of light-harvesting genes is deciphered by Henry and Cogdell (2013) in Chapter 7, and Willison and Magnin (2013) review the function and evolution of endogenous plasmids in Chapter 8. The evolution of bacteriophytochromes is the focus of Papiz and Bellini's (2013) Chapter 9, whereas in Chapter 10, Zappa and Bauer (2013) analyse three *Rhodobacter* genomes to deduce strategies for maintaining iron homeostasis.

The last three chapters are devoted to the interesting group of purple bacteria known as the aerobic anoxygenic phototrophs. Yurkov and Hughes (2013) provide a broad introduction to the group in Chapter 11, and in Chapter 12, Zheng, Koblížek, Beatty, and Jiao (2013) use genome sequences to suggest that photosynthesis genes have been acquired by horizontal gene transfer in one species, and lost from the phototrophic ancestor of a present-day chemotroph. In the final chapter, Koblížek, Zeng, Horák, and Oborník (2013) perform an extensive analysis of the genomes of several members of the *Roseobacter* clade, and conclude that in progenitors of these species there has been regressive loss of carbon dioxide fixation genes, followed by loss of photosynthesis genes, in pathways leading from photoautotrophs through photoheterotrophs to chemoheterotrophs.

REFERENCES

Blankenship, R. E. (2010). Early evolution of photosynthesis. *Plant Physiology*, *154*, 434–438.

Bryant, D. A., & Liu, Z. (2013). Green bacteria: insights into green bacterial evolution through genomic analyses. *Advances in Botanical Research*, *66*, 99–150.

Gomelsky, M., & Zeilstra-Ryalls, J. H. (2013). The living genome of a purple nonsulfur photosynthetic bacterium: overview of the *Rhodobacter sphaeroides* transcriptome landscapes. *Advances in Botanical Research*, *66*, 179–203.

Gupta, R. S. (2013). Molecular markers for photosynthetic bacteria and insights into the origin and spread of photosynthesis. *Advances in Botanical Research, 66*, 37–66.

Henry, S. L., & Cogdell, R. J. (2013). The evolution of the purple photosynthetic bacterial light-harvesting system. *Advances in Botanical Research, 66*, 205–226.

Koblížek, M., Zeng, Y., Horák, A., & Oborník, M. (2013). Regressive evolution of photosynthesis in the *Roseobacter* clade. *Advances in Botanical Research, 66*, 385–405.

Mulkidjanian, A. Y., & Galperin, M. Y. (2013). A time to scatter genes and a time to gather them: evolution of photosynthesis genes in bacteria. *Advances in Botanical Research, 66*, 1–35.

Nagashima, S., & Nagashima, K. V. P. (2013). Comparison of photosynthesis gene clusters retrieved from total genome sequences of purple bacteria. *Advances in Botanical Research, 66*, 151–178.

Papiz, M., & Bellini, D. (2013). Evolution of bacteriophytochromes in photosynthetic bacteria. *Advances in Botanical Research, 66*, 267–288.

Sattley, W. M., & Swingley, W. D. (2013). Properties and evolutionary implications of the heliobacterial genome. *Advances in Botanical Research, 66*, 67–97.

Willison, J. C., & Magnin, J.-P. (2013). Role and evolution of endogenous plasmids in photosynthetic bacteria. *Advances in Botanical Research, 66*, 227–265.

Yurkov, V., & Hughes, E. (2013). Genes associated with the peculiar phenotypes of the aerobic anoxygenic phototrophs. *Advances in Botanical Research, 66*, 327–358.

Zappa, S., & Bauer, C. E. (2013). Iron homeostasis in the *Rhodobacter* genus. *Advances in Botanical Research, 66*, 289–326.

Zheng, Q., Koblížek, M., Beatty, J. T., & Jiao, N. (2013). Evolutionary divergence of marine aerobic anoxygenic phototrophic bacteria as seen from diverse organisations of their photosynthesis gene clusters. *Advances in Botanical Research, 66*, 359–383.

CONTRIBUTORS

Carl E. Bauer
Department of Molecular and Cellular Biochemistry, Indiana University, Bloomington, IN, USA

J. Thomas Beatty
Department of Microbiology and Immunology, University of British Columbia, Vancouver, BC, Canada

Dom Bellini
Institute of Integrative Biology, University of Liverpool, Liverpool, UK

Donald A. Bryant
Department of Biochemistry and Molecular Biology, The Pennsylvania State University, University Park, PA, USA; Department of Chemistry and Biochemistry, Montana State University, Bozeman, MT, USA

Richard J. Cogdell
Institute of Molecular, Cell and Systems Biology, College of Medical, Veterinary and Life Sciences, Glasgow Biomedical Research Centre, University of Glasgow, Glasgow, UK

Michael Y. Galperin
National Center for Biotechnology Information, National Library of Medicine, National Institutes of Health, Bethesda, MD, USA

Mark Gomelsky
Department of Molecular Biology, University of Wyoming, Laramie, WY, USA

Radhey S. Gupta
Department of Biochemistry and Biomedical Sciences, McMaster University, Hamilton, ON, Canada

Sarah L. Henry
Institute of Molecular, Cell and Systems Biology, College of Medical, Veterinary and Life Sciences, Glasgow Biomedical Research Centre, University of Glasgow, Glasgow, UK

Aleš Horák
Institute of Parasitology CAS, České Budějovice, Czech Republic

Elizabeth Hughes
Department of Microbiology, University of Manitoba, Winnipeg, MB, Canada

Nianzhi Jiao
State Key Laboratory of Marine Environmental Science, Xiamen University, Xiamen, PR China

Michal Koblížek
Institute of Microbiology CAS, Opatovický mlýn, Třeboň, Czech Republic

Michal Koblížek
Department of Phototrophic Microorganisms – Algatech, Institute of Microbiology CAS, Třeboň, Czech Republic; Faculty of Science, University of South Bohemia, České Budějovice, Czech Republic

Zhenfeng Liu
Department of Biochemistry and Molecular Biology, The Pennsylvania State University, University Park, PA, USA; Department of Biological Sciences, University of Southern California, Los Angeles, CA, USA

Jean-Pierre Magnin
Laboratoire d'Electrochimie et de Physicochimie des Matériaux et des Interfaces, St Martin d'Hères, Grenoble, France

Armen Y. Mulkidjanian
School of Physics, University of Osnabrück, Osnabrück, Germany; School of Bioengineering and Bioinformatics, Moscow State University, Moscow, Russia; Belozersky Institute of Physico-Chemical Biology, Moscow State University, Moscow, Russia

Sakiko Nagashima
Research Institute for Photosynthetic Hydrogen Production, Kanagawa University, Hiratsuka, Kanagawa, Japan; Department of Biological Science, Tokyo Metropolitan University, Hachioji, Tokyo, Japan

Kenji V. P. Nagashima
Research Institute for Photosynthetic Hydrogen Production, Kanagawa University, Hiratsuka, Kanagawa, Japan; Precursory Research for Embryonic Science and Technology (PRESTO), Japan Science and Technology Agency (JST), Kawaguchi, Saitama, Japan

Miroslav Oborník
Department of Phototrophic Microorganisms – Algatech, Institute of Microbiology CAS, Třeboň, Czech Republic; Faculty of Science, University of South Bohemia, České Budějovice, Czech Republic; Institute of Parasitology CAS, České Budějovice, Czech Republic

Miroslav Papiz
Institute of Integrative Biology, University of Liverpool, Liverpool, UK

W. Matthew Sattley
Indiana Wesleyan University, Division of Natural Sciences, Marion, IN, USA

Wesley D. Swingley
Northern Illinois University, Department of Biological Sciences, DeKalb, IL, USA

John C. Willison
Laboratoire de Chimie et Biologie des Métaux, CEA Grenoble, Grenoble-INP, France

Vladimir Yurkov
Department of Microbiology, University of Manitoba, Winnipeg, MB, Canada

Sébastien Zappa
Department of Molecular and Cellular Biochemistry, Indiana University, Bloomington, IN, USA

Jill H. Zeilstra-Ryalls
Department of Biological Sciences, Bowling Green State University, Bowling Green, OH, USA

Yonghui Zeng
Department of Phototrophic Microorganisms – Algatech, Institute of Microbiology CAS, Třeboň, Czech Republic

Qiang Zheng
State Key Laboratory of Marine Environmental Science, Xiamen University, Xiamen, PR China

CHAPTER ONE

A Time to Scatter Genes and a Time to Gather Them: Evolution of Photosynthesis Genes in Bacteria

Armen Y. Mulkidjanian[*,**,†,1], **Michael Y. Galperin**[‡]
[*]School of Physics, University of Osnabrück, Osnabrück, Germany
[**]School of Bioengineering and Bioinformatics, Moscow State University, Moscow, Russia
[†]Belozersky Institute of Physico-Chemical Biology, Moscow State University, Moscow, Russia
[‡]National Center for Biotechnology Information, National Library of Medicine, National Institutes of Health, Bethesda, MD, USA
[1]Corresponding author: E-mail: amulkid@uos.de

Contents

Abstract

Genome sequencing opened entirely new avenues for studying the evolution of photosynthesis. By systematically comparing sequences of photosynthesis-related genes and their products in phototrophic members of diverse bacterial lineages, it has become possible to delineate their common and distinct traits, analyse their evolutionary relationships, and reconstruct the likely scenarios for the overall evolution of the photosynthetic machinery. We consider here the comparative genomics data on the distribution of photosynthesis genes among certain representatives of six bacterial phyla, Acidobacteria, Chlorobi, Chloroflexi, Cyanobacteria, Firmicutes, and Proteobacteria and put these data in a broader geochemical context. We address the tentative nature of the first photosynthetic organisms, the driving forces behind their origin, and review the evidence for the early origin of abiogenic photosynthesis.

1. INTRODUCTION

Photosynthesis is a key biological process that may have emerged even before the origin of life on the Earth and played a key role in shaping the

Advances in Botanical Research, Volume 66
ISSN 0065-2296, http://dx.doi.org/10.1016/B978-0-12-397923-0.00001-1

planet and its atmosphere. Indeed, abiogenic photosynthesis – synthesis of the first organic molecules on the Hadean Earth from CO_2 and H_2O driven by the energy of solar UV radiation – most likely contributed the necessary precursors for the formation of the first cells (Guzman & Martin, 2009, 2010; Moore & Webster, 1913; Mulkidjanian, 2009; Mulkidjanian, Bychkov, Dibrova, Galperin, & Koonin, 2012a; Mulkidjanian, Bychkov, Dibrova, Galperin, & Koonin, 2012b; Mulkidjanian & Galperin, 2009; Schoonen, Smirnov, & Cohn, 2004; Zhang, Martin, Friend, Schoonen, & Holland, 2004; Zhang et al., 2007). At the next step, anoxygenic photosynthesis provided a ready way for harnessing the energy of the Sun into the accumulation of bacterial biomass (Sleep, 2010; Sleep & Bird, 2007, 2008), which ultimately allowed the gradual emergence of complex multicellular organisms.

The subsequent emergence of oxygenic photosynthesis dramatically changed the conditions on the planet by providing the readily available acceptor of electrons for the electron-transport chains of the increasingly complex organisms, and by creating the ozone shield that protected these organisms from the damaging short-wave UV radiation (Garcia-Pichel, 1998). Finally, through acquisition of cyanobacterial symbionts, the ability to conduct photosynthesis was conferred to several lineages of eukaryotic cells, which led to the emergence of apicomplexans, diatoms, red and brown algae, and green plants (Green, 2011; Keeling, 2009, 2010).

Accordingly, the problem of origin and evolution of photosynthesis is a core element in any concept of the origin and evolution of life on Earth. The complexity of this problem is exacerbated by a certain degree of confusion regarding the fossil data. The early naive reports of full-fledged fossils of trichomic cyanobacteria-like microorganisms in the Early Archaean (Schopf, 1993; Schopf & Packer, 1987) have been disputed (Brasier et al., 2002) and are largely being neglected. However, later findings of carbon (graphite) deposits associated with likely microbial mats (Tice & Lowe, 2004, 2006) have again pushed the time of emergence of photosynthetic microbes back to the ~3.5 billion years ago mark. Furthermore, geochemical analyses explained the origin of 3.8 Gy old carbon-rich deposits (black shales) as originating from anoxygenic photosynthesis (Sleep & Bird, 2007, 2008). That said, biochemical characterization of those fossils still remains out of reach, forcing researchers to seek alternative ways to study the origins of photosynthesis.

Genome sequencing opened an entirely new avenue for studying the evolution of photosynthesis. By systematically comparing sequences of photosynthesis-related genes – and their products – in phototrophic members of diverse bacterial lineages, it has become possible to delineate their common and distinct traits, analyse their evolutionary relationships, and reconstruct the likely scenarios for the overall evolution of the photosynthetic machinery.

Some time ago, we used comparative genomics to analyse the distribution of photosynthesis-related genes in different lineages and, specifically, delineated cyanobacterial clusters of orthologous groups of proteins (Cyanobacterial clusters of Orthologous Groups of proteins (CyOGs)) (Mulkidjanian et al., 2006). As part of that work, we have noticed that 84 CyOGs were exclusively shared by cyanobacteria and plants and/or other plastid-carrying eukaryotes, such as diatoms or apicomplexans. That set included 49 CyOGs with known functions, which were all involved in photosynthesis, and 35 families of uncharacterized proteins that could also be involved in photosynthesis. In the same article we compared the distribution of photosynthesis-related genes in cyanobacteria with that in other phototrophic prokaryotes. Based on this analysis, we suggested that photosynthesis originated among the direct ancestors of cyanobacteria – anoxygenic procyanobacteria – and that members of other phyla obtained their photosynthesis genes via lateral gene transfer.

Given several recently published comprehensive reviews on the evolution of photosynthesis (Bryant & Frigaard, 2006; Bryant et al., 2012; Gupta, 2012; Hohmann-Marriott & Blankenship, 2011), in this chapter we provide an update of our earlier genome analysis and also attempt to consider the problem of the evolution of photosynthesis in a broader geochemical context. We check to what extent our predictions on the photosynthetic function of 35 uncharacterized enzymes were correct, briefly address the tentative nature of the first photosynthetic organisms, and review the evidence for an early origin of abiogenic photosynthesis.

2. PHOTOSYNTHESIS GENES OF CYANOBACTERIA AND PLANTS

A comparative study of cyanobacterial genomes several years ago described a set of 1054 protein families, referred to as core CyOGs, that had been encoded in at least 14 of the 15 complete cyanobacterial genomes available at that time (Mulkidjanian et al., 2006). Of those 1054 core CyOGs, 84 protein families were found exclusively in cyanobacteria and plants (*Arabidopsis thaliana*, rice, the red algae *Cyanidioschyzon merolae* and *Porphyra purpurea*, and/or the diatom *Thalassiosira pseudonana*).

Some of those 84 proteins had been previously shown to participate in photosynthesis as components of photosystems I and II, light-harvesting antennas, and so on. However, members of 35 protein families with the same phylogenetic distribution had no known function (Mulkidjanian et al., 2006). We have reasoned that the proteins encoded in (nearly) all cyanobacterial genomes and at least some chloroplast-containing eukaryotes, but not in any other bacterial or eukaryotic genomes were likely to either directly participate in photosynthesis or have photosynthesis-related functions.

This proposal had been partly verified by the analysis of the proteins with similar phylogenetic profiles that, although not yet properly annotated in the public databases, had been experimentally characterized by that time. One of these was GENOMES UNCOUPLED4 (GUN4) protein, a cofactor of Mg-chelatase, which had been proposed to regulate chlorophyll biosynthesis and intracellular signalling (Larkin, Alonso, Ecker, & Chory, 2003). This protein is encoded in *Synechocystis* sp. PCC 6803 by three paralogous genes, *sll0558*, *sll1380*, and *slr1958*, and crystal structures of Sll0558 and its orthologue from *Thermosynechococcus elongatus* have been solved (PDB entries 1Y6I and 1Z3X, respectively (Davison et al., 2005; Verdecia et al., 2005)). While this protein is currently the subject of intensive research (Adhikari et al., 2011), the corresponding entries in public databases (e.g. P72583 in UniProt (The UniProt Consortium, 2012)) are still annotated as 'Ycf53-like' proteins, although protein domain databases, such as Pfam (Punta et al., 2012) already identify them as members of the GUN4 family.

In another interesting case, a *Chlamydomonas reinhardtii* protein Tab2, an orthologue of the *Synechocystis* sp. PCC 6803 protein Sll2002, has been characterized as an RNA-binding protein that specifically interacts with an upstream region of the mRNA of the *psaB* gene and is required for translation of the *psaB* product, photosystem I reaction centre (RC) protein, and the assembly of the photosystem I complex (Dauvillee, Stampacchia, Girard-Bascou, & Rochaix, 2003). Shortly after that, an *A. thaliana* orthologue of Tab2 (At3g08010, designated ATAB2) was shown to bind to the 5′-untranslated regions in the mRNA of *psaB* and several other chloroplast genes, including *psbA*, *psbB*, and *psbD/C* (Barneche, Winter, Crevecoeur, & Rochaix, 2006). Curiously, while some plant Tab2 proteins are marked as such, cyanobacterial members of the Tab2 family are still listed as uncharacterized proteins, and the corresponding protein family (PF06485 in Pfam) is referred to as domain of unknown function, DUF1092 (Punta et al., 2012).

In the past several years, some of these 35 proteins predicted to have photosynthesis-related functions have been experimentally characterized, either in cyanobacteria, or in plants, and in some cases in both groups. Table 1.1 shows 11 such proteins, listing their locus tags in *Synechocystis* sp. PCC 6803, assigned gene names, their current annotations, orthologs in *A. thaliana* (where available) or in red algae, and the respective entries in the public databases, UniProt and Pfam (Punta et al., 2012; The UniProt Consortium, 2012). These data clearly demonstrate the predictive power of the phylogenetic patterns: all experimentally characterized proteins indeed turned out to be involved in photosynthesis, either as auxiliary or regulatory subunits of the photosynthetic reaction complexes (Table 1.1). Nine of the 11 proteins have been found in all cyanobacterial and plant genomes. One of the remaining two proteins, Ycf34, was encoded in all cyanobacteria and in chloroplasts of diatoms and red and brown algae but apparently lost among green plants. Finally, Ycf86 has been found so far only in cyanobacteria and red algae.

These recent data provide additional support to the original prediction that proteins with the same phylogenetic pattern (encoded in the genomes of photosynthetic organisms but not in the genomes of non-photosynthetic organisms) should have photosynthesis-related functions. In Table 1.2, we list 23 such proteins that have not yet been experimentally characterized. They all represent widespread protein families that have been annotated as domain of unknown function, DUFs, in Pfam (Punta et al., 2012). Again, several of these proteins are encoded in chloroplasts of diatoms and red and brown algae, but seem to have been lost from green plants. Others are found in (nearly) all photosynthetic organisms and represent attractive targets for future experimental research.

3. EVOLUTION OF THE PHOTOSYNTHESIS GENE SET

3.1. The Common Photosynthesis Gene Set

Phototrophic organisms depend on chlorophyll-containing photosynthetic RCs of type I and/or of type II (RC1 and RC2, respectively). Photosynthetic RCs are found in organisms that belong to several distinct prokaryotic and eukaryotic lineages. However, all eukaryotic phototrophs appear to have inherited their photosynthetic organelles, plastids, from cyanobacteria. In contrast, among prokaryotes, phototrophy has been found in representatives of several different phyla. Photosynthesis is found, in addition to the Cyanobacteria, in the Bacteroidetes/Chlorobi group (e.g. *Chlorobium tepidum*), Firmicutes (e.g. *Heliobacillus mobilis*), Acidobacteria (*Candidatus Chloracidobacterium thermophilum*), Chloroflexi (e.g. *Chloroflexus aurantiacus*),

Table 1.1 Recently characterized conserved cyanobacterial and plant proteins

Locus	Length, aa	Gene name	Updated annotation	Plant homologue	UniProt entry	Pfam domain	References
ssl3364	73	CP12	Thioredoxin-regulated chloroplast protein CP12	At2g47400	P73654	CP12	(Erales, Lignon, & Gontero, 2009; Gontero & Maberly, 2012; Howard et al., 2011)
sll1414	215	psb29 (THF1)	Photosystem II biogenesis protein psb29 (THF1)	At2g20890	P73956	Thylakoid Format	(Keren, Ohkawa, Welsh, Liberton, & Pakrasi, 2005)
sll1509	112	Ycf20	Chloroplast protein Ycf20 involved in dissipation of absorbed light energy	At5g43050	P72983	DUF565	(Jung & Niyogi, 2010)
ssl1417	69	Ycf33	Photosystem I protein Ycf33 involved in cyclic electron transport	At4g16410	P74788	DUF751	(Ohtsuka, Oyabu, Kashino, Satoh, & Koike, 2004)

sll0933	126	PAM68	Photosystem II biogenesis protein Sll0933	At5g52780	P72865	DUF3464	(Armbruster et al., 2010; Rengstl, Oster, Stengel, & Nickelsen, 2011)
slr0815	111	CCR2	Thylakoid membrane protein involved in photosystem II response to cold stress	At3g17930	P74048	DUF3007	(Li, Gao, Yin, & Xu, 2012)
ssl0352	62	NdhS	NADPH: plastoquinone oxidoreductase subunit NdhS (CRR31)	At4g23890	P74795	DUF3252	(Battchikova et al., 2011; Yamamoto, Peng, Fukao, & Shikanai, 2011)
ssl3451	77	SipA	Regulator of cyanobacterial sensor kinase NblS (Hik33)	At5g20935	P73286	DUF3148	(Espinosa, Fuentes, Burillo, Rodriguez-Mateos, & Contreras, 2006; Sakayori, Shiraiwa, & Suzuki, 2009)

Continued

Table 1.1 Recently characterized conserved cyanobacterial and plant proteins—cont'd

Locus	Length, aa	Gene name	Updated annotation	Plant homologue	UniProt entry	Pfam domain	References
slr0575	184	APE1	Acclimation of photosynthesis to environment (APE1) protein, affects chlorophyll fluorescence	At5g38660	Q55403	DUF2854	(Walters, Shephard, Rogers, Rolfe, & Horton, 2003)
ssr1425	84	Ycf34	Fe-S cluster chloroplast protein Ycf34, regulator of the photosynthetic electron transport	CypaCp008	P74777	Ycf34	(Wallner et al., 2012)
ssr2998	78	petP (Ycf86)	Cytochrome b_6f complex subunit petP (CP19, Ycf86)	PopuCp097*	P72798	DUF2862	(Volkmer et al., 2007)

*Found only in cyanobacteria and in red algae.

Table 1.2 Uncharacterized conserved cyanobacterial and plant proteins

Locus	Length, aa	Gene name	Plant homolog	UniProt entry	Pfam domain
slr1638	117	PM23	At1g63610	P74354	DUF760
sll0661	131	Ycf35	CypaCp141	Q55981	DUF1257
sll0584	169	Ycf36	At5g67370	Q55866	DUF1230
sll1702	198	Ycf51	CypaCp078	P73690	DUF2518
sll1879	543	Ycf55	PopuCp018	P74126	DUF3685
sll1737	153	Ycf60	At2g47840	P73387	–
slr0503	353	Ycf66	MapoCp005*	F7UTS9	Ycf66_N
slr1699	244	–	At5g47860	P73194	DUF1350
sll1656	189	–	At2g15290	P72815	DUF3611
slr0438	119	–	At3g15110	Q55125	DUF3082
slr0589	185	–	At3g26710	P74727	DUF3529
slr1470	134	–	At1g14345	P74154	DUF304
slr1052	367	–	At3g26580	P73017	TPR_16
sll1071	264	–	At5g52970	P73281	Repair_PSII
slr0948	190	–	At1g59840	P74315	DUF2930
sll0272	156	–	At2g04039	P74394	DUF2996
sll2013	179	–	At5g39520	P73665	DUF1997
slr1195	154	–	At5g08400	P73343	DUF3531
slr1660	214	–	CR066†	P74661	DUF3172
slr1702	214	–	At5g27560	P73200	DUF1995
ssr3188	89	–	At5g52960	P73653	DUF3143
slr0598	118	–	At3g19900	P74744	DUF3067
ssl3829	88	–	At5g39210	P73675	DUF3571

**Marchantia polymorpha* and some other green plants.
†*Chlamydomonas reinhardtii* and other green algae.

and in three different classes of Proteobacteria: α-Proteobacteria (e.g. *Rhodopseudomonas palustris*), β-Proteobacteria (e.g. *Rubrivivax gelatinosus*), and γ-Proteobacteria (e.g. *Chromatium vinosum*). The first three phyla have photosynthetic RCs that are similar to the cyanobacterial PSI and use low-potential FeS clusters as electron acceptors (RC1 type). The RCs of members of Proteobacteria and Chloroflexi (RC2 type) use bound quinones as ultimate electron acceptors and are similar to the cyanobacterial PSII (although lacking the oxygen-evolving complex) (Bryant et al., 2012; Hohmann-Marriott & Blankenship, 2011).

While in 2006 it was not clear whether non-phototrophic Chlorobi exist, they have now been found and characterized (Iino et al., 2010; Liu et al., 2012). Hence, now only Cyanobacteria are left without non-phototrophic members (Table 1.3).

Table 1.3 Phototrophic bacteria with completely sequenced genomes and their heterotrophic relatives

Taxonomy*	Representative organism (GenBank genome entry)	Proteins	Photo-system	CO_2 assimilation	Photoautotrophic growth	References
Phylum: *Acidobacteria*						
	Candidatus Chloracidobacteriumthermophilum (CP002514, CP002515)	3054	RC1	N/A	No	(Bryant et al., 2007; Garcia Costas et al., 2012)
	Terriglobus saanensis (CP002467)	4180	–	N/A	No	(Mannisto, Rawat, Starovoytov, & Haggblom, 2011; Rawat, Mannisto, Bromberg, & Haggblom, 2012)
Phylum: *Chlorobi*						
Class: *Chlorobia* Order: *Chlorobiales* Family: *Chlorobiaceae*	*Chlorobium tepidum* (AE006470)	2245	RC1	Reverse TCA	Yes	(Eisen et al., 2002; Imhoff, 2003; Li, Sawaya, Tabita, & Eisenberg, 2005; Wahlund, Woese, Castenholz, & Madigan, 1991)
	Chlorobaculum parvum (CP001099)	2043	RC1	Reverse TCA	Yes	
	Chlorobium limicola (CP001097)	2434	RC1	Reverse TCA	Yes	
	Chloroherpeton thalassium (CP001100)	2710	RC1	Reverse TCA	Yes	
	Prosthecochloris aestuarii (CP001108)	2327	RC1	Reverse TCA	Yes	

Class: *Ignavibacteriae*	*Ignavibacterium album* (CP003418)	3195	–	N/A	No	(Iino et al., 2010; Liu et al., 2012; Podosokorskaya et al., in press)
Phylum: *Chloroflexi*						
Class: *Chloroflexi* **Order:** *Chloroflexales* **Family:** *Chloroflexaceae*	*Chloroflexus aurantiacus* (CP000909)	3853	RC2	3-Hydroxypropionate cycle	Yes	(Klatt, Bryant, & Ward, 2007)
	Roseiflexus castenholzii (CP000804)	4330	RC2	3-Hydroxypropionate cycle	No	(Tang et al., 2011) (Gupta, Chander, & George, in press; Hanada et al., 2002; Herter et al., 2001; Klatt et al., 2007)
Family: *Oscillochloridaceae*	*Oscillochloris trichoides* (ADVR00000000)	3231	RC2	CBB cycle	Yes	(Berg, Keppen, Krasil'nikova, Ugol'kova, & Ivanovskii, 2005; Keppen, Baulina, Lysenko, & Kondratieva, 1993; Kuznetsov et al., 2011)
Order: *Herpetosiphonales*	*Herpetosiphon aurantiacus* (CP000875)	5279	–	N/A	No	(Kiss et al., 2011; Klatt et al., 2007)

Continued

Table 1.3 Phototrophic bacteria with completely sequenced genomes and their heterotrophic relatives—cont'd

Taxonomy*	Representative organism (GenBank genome entry)	Proteins	Photo-system	CO_2 assimilation	Photoautotrophic growth	References
Phylum: *Firmicutes*						
Class: *Clostridia* Order: *Clostridiales* Family: *Heliobacteriaceae*	*Heliobacterium modesticaldum* (CP000930)	2999	RC1	PEP carboxykinase	No	(Kimble, Mandelco, Woese, & Madigan, 1995; Sarrou et al., 2012; Sattley et al., 2008; Tang, Yue, & Blankenship, 2010)
Family: *Peptococcaceae*	*Desulfotomaculum reducens* (CP000612)	3276	–	N/A	No	(Junier et al., 2010)
Phylum: *Proteobacteria*						
Class: *α-Proteobacteria*	*Rhodobacter capsulatus* (CP001312)	3642	RC2	CBB cycle	Yes	(Imhoff, Truper, & Pfennig, 1984)
Order: *Rhodobacterales*	*Rhodobacter sphaeroides* (CP000143)	4242	RC2	CBB cycle	Yes	
Family: *Rhodobacteraceae*	*Paracoccus denitrificans* (CP000490)	5077	–	CBB cycle	No	
Order: *Rhodospirillales*	*Rhodospirillum rubrum* (CP000230)	3838	RC2	CBB cycle	Yes	
Family *Rhodospirillaceae*	*Magnetospirillum magneticum* (AP007255)	4561	–	CBB cycle	No	(Geelhoed, Kleerebezem, Sorokin, Stams, & van Loosdrecht, 2010)

Class: *β-Proteobacteria*	*Rubrivivax gelatinosus* (AP012320)	4693	RC2	CBB cycle	No	(Nagashima et al., 2012)
Order: *Burkholderiales*	*Methylibium petroleiphilum* (CP000555)		–	N/A	No	(Kane et al., 2007)
Class: *γ-Proteobacteria*	*Allochromatium vinosum* (CP001896)	3220	RC2	CBB cycle	Yes	(Imhoff, Suling, & Petri, 1998;
Order: *Chromatiales*	*Thiocystis violascens* (CP003154)	4330	RC2	CBB cycle	Yes	Weissgerber et al., 2011)
Order: *Methylococcales*	*Methylococcus capsulatus* (AE017282)	2956	–	N/A	No	(Ward et al., 2004)
Phylum: *Cyanobacteria*						
Order: *Chroococcales*	*Acaryochloris marina* (CP000828)	8383	RC1, RC2	CBB cycle	Yes	(Pfreundt, Stal, Voss, & Hess, 2012)
	Microcystis aeruginosa (AP009552)	6312	RC1, RC2	CBB cycle	Yes	(Kaneko et al., 2007)
	Synechococcus elongatus (CP000100)	2662	RC1, RC2	CBB cycle	Yes	
	Synechococcus sp. PCC 7002 (CP000951)	3187	RC1, RC2	CBB cycle	Yes	
	Synechocystis sp. PCC 6803 (BA000022)	3575	RC1, RC2	CBB cycle	Yes	(Kaneko et al., 1996)
	Candidatus Atelocyanobacterium thalassa (UCYN-A, CP001842)	1199	RC1	N/A	No	(Thompson et al., 2012; Tripp et al., 2010; Zehr et al., 2008)
Order: *Gloeobacterales*	*Gloeobacter violaceus* (BA000045)	4430	RC1, RC2	CBB cycle	Yes	(Nakamura et al., 2003)

Continued

Table 1.3 Phototrophic bacteria with completely sequenced genomes and their heterotrophic relatives—cont'd

Taxonomy[*]	Representative organism (GenBank genome entry)	Proteins	Photo-system	CO_2 assimilation	Photoautotrophic growth	References
Order: *Nostocales*	*Anabaena variabilis* (CP000117)	5710	RC1, RC2	CBB cycle	Yes	
	Nostoc sp. PCC 7120 (BA000019)	6129	RC1, RC2	CBB cycle	Yes	
Order: *Oscillatoriales*	*Arthrospira platensis* (CM001632)	6108	RC1, RC2	CBB cycle	Yes	
Order: *Prochlorales*	*Prochlorococcus marinus* (AE017126)	1883	RC1, RC2	CBB cycle	Yes	(Chisholm et al., 1992; Dufresne et al., 2003; Rocap et al., 2003)

[*]Taxonomy of phototrophic strains, according to the List of Prokaryotic Names with Standing in Nomenclature (Munoz et al., 2011) and the NCBI Taxonomy database (Federhen, 2012). Closely related non-phototrophic bacteria (selected based on the 16S rRNA sequence similarity) are listed with the highest taxon that is distinct from that of the respective phototroph(s).

The existence of free-living heterotrophic relatives of known phototrophs in Acidobacteria, Chlorobi, Chloroflexi, Firmicutes, and Proteobacteria (Table 1.3) can be contrasted with the secondary loss of the photosynthetic ability among degenerate plastids, e.g. in apicomplexans, which lack both RC1 and RC2. A symbiotic cyanobacterium, UCYN-A (*Candidatus Atelocyanobacterium thalassa*), with streamlined metabolism has been described that lacks the RC2 but still retains a functional RC1 (Thompson et al., 2012; Tripp et al., 2010; Zehr et al., 2008). In contrast, the free-living heterotrophic members of the other five phyla typically encode relatively large protein sets (Table 1.3) and do not display any signs of genome degradation.

The 'patchy' distribution of the ability to conduct photosynthesis could be explained either by a massive loss of photosynthesis genes by non-photosynthetic members of the respective phyla or by an acquisition of this ability by a handful of selected genera (Table 1.3) through lateral gene transfer of certain photosynthesis genes. Given the obvious evolutionary advantage of having solar radiation as a source of energy, the first scenario appears extremely unlikely. As a result, there is a general consensus that photosynthesis genes are being spread through lateral gene transfer (Blankenship, 1992; Bryant & Frigaard, 2006; Bryant et al., 2012; Gupta, 2012; Hohmann-Marriott & Blankenship, 2011; Mulkidjanian et al., 2006; Olson & Blankenship, 2004). This consensus has been further supported by the findings that photosynthesis genes could be transduced by phages (Alperovitch-Lavy et al., 2011; Mann, Cook, Millard, Bailey, & Clokie, 2003; Sharon et al., 2009; Sullivan et al., 2006) and expressed in the infected host cells (Lindell et al., 2004).

Despite the obvious propensity of photosynthesis genes to lateral gene transfer, a direct comparison of the completely sequenced genomes of phototrophic bacteria from different lineages revealed a surprisingly little overlap between the respective gene sets (Table 1.4). This circumstance greatly affected the evolutionary analyses of photosynthesis genes. Indeed, the typical approaches to such analyses involve identification of shared traits and construction of phylogenetic trees from sequences of the genes (proteins) that are responsible for these shared traits. In this case, several attempts at delineation of the 'photosynthesis gene set' of the genes shared by all photosynthetic organisms revealed that: (1) there are very few such genes and (2) most of these genes are involved in biosynthesis of (bacterio)chlorophyll and related processes, rather than in photosynthesis per se (Mulkidjanian et al., 2006; Raymond, Zhaxybayeva, Gogarten, & Blankenship, 2003; Raymond, Zhaxybayeva, Gogarten, Gerdes, & Blankenship, 2002; Sato, 2002;

Table 1.4 Distribution of the core photosynthesis genes in various phototrophic lineages*

		Anoxygenic phototrophs					Oxygenic phototrophs	
System or pathway	**Genes**	**Acido**	**Chlorobium**	**Chloroflexus**	**Helio**	**Purple**	**Cyano**	**Plants**
Chlorophyll biosynthesis	*chlB, chlD, chlG, chlH, chlI, chlL, chlM, chlN, chlP*	+	+	+	+	+	+	+
Photosystem I								
Core RC1 subunit	*psaB/pshA*	+	+	–	+	–	+	+
Iron-sulfur subunit	*psaC*	+	±†	–	+	–	+	+
Photosystem I subunits	*psaD, psaE, psaF, psaI, psaJ, psaK, psaL, psaM*	–	–	–	–	–	+	+
Photosystem II								
Core RC2 subunit D1/D2	*psbA/psbD*	–	–	+	–	+	+	+
Photosystem II subunits	*psbB, psbC, psbE, psbF, psbH, psbI, psbJ, psbK, psbL, psbM, psbN, psbO, psbP, psbQ, psbT, psbU, psbV, psbW, psbX, psbY(Ycf32), psbZ(Ycf9), psb27*	–	–	–	–	–	+	+

Cytochrome b_6f complex								
Cytochrome b_6 with fused or separate subunit IV, Rieske iron-sulfur protein	*petB* (±*petD*), *petC*	+	+	−	+	+	+	+
Cytochrome f, other subunits	*petA*, *petD*, *petG*, *petL*, *petM*, *petN*	−	−	−	−	−	+	+
Cytochrome *c* (c_6, c_{553})	*petJ*, *cytM*	+	+	+	+	+	+	+
Plastocyanin	*petE*	−	−	±	−	−	+	+
CBB cycle								
RuBisCO	*rbcS*, *rbcL*	−	−/+	−	−	+	+	+
Common enzymes	*pgk*, *gapA*, *rpe*, *tpiA*, *tktA*	+	+	+	+	+	+	+

[*]Presence or absence of orthologs of the respective cyanobacterial genes in the genomes of phototrophic representatives of Acidobacteria (*Candidatus Chloracidobacterium thermophilum*, Genbank entry CP002514, CP002515), Chlorobi (*Chlorobium tepidum* TLS, AE006470), Chloroflexi (*Chloroflexus aurantiacus* J-10-fl, CP000909), Firmicutes (*Heliobacterium modesticaldum* Ice1, CP000930), Proteobacteria (*Rhodopseudomonas palustris* CGA009, BX571963), Cyanobacteria (*Synechocystis* sp. PCC 6803, BA000022), and plants (*Arabidopsis thaliana*, NC_003070, NC_003076, NC_000932). This table has been originally compiled for reference Mulkidjanian et al. (2006) and updated based on the analysis of the complete genomes of *Candidatus* Chloracidobacterium thermophilum and *H. modesticaldum* (Garcia Costas et al., 2012; Sattley et al., 2008).
[†]Iron-sulfur protein PscB of *C. tepidum* is not homologous to PsaC-like FeS-subunits of other groups of phototrophs.

Zhaxybayeva, Hamel, Raymond, & Gogarten, 2004). The major additions to the genomic analysis after 2006, as included in Table 1.4, was the discovery of a photosynthetic machinery, namely a type I RC and the accompanying set of protein-coding genes, in a representative of *Acidobacteria*, *Candidatus Chloracidobacterium thermophilum* (Bryant et al., 2007; Garcia Costas et al., 2012; Tsukatani, Romberger, Golbeck, & Bryant, 2012). Another important new result was the discovery of a cytochrome b_6f complex in a non-phototrophic (Table 1.3) representative of Chloroflexi, *Herpetosiphon aurantiacus* (Kiss et al., 2011). Before that, in Chloroflexi, only the alternative complex III had been identified as an oxidoreductase that would connect the membrane menaquinol pool with high-potential electron acceptors (Yanyushin, 2002; Yanyushin, del Rosario, Brune, & Blankenship, 2005). Thus, cytochrome *bc* complexes have now been found in all phototroph-containing phyla. Still, the principal observation that the shared set of photosynthesis genes is related to the (bacterio)chlorophyll biosynthesis and not to the photosynthetic machinery per se still stands.

3.2. Who Were the Ancestral Phototrophs?

The inability to draw conclusions on the evolution of photosynthesis and the nature of ancestral form of the RC from genome comparisons alone prompted us to bring the genomic results into a broader context.

Previously, we have argued that since all free-living cyanobacteria contain almost 100 photosynthesis-related genes and are obligate autotrophs, their anoxygenic ancestors, procyanobacteria, could be the first phototrophic organisms. In support, we have invoked the geological evidence, as obtained by Tice and Lowe at Buck Reef Chert, a 250–400 m-thick rock running along the South African coast (Tice & Lowe, 2004, 2006). Tice and Lowe have identified traces of a primordial phototrophic community within this >3.4 Gy-old chert and defined the inhabitants of this community as partially filamentous phototrophs, which, according to the carbon isotopic composition, used the Calvin–Benson–Bassham (CBB) cycle to fix CO_2. Overall, this set of features most closely resembles cyanobacteria, so that the Buck Reef Chert may have been inhabited by their direct ancestors. We have suggested that phototrophic organisms belonging to phyla other than cyanobacteria could have obtained their photosynthesis genes via lateral gene transfer from the (pro)cyanobacterial lineage at different steps of evolution.

According to the proposed scenario, the ancestors of *Chlorobium*, *Heliobacterium*, and *Chloracidobacterium* must have acquired the primordial, homodimeric form of the RC1, whereas proteobacterial phototrophic lineages and *Chloroflexus* acquired their RC2 before it 'learned' to oxidize water. Anoxygenic phototrophs usually dwell in the depth of microbial mats. Perhaps therefore they were subject to a weaker selective pressure from light and oxygen than those (ancestors of modern cyanobacteria) that remained on the surface, resulting in preservation of ancestral features of their photosynthetic machinery. Thus, photosynthetic enzymes of anaerobic bacteria can be considered snapshots of the ancient RCs: the homodimeric RC1 of *Heliobacillus mobilis* (PshA) and *Chlorobium tepidum* (PscA) are probably more similar to the ancient homodimeric RC1 than the highly evolved heterodimeric PSI (PsaA/PsaB) of modern cyanobacteria.

This scenario is in a good correspondence with the most recent work of Gupta (2012), who has analysed conserved signature insertions and deletions in key proteins involved in bacteriochlorophyll biosynthesis. In this work the bacteriochlorophyll synthesis enzymes from Heliobacteriaceae were identified as primitive in comparison to all other photosynthetic lineages, and some ancient Firmicutes were suggested as first phototrophic organisms (Gupta, 2012). Heliobacteriaceae, however, form a very narrow group of phototrophs within the Firmicutes phylum. Furthermore, the photosynthetic apparata of *Heliobacillus mobilis* and *Heliobacterium modesticaldum* are harboured on large operons (Sattley et al., 2008; Xiong, Inoue, & Bauer, 1998), potential subjects of lateral gene transfer. Hence, it seems very likely that heliobacteria, indeed, obtained their photosynthesis genes via lateral gene transfer. The same operon in *H. modesticaldum* also carries the genes of a menaquinone-dependent cytochrome b_6f complex, which, arguably, also represents a primitive form of cytochrome *bc* complexes (Dibrova, Cherepanov, et al., 2013, in press; Dibrova, Chudetsky, et al., 2012b). Most likely, all these proteins emerged and were shaped not within Heliobacteriaceae, but within some other anoxygenic, menaquinone-containing phototrophic lineage, which, we believe, directly preceded cyanobacteria. This lineage later also invented oxygenic photosynthesis, and underwent dramatic changes in response to the oxygenation of the biosphere (Mulkidjanian et al., 2006; Raymond & Blankenship, 2004; Rutherford, Osyczka, & Rappaport, 2012), which they could not evade. In contrast, the strictly anaerobic heliobacteria retained the low-potential menaquinone and, correspondingly, the ancestral versions of the homodimeric photosynthetic reaction complex and of the cytochrome b_6f complex.

The proposed scenario is illustrated in Fig. 1.1, which we have upgraded as compared to a previous publication (Mulkidjanian et al., 2006). Here, the phototrophic phyla are depicted in accordance with the depth of their

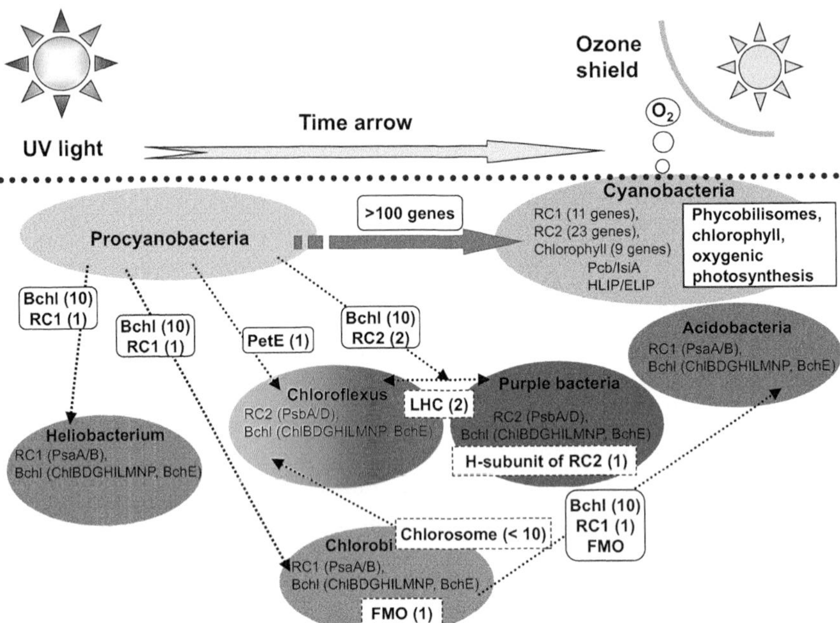

Figure 1.1 Distribution of the photosynthesis gene contents in different lineages of phototrophs and the directions of proposed lateral gene transfer. *(Modified from Mulkidjanian, Bychkov, Dibrova, Galperin, & Koonin (2012), Mulkidjanian, Bychkov, Dibrova, Galperin, & Koonin (2006), see text for details).* (For colour version of this figure, the reader is referred to the online version of this book.)

location in modern (and, perhaps, primordial) microbial mats (Nisbet & Sleep, 2001). Rounded text boxes show the extent of photosynthesis gene lateral transfer between the phyla; the numbers of CyOGs transferred are indicated in parentheses. Rectangular text boxes show major photosynthesis-relevant 'inventions' that occurred inside (solid box) or outside (dashed boxes) the (pro)cyanobacterial lineage. The colouring of the Chloroflexi reflects the presence in this group of both green chlorosome-containing organisms (Pierson & Castenholz, 1974) and the pink ones, lacking the chlorosomes (Hanada, Takaichi, Matsuura, & Nakamura, 2002). The chlorosome-less Chloroflexi carry their photosynthesis genes in an operon that is similar in gene content and even in the gene order to the RC2-encoding operon of purple bacteria (Yamada et al., 2005). Since the RC2 of Chloroflexi is menaquinone-dependent and operates at lower redox potentials than the ubiquinone-dependent RC2 of proteobacteria, we suggest that the latter have attained their RC2 and, perhaps, the light-harvesting proteins via Chloroflexi. In proteobacteria, a new protein was recruited, the 'heavy'

H-subunit that cups/protects RC2 from the cytoplasmic side. In other phototrophs, the RCs are cupped either by phycobilisomes (cyanobacteria) or by chlorosomes (Chlorobi, Chloroflexi, and *Cand.* Chloracidobacterium), so that part of the light excitation energy pours into the RC2 through this interface. The gene for the H-subunit, although present in the photosynthesis gene cluster, is located separately from the genes of the other two RC subunits (Suwanto & Kaplan, 1989). In Chlorobi and *Ca.* Chloracidobacterium, a bacteriochlorophyll-binding Fenna–Matthews–Olson protein (FMO) protein (Matthews, Fenna, Bolognesi, Schmid, & Olson, 1979) is used to mediate the excitation transfer from the chlorosome to the RC1.

The picture emphasizes that while the surface phototrophs, moving along the time arrow from the anoxic into the self-made oxygenated world, underwent a major transformation from anoxygenic procyanobacteria to cyanobacteria (Raymond & Segre, 2006; Rutherford et al., 2012), the inhabitants of the lower layers of microbial communities (mats), better protected from oxygen, could retain their traits in the course of evolution.

4. PHOTOSYNTHESIS AND THE EMERGENCE OF LIFE

The evolutionary scenario, as put forward in the previous section, implies that the first phototrophic organisms depended on solar light and dwelled in illuminated habitats. An alternative hypothesis, proposed by Nisbet and co-workers (Nisbet, Cann, & Dover, 1995), has suggested that anoxygenic photosynthesis could have evolved from the infrared phototaxis systems of bacteria that dwelled around deep-sea hydrothermal vents. Geologically, this hypothesis joined the popular line of thinking, according to which life emerged around deep-sea hydrothermal vents, where it was protected from the damaging impact of the solar UV radiation, which had been orders of magnitude stronger in the absence of ozone layer (Russell, Hall, Cairns-Smith, & Braterman, 1988; Sagan, 1973). Biochemically, this hypothesis provided backing to the phylogenetic reconstruction, according to which the proteobacterial bacteriochlorophylls, which absorb in the infrared part of the solar spectrum, were the first photosynthetic pigments and proteobacteria, accordingly, could be the first phototrophs (Xiong, Fischer, Inoue, Nakahara, & Bauer, 2000).

The scenario of proteobacteria as first phototrophs, as well as the underlying phylogenetic analysis, has been criticized by many authors and from different viewpoints (Bryant et al., 2012; Green & Gantt, 2000; Gupta, 2012; Mix, Haig, & Cavanaugh, 2005). As one more argument against the ancestral status of phototrophic proteobacteria, it could be noted that high-potential quinones are found exactly in those lineages of proteobacteria that harbour photosynthetic enzymes, i.e. α-, β-, and γ-proteobacteria. As argued by Nitschke and co-workers, the replacement of menaquinone by a high-potential ubiquinone took place in these lineages (Schoepp-Cothenet et al., 2009; Schoepp-Cothenet et al., 2013); accordingly, the very emergence of α-, β-, and γ-proteobacteria should have followed the oxygenation of the atmosphere only some 2.5 Gy ago (Hazen et al., 2011).

The geological viewpoint on the emergence of life around deep-sea hydrothermal vents has also been challenged. It has gradually become clear that the emergence of the first biopolymers could hardly happen without the participation of solar UV radiation as a selective factor. The common property of native nucleobases, which discriminates them from other molecules of comparable complexity, is their exceptional photostability (Mulkidjanian, Cherepanov, & Galperin, 2003; Serrano-Andres & Merchan, 2009; Sobolewski & Domcke, 2006). We have argued earlier that because of this property, nucleotides could have been photo-selected by solar UV radiation – in the absence of an ozone layer – from a plethora of abiotically (photo)synthesized organic compounds (Mulkidjanian et al., 2003). It has been shown that nucleobases and nucleotides can specifically form in formamide-containing solutions, particularly under UV irradiation and in the presence of phosphorous compounds (Barks et al., 2010; Costanzo, Saladino, Crestini, Ciciriello, & Di Mauro, 2007; Schoffstall, 1976). More recently, it has been found that after a prolonged UV illumination of complex mixtures of ribonucleotides and diverse by-products of nucleotide synthesis, only 2′,3′-cyclic nucleotides remained in the solution as the most photostable of the produced molecules (Powner, Gerland, & Sutherland, 2009). The 2′,3′-cyclic ribonucleotides can polymerize into oligomers even in the absence of templates (Verlander, Lohrmann, & Orgel, 1973); this polymerization is driven by the cleavage of one of the two phosphoester bonds (transesterification). Hence, cyclic nucleotides, which could form abiotically at high concentrations of formamide and phosphate (Costanzo, Pino, Botta, Saladino, & Di Mauro, 2011; Costanzo et al., 2007; Saladino, Botta, Pino, Costanzo, & Di Mauro, 2012a; Saladino, Crestini, Pino, Costanzo, & Di Mauro, 2012b), could serve as both monomers and the energy source for the abiotic formation of RNA replicators and ribozymes.

Independently, the 'hatcheries' of the first cells were reconstructed by combining a geochemical analysis with phylogenomic scrutiny of the inorganic ion requirements of universal components of modern cells

(Mulkidjanian, Bychkov, Dibrova, Galperin, & Koonin, 2012a; Mulkidjanian, Bychkov, Dibrova, Galperin, & Koonin, 2012b). These ubiquitous, and by inference primordial, proteins and functional systems show affinity to and functional requirement for K^+, Zn^{2+}, Mn^{2+}, and phosphate. Thus, protocells must have evolved in habitats with a high K^+/Na^+ ratio and relatively high concentrations of Zn, Mn, and phosphorous compounds. Geochemical reconstruction shows that the ionic composition conducive to the origin of cells is compatible with emissions of vapour-dominated zones of inland geothermal systems. A major distinctive feature of such systems is the separation of the vapour phase from the liquid phase due to the boiling of the ascending hot hydrothermal fluids. The ascending vapour, after reaching the surface of the rock, discharges via numerous fumaroles and mud pots, which make a geothermal field. The chemical composition of the two phases differs dramatically: the liquid phase contains large amounts of Na and Cl whereas the vapour phase is specifically enriched in potassium ions, H_2S, CO_2, and NH_3 (Aver'ev, 1961; Fournier, 2004; Mulkidjanian, Bychkov, Dibrova, Galperin, & Koonin, 2012a; Mulkidjanian, Bychkov, Dibrova, Galperin, & Koonin, 2012b; White, Muffler, & Truesdell, 1971).

As argued elsewhere, anoxic geothermal fields should have been particularly conducive for abiogenic synthesis of ribonucleotides and their polymerization (Dibrova, Cherepanov, et al., submitted for publication; Dibrova, Chudetsky, et al., 2012; Mulkidjanian, Bychkov, Dibrova, Galperin, & Koonin, 2013; Mulkidjanian, Bychkov, Dibrova, Galperin, & Koonin, 2012a, 2012b), unlike any marine habitats that could never been enriched in simple amides, phosphorous compounds and borate, all of which are needed for abiotic formation of nucleobases and nucleotides (Benner, Carrigan, Ricardo, & Frye, 2006; Saladino, Botta, et al., 2012a; Saladino, Crestini, et al., 2012b). Two types of environments relevant for the early stages of evolution can be expected at anoxic geothermal fields, namely: (1) periodically wetted, illuminated mineral surfaces that could serve as templates and (photo)catalysts for diverse abiotic syntheses and (2) puddles and pools of cooled, condensed vapour that would function as concentrators of prebiotic organic molecules. Each such pool would 'harvest' substrates from its catchment area and should have contained mixture of water, simple amides, silica, metal sulfides, and amphiphilic molecules (which could be present as micelles). These pools could have served as hatcheries of the first replicating organisms. Under anoxic, CO_2-dominated atmosphere, the ionic composition of pools of cool, condensed vapour at anoxic geothermal fields would resemble the internal milieu of modern cells. Such pools would be lined with porous silicate minerals mixed with metal sulfides, and enriched in K^+ ions and phosphorous compounds.

Concerning the mentioned clear preference of the ancient proteins for Zn and Mn as transition metal cofactors (Mulkidjanian, Bychkov, Dibrova, Galperin, & Koonin, 2012a; Mulkidjanian, Bychkov, Dibrova, Galperin, & Koonin, 2012b; Mulkidjanian & Galperin, 2009), it is noteworthy that these two metals precipitate together at the outlets of the hydrothermal and geothermal systems, forming Zn- and Mn-enriched, ring-like deposits around them (Reed & Palandri, 2006; Tivey, 2007). Currently, these metals can precipitate both as sulfides (at the sea floor) and oxides (at terrestrial systems). At the primordial Earth, however, only sulfides could precipitate at inland geothermal fields. Sulfides of Zn and Mn are among the most potent photocatalysts (Fox & Dulay, 1993; Henglein, 1984; Mulkidjanian, 2009 and references therein). Under solar light that contained an essential UV component and in the presence of high levels of CO_2 in the primordial atmosphere (Sleep, 2010), ZnS and MnS would reduce carbon dioxide to diverse organic molecules. Numerous experiments have shown that this kind of photosynthesis proceeds with a high yield, reaching 80% in the case of ZnS particles (Guzman & Martin, 2009, 2010; Henglein, 1984; Kisch & Twardzik, 1991; Reber & Meier, 1984; Yanagida, Azuma, Midori, Pac, & Sakurai, 1985a; Yanagida, Kizumoto, Ishimaru, Pac, & Sakurai, 1985b; Zhang et al., 2004, 2007). Hence, the preference of the ancient cellular systems for Zn and Mn as transition metal cofactors might simply reflect the fact that the first life forms dwelled among ZnS and MnS-enriched photosynthesizing precipitates and recruited the Zn and Mn ions, which were released upon photosynthesis, for stabilizing their proteins and RNA polymers (Mulkidjanian, 2009; Mulkidjanian & Galperin, 2009).

Hence, the porous sediments, enriched in sulfides of Zn and Mn, could serve as photosynthesizing habitats of first heterotrophic cells (Mulkidjanian, Bychkov, Dibrova, Galperin, & Koonin, 2012a; Mulkidjanian, Bychkov, Dibrova, Galperin, & Koonin, 2012b). Apparently, such organisms did not initially need photosynthetic systems of their own, because they could get organic molecules for free, from abiotic photosynthesis (Zhang et al., 2004, 2007) and also from geothermal vapour, which carries diverse organic molecules (Sleep, Meibom, Fridriksson, Coleman, & Bird, 2004).

The ZnS- and MnS-containing sediments could also provide shelter from the UV irradiation for the first organisms. Even a thin, 5 μm-layer of ZnS would attenuate the UV light by a factor of 10^{10} (Mulkidjanian, Bychkov, Dibrova, Galperin, & Koonin, 2012a; Mulkidjanian, Bychkov, Dibrova,

Galperin, & Koonin, 2012b). A stratified system could have been established within geothermal ponds where the illuminated upper layers were involved in 'light harvesting' and production of reduced organic compounds, whereas the deeper, less productive but better protected layers would have provided shelter for the replicating organisms (Mulkidjanian, Bychkov, Dibrova, Galperin, & Koonin, 2012a; Mulkidjanian, Bychkov, Dibrova, Galperin, & Koonin, 2012b). The light gradient and the interlayer metabolite exchange are typical of modern stratified phototrophic microbial communities (Nold & Ward, 1996).

It is noteworthy that for a particular organism it might be beneficial to get closer to the surface of a sediment and, hence, closer to the source of abiotically produced organic molecules. Therefore organisms that could synthesize or recruit UV-absorbing compounds, such as porphyrins, might have an evolutionary advantage (Mulkidjanian & Junge, 1997). Porphyrin-carrying membrane proteins may have eventually evolved later into the first biogenic photosynthetic apparata.

Halmann and colleagues (Halmann, Aurian-Blajeni, & Bloch, 1980) have noted the similarity between physical mechanisms of chlorophyll-based and semiconductor-based photosyntheses, which both include light-induced charge separation followed by the stabilization of the low-energy, reduced states, as shown in Fig. 1.2, where the energy diagrams for a ZnS crystal and a sulfide-oxidizing RC1 of Chlorobi are compared. Even the

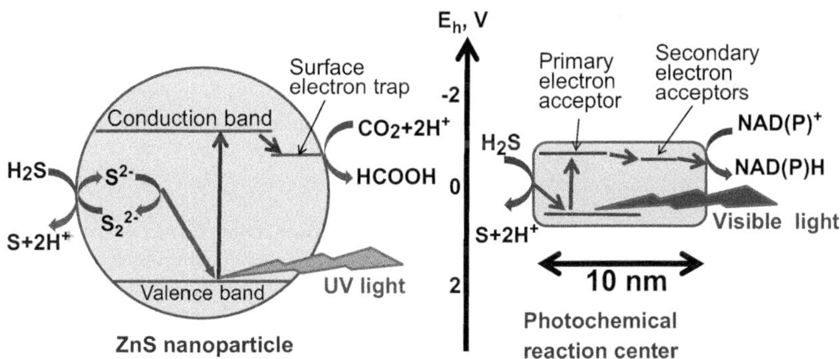

Figure 1.2 ***A comparison of energy diagrams for a photosynthesizing ZnS nanoparticle and a photosynthetic reaction centre.*** Left, the energy diagram for a ZnS crystal, based on references Henglein, Gutierrez, and Fischer (1984), Kisch and Künneth (1991), Yoneyama (1997). Right, an energy diagram for a simple, sulfide-oxidizing reaction centre complex of green sulfur bacteria is shown as an example, see Bryant et al. (2012), Frigaard and Bryant (2004), Jagannathan and Golbeck (2008) for reviews on this type of reaction centre. *(The figure is taken from Mulkidjanian and Galperin (2009)).* (For colour version of this figure, the reader is referred to the online version of this book.)

same reaction of sulfide oxidation is utilized to re-fill the photo-generated electron vacancies (holes).

The emergence of biogenic photosynthesis seems to have happened after the separation of bacteria from archaea, so it was not a very early event. Therefore it cannot be fully excluded that biogenic photosynthesis may have emerged after some ancestral bacteria, which possessed porphyrin-carrying membrane proteins, had colonized deep-sea hydrothermal vents.

Still, we consider it more plausible that porphyrin/chlorophyll photosynthesis was 'invented' by terrestrial life forms, which moved away from geothermal fields and invaded new habitats that were depleted in ZnS, MnS, and abiogenically produced organic molecules. Upon getting away from geothermal fields, the use of biogenic photosynthesis could have initially complemented the gradually diminishing ZnS-mediated photosynthesis; its contribution, however, should have increased with the departure from geothermal fields and invading, for example, terrestrial fresh-water basins. In this framework, the emergence of biogenic photosynthesis might represent a clear-cut case of functional takeover – with the primeval photochemical RCs and primordial CBB cycle accomplishing together the function that ZnS- and MnS-rich precipitates carried out at the geothermal fields: namely, utilization of solar and geothermal energy for producing organic compounds from CO_2.

ACKNOWLEDGEMENTS

This study was supported by grants from the Deutsche Forschungsgemeinschaft (DFG-Mu-1285/1-10, DFG-436-RUS 113/963/0-1), the COST Action CM0902 of the EU, the Russian Government grant no. 02.740.11.5228 (A.Y.M.), and by the Intramural Research Program of the NIH, National Library of Medicine (M.Y.G.).

REFERENCES

Adhikari, N. D., Froehlich, J. E., Strand, D. D., Buck, S. M., Kramer, D. M., & Larkin, R. M. (2011). GUN4-porphyrin complexes bind the ChlH/GUN5 subunit of Mg-chelatase and promote chlorophyll biosynthesis in *Arabidopsis*. *Plant Cell*, *23*, 1449–1467.

Alperovitch-Lavy, A., Sharon, I., Rohwer, F., Aro, E. M., Glaser, F., Milo, R., et al. (2011). Reconstructing a puzzle: existence of cyanophages containing both photosystem-I and photosystem-II gene suites inferred from oceanic metagenomic datasets. *Environmetal Microbiology*, *13*, 24–32.

Armbruster, U., Zuhlke, J., Rengstl, B., Kreller, R., Makarenko, E., Ruhle, T., et al. (2010). The *Arabidopsis* thylakoid protein PAM68 is required for efficient D1 biogenesis and photosystem II assembly. *Plant Cell*, *22*, 3439–3460.

Aver'ev, V. V. (1961). Conditions for the discharge of the Pauzhetka high-temperature waters in Southern Kamchatka. In *Proceedings of the volcanology* (pp. 90–98). Moscow: Laboratory of the Russian Academy of Sciences. Issue 19.

Barks, H. L., Buckley, R., Grieves, G. A., Di Mauro, E., Hud, N.V., & Orlando, T. M. (2010). Guanine, adenine, and hypoxanthine production in UV-irradiated formamide solutions: relaxation of the requirements for prebiotic purine nucleobase formation. *Chembiochem: A European Journal of Chemical Biology*, *11*, 1240–1243.

Barneche, F., Winter, V., Crevecoeur, M., & Rochaix, J. D. (2006). ATAB2 is a novel factor in the signalling pathway of light-controlled synthesis of photosystem proteins. *EMBO Journal*, *25*, 5907–5918.

Battchikova, N., Wei, L., Du, L., Bersanini, L., Aro, E. M., & Ma, W. (2011). Identification of novel Ssl0352 protein (NdhS), essential for efficient operation of cyclic electron transport around photosystem I, in NADPH:plastoquinone oxidoreductase (NDH-1) complexes of *Synechocystis* sp. PCC 6803. *Journal of Biological Chemistry*, *286*, 36992–37001.

Benner, S. A., Carrigan, M. A., Ricardo, A., & Frye, F. (2006). Setting the stage: the history, chemistry and geobiology behind RNA. In R. F. Gesteland, T. R. Cech & J. Atkins (Eds.), *The RNA world* (3rd ed.). Cold Spring Harbor, NY: Cold Spring Harbor Laboratory Press.

Berg, I. A., Keppen, O. I., Krasil'nikova, E. N., Ugol'kova, N.V., & Ivanovskii, R. N. (2005). Carbon metabolism of filamentous anoxygenic phototrophic bacteria of the family *Oscillochloridaceae*. *Microbiology*, *74*, 258–264.

Blankenship, R. E. (1992). Origin and early evolution of photosynthesis. *Photosynthesis Research*, *33*, 91–111.

Brasier, M. D., Green, O. R., Jephcoat, A. P., Kleppe, A. K., Van Kranendonk, M. J., Lindsay, J. F., et al. (2002). Questioning the evidence for Earth's oldest fossils. *Nature*, *416*, 76–81.

Bryant, D. A., Costas, A. M., Maresca, J. A., Chew, A. G., Klatt, C. G., Bateson, M. M., et al. (2007). *Candidatus* Chloracidobacterium thermophilum: an aerobic phototrophic Acidobacterium. *Science*, *317*, 523–526.

Bryant, D. A., & Frigaard, N. U. (2006). Prokaryotic photosynthesis and phototrophy illuminated. *Trends in Microbiology*, *14*, 488–496.

Bryant, D. A., Liu, Z., Li, T., Zhao, F., Costas, A. M.G., Klatt, C. G., et al. (2012). Comparative and functional genomics of anoxygenic green bacteria from the taxa *Chlorobi*, *Chloroflexi*, and Acidobacteria. In R. L. Burnap & W. F. L. Vermaas (Eds.), *Functional genomics and evolution of photosynthetic systems* (pp. 47–102). New York: Springer.

Chisholm, S. W., Frankel, S. L., Goericke, R., Olson, R. J., Palenik, B., Waterbury, J. B., et al. (1992). *Prochlorococcus marinus* nov. gen. nov. sp.: an oxyphototrophic marine prokaryote containing divinyl chlorophyll a and b. *Archives of Microbiology*, *157*, 297–300.

Costanzo, G., Pino, S., Botta, G., Saladino, R., & Di Mauro, E. (2011). May cyclic nucleotides be a source for abiotic RNA synthesis? *Origins of Life and Evolution of the Biosphere: The Journal of the International Society for the Study of the Origin of Life*, *41*, 559–562.

Costanzo, G., Saladino, R., Crestini, C., Ciciriello, F., & Di Mauro, E. (2007). Nucleoside phosphorylation by phosphate minerals. *Journal of Biological Chemistry*, *282*, 16729–16735.

Dauvillee, D., Stampacchia, O., Girard-Bascou, J., & Rochaix, J. D. (2003). Tab2 is a novel conserved RNA binding protein required for translation of the chloroplast *psaB* mRNA. *EMBO Journal*, *22*, 6378–6388.

Davison, P. A., Schubert, H. L., Reid, J. D., Iorg, C. D., Heroux, A., Hill, C. P., et al. (2005). Structural and biochemical characterization of Gun4 suggests a mechanism for its role in chlorophyll biosynthesis. *Biochemistry*, *44*, 7603–7612.

Dibrova, D. V., Cherepanov, D. A., Galperin, M. Y., Skulachev, V. P., & Mulkidjanian, A. Y. (2013). Evolution of cytochrome bc complexes: from membrane electron translocases to triggers of apoptosis. *Biochimica et Biophysica Acta*, in press.

Dibrova, D.V., Chudetsky, M.Y., Galperin, M.Y., Koonin, E.V., & Mulkidjanian, A.Y. (2012b). The role of energy in the emergence of biology from chemistry. *Origins of Life and Evolution of the Biosphere: The Journal of the International Society for the Study of the Origin of Life*, *45*, 459–468.

Dufresne, A., Salanoubat, M., Partensky, F., Artiguenave, F., Axmann, I. M., Barbe, V., et al. (2003). Genome sequence of the cyanobacterium *Prochlorococcus marinus* SS120, a nearly minimal oxyphototrophic genome. *Proceedings of the National Academy of Sciences of the United States of America, 100,* 10020–10025.

Eisen, J. A., Nelson, K. E., Paulsen, I. T., Heidelberg, J. F., Wu, M., Dodson, R. J., et al. (2002). The complete genome sequence of *Chlorobium tepidum* TLS, a photosynthetic, anaerobic, green-sulfur bacterium. *Proceedings of the National Academy of Sciences of the United States of America, 99,* 9509–9514.

Erales, J., Lignon, S., & Gontero, B. (2009). CP12 from *Chlamydomonas reinhardtii,* a permanent specific "chaperone-like" protein of glyceraldehyde-3-phosphate dehydrogenase. *Journal of Biological Chemistry, 284,* 12735–12744.

Espinosa, J., Fuentes, I., Burillo, S., Rodriguez-Mateos, F., & Contreras, A. (2006). SipA, a novel type of protein from *Synechococcus* sp. PCC 7942, binds to the kinase domain of NblS. *FEMS Microbiology Letters, 254,* 41–47.

Federhen, S. (2012). The NCBI taxonomy database. *Nucleic Acids Res, 40,* D136–D143.

Fournier, R. O. (2004). *Geochemistry and dynamics of the Yellowstone National Park hydrothermal system.* Menlo Park, CA: US Geological Survey.

Fox, M. A., & Dulay, M. T. (1993). Heterogeneous photocatalysis. *Chemical Reviews, 93,* 341–357.

Frigaard, N. U., & Bryant, D. (2004). Seeing green bacteria in a new light: genomics-enabled studies of the photosynthetic apparatus in green sulfur bacteria and filamentous anoxygenic phototrophic bacteria. *Archives of Microbiology, 182,* 265–276.

Garcia Costas, A. M., Liu, Z., Tomsho, L. P., Schuster, S. C., Ward, D. M., & Bryant, D. A. (2012). Complete genome of *Candidatus* Chloracidobacterium thermophilum, a chlorophyll-based photoheterotroph belonging to the phylum Acidobacteria. *Environmetal Microbiology, 14,* 177–190.

Garcia-Pichel, F. (1998). Solar ultraviolet and the evolutionary history of cyanobacteria. *Origins of Life and Evolution of the Biosphere: The Journal of the International Society for the Study of the Origin of Life, 28,* 321–347.

Geelhoed, J. S., Kleerebezem, R., Sorokin, D.Y., Stams, A. J., & van Loosdrecht, M. C. (2010). Reduced inorganic sulfur oxidation supports autotrophic and mixotrophic growth of *Magnetospirillum* strain J10 and *Magnetospirillum gryphiswaldense. Environmetal Microbiology, 12,* 1031–1040.

Gontero, B., & Maberly, S. C. (2012). An intrinsically disordered protein, CP12: jack of all trades and master of the Calvin cycle. *Biochemical Society Transactions, 40,* 995–999.

Green, B. R. (2011). Chloroplast genomes of photosynthetic eukaryotes. *Plant Journal, 66,* 34–44.

Green, B. R., & Gantt, E. (2000). Is photosynthesis really derived from purple bacteria? *Journal of Phycology, 36,* 983–985.

Gupta, R. S. (2012). Origin and spread of photosynthesis based upon conserved sequence features in key bacteriochlorophyll biosynthesis proteins. *Molecular Biology and Evolution, 29,* 3397–3412.

Gupta, R. S., Chander, P., & George, S. (2013). Phylogenetic framework and molecular signatures for the class *Chloroflexi* and its different clades; proposal for division of the class *Chloroflexi* class. nov. into the suborder *Chloroflexineae* subord. nov., consisting of the emended family *Oscillochloridaceae* and the family *Chloroflexaceae* fam. nov., and the suborder *Roseiflexineae* subord. nov., containing the family *Roseiflexaceae* fam. nov. *Antonie Van Leeuwenhoek, 103,* 99–119.

Guzman, M. I., & Martin, S. T. (2009). Prebiotic metabolism: production by mineral photoelectrochemistry of alpha-ketocarboxylic acids in the reductive tricarboxylic acid cycle. *Astrobiology, 9,* 833–842.

Guzman, M. I., & Martin, S. T. (2010). Photo-production of lactate from glyoxylate: how minerals can facilitate energy storage in a prebiotic world. *Chemical Communications (Cambridge, England), 46*, 2265–2267.

Halmann, M., Aurian-Blajeni, B., & Bloch, S. (1980). Photoassisted carbon dioxide reduction and formation of two and three-carbon compounds. In *Origin of life: Proceedings of the third ISSOL meeting and sixth ICOL meeting* (pp. 143–150). Dordrecht, Jerusalem, Israel: D. Reidel Publishing Co.

Hanada, S., Takaichi, S., Matsuura, K., & Nakamura, K. (2002). *Roseiflexus castenholzii* gen. nov., sp. nov., a thermophilic, filamentous, photosynthetic bacterium that lacks chlorosomes. *International Journal of Systematic and Evolutionary Microbiology, 52*, 187–193.

Hazen, R. M., Bekker, A., Bish, D. L., Bleeker, W., Downs, R. T., Farquhar, J., et al. (2011). Needs and opportunities in mineral evolution research. *American Mineralogist, 96*, 953–963.

Henglein, A. (1984). Catalysis of photochemical reactions by colloidal semiconductors. *Pure and Applied Chemistry. Chimie pure et appliquée, 56*, 1215–1224.

Henglein, A., Gutierrez, M., & Fischer, C. H. (1984). Photochemistry of colloidal metal sulfides. 6. Kinetics of interfacial reactions at ZnS particles. *Berichte Der Bunsen-Gesellschaft-Physical Chemistry and Chemical Physics, 88*, 170–175.

Herter, S., Farfsing, J., Gad'On, N., Rieder, C., Eisenreich, W., Bacher, A., et al. (2001). Autotrophic CO_2 fixation by *Chloroflexus aurantiacus*: study of glyoxylate formation and assimilation via the 3-hydroxypropionate cycle. *Journal of Bacteriology, 183*, 4305–4316.

Hohmann-Marriott, M. F., & Blankenship, R. E. (2011). Evolution of photosynthesis. *Annual Review of Plant Biology, 62*, 515–548.

Howard, T. P., Fryer, M. J., Singh, P., Metodiev, M., Lytovchenko, A., Obata, T., et al. (2011). Antisense suppression of the small chloroplast protein CP12 in tobacco alters carbon partitioning and severely restricts growth. *Plant Physiology, 157*, 620–631.

Iino, T., Mori, K., Uchino, Y., Nakagawa, T., Harayama, S., & Suzuki, K. (2010). Ignavibacterium album gen. nov., sp. nov., a moderately thermophilic anaerobic bacterium isolated from microbial mats at a terrestrial hot spring and proposal of Ignavibacteria classis nov., for a novel lineage at the periphery of green sulfur bacteria. *International Journal of Systematic and Evolutionary Microbiology, 60*, 1376–1382.

Imhoff, J. F. (2003). Phylogenetic taxonomy of the family *Chlorobiaceae* on the basis of 16S rRNA and *fmo* (Fenna–Matthews–Olson protein) gene sequences. *International Journal of Systematic and Evolutionary Microbiology, 53*, 941–951.

Imhoff, J. F., Suling, J., & Petri, R. (1998). Phylogenetic relationships among the *Chromatiaceae*, their taxonomic reclassification and description of the new genera *Allochromatium, Halochromatium, Isochromatium, Marichromatium, Thiococcus, Thiohalocapsa* and *Thermochromatium*. *International Journal of Systematic Bacteriology, 48*(Pt 4), 1129–1143.

Imhoff, J. F., Truper, H. G., & Pfennig, N. (1984). Rearrangement of the species and genera of the phototrophic 'purple nonsulfur bacteria'. *International Journal of Systematic Bacteriology, 34*, 340–343.

Jagannathan, B., & Golbeck, J. H. (2008). Unifying principles in homodimeric type I photosynthetic reaction centers: properties of PscB and the F-A, F-B and F-X iron-sulfur clusters in green sulfur bacteria. *Biochimica et Biophysica Acta, 1777*, 1535–1544.

Jung, H. S., & Niyogi, K. K. (2010). Mutations in *Arabidopsis* YCF20-like genes affect thermal dissipation of excess absorbed light energy. *Planta, 231*, 923–937.

Junier, P., Junier, T., Podell, S., Sims, D. R., Detter, J. C., Lykidis, A., et al. (2010). The genome of the Gram-positive metal- and sulfate-reducing bacterium *Desulfotomaculum reducens* strain MI-1. *Environmetal Microbiology, 12*, 2738–2754.

Kane, S. R., Chakicherla, A. Y., Chain, P. S., Schmidt, R., Shin, M. W., Legler, T. C., et al. (2007). Whole-genome analysis of the methyl *tert*-butyl ether-degrading beta-proteobacterium *Methylibium petroleiphilum* PM1. *Journal of Bacteriology, 189*, 1931–1945.

Kaneko, T., Nakajima, N., Okamoto, S., Suzuki, I., Tanabe, Y., Tamaoki, M., et al. (2007). Complete genomic structure of the bloom-forming toxic cyanobacterium *Microcystis aeruginosa* NIES-843. *DNA Research, 14*, 247–256.

Kaneko, T., Sato, S., Kotani, H., Tanaka, A., Asamizu, E., Nakamura, Y., et al. (1996). Sequence analysis of the genome of the unicellular cyanobacterium *Synechocystis* sp. strain PCC6803. II. Sequence determination of the entire genome and assignment of potential protein-coding regions. *DNA Research, 3*, 109–136.

Keeling, P. J. (2009). Chromalveolates and the evolution of plastids by secondary endosymbiosis. *Journal of Eukaryotic Microbiology, 56*, 1–8.

Keeling, P. J. (2010). The endosymbiotic origin, diversification and fate of plastids. *Philosophical Transactions of the Royal Society of London. Series B, Biological Sciences, 365*, 729–748.

Keppen, O. I., Baulina, O. I., Lysenko, A. M., & Kondratieva, E. N. (1993). A new green bacterium of the *Chloroflexaceae* family. *Mikrobiologiya, 62*, 267–275.

Keren, N., Ohkawa, H., Welsh, E. A., Liberton, M., & Pakrasi, H. B. (2005). Psb29, a conserved 22-kD protein, functions in the biogenesis of photosystem II complexes in *Synechocystis* and *Arabidopsis*. *Plant Cell, 17*, 2768–2781.

Kimble, L. K., Mandelco, L., Woese, C. R., & Madigan, M. T. (1995). *Heliobacterium modesticaldum*, sp. nov., a thermophilic heliobacterium of hot springs and volcanic soils. *Archives of Microbiology, 163*, 259–267.

Kisch, H., & Künneth, R. (1991). Photocatalysis by semiconductor powders: preparative and mechanistic aspects. In J. Rabek (Ed.), *Photochemistry and photophysics* (pp. 131–175). CRC Press Inc.

Kisch, H., & Twardzik, G. (1991). Heterogeneous photocatalysis 9. Zinc-sulfide catalyzed photoreduction of carbon dioxide. *Chemische Berichte, 124*, 1161–1162.

Kiss, H., Nett, M., Domin, N., Martin, K., Maresca, J. A., Copeland, A., et al. (2011). Complete genome sequence of the filamentous gliding predatory bacterium *Herpetosiphon aurantiacus* type strain (114-95^{T}). *Standards in Genomic Sciences, 5*, 356–370.

Klatt, C. G., Bryant, D. A., & Ward, D. M. (2007). Comparative genomics provides evidence for the 3-hydroxypropionate autotrophic pathway in filamentous anoxygenic phototrophic bacteria and in hot spring microbial mats. *Environmetal Microbiology, 9*, 2067–2078.

Kuznetsov, B. B., Ivanovsky, R. N., Keppen, O. I., Sukhacheva, M. V., Bumazhkin, B. K., Patutina, E. O., et al. (2011). Draft genome sequence of the anoxygenic filamentous phototrophic bacterium *Oscillochloris trichoides* subsp. DG-6. *Journal of Bacteriology, 193*, 321–322.

Larkin, R. M., Alonso, J. M., Ecker, J. R., & Chory, J. (2003). GUN4, a regulator of chlorophyll synthesis and intracellular signaling. *Science, 299*, 902–906.

Li, W., Gao, H., Yin, C., & Xu, X. (2012). Identification of a novel thylakoid protein gene involved in cold acclimation in cyanobacteria. *Microbiology, 158*, 2440–2449.

Lindell, D., Sullivan, M. B., Johnson, Z. I., Tolonen, A. C., Rohwer, F., & Chisholm, S. W. (2004). Transfer of photosynthesis genes to and from *Prochlorococcus* viruses. *Proceedings of the National Academy of Sciences of the United States of America, 101*, 11013–11018.

Li, H., Sawaya, M. R., Tabita, F. R., & Eisenberg, D. (2005). Crystal structure of a RuBisCO-like protein from the green sulfur bacterium *Chlorobium tepidum*. *Structure, 13*, 779–789.

Liu, Z., Frigaard, N. U., Vogl, K., Iino, T., Ohkuma, M., Overmann, J., et al. (2012). Complete genome of *Ignavibacterium album*, a metabolically versatile, flagellated, facultative anaerobe from the phylum Chlorobi. *Frontiers in Microbiology, 3*, 185.

Mann, N. H., Cook, A., Millard, A., Bailey, S., & Clokie, M. (2003). Marine ecosystems: bacterial photosynthesis genes in a virus. *Nature, 424*, 741.

Mannisto, M. K., Rawat, S., Starovoytov, V., & Haggblom, M. M. (2011). *Terriglobus saanensis* sp. nov., an acidobacterium isolated from tundra soil. *International Journal of Systematic and Evolutionary Microbiology, 61*, 1823–1828.

Matthews, B. W., Fenna, R. E., Bolognesi, M. C., Schmid, M. F., & Olson, J. M. (1979). Structure of a bacteriochlorophyll *a*-protein from the green photosynthetic bacterium *Prosthecochloris aestuarii*. *Journal of Molecular Biology*, *131*, 259–285.

Mix, L. J., Haig, D., & Cavanaugh, C. M. (2005). Phylogenetic analyses of the core antenna domain: investigating the origin of photosystem I. *Journal of Molecular Evolution*, *60*, 153–163.

Moore, B., & Webster, T. A. (1913). Synthesis by sunlight in relationship to the origin of life. Synthesis of formaldehyde from carbon dioxide and water by inorganic colloids acting as transformers of light energy. *Proceedings of the Royal Society of London. Series B, Containing Papers of a Biological Character. Royal Society (Great Britain)*, *87*, 163–176.

Mulkidjanian, A. Y. (2009). On the origin of life in the zinc world: 1. Photosynthesizing, porous edifices built of hydrothermally precipitated zinc sulfide as cradles of life on Earth. *Biology Direct*, *4*, 26.

Mulkidjanian, A. Y., Bychkov, A. Y., Dibrova, D. V., Galperin, M. Y., & Koonin, E. V. (2012a). Origin of first cells at terrestrial, anoxic geothermal fields. *Proceedings of the National Academy of Sciences of the United States of America*, *109*, E821–E830.

Mulkidjanian, A. Y., Bychkov, A. Y., Dibrova, D. V., Galperin, M. Y., & Koonin, E. V. (2012b). Open questions on the origin of life at anoxic geothermal fields. *Origins of Life and Evolution of the Biosphere: The Journal of the International Society for the Study of the Origin of Life*, *42*, 507–516.

Mulkidjanian, A. Y., Cherepanov, D. A., & Galperin, M. Y. (2003). Survival of the fittest before the beginning of life: selection of the first oligonucleotide-like polymers by UV light. *BMC Evolutionary Biology*, *3*, 12.

Mulkidjanian, A. Y., & Galperin, M. Y. (2009). On the origin of life in the zinc world. 2. Validation of the hypothesis on the photosynthesizing zinc sulfide edifices as cradles of life on Earth. *Biology Direct*, *4*, 27.

Mulkidjanian, A. Y., & Junge, W. (1997). On the origin of photosynthesis as inferred from sequence analysis. A primordial UV-protector as common ancestor of reaction centers and antenna proteins. *Photosynthesis Research*, *51*, 27–42.

Mulkidjanian, A. Y., Koonin, E. V., Makarova, K. S., Mekhedov, S. L., Sorokin, A., Wolf, Y. I., et al. (2006). The cyanobacterial genome core and the origin of photosynthesis. *Proceedings of the National Academy of Sciences of the United States of America*, *103*, 13126–13131.

Munoz, R., Yarza, P., Ludwig, W., Euzeby, J., Amann, R., Schleifer, K. H., et al. (2011). Release LTPs104 of the all-species living tree. *Systematic and Applied Microbiology*, *34*, 169–170.

Nagashima, S., Kamimura, A., Shimizu, T., Nakamura-Isaki, S., Aono, E., Sakamoto, K., et al. (2012). Complete genome sequence of phototrophic betaproteobacterium *Rubrivivax gelatinosus* IL144. *Journal of Bacteriology*, *194*, 3541–3542.

Nakamura, Y., Kaneko, T., Sato, S., Mimuro, M., Miyashita, H., Tsuchiya, T., et al. (2003). Complete genome structure of *Gloeobacter violaceus* PCC 7421, a cyanobacterium that lacks thylakoids. *DNA Research*, *10*, 137–145.

Nisbet, E. G., Cann, J. R., & Dover, C. L. (1995). Did photosynthesis begin from thermotaxis? *Nature*, *373*, 479–480.

Nisbet, E. G., & Sleep, N. H. (2001). The habitat and nature of early life. *Nature*, *409*, 1083–1091.

Nold, S. C., & Ward, D. M. (1996). Photosynthate partitioning and fermentation in hot spring microbial mat communities. *Applied and Environmental Microbiology*, *62*, 4598–4607.

Ohtsuka, M., Oyabu, J., Kashino, Y., Satoh, K., & Koike, H. (2004). Inactivation of *ycf33* results in an altered cyclic electron transport pathway around photosystem I in *Synechocystis* sp. PCC6803. *Plant Cell Physiology*, *45*, 1243–1251.

Olson, J. M., & Blankenship, R. E. (2004). Thinking about the evolution of photosynthesis. *Photosynthesis Research*, *80*, 373–386.

Pfreundt, U., Stal, L. J., Voss, B., & Hess, W. R. (2012). Dinitrogen fixation in a unicellular chlorophyll d-containing cyanobacterium. *ISME Journal*, *6*, 1367–1377.

Pierson, B. K., & Castenholz, R.W. (1974). A phototrophic gliding filamentous bacterium of hot springs, *Chloroflexus aurantiacus*, gen. and sp. nov. *Archives of Microbiology, 100*, 5–24.

Podosokorskaya, O., Kadnikov, V., Gavrilov, S., Mardanov, A., Merkel, A., Karnachuk, O., et al. (2013). Characterization of *Melioribacter roseus* gen. nov., sp. nov., a novel facultatively anaerobic thermophilic cellulolytic bacterium from the class Ignavibacteria, and a proposal of a novel bacterial phylum *Ignavibacteriae*. *Environmental Microbiology*, published online, doi: 10.1111/1462-2920.12067.

Powner, M. W., Gerland, B., & Sutherland, J. D. (2009). Synthesis of activated pyrimidine ribonucleotides in prebiotically plausible conditions. *Nature, 459*, 239–242.

Punta, M., Coggill, P. C., Eberhardt, R.Y., Mistry, J., Tate, J., Boursnell, C., et al. (2012). The Pfam protein families database. *Nucleic Acids Research, 40*, D290–D301.

Rawat, S. R., Mannisto, M. K., Bromberg, Y., & Haggblom, M. M. (2012). Comparative genomic and physiological analysis provides insights into the role of *Acidobacteria* in organic carbon utilization in Arctic tundra soils. *FEMS Microbiology Ecology, 82*, 341–355.

Raymond, J., & Blankenship, R. E. (2004). The evolutionary development of the protein complement of photosystem 2. *Biochimica et Biophysica Acta, 1655*, 133–139.

Raymond, J., & Segre, D. (2006). The effect of oxygen on biochemical networks and the evolution of complex life. *Science, 311*, 1764–1767.

Raymond, J., Zhaxybayeva, O., Gogarten, J. P., & Blankenship, R. E. (2003). Evolution of photosynthetic prokaryotes: a maximum-likelihood mapping approach. *Philosophical Transactions of the Royal Society of London. Series B, Biological Sciences, 358*, 223–230.

Raymond, J., Zhaxybayeva, O., Gogarten, J. P., Gerdes, S. Y., & Blankenship, R. E. (2002). Whole-genome analysis of photosynthetic prokaryotes. *Science, 298*, 1616–1620.

Reber, J. F., & Meier, K. (1984). Photochemical production of hydrogen with zinc-sulfide suspensions. *Journal of Physical Chemistry, 88*, 5903–5913.

Reed, M. H., & Palandri, J. (2006). Sulfide mineral precipitation from hydrothermal fluids. In D. Vaughan (Ed.), *Sulfide mineralogy and geochemistry* (pp. 609–631). Geochemical Society, Mineralogical Society of America.

Rengstl, B., Oster, U., Stengel, A., & Nickelsen, J. (2011). An intermediate membrane subfraction in cyanobacteria is involved in an assembly network for photosystem II biogenesis. *Journal of Biological Chemistry, 286*, 21944–21951.

Rocap, G., Larimer, F. W., Lamerdin, J., Malfatti, S., Chain, P., Ahlgren, N. A., et al. (2003). Genome divergence in two *Prochlorococcus* ecotypes reflects oceanic niche differentiation. *Nature, 424*, 1042–1047.

Russell, M. J., Hall, A. J., Cairns-Smith, A. G., & Braterman, P. S. (1988). Submarine hot springs and the origin of life. *Nature, 336*, 117.

Rutherford, A. W., Osyczka, A., & Rappaport, F. (2012). Back-reactions, short-circuits, leaks and other energy wasteful reactions in biological electron transfer: redox tuning to survive life in O_2. *FEBS Letters, 586*, 603–616.

Sagan, C. (1973). Ultraviolet selection pressure on earliest organisms. *Journal of Theoretical Biology, 39*, 195–200.

Sakayori, T., Shiraiwa, Y., & Suzuki, I. (2009). A *Synechocystis* homolog of SipA protein, Ssl3451, enhances the activity of the histidine kinase Hik33. *Plant Cell Physiology, 50*, 1439–1448.

Saladino, R., Botta, G., Pino, S., Costanzo, G., & Di Mauro, E. (2012a). From the one-carbon amide formamide to RNA all the steps are prebiotically possible. *Biochimie, 94*, 1451–1456.

Saladino, R., Crestini, C., Pino, S., Costanzo, G., & Di Mauro, E. (2012b). Formamide and the origin of life. *Physics of Life Reviews, 9*, 84–104.

Sarrou, I., Khan, Z., Cowgill, J., Lin, S., Brune, D., Romberger, S., et al. (2012). Purification of the photosynthetic reaction center from *Heliobacterium modesticaldum*. *Photosynthesis Research, 111*, 291–302.

Sato, N. (2002). Comparative analysis of the genomes of cyanobacteria and plants. *Genome Informatics, 13*, 173–182.

Sattley, W. M., Madigan, M. T., Swingley, W. D., Cheung, P. C., Clocksin, K. M., Conrad, A. L., et al. (2008). The genome of *Heliobacterium modesticaldum*, a phototrophic representative of the Firmicutes containing the simplest photosynthetic apparatus. *Journal of Bacteriology, 190*, 4687–4696.

Schoepp-Cothenet, B., Lieutaud, C., Baymann, F., Vermeglio, A., Friedrich, T., Kramer, D. M., et al. (2009). Menaquinone as pool quinone in a purple bacterium. *Proceedings of the National Academy of Sciences of the United States of America, 106*, 8549–8554.

Schoepp-Cothenet, B., van Lis, R., Atteia, A., Baymann, F., Capowiez, L., Ducluzeau, A. L., et al. (2013). On the universal core of bioenergetics. *Biochimica et Biophysica Acta, 1827*, 79–93.

Schoffstall, A. M. (1976). Prebiotic phosphorylation of nucleosides in formamide. *Origins of Life, 7*, 399–412.

Schoonen, M., Smirnov, A., & Cohn, C. (2004). A perspective on the role of minerals in prebiotic synthesis. *Ambio, 33*, 539–551.

Schopf, J. W. (1993). Microfossils of the Early Archean Apex chert: new evidence of the antiquity of life. *Science, 260*, 640–646.

Schopf, J. W., & Packer, B. M. (1987). Early Archean (3.3-billion to 3.5-billion-year-old) microfossils from Warrawoona Group, Australia. *Science, 237*, 70–73.

Serrano-Andres, L., & Merchan, M. (2009). Are the five natural DNA/RNA base monomers a good choice from natural selection? A photochemical perspective. *Journal of Photochemistry and Photobiology C-photochemistry Reviews, 10*, 21–32.

Sharon, I., Alperovitch, A., Rohwer, F., Haynes, M., Glaser, F., Atamna-Ismaeel, N., et al. (2009). Photosystem I gene cassettes are present in marine virus genomes. *Nature, 461*, 258–262.

Sleep, N. H. (2010). The Hadean–Archaean environment. *Cold Spring Harbor Perspectives in Biology, 2*, a002527.

Sleep, N. H., & Bird, D. K. (2007). Niches of the pre-photosynthetic biosphere and geologic preservation of Earth's earliest ecology. *Geobiology, 5*, 101–117.

Sleep, N. H., & Bird, D. K. (2008). Evolutionary ecology during the rise of dioxygen in the Earth's atmosphere. *Philosophical Transactions of the Royal Society of London. Series B, Biological Sciences, 363*, 2651–2664.

Sleep, N. H., Meibom, A., Fridriksson, T., Coleman, R. G., & Bird, D. K. (2004). H_2-rich fluids from serpentinization: geochemical and biotic implications. *Proceedings of the National Academy of Sciences of the United States of America, 101*, 12818–12823.

Sobolewski, A. L., & Domcke, W. (2006). The chemical physics of the photostability of life. *Europhysics News, 37*, 20–23.

Sullivan, M. B., Lindell, D., Lee, J. A., Thompson, L. R., Bielawski, J. P., & Chisholm, S. W. (2006). Prevalence and evolution of core photosystem II genes in marine cyanobacterial viruses and their hosts. *PLoS Biology, 4*, e234.

Suwanto, A., & Kaplan, S. (1989). Physical and genetic mapping of the *Rhodobacter sphaeroides* 2.4.1 genome: genome size, fragment identification, and gene localization. *Journal of Bacteriology, 171*, 5840–5849.

Tang, K. H., Barry, K., Chertkov, O., Dalin, E., Han, C. S., Hauser, L. J., et al. (2011). Complete genome sequence of the filamentous anoxygenic phototrophic bacterium. *Chloroflexus Aurantiacus. BMC Genomics, 12*, 334.

Tang, K. H., Yue, H., & Blankenship, R. E. (2010). Energy metabolism of *Heliobacterium modesticaldum* during phototrophic and chemotrophic growth. *BMC Microbiology, 10*, 150.

The UniProt Consortium. (2012). Reorganizing the protein space at the Universal Protein Resource (UniProt). *Nucleic Acids Research., 40*, D71–D75.

Thompson, A. W., Foster, R. A., Krupke, A., Carter, B. J., Musat, N., Vaulot, D., et al. (2012). Unicellular cyanobacterium symbiotic with a single-celled eukaryotic alga. *Science, 337*, 1546–1550.

Tice, M. M., & Lowe, D. R. (2004). Photosynthetic microbial mats in the 3,416-Myr-old ocean. *Nature, 431*, 549–552.

Tice, M. M., & Lowe, D. R. (2006). Hydrogen-based carbon fixation in the earliest known photosynthetic organisms. *Geology, 34*, 37–40.

Tivey, M. K. (2007). Generation of seafloor hydrothermal vent fluids and associated mineral deposits. *Oceanography, 20*, 50–65.

Tripp, H. J., Bench, S. R., Turk, K. A., Foster, R. A., Desany, B. A., Niazi, F., et al. (2010). Metabolic streamlining in an open-ocean nitrogen-fixing cyanobacterium. *Nature, 464*, 90–94.

Tsukatani, Y., Romberger, S. P., Golbeck, J. H., & Bryant, D. A. (2012). Isolation and characterization of homodimeric type-I reaction center complex from *Candidatus* Chloracidobacterium thermophilum, an aerobic chlorophototroph. *Journal of Biological Chemistry, 287*, 5720–5732.

Verdecia, M. A., Larki, R. M., Ferrer, J. L., Riek, R., Chory, J., & Noel, J. P. (2005). Structure of the Mg-chelatase cofactor GUN4 reveals a novel hand-shaped fold for porphyrin binding. *PLoS Biology, 3*, e151.

Verlander, M. S., Lohrmann, R., & Orgel, L. E. (1973). Catalysts for the self-polymerization of adenosine cyclic 2', 3'-phosphate. *Journal of Molecular Evolution, 2*, 303–316.

Volkmer, T., Schneider, D., Bernat, G., Kirchhoff, H., Wenk, S. O., & Rogner, M. (2007). Ssr2998 of *Synechocystis* sp. PCC 6803 is involved in regulation of cyanobacterial electron transport and associated with the cytochrome b_6f complex. *Journal of Biological Chemistry, 282*, 3730–3737.

Wahlund, T. M., Woese, C. R., Castenholz, R. W., & Madigan, M. T. (1991). A thermophilic green sulfur bacterium from New Zealand hot springs, *Chlorobium tepidum* sp. nov. *Archives of Microbiology, 159*, 81–90.

Wallner, T., Hagiwara, Y., Bernat, G., Sobotka, R., Reijerse, E. J., Frankenberg-Dinkel, N., et al. (2012). Inactivation of the conserved open reading frame *ycf34* of *Synechocystis* sp. PCC 6803 interferes with the photosynthetic electron transport chain. *Biochimica et Biophysica Acta, 1817*, 2016–2026.

Walters, R. G., Shephard, F., Rogers, J. J., Rolfe, S. A., & Horton, P. (2003). Identification of mutants of *Arabidopsis* defective in acclimation of photosynthesis to the light environment. *Plant Physiology, 131*, 472–481.

Ward, N., Larsen, O., Sakwa, J., Bruseth, L., Khouri, H., Durkin, A. S., et al. (2004). Genomic insights into methanotrophy: the complete genome sequence of *Methylococcus capsulatus* (Bath). *PLoS Biology, 2*, e303.

Weissgerber, T., Zigann, R., Bruce, D., Chang, Y. J., Detter, J. C., Han, C., et al. (2011). Complete genome sequence of *Allochromatium vinosum* DSM 180^T. *Standards in Genomic Sciences, 5*, 311–330.

White, D. E., Muffler, L. J.P., & Truesdell, A. N. (1971). Vapor-dominated hydrothermal systems compared with hot-water systems. *Economic Geology and the Bulletin of the Society of Economic Geologists, 66*, 75–97.

Xiong, J., Fischer, W. M., Inoue, K., Nakahara, M., & Bauer, C. E. (2000). Molecular evidence for the early evolution of photosynthesis. *Science, 289*, 1724–1730.

Xiong, J., Inoue, K., & Bauer, C. E. (1998). Tracking molecular evolution of photosynthesis by characterization of a major photosynthesis gene cluster from *Heliobacillus mobilis*. *Proceedings of the National Academy of Sciences of the United States of America, 95*, 14851–14856.

Yamada, M., Zhang, H., Hanada, S., Nagashima, K.V., Shimada, K., & Matsuura, K. (2005). Structural and spectroscopic properties of a reaction center complex from the chlorosome-lacking filamentous anoxygenic phototrophic bacterium *Roseiflexus castenholzii*. *Journal of Bacteriology, 187*, 1702–1709.

Yamamoto, H., Peng, L., Fukao, Y., & Shikanai, T. (2011). An Src homology 3 domain-like fold protein forms a ferredoxin binding site for the chloroplast NADH dehydrogenase-like complex in *Arabidopsis*. *Plant Cell, 23*, 1480–1493.

Yanagida, S., Azuma, T., Midori, Y., Pac, C., & Sakurai, H. (1985a). Semiconductor photocatalysis. Part 4. Hydrogen evolution and photoredox reactions of cyclic ethers catalyzed by zinc-sulfide. *Journal of the Chemical Society—Perkin Transactions, 2*, 1487–1493.

Yanagida, S., Kizumoto, H., Ishimaru, Y., Pac, C., & Sakurai, H. (1985b). Zinc sulfide catalyzed photochemical conversion of primary amines to secondary amines. *Chemistry Letters*, 141–144.

Yanyushin, M. F. (2002). Fractionation of cytochromes of phototrophically grown *Chloroflexus aurantiacus*. Is there a cytochrome *bc* complex among them? *FEBS Letters, 512*, 125–128.

Yanyushin, M. F., del Rosario, M. C., Brune, D. C., & Blankenship, R. E. (2005). New class of bacterial membrane oxidoreductases. *Biochemistry, 44*, 10037–10045.

Yoneyama, H. (1997). Photoreduction of carbon dioxide on quantized semiconductor nanoparticles in solution. *Catalysis Today, 39*, 169–175.

Zehr, J. P., Bench, S. R., Carter, B. J., Hewson, I., Niazi, F., Shi, T., et al. (2008). Globally distributed uncultivated oceanic N_2-fixing cyanobacteria lack oxygenic photosystem II. *Science, 322*, 1110–1112.

Zhang, X. V., Ellery, S. P., Friend, C. M., Holland, H. D., Michel, F. M., Schoonen, M. A.A., et al. (2007). Photodriven reduction and oxidation reactions on colloidal semiconductor particles: implications for prebiotic synthesis. *Journal of Photochemistry and Photobiology A: Chemistry, 185*, 301–311.

Zhang, X. V., Martin, S. T., Friend, C. M., Schoonen, M. A.A., & Holland, H. D. (2004). Mineral-assisted pathways in prebiotic synthesis: photoelectrochemical reduction of carbon(+IV) by manganese sulfide. *Journal of the American Chemical Society., 126*, 11247–11253.

Zhaxybayeva, O., Hamel, L., Raymond, J., & Gogarten, J. P. (2004). Visualization of the phylogenetic content of five genomes using dekapentagonal maps. *Genome Biology, 5*, R20.

CHAPTER TWO

Molecular Markers for Photosynthetic Bacteria and Insights into the Origin and Spread of Photosynthesis

Radhey S. Gupta
Department of Biochemistry and Biomedical Sciences, McMaster University, Hamilton, ON, Canada
E-mail: gupta@mcmaster.ca

Contents

Abstract

Microbial genomes provide a rich resource for identifying molecular markers that can reliably distinguish different groups of photosynthetic organisms and provide insights into the origin and spread of photosynthesis. This review summarises work on molecular markers consisting of conserved signature indels (CSIs) and signature proteins that are specific for different phyla of photosynthetic bacteria and their major subclades. Based upon these markers and phylogenetic analyses, photosynthetic

Advances in Botanical Research, Volume 66
ISSN 0065-2296, http://dx.doi.org/10.1016/B978-0-12-397923-0.00002-3

Chloroflexi are divided into two suborders viz. *Chloroflexineae* and *Roseiflexineae* consistent with their biochemical, morphological and molecular characteristics. Within Cyanobacteria three major clades (referred to as Clade A, B and C) are distinguished at the highest phylogenetic/taxonomic levels. Phylogenetic analyses and several CSIs in key bacteriochlorophyll (BChl) biosynthesis proteins, BchL, BchN and BchB, provide evidence that their homologues in Proteobacteria have been acquired from Clade C Cyanobacteria, whereas those in the Chlorobi are derived from Chloroflexi by lateral gene transfers. Phylogenetic analyses and other CSIs in the sequentially duplicated proteins BchL, BchX and NifH provide evidence that the BchX homologues required for anoxygenic photosynthesis originated before the BchL homologues needed for oxygenic photosynthesis. Another CSI in the duplicated BchX–BchL proteins provides evidence that the BchL homologues of *Heliobacteriaceae* (*Firmicutes* phylum) are primitive in comparison to other photosynthetic bacteria.

1. INTRODUCTION

An understanding of the origin of photosynthesis, which sustains most life on earth, is an important unresolved problem in the evolutionary history of life (Blankenship, 1992; Blankenship & Hartman, 1998; Dismukes et al., 2001; Hartman, 1998; Hohmann-Marriott & Blankenship, 2011). Except for plants and algae that are secondarily photosynthetic due to endosymbiotic acquisition of Cyanobacteria (Margulis, 1993; Morden, Delwiche, Kuhsel, & Palmer, 1992), the bacteriochlorophyll (BChl)-based photosynthesis is found in five discontinuous phyla of cultured bacteria viz. Cyanobacteria, Chloroflexi, Bacteroidetes/Chlorobi, Firmicutes (*Heliobacteriaceae*) and Proteobacteria (Blankenship, 1992; Bryant & Frigaard, 2006; Gest & Favinger, 1983; Hohmann-Marriott & Blankenship, 2011; Olson & Pierson, 1987). Additionally, an uncultured bacterium belonging to the phylum Acidobacteria is also inferred to be photosynthetic (Bryant et al., 2007; Raymond, 2008). Of these groups, Cyanobacteria and Chlorobi are comprised entirely of photosynthetic bacteria, whereas the remaining three phyla contain both photosynthetic as well as non-photosynthetic members (Bryant & Frigaard, 2006; Castenholz, 2001; Overmann, 2003). Despite their phylogenetic discontinuity (Ciccarelli et al., 2006; Gest & Blankenship, 2004; Gupta, 1998; Ludwig & Klenk, 2005; Olsen, Woese, & Overbeek, 1994), the observed similarities in the photosynthetic pigments and the charge transfer mechanisms in the reaction centres (RCs) of various phototrophic bacteria indicate that photosynthesis has evolved only once (Blankenship, 1994; Golbeck, 1993; Nelson & Ben Shem, 2005; Nitschke & Rutherford, 1991; Olson & Blankenship, 2004; Sadekar, Raymond, & Blankenship, 2006; Schubert et al., 1998). However, based upon the known characteristics of different photosynthetic organisms, or by means of commonly used phylogenetic approaches, it has proven difficult to determine in which of these bacterial phyla photosynthetic ability first evolved and how this ability has spread to other bacterial phyla (Dismukes et al., 2001; Gupta, 2003; Gupta, Mukhtar, & Singh, 1999; Mulkidjanian et al., 2006; Pierson, 1994; Raymond, Zhaxybayeva, Gogarten, Gerdes, & Blankenship, 2002; Raymond, Zhaxybayeva, Gogarten, & Blankenship, 2003; Vermaas, 1994; Xiong & Bauer, 2002; Xiong, Fischer, Inoue, Nakahara, & Bauer, 2000). Important insights in these regards are now provided by novel and reliable molecular markers, whose discovery has become possible due to the availability of genomic sequences.

Comparative analysis of genomic sequences provide a powerful means for identifying molecular markers that are distinctive characteristics of either different phyla of photosynthetic bacteria or provide information regarding evolutionary relationships among them (Gupta, 1998, 2000b; Koonin, Aravind, & Kondrashov, 2000; Nobrega & Pennacchio, 2004). Based upon genome sequences, two different kinds of molecular markers for different bacterial groups are being identified. The first of these markers consists of conserved signature indels (CSIs) (i.e. inserts or deletions) in protein sequences that are uniquely found in particular groups of organisms (Gupta, 1998, 2000b; Gupta & Griffiths, 2002). The CSIs that provide useful molecular markers for evolutionary studies are of defined lengths, present at the same location in different homologues, and they are flanked on both sides by conserved regions to ensure that they are reliable genetic characteristics (Gao & Gupta, 2012; Gupta, 1998, 2010a; Gupta & Griffiths, 2002). Because most CSIs in protein sequences (even a 1 aa indel) are the products of rare and highly specific genetic changes, their presence or absence in gene/protein sequences is generally not affected by factors such as differences in evolutionary rates at different sites or among different species that greatly influence branching patterns of species in phylogenetic trees (Felsenstein, 1988, 2004; Gupta, 1998; Moreira & Philippe, 2000). Hence, when a discovered CSI is restricted to a phylogenetically defined group of organisms, its most parsimonious explanation is that the genetic change that gave rise to it first occurred in a common ancestor of that group and then vertically passed onto various descendant species (Gupta, 1998; Rivera & Lake, 1992; Rokas & Holland, 2000). Therefore, the shared presence of such markers or 'molecular synapomorphies' in different group(s) of organisms provides powerful means to establish ancestral evolutionary relationships, as well as to identify cases of lateral gene transfer (LGT) among unrelated taxa.

Additionally, depending upon the presence or absence of these indels in outgroup species, it is possible to determine which of the two character states of the gene/protein (i.e. indel-containing or indel-lacking) is ancestral (Baldauf & Palmer, 1993; Gupta, 1998, 2003, 2010b).

The second kind of molecular markers that have proven useful for taxonomic and evolutionary studies are whole proteins that are limited to different monophyletic clades of organisms (Dutilh, Snel, Ettema, & Huynen, 2008; Gupta & Lorenzini, 2007; Gupta & Mok, 2007; Lerat, Daubin, Ochman, & Moran, 2005). Blast searches with these proteins indicate that they are distinctive characteristics of various species/strains from particular clades. Although the mechanisms responsible for the origin/evolution of the genes for these proteins are unclear (Dutilh et al., 2008; Kuo & Ochman, 2009), their presence in a conserved state in all or most species/strains from particular clades, but nowhere else, indicates that the genes for these proteins first evolved in a common ancestor of these clades followed by their retention by various descendants (Dutilh et al., 2008; Fang, Rocha, & Danchin, 2008; Narra, Cordes, & Ochman, 2008). Because of their clade specificity, these conserved signature proteins (CSPs) again provide valuable molecular markers for distinguishing different clades of organisms (Dutilh et al., 2008; Gupta & Lorenzini, 2007; Gupta & Mok, 2007). The studies based on both CSIs and CSPs are generally carried out in conjunction with detailed phylogenetic studies based upon concatenated sequences for large numbers of proteins. The vast majority of the clades that are identified based upon discovered CSIs and CSPs generally correspond to well-supported clades in phylogenetic trees, thereby providing evidence that these signatures provide reliable molecular markers for these groups of organisms (Gao, Mohan, & Gupta, 2009; Gupta & Mathews, 2010; Gupta & Mok, 2007). Extensive work on both CSIs and CSPs indicates that both kinds of molecular markers are present at different phylogenetic depths and the results obtained using both these approaches show excellent agreement with each other (Gupta, 2009; Gupta & Lorenzini, 2007; Gupta & Mathews, 2010; Gupta & Mok, 2007). In addition to the signatures that are specific for the monophyletic clades of organisms, some signatures are found to be commonly shared by phylogenetically unrelated organisms such as Chlamydiae and Actinobacteria (Griffiths & Gupta, 2006), or Chlorobi and Chloroflexi (Gupta, 2012), or Cyanobacteria and Deinococcus-Thermus (Gupta & Johari, 1998). The shared presence of such CSIs or CSPs in several of these cases has been shown to be due to lateral transfer of genes between these phyla (Griffiths & Gupta, 2006). The work on CSIs and CSPs that is specific for different phyla of photosynthetic bacteria, or those that are shared among some of them, and the implications of these signatures for the origin and spread of photosynthesis are discussed below.

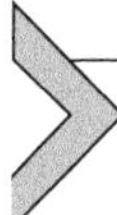

2. MOLECULAR SIGNATURES FOR DIFFERENT PHYLA OF PHOTOSYNTHETIC BACTERIA

2.1. Molecular Signatures for Chloroflexi and Their Revised Taxonomy

The phylum Chloroflexi is comprised of diverse group of organisms that include anoxygenic photoautotrophs, aerobic chemoheterotrophs, thermophilic organisms as well as anaerobic organisms that obtain energy by reductive dehalogenation of organic chlorinated compounds (Garrity & Holt, 2001a; Hugenholtz & Stackebrandt, 2004). Until 2001, this phylum consisted of a single class Chloroflexia containing two orders *Chloroflexales* and *Herpetosiphonales*, each containing only a single family (viz. *Chloroflexaceae* and *Herpetosiphonaceae*) and together made up of five genera (Garrity & Holt, 2001a; Gupta et al., 2013). However, in the past 10 years, this phylum has undergone enormous expansion both by inclusion of newly discovered species and by amalgamation of species that were previously part of other bacterial phyla (e.g. Thermomicrobia, Actinobacteria) (Garrity & Holt, 2001b; Hugenholtz & Stackebrandt, 2004). As a result, the phylum is currently made up of six classes viz. Chloroflexia, Thermomicrobia, Dehalococcoidetes, Anaerolineae, Caldilineae and Ktedonobacteria (Euzeby, 2011; Hugenholtz & Stackebrandt, 2004; Moe, Yan, Nobre, da Costa, & Rainey, 2009; Yabe, Aiba, Sakai, Hazaka, & Yokota, 2010; Yamada et al., 2006; Yarza et al., 2010). Currently, no morphological, physiological, biochemical or molecular trait is known that is uniquely shared by different species from this phylum or its different constituent classes (Hanada & Pierson, 2006; Hugenholtz & Stackebrandt, 2004; Yabe et al., 2010; Yamada et al., 2006).

The photosynthetic ability within the phylum Chloroflexi is found only within the species belonging to the class Chloroflexia (Garrity & Holt, 2001a; Hanada & Pierson, 2006; Pierson & Castenholz, 1992). Of the two orders that are part of this class, Herpetosiphonales contains a single genus *Herpetosiphon*, which is non-photosynthetic, whereas all photosynthetic genera, i.e. *Oscillochloris*, *Chloroflexus*, *Chloronema*, *Roseiflexus* and *Heliothrix*, are part of the order Chloroflexales. Different genera within the order Chloroflexales are presently divided into two families viz. *Chloroflexaceae* and *Oscillochloridaceae* (Euzeby, 2011; Garrity &

Holt, 2001a; Hanada & Pierson, 2006; Keppen, Tourova, Kuznetsov, Ivanovsky, & Gorlenko, 2000). However, the grouping of these genera into these two families is not supported by their biochemical and morphological characteristics (Hanada & Pierson, 2006). Until recently, no molecular markers were known for the class Chloroflexia or its different constituent orders and families.

The genome sequences for 18 species/strains from the phylum Chloroflexi are now available (Gupta, Chander, & George, 2013). Based upon these sequences, robust phylogenetic trees for the species from the phylum Chloroflexi have been constructed and large numbers of CSIs that are specific for different clades of Chloroflexi have been identified (Gupta et al., 2013). A summary of these results is presented in Fig. 2.1. These studies have identified five CSIs in widely distributed proteins that are uniquely found in various species from the class Chloroflexia and another nine CSIs that are specific for the species from the order Chloroflexales (Gupta et al., 2013). Due to the specificities of these CSIs for either the class Chloroflexia or the order Chloroflexales, they provide novel (genetic) molecular markers for distinguishing these groups of bacteria from all others.

An important finding from these studies was that the species *Oscillochloris trichoides*, which is a part of the family *Oscillochloridaceae*, showed a strong and specific relationship to the species from the genus *Chloroflexus*, both in phylogenetic trees and based upon discovered molecular signatures (Gupta et al., 2013). Seven CSIs identified in this work were uniquely shared by *O. trichoides* and various *Chloroflexus* spp., but they were not found in any *Roseiflexus* spp. or other bacteria. In addition to their phylogenetic clustering and the shared presence of many novel CSIs, the species from the genera *Chloroflexus* and *Oscillochloris* (and also *Chloronema*) also differ from species of the genera *Roseiflexus* (and *Heliothrix*) by their green colour, shared presence of the chlorosomes and BChl *c* (in addition to BChl *a* and *d* in some species), by their fatty acid profiles, and by the presence of β- and γ-carotenes and quinone MK-10 (Hanada & Pierson, 2006; Hanada, Takaichi, Matsuura, & Nakamura, 2002). In contrast, species from the genus *Roseiflexus* (and also *Heliothrix*) are orange-red bacteria, which lack chlorosomes and BChl *c* and differ from the species of *Chloroflexus* and *Oscillochloris* genera in their carotenoids, quinones and fatty acid profiles (Hanada & Pierson, 2006; Hanada et al., 2002). The species from these latter genera consistently branch more deeply in phylogenetic trees (Gupta, 2012; Hanada & Pierson, 2006) and three CSIs that are specific for the *Roseiflexus* spp. have been identified (Gupta et al., 2013). Based upon the above mentioned molecular, biochemical, morphological and phylogenetic observations, the order Chloroflexales

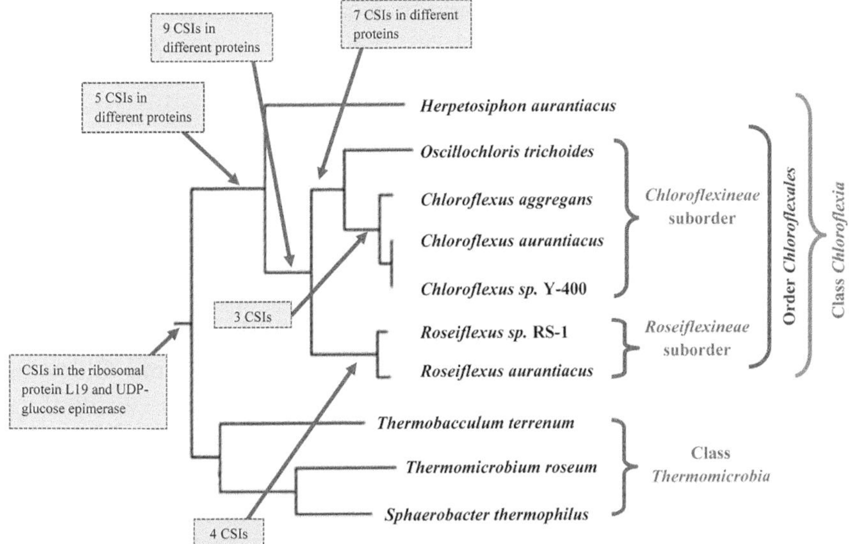

Figure 2.1 A summary diagram showing the evolutionary relationships among sequenced species belonging to the classes Chloroflexia and Thermomicrobia and different taxonomic groups within the class Chloroflexia that can now be clearly distinguished based upon the discovered molecular signatures (Gupta et al., 2013). (For colour version of this figure, the reader is referred to the online version of this book.)

was recently divided into two new suborders (Gupta et al., 2013). The first of these suborders Chloroflexineae consists of the family *Oscillochloridaceae* (emended to include the genus *Chloronema*) and a new family *Chloroflexaceae* containing the genus *Chloroflexus*. The second suborder Roseiflexineae contains a single family *Roseiflexaceae* that includes the genera *Roseiflexus* and *Heliothrix*.

Phylogenetic analyses of the sequenced species from the phylum Chloroflexi also show that different classes that are currently part of this phylum do not form a monophyletic clade in phylogenetic trees (Gupta et al., 2013). Additionally, no CSI or other molecular signature was identified that is uniquely shared by all of the species from this phylum (Gupta et al., 2013). Nonetheless, a specific grouping of the classes Chloroflexia and Thermomicrobia is supported by both phylogenetic means and the identified CSIs. Based upon these results, it has been proposed that the phylum Chloroflexi '*sensu stricto*' should be limited to only the classes Chloroflexia and Thermomicrobia and the other four classes (viz. Dehalococcoidetes, Anaerolineae, Caldilineae and Ktedonobacteria) that are presently grouped with it should be regarded as related taxa awaiting more detailed investigations to clarify their relationships (Gupta et al., 2013).

2.2. Molecular Signatures for the Chlorobi

The Chlorobi species are mainly found in anoxic aquatic settings, where sunlight is able to penetrate (Bryant & Frigaard, 2006; Overmann, 2003). In phylogenetic trees, they exhibit close affinity to the non-photosynthetic Bacteroidetes species that are found in diverse habitats (Bryant & Frigaard, 2006; Gupta, 2004; Olsen et al., 1994; Overmann, 2003). A specific relationship between the Bacteroidetes and Chlorobi is further established by three CSIs in important proteins (viz. 8–9 aa insert in FtsK, 1 aa insert in UvrB and 18 aa inset in ATP synthase α subunit) as well as three CSPs (PG0081, PG0649 and PG2432) that are uniquely present in different sequenced species from these two groups (Fig. 2.2) (Gupta, 2004; Gupta & Lorenzini, 2007). Additionally, these two groups of bacteria also exhibit a close relationship to the *Fibrobacter succinogenes* (Gupta, 2004; Griffiths & Gupta, 2001), which is currently placed in a separate phylum (Fibrobacteres) (Ludwig & Klenk, 2005). A specific relationship of the Chlorobi and Bacteroidetes phyla to the Fibrobacteres is strongly supported by two large CSIs in the proteins, RNA polymerase β′ subunit and serine hydroxymethyl transferase, that are uniquely found in various genome-sequenced species from these phyla (Fig. 2.2) (Gupta, 2004). Additionally, a signature protein PG0081 (accession number NP_904430) is also uniquely present in various species from these three bacteria phyla (Gupta & Lorenzini, 2007). These results provide compelling evidence that the species from these three taxa have shared a common ancestor exclusive of all other bacteria and they should be recognised as part of a single phylum or superphylum viz. 'FCB phylum' (for Fibrobacteres–Chlorobi–Bacteroidetes) (Gupta, 2004; Gupta & Lorenzini, 2007).

The analysis of genomic sequences from Chlorobi species has identified 50 CSPs as well as two prominent CSIs (viz. 28 aa insert in the DNA polymerase III alpha subunit and 12–14 aa insert in alanyl-tRNA synthetase) that are specific for the species from this phylum (Gupta & Lorenzini, 2007). While most of the Chlorobi-specific proteins are of unknown functions, some are involved in photosynthesis-related functions such as the chlorosome envelope A (Plut_0265) and C (Plut_0264) proteins, the BChl A protein (Plut_1500) (Blankenship, Olson, & Miller, 1995), and

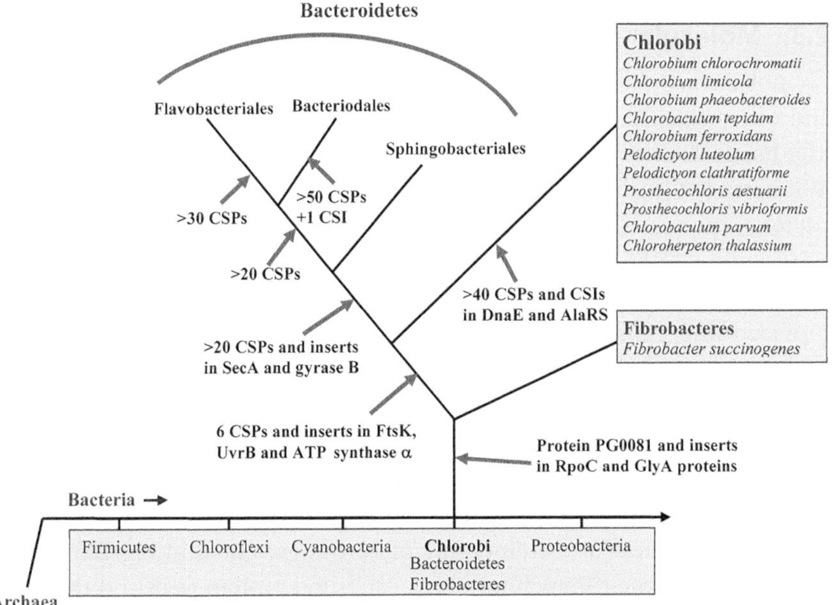

Figure 2.2 ***A summary diagram showing the evolutionary relationships among the FCB phylum of bacteria based on different identified CSIs and CSPs (Gupta, 2004; Gupta & Lorenzini, 2007).*** The arrows mark the evolutionary stages where the genetic changes responsible for indicated CSIs and CSPs were likely introduced. The placement of the FCB phylum in relation to other phyla of cultured photosynthetic bacteria (rooted in relation to Archaea) is based upon earlier work (Griffiths & Gupta, 2004, 2007; Gupta, 2003, 2005a). (For colour version of this figure, the reader is referred to the online version of this book.)

two other proteins (Plut_0620:PscD and Plut_1628) that are annotated as part of the photosystem P840 RC (Eisen et al., 2002; Frigaard, Chew, Li, Maresca, & Bryant, 2003). Three additional Chlorobi-specific proteins, Plut_1714–Plut_1716, are also clustered together in the genome indicating that they may form a functional unit. Sixty-five additional proteins are also specific for the Chlorobi species, but they are missing in some species (Gupta & Lorenzini, 2007). However, eight of these proteins are only found in *Chlorobium luteolum* and *Chlorobium phaeovibrioides*, which form a strongly supported clade in phylogenetic trees (Gupta & Lorenzini, 2007; Imhoff, 2003). These CSPs and CSIs provide novel tools for genetic and biochemical investigations on the Chlorobi species.

In addition to these CSIs, several other CSIs are uniquely shared by various Chloroflexales and Chlorobi species. The significance of these CSIs will be discussed in Section 3.

2.3. Molecular Signatures for Cyanobacteria

Cyanobacteria differ greatly in terms of their morphology, physiology and other characteristics (Castenholz, 2001; Rippka, Deruelles, Waterbury, Herdman, & Stanier, 1979; Sanchez-Baracaldo, Hayes, & Blank, 2005; Wilmotte & Golubic, 1991). In 16S ribosomal RNA gene trees, the cyanobacterial species/strains form 14 unresolved clusters and their evolutionary relationships or taxonomy are not resolved (Wilmotte & Herdman, 2001). Hence, the availability of genome sequences for >40 Cyanobacteria have provided important means to clarify their phylogeny and taxonomy. Based upon concatenated protein trees and discovery of large numbers of molecular signatures that are specific for either all Cyanobacteria or their different clades, a reliable picture of cyanobacterial phylogeny and taxonomy has begun to emerge (Gupta, 2009; Gupta & Mathews, 2010; Sanchez-Baracaldo et al., 2005; Shi & Falkowski, 2008; Swingley, Blankenship, & Raymond, 2008). An overview of the emerging picture is presented in Fig. 2.3.

Based upon their branching in concatenated protein trees and the identified signatures, the sequenced cyanobacterial species at the highest level form three main clades. The first of these clades consisting of *Gloeobacter* and the *Synechococcus* strains JA-3-3Ab and JA2-3-B′a (Clade A) forms the deepest branching lineage within Cyanobacteria. The species from this clade can be distinguished from all other Cyanobacteria by means of a number of unique CSIs and CSPs (Gupta, 2009, 2010b; Gupta & Mathews, 2010) (Fig. 2.3). The deep branching of Clade A species in comparison to all other Cyanobacteria is independently supported by several CSIs in important proteins, where the particular CSIs are commonly shared by all other Cyanobacteria, but they are lacking in the Clade A Cyanobacteria as well as all other bacteria (Gupta, 2009). The Clade B of Cyanobacteria contains the majority of known Cyanobacteria except the Clade A Cyanobacteria and the unicellular marine Cyanobacteria (Clade C) (Fig. 2.3). This clade, as currently defined, includes all of the species/strains from the orders Chroococcales, Nostocales and Oscillatoriales as well as certain deeper branching cyanobacterial species such as *Acaryochloris marina* and *Thermosynechococcus elongatus*. Within Clade B, many signatures that are specific for either the order Chroococcales, or the order Nostocales, or those indicating interrelationship among the orders Nostocales, Oscillatoriales and Chroococcales have also been identified (Fig. 2.3) (Gupta, 2009, 2010b; Gupta & Mathews, 2010).

The third important clade of Cyanobacteria, referred to as Clade C, is comprised of the marine unicellular *Synechococcus* and *Prochlorococcus*

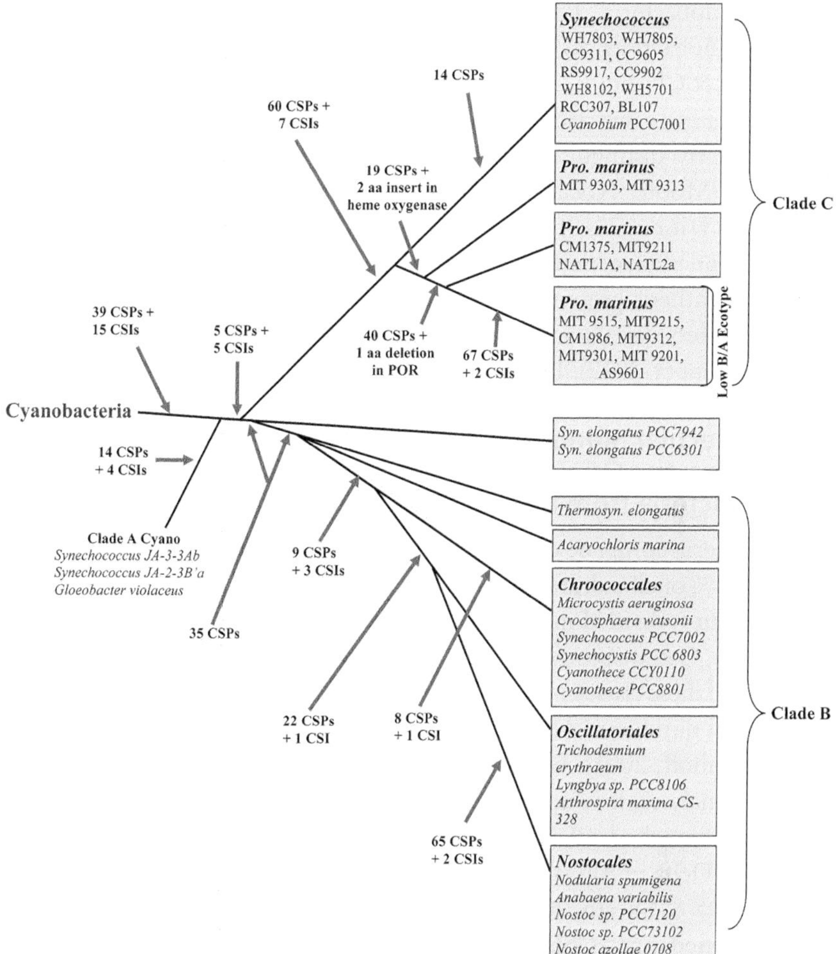

Figure 2.3 ***A summary diagram showing different molecular signatures consisting of CSIs and CSPs that have been identified for various cyanobacterial species/clades (Gupta, 2009; Gupta & Mathews, 2010; Gupta, Pereira, Chandrasekera, & Johari, 2003).*** All of the clades identified by these signatures are also supported by phylogenetic analyses (Gupta, 2009; Gupta & Mathews, 2010; Swingley et al., 2008). (For colour version of this figure, the reader is referred to the online version of this book.)

species/strains. This clade is separated from all other Cyanobacteria by a long branch in phylogenetic trees (Gupta, 2009, 2010b; Gupta & Mathews, 2010). Large numbers of CSIs and CSPs that are specific for this clade have been identified (Gupta, 2009, 2010b; Gupta & Mathews, 2010). Within this clade, many signatures distinguish the *Prochlorococcus* strains/isolates from other Cyanobacteria. The *Prochlorococcus* strains/isolates also differ from other Cyanobacteria in their photosynthetic pigments (Rocap, Distel, Waterbury, & Chisholm, 2002; Rocap et al., 2003). Among the *Prochlorococcus* strains/isolates, a subclade corresponding to the low B/A ecotype can also be clearly distinguished based upon many identified CSIs and CSPs (Fig. 2.3) (Moore, Rocap, & Chisholm, 1998; Rocap et al., 2002, 2003 Gupta and Wilson, 2010). The Clade C of Cyanobacteria is of much interest as recent phylogenetic studies and signature sequences in several key BChl biosynthesis proteins provide evidence that the genes for these proteins in Proteobacteria are derived from Clade C Cyanobacteria by means of LGTs (Gupta, 2012). This work is described below.

2.4. Molecular Signatures for Proteobacteria: Evidence Supporting the Origin of BChl Biosynthesis Genes in Proteobacteria from the Clade C Cyanobacteria

Proteobacteria comprise one of the largest phyla within bacteria accounting for over 45% of all cultured bacteria (Cole et al., 2009; Kersters et al., 2006). The phylum is divided into five main classes: Alphaproteobacteria, Betaproteobacteria, Gammaproteobacteria, Deltaproteobacteria and Epsilonproteobacteria. Of these classes, photosynthetic ability is distributed sporadically only in a limited number of Alpha (α)-, Beta (β)- and Gamma (γ)-proteobacteria species (Imhoff, 2001). Based upon phylogenetic analyses as well as other forms of analyses, it is now established that different classes of Proteobacteria have branched off from a common ancestor in the following order: Epsilon → Delta → Alpha → Beta → Gamma-proteobacteria (Gupta, 2000b, 2001; Gupta & Sneath, 2007). Comparative analyses of genomic sequences have identified many CSIs and CSPs that are distinctive characteristics of the Alpha-, Gamma- and Epsilonproteobacteria (Gao et al., 2009; Gupta, 2005b; Gupta & Mok, 2007; Kainth & Gupta, 2005). However, the most extensive work in this regard has been carried out with Alphaproteobacteria, which contains the majority of photosynthetic bacteria belonging to the phylum Proteobacteria (Gupta, 2005b; Gupta & Mok, 2007; Kainth & Gupta, 2005).

Within α-proteobacteria, phototrophs are present in four of the six orders: Rhodobacterales, Rhodospirillales, Rhizobiales (*Bradyrhizobiaceae*) and Sphingomonadales (Imhoff, 2001). However, these four orders do not form a monophyletic clade within the α-proteobacteria (Gupta, 2005b; Gupta & Mok, 2007; Williams, Sobral, & Dickerman, 2007) and each of them contains both phototrophs and non-phototrophs. Comparative analyses of the genomes from α-proteobacteria have identified large numbers of CSIs and CSPs that are specific for either all α-proteobacteria or different orders of α-proteobacteria. (Gupta & Mok, 2007; Kainth & Gupta, 2005). A summary of these molecular signatures is provided in Fig. 2.4. These signatures include 72 CSPs and a 3 aa conserved insert in seryl-tRNA synthetase that are specific for the *Bradyrhizobiaceae*, 35 CSPs that are specific for Rhodobacterales, 4 CSPs each for the *Rhodospirillaceae* and *Acetobacteraceae* families and 31 CSPs that are specific for the order Sphingomonadales (Gupta, 2005b; Gupta & Mok, 2007). Additionally, a 25 aa insert in the RNA polymerase β subunit and a 4 aa insert in DNA gyrase B, which are specific for the orders Rhodospirillales and Sphingomonadales, respectively, have also been identified (Gupta & Mok, 2007).

Earlier phylogenetic studies based on BchL, BchN and BchB proteins, the three subunits of the light-independent (or dark-operative) protochlorophyllide oxidoreductase (DPOR), have led to the inference that Proteobacteria was the earliest branching lineage in which photosynthetic ability first evolved (Xiong & Bauer, 2002; Xiong et al., 2000). However, recent work on the same protein sequences strongly indicates that the deep branching of the proteobacterial homologues in these earlier studies was caused by the highly divergent nature of the proteobacterial homologues and lack of any close relatives to them in the dataset that was used in earlier phylogenetic studies (Green & Gantt, 2000 Gupta, 2013). Recent studies on the BchL, BchN and BchB proteins now provide compelling evidence that the proteobacterial homologues of these proteins are closely related to sequences found in the Clade C Cyanobacteria. The observations supporting this inference include: (i) branching of the proteobacterial homologues with the Clade C Cyanobacteria in phylogenetic trees based upon BchL, BchN and BchB sequences (Gupta, 2012); (ii) For all three of these proteins, the proteobacterial homologues exhibit maximal sequence similarity (~60–70% identity) to the Clade C Cyanobacteria in comparison to all other phototrophic lineage (~32–36% identity) (Gupta, 2012); (iii) In a sequence alignment of BchN proteins, three CSIs have been identified that are uniquely shared by all proteobacterial homologues and those from the Clade C Cyanobacteria. Partial sequence alignments of the BchN homologues showing two of these CSIs (① and ③) are presented in Fig. 2.5. These observations strongly indicate that the genes for these proteins have undergone LGTs between these two groups of photosynthetic bacteria.

While the photosynthetic ability within the Proteobacteria is sporadically distributed in a limited number of species belonging to the α-,

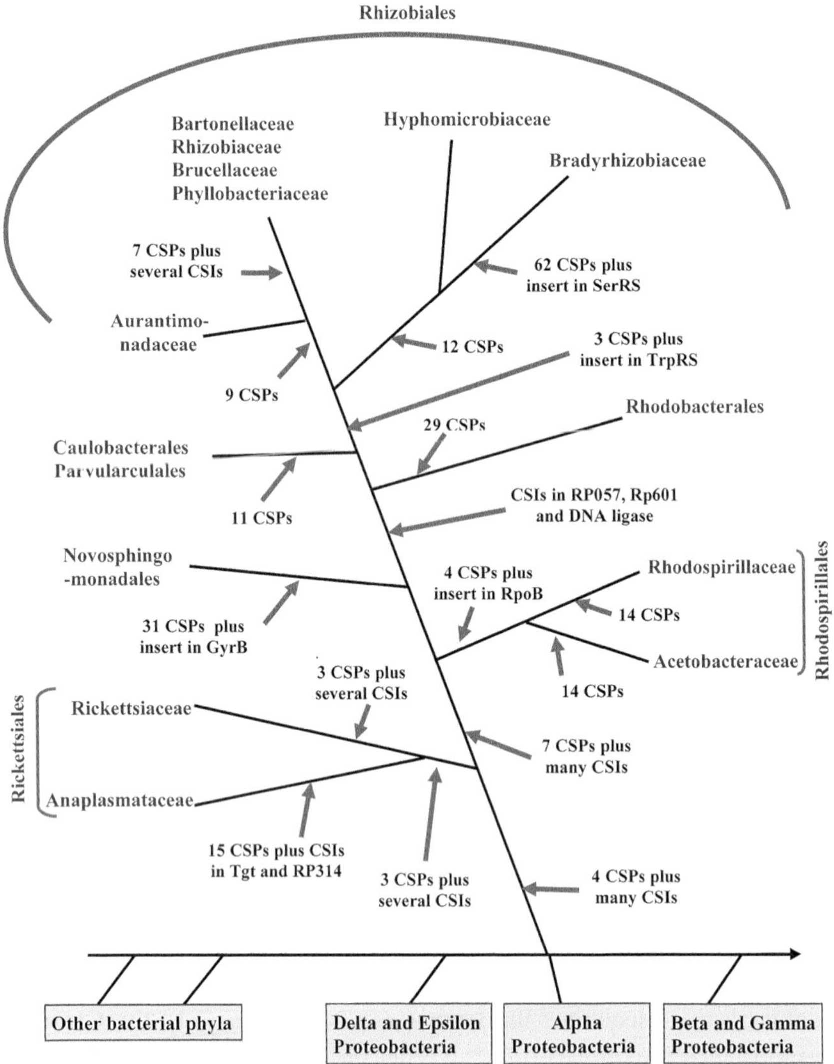

Figure 2.4 ***A summary diagram showing the evolutionary relationships among the Alpha(α)-proteobacteria based upon the species distribution patterns of different discovered CSPs and CSIs (Gupta, 2005b; Gupta & Mok, 2007).*** The placement of α-proteobacteria in between the δ–ε-proteobacteria and the β- and γ-proteobacteria is based on earlier studies (Gupta, 2000b; Gupta & Sneath, 2007). (For colour version of this figure, the reader is referred to the online version of this book.)

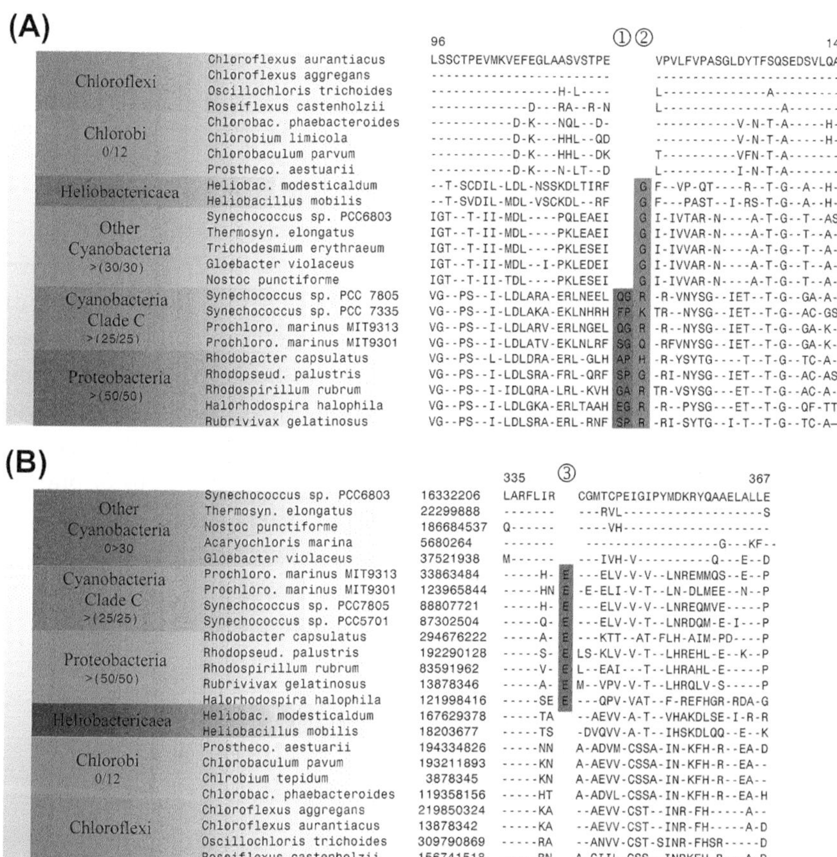

Figure 2.5 ***Excerpts from the sequence alignment for BchN homologues showing two CSIs (① and ③) that are commonly shared by all Clade C Cyanobacteria and Proteobacteria, and an additional CSI ② consisting of 1 aa deletion that is uniquely present in different Chlorobi and Chloroflexi homologues (Gupta, 2012).*** The dashes in the alignments show identity with the amino acid on the top line. The numbers on the top indicate the position of the sequence in the species on the top line. The numbers below the group names indicate the presence or absence these CSIs in different homologues from these groups. The Genebank identification numbers of different protein sequences are shown in the second column. (For colour version of this figure, the reader is referred to the online version of this book.)

β- and γ-classes, the phylum Cyanobacteria is monophyletic and it is made up entirely of photosynthetic bacteria (Blank & Sanchez-Baracaldo, 2010; Castenholz, 2001; Gupta, 2010b; Mulkidjanian et al., 2006). Additionally, within this phylum, the Clade C is indicated to be a derived clade based upon its branching position in phylogenetic trees and several CSIs in important proteins (Fig. 2.3) (Gupta, 2009; Gupta & Mathews, 2010). These observations indicate that different CSIs and other characteristics that distinguish the BchL, BhcN and BchB homologues of Clade C Cyanobacteria from other members of the Cyanobacteria phylum, they initially originated in a common ancestor of the Clade C Cyanobacteria and then these genes were laterally acquired by the Proteobacteria (Gupta, 2012). The alternative possibility that these genetic changes first occurred in a proteobacterium and subsequently transferred to the Clade C Cyanobacteria as well as other Proteobacteria would require numerous gene transfers, gene losses and gene replacement events, and is highly unlikely (Gupta, 2012). The observed close similarities in the components of the DPOR complex between the Clade C Cyanobacteria and Proteobacteria suggest that these two groups of photosynthetic bacteria should possess and commonly share certain unique aspect of photosynthesis. However, no unique photosynthetic characteristic is presently known that is commonly shared by these two groups of phototrophs and studies in this regard should be of much interest.

3. ORIGIN AND SPREAD OF PHOTOSYNTHESIS BASED UPON MOLECULAR SIGNATURES IN KEY BChl BIOSYNTHESIS PROTEINS

As noted earlier, the BChl-based photosynthesis within prokaryotes is found in five discontinuous phyla of cultured bacteria. However, in which particular bacterial group photosynthesis first originated and how this ability was acquired by other bacterial phyla has proven difficult to resolve. The main difficulty in this regard is that photosynthesis-related genes, which are clustered in genomes (Choudhary & Kaplan, 2000; Xiong, Inoue, & Bauer, 1998), are prone to LGTs (Raymond, 2009; Raymond et al., 2002, 2003; Zhaxybayeva, Gogarten, Charlebois, Doolittle, & Papke, 2006), making it difficult to interpret the results of phylogenetic analyses based on such genes/proteins (Green & Gantt, 2000; Xiong et al., 1998, 2000). Analyses based on other genes/proteins do not necessarily reflect the evolution of photosynthesis process (Ciccarelli et al., 2006; Gupta, 2000a, 2003; Gupta et al., 1999; Olsen et al., 1994). Thus, novel approaches are required to gain insights into this important problem.

Recent work on several key proteins involved in BChl biosynthesis has identified a number of CSIs that provide important insights in these regards. The proteins BchL, BchN and BchB (referred to as BchL–N–B) are part of an enzyme complex viz. light-independent (DPOR) that converts protochlorophyllide to chlorophyllide *a* (chlorin) (Beale, 1999; Burke, Hearst, & Sidow, 1993; Chew & Bryant, 2007; Mulkidjanian et al., 2006; Raymond et al., 2002; Raymond, Siefert, Staples, & Blankenship, 2004). Importantly, these three proteins, which are uniquely found in all phototrophs, exhibit significant sequence similarity to three other proteins BchX, BchY and BchZ (referred to as BchX–Y–Z) that form a second enzyme complex viz. chlorin reductase, which reduces chlorin to bacteriochlorin that serves as the direct precursor for the BChls (Beale, 1999; Chew & Bryant, 2007). The latter three proteins are found in different prokaryotic phototrophs except Cyanobacteria. The observed similarities in the sequences and structures of these two sets of proteins indicate that they have evolved from an ancient gene duplication in a common ancestor of all phototrophs (Burke et al., 1993; Chew & Bryant, 2007; Raymond et al., 2004; Xiong & Bauer, 2002). Additionally, these two sets of proteins also exhibit significant sequence and structural similarity to the three subunits (viz. NifH, NifD and NifK) of the nitrogenase complex (Muraki et al., 2010; Sarma et al., 2008), which plays a central role in nitrogen fixation (Burke et al., 1993; Haselkorn, 1986; Raymond et al., 2004; Xiong et al., 2000). These duplicated sets of photosynthesis-related proteins provide valuable means for investigating the origin of photosynthesis.

3.1. Conserved Indels in the NifH, BchX and BchL Proteins Provide Evidence that BchX Homologues Originated Prior to the BchL Homologues

One important question concerning the origin of photosynthesis is which of the two forms of photosynthesis (i.e. oxygenic photosynthesis carried out by Cyanobacteria or the anoxygenic photosynthesis carried out by other bacterial phyla) originated first (Blankenship, 1992, 2010; Burke et al., 1993; Hohmann-Marriott & Blankenship, 2011; Mulkidjanian et al., 2006; Olson & Blankenship, 2004). In the pathway leading to the biosynthesis of BChl/Chl, the DPOR enzyme complex (BchL–N–B) involved in the production of chlorin (a precursor to Chl), precedes the chlorin reductase complex (BchX–Y–Z), which produces bacteriochlorin – a direct precursor of BChl (Beale, 1999; Burke et al., 1993; Chew & Bryant, 2007). According to a hypothesis proposed by Granick (1965), in a given biochemical pathway the enzymes/proteins which carry out an earlier biochemical step have likely evolved earlier than those carrying out later steps. This hypothesis suggests that oxygenic photosynthesis based on Chl probably evolved earlier than the anoxygenic photosynthesis requiring BChl (Mauzerall, 1978; Olson & Blankenship, 2004; Olson & Pierson, 1987). However, phylogenetic studies based on NifH, BchX and BchL proteins have not supported this prediction (Burke et al., 1993; Raymond et al., 2003; Xiong et al., 2000).

More definitive evidence settling this question has come from identification of two CSIs (④ and ⑤) in the sequence alignments of NifH, BchX and BchL sequences that are uniquely shared by all NifH and BchX homologues, but not found in any BchL homologues (Fig. 2.6) (Gupta, 2012). Due to the earlier divergence of the NifH protein from the BchX and BchL protein pair, the NifH sequences can be used to determine whether any CSI present in the latter two proteins is an insert or a deletion. The fact that both the above CSIs are present in all NifH and BchX homologues, but are lacking in the BchL homologues, provides strong evidence that the presence of these indels is the ancestral state of the BchL–BchX protein, and so the BchX homologues containing these CSIs are primitive in comparison to the BchL homologues (Gupta, 2012). Thus, it can be inferred that these CSIs were caused by highly specific deletions in a common ancestor of the BchL gene after its divergence from the BchX gene by duplication. Due to the presence of these CSIs in conserved regions and their presence in all NifH and BchX homologues, but none of the BchL homologues, this is the simplest and most parsimonious explanation of these results. The species distribution patterns of these CSIs thus provide strong evidence that the anoxygenic photosynthesis supported by BchX homologues originated before the oxygenic photosynthesis requiring BchL homologues. This inference is independently supported by phylogenetic studies based on NifH, BchX and BchL sequences (Burke et al., 1993; Gupta, 2012; Xiong et al., 2000).

3.2. A Conserved Indel in the BchL Protein Provides Evidence that the BchL Homologues from *Heliobacteriaceae* are Primitive in Comparison to Sequences from Other Phototrophs

A central question in photosynthesis is to determine in which bacterial group photosynthetic ability first evolved. The sequence alignments of BchL, BchB and BchN proteins have also identified several CSIs that provide important insights in this regard. Two CSIs in the BchL homologues that have proven particularly useful in this regard are shown in Fig. 2.7 (Gupta, 2012). Both these CSIs are of defined lengths and they are flanked on both sides by conserved residues indicating that they are reliable molecular markers. Of these two CSIs, the first consists of a 1 aa indel (CSI ⑥) that is specifically present in the two *Heliobacteriaceae* species (Fig. 2.7A). The absence of this indel

Group	Species	38	④	66	127	⑤	164
NifH homologues (>200/200)	Methanosarcina barkeri	GCDPKADCT	R	LVLGGVAQTTIMDTLRELG	LGDVVCGGFAMPIR	EG	KAQEVYIVASGEMMATYAANNI
	Methanosarcina acetivorans	-----R-S-	-	ILAE-KFIPAVLEEH--QL	---------S----	--	F-E-I-LIC--GF-SI------
	Rhodopseudomonas palustris	-----S-S-	T	ILR--EDLP-VL-S--DS-	----------V---	N-	I-ESAFV-T-SDF--IF----L
	Desulfitobacterium hafniense	-----S-S-	N	TLR--KYIP-VL-----KS	----------I---	--	I-EH-FT-S-SDF-SI--S--L
	Rhodospirillum rubrum	-------S-	-	-I---KP-E-L--V---Q-	--------------	D-	---------------V------
	Chlorobium tepidum	-------S-	-	-L---LQ-K-VL-----E-	--------------	D-	--E-I---C------M------
	Ch. her. II	-------S-	-	-L---LI-K-VL-----E-	--------------	D-	--E-I---V------M------
	Clostridium acetobutylicum	-------S-	-	-L---L--K-VL-----E-	--------------	--	--K-I----------M------
	Azotobacter vinelandii	-------S-	-	-I-HSK--G-V-EMAASA-	--------------	-N	----I---C------M------
	Heliobacterium modesticaldum	-------S-	-	-I-HSK--A-V--LA--K-	--------------	-N	----I---T------M------
	He. chlorum	-------S-	-	-I-HSK--A-V--LA--K-	--------------	-N	----I---T------M------
	Sinorhizobium meliloti	-------S-	-	-I-NAK--D-VLHLAATE-	--------------	-N	----I---M------L------
	Rhodobacter sphaeroides	-------S-	-	-I-NTKL-D-VLHLAA-A-	--------------	-N	----I---M------L------
	Nostoc punctiforme	-------S-	-	-M-HSK----VLHLAA-R-	--------------	--	----I---T------M------
	Trichodesmium erythraeum	-------S-	-	-I-DAK----VLHVAA---	--------------	-N	----I---C------M------
BchX homologues (>70/70)	Heliobacterium modesticaldum	-----H-S-	V	ILFN--NPP-LLEYWA--N	---------GV--S	KS	I-KSIIL--GNDHQSL-V----
	Chloroflexus aurantiacus	-----H-SC	N	ALF--ISLP-LG-VW--FK	----------T-LS	RS	L-E--I-LCGNDRQSL------
	Chloroflexus aggregans	-----H-SC	N	ALF---SLP-LG-VW--FK	----------T-LS	RS	L-E--I-LCGNDRQSL------
	Roseiflexus castenholzii	-----H-SC	N	TIF--HSLP-LG-QW-LFR	----------T-LA	RS	L-EQ-I-LVGHDRQSL------
	Ro. sp. RS-1	-----H-SC	N	TIF--HSLP-LG-QW-LFK	----------T-LA	RS	L-EQ-I-LVGHDRQSL------
	Chlorobium phaeobacteroides	-----H-S-	T	SLF---SLP-VTEVFA-KN	----------T-LA	RS	LSE--ILLTNNDRQSIFT----
	Chlorobium tepidum	-----H-S-	T	SLF--ISLP-VTEVFA-KN	----------T-LA	RS	LSE--LL-T-NDRQSIFTS---
	Prosthecochloris vibrioformis	-----H-S-	T	SLF--MSLP-LT-VFS-KN	----------T-LA	RS	LSE--IL-T-NDRQSIFT----
	Chloroherpeton thalassium	-----H-S-	T	SLF---SLP-VTEVFAKKN	----------T-LS	RS	LCE--IL-V-NDRQSIF-----
	Rhodobacter sphaeroides	-----S-T-	S	-LF--K-CP--IE-SARKK	---------GL--A	RD	M--K-IL-G-NDLQSL-VT--V
	Rhodospirillum rubrum	-----S-T-	S	-LF--R-CP--IE-SSARK	---------GL--A	RD	LC-K-IV-G-NDLQSL-VV--V
	Methylobacterium extorquens	-----S-T-	S	-LF--R-CP--IE-STKKK	---------GL--A	RD	MC-K-IV-G-NDLQSL-V---V
	Rubrivivax gelatinosus	-----S-T-	S	-LF--K-TP--IE-SAKKK	---------GL--A	RD	MC-K-IV-G-NDLQSL-V---V
	Halorhodospira halophila	-----S-T-	S	-LF--R-CP--I--SSRKK	---------GL--A	RD	LC-K-IL-GANDLQSL-VV--V
BchL homologues (0>130)	Synechococcus sp. RC307	-----H-S-		FT-THKMVP-VI-I-E-VD	----------A-LQ		H-NYCL--TANDFDSIF-M-R-
	Prochlorococcus marinus MIT9303	-----H-S-		FT-THRMVP-VI-I-E-VD	----------A-LQ		H-NYCL--TANDFDSIF-M-R-
	Rhodobacter sphaeroides	-----H-S-		FT-T-SLVP-VI-V-KDVD	----------A-LQ		H-DQAVV-TANDFDSI--M-R-
	Methylobac.	-----H-S-		FT-TKRLAP-VI-A-EAVK	----------S-LQ		H-DRAL--TANDFDSIF-M-R-
	Chlorobium phaeobacteroides	-----H-S-		FPIT-KL-K-VIEA-E-VD	---------SA-LN		Y-DYAI-I-TNDFDSIF---RL
	Chloroflexus aurantiacus	-----H-S-		FP-T-HL-P-VI-V-DSVN	---------SA-LN		Y-DYGL-I-CNDFDSIF---RL
	Trichodesmium erythraeum	-----H-S-		FT-T-FLIP--I---Q-KD	----------A-LN		YSDYCM--TDNGFD-LF---R-
	Synechocystis sp. PCC6803	-----H-S-		FT-T-FLIP--I---Q-KD	----------A-LN		Y-DYCL--TDNGFD-LF---R-
	Heliobacterium modesticaldum	-----S-S-		FTIA-RMIP-VVEI-DKFN	----------T-LQ		Y-DLAC--S-NDFD-LF---R-
	Heliobacterium mobilis	-----S-S-		FTIA-KMIP-VVEI-DKFN	----------T-LQ		Y-DLACV-S-NDFD-LF---R-

Figure 2.6 ***Partial sequence alignments of the NifH, BchX and BchL homologues showing two different CSIs in these proteins that are commonly shared by the NifH and BchX homologues, but not found in any of the BchL homologues.*** The numbers below the group names indicate that, of the available homologues from these groups, how many contained or lacked these CSIs. The dashes (-) indicate identity with the amino acid on the top line. See the color plate.

in the BchL homologues from all other phototrophic bacteria indicates that it is a specific characteristic of the *Heliobacteriaceae* BChl (Gupta, 2012). The second CSI (CSI ⑦), present in an adjoining region, is a 5 aa indel that is uniquely found in different Chlorobi and Chloroflexi homologues, but absent in all other bacteria. These CSIs could be either inserts in the genes from these particular taxa or alternatively they could result from deletion(s) in the BchL homologues from other phototrophic lineages. Insights in these regards are provided by a multiple sequence alignment of diverse BchL and BchX homologues for the regions where these CSIs are found (Fig. 2.7B). From the sequence alignment of these two proteins, which shows sufficient conservation in this region, it is clear that the 1 aa CSI (CSI ⑥) that is uniquely found in the BchL homologues of *Heliobacteriaceae* is also present in all BchX homologues. Because, BchX and BchL homologues are derived by gene duplication in a common ancestor of all phototrophs (Burke et al., 1993; Xiong et al., 2000), the presence of this CSI in all BchX homologues as well as the BchL homologue from *Heliobacteriaceae* suggests that the BchL homologues from *Heliobacteriaceae* are primitive in comparison to those from other

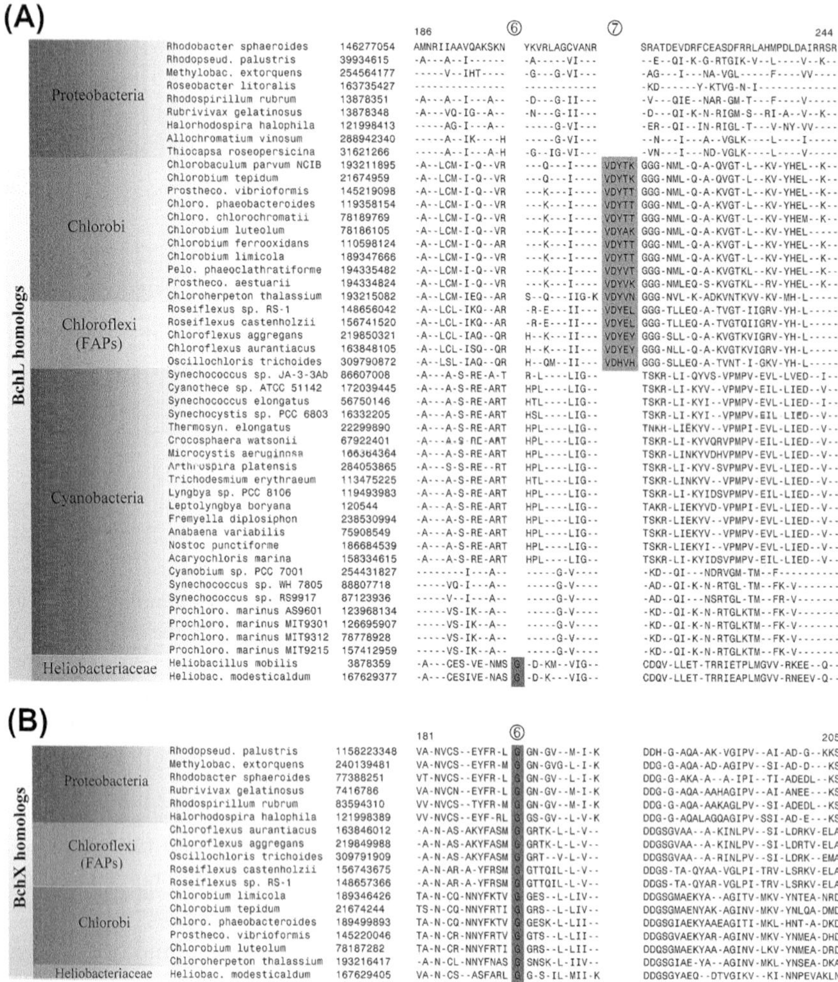

Figure 2.7 ***(A) Excerpts from the sequence alignment of BchL proteins showing two conserved signature indels that are specific for different lineages of phototrophic bacteria.*** The CSI ⑥ is specific for the *Heliobacteriaceae*, whereas CSI ⑦ is commonly shared by different Chlorobi and Chloroflexi homologues. The dashes (-) in these as well as other sequence alignments indicate identity with the amino acid on the top line. (B) A sequence alignment of the BchX homologues from different phototrophic lineages for the same region as shown in part A for the BchL protein sequences. See the color plate.

phototrophic lineages (Gupta, 2012). The absence of this CSI in the BchL homologues from other phototrophs is thus due to a deletion that occurred in a common ancestor of the other lineages. Due to the absence of this indel in all other BchL homologues and the distinct branching of the BchX and BchL homologues in phylogenetic tress (Gupta, 2012), the chance occurrence of this indel in the *Heliobacteriaceae* BchL homologues, or their acquisition of a gene containing this CSI by means of LGT is considered unlikely (Gupta, 2012). The species distribution of CSI ⑥ thus strongly suggests that the gene for the BchL protein first originated within the *Heliobacteriaceae* (Firmicutes) phylum, which is also indicated to be the earliest branching phylum within the bacteria (Ciccarelli et al., 2006; Gupta, 2001, 2003, 2011). This in turn suggests that photosynthesis evolved very early in the evolutionary history of life.

In contrast to the CSI ⑥, the CSI ⑦ is specifically present in the BchL homologues of Chlorobi and Chloroflexi. However, this CSI is absent in all of the BchX homologues indicating that the absence of this indel is the ancestral character state of the BchX–BchL protein. Therefore, this CSI represents an insert in the BchL homologues of Chlorobi and Chloroflexi and its shared presence in these two phylogenetically distinct lineages could be due to LGTs. In addition to this CSI, a number of other discovered CSIs in the BchL, BchB and BchN proteins are also uniquely shared by the Chlorobi and Chloroflexi homologues (Fig. 2.5; see CSI ② in the BchN homologues) suggesting that the genes for all three DPOR subunits have undergone LGTs between these two groups of photosynthetic organisms (Gupta, 2012). This inference is also strongly supported by phylogenetic studies and pairwise sequence similarity studies on the BchL, BchB and BchN protein sequences (Gupta, 2012). In cases of LGT, it is generally difficult to determine the direction in which LGT has occurred. However, based upon a number of other CSIs in the BchB protein, it is possible to infer that the gene transfer in these cases has occurred from Chloroflexi to a common ancestor of the Chlorobi (Gupta, 2012). Further, based upon these other CSIs, it is also possible to infer that the Chloroflexi species from which this gene transfer occurred was very likely a *Chloroflexus* spp. (i.e. excluding the genera *Roseiflexus* and *Oscillochloris*) (Gupta, 2012).

3.3. Implications of the Identified CSIs for the Origin and Spread of Photosynthesis

The identified CSIs in the BchL–N–B and BchX proteins indicate that the homologues of the BchL–N–B proteins (i.e. the DPOR enzyme complex) first originated in the *Heliobacteriaceae* species, which are part of the phylum

Firmicutes. The *Heliobacteriaceae* species are known to contain a primitive photosynthetic RC in which the antenna and RC complexes are both part of a single protein (Blankenship, 1992; Heinnickel & Golbeck, 2007; Sattley et al., 2008; Trost & Blankenship, 1989; Vassiliev, Antonkine, & Golbeck, 2001; Vermaas, 1994). Further, unlike other photosynthetic prokaryotes, thus far no photoautotrophic growth has been observed for any *Heliobacteriaceae* species (Bryant & Frigaard, 2006; Gest & Favinger, 1983; Madigan, 2006; Sattley & Blankenship, 2010). These observations raise the possibility that although the genes for some of the key photosynthesis proteins (viz. the DPOR complex) and a primitive photosynthetic RC first evolved in the *Heliobacteriaceae*, functional photosynthetic ability that could support photoautotrophic growth was not developed in this phylum.

The results of pairwise sequence similarities of BchL–N–B proteins indicate that their homologues from *Heliobacteriaceae* show greater similarity to those from Cyanobacteria (except Clade C) and Chloroflexi/Chlorobi. Earlier work based on several CSIs in universally distributed proteins provides evidence that the phylum Chloroflexi branched after the Firmicutes but prior to Cyanobacteria (Ciccarelli et al., 2006; Gupta, 2001, 2003). These observations suggest that either Chloroflexi or Cyanobacteria were the earliest recipients of the genes for photosynthesis proteins from *Heliobacteriaceae*. Geological and fossil evidence indicate that the earliest phototrophic microbial communities that existed ~3.4 Ga ago used the Calvin–Benson–Bassham cycle for CO_2 fixation, and they were comprised of filamentous anoxygenic bacteria (Dismukes et al., 2001; Tice & Lowe, 2004, 2006). In contrast, oxygenic photosynthesis attributable to Cyanobacteria is indicated to have evolved about 2.2–2.6 Ga ago (Blank & Sanchez-Baracaldo, 2010; Kazmierczak & Altermann, 2002; Olson, 2006; Olson & Blankenship, 2004). Because Chloroflexi have filamentous morphology and they are capable of carrying out anoxygenic photosynthesis by a variety of mechanisms including the Calvin–Benson–Bassham cycle (Hanada & Pierson, 2006), it is likely that they were the earliest phototrophic lineage in which photosynthetic ability was fully developed (Olson, 2006; Tice & Lowe, 2006). In contrast to all other photosynthetic bacteria, which contain only a single photosynthetic RC, Cyanobacteria possess two different types of RCs viz. RC-1 and RC-2 (or PSI and PSII) in order to carry out oxygenic photosynthesis (Blankenship & Hartman, 1998; Golbeck, 1993). One of these RCs, RC-1 is similar to that found in the *Heliobacteriaceae* species, whereas RC-2 is similar to that present in Chloroflexi (Blankenship, 1994; Olson & Pierson, 1987). Based upon these observations, it is likely that photosynthetic organisms that individually possessed either RC-1 or RC-2 evolved prior to the evolution of Cyanobacteria that contain both these RCs. The inferences from other CSIs discussed in this review provide evidence that the genes for the BchL–N–B proteins in Proteobacteria are derived from the Clade C Cyanobacteria, whereas those in Chlorobi were acquired from *Chloroflexus* or a related bacterium by means of LGTs (Gupta, 2012).

The work on different CSIs in the BchL, BchN, BchB and BchX proteins reviewed here has significantly advanced our understanding of origin and spread of photosynthesis, which represents one of the major evolutionary innovations in the history of life. Nonetheless, it should be acknowledged that these proteins represent only a small fraction of the total protein complement necessary for photosynthesis processes. Because photosynthesis gene repertoires in different lineages have been shaped by different processes including LGT, acquisition of novel genes, as well as gene losses, the overall evolution of photosynthesis in different lineages is apparently very complex (Hohmann-Marriott & Blankenship, 2011; Xiong & Bauer, 2002). Therefore, it is possible, and in fact quite likely, that not all components of this complex process will exhibit similar evolutionary histories and in most cases it will prove difficult to determine or resolve their relationships. In this context, the evolutionary histories of the BchL, BchN, BchB and BchX genes/proteins, which are unique and central components of photosynthesis, which we have been able to deduce by means of the discovered CSIs, represent an important advancement.

ABBREVIATIONS

CSI conserved signature indel (insert or deletion)
CSP conserved signature protein
BChl bacteriochlorophyll
FCB Fibrobacter–Chlorobi and Bacteroidetes
LGT(s) lateral gene transfer(s)
RC(s) reaction centre(s)
DPOR dark-operative protochlorophyllide oxidoreductase

ACKNOWLEDGEMENTS

The research work from the author's lab was supported by a research grant from the National Science and Engineering Research Council of Canada. I thank Mobolaji Adeolu and Chirayu Chokshi for their assistance in the preparation of this manuscript.

REFERENCES

Baldauf, S. L., & Palmer, J. D. (1993). Animals and fungi are each other's closest relatives: congruent evidence from multiple proteins. *Proceedings of the National Academy of Sciences of the United States of America, 90*, 11558–11562.

Beale, S. I. (1999). Enzyme of chlorophyll biosynthesis. *Photosynthesis Research, 60*, 43–73.

Blankenship, R. E. (1992). Origin and early evolution of photosynthesis. *Photosynthesis Research, 33*, 91–111.

Blankenship, R. E. (1994). Protein structure, electron transfer and evolution of prokaryotic photosynthetic reaction centers. *Antonie van Leeuwenhoek, 65*, 311–329.

Blankenship, R. E. (2010). Early evolution of photosynthesis. *Plant Physiology, 154*(2), 434–438.

Blankenship, R. E., & Hartman, H. (1998). The origin and evolution of oxygenic photosynthesis. *Trends Biochemical Sciences, 23*, 94–97.

Blankenship, R. E., Olson, J. M., & Miller, M. (1995). Antenna complexes from green photosynthetic bacteria. In R. E. Blankenship, M. T. Madigan & C. E. Bauer (Eds.), *Anoxygenic photosynthetic bacteria* (pp. 399–435). Dordrecht: Kluwer.

Blank, C. E., & Sanchez-Baracaldo, P. (2010). Timing of morphological and ecological innovations in the cyanobacteria – a key to understanding the rise in atmospheric oxygen. *Geobiology, 8*(1), 1–23.

Bryant, D. A., Costas, A. M., Maresca, J. A., Chew, A. G., Klatt, C. G., Bateson, M. M., et al. (2007). *Candidatus Chloracidobacterium thermophilum*: an aerobic phototrophic acidobacterium. *Science, 317*(5837), 523–526.

Bryant, D. A., & Frigaard, N. U. (2006). Prokaryotic photosynthesis and phototrophy illuminated. *Trends in Microbiology, 14*(11), 488–496.

Burke, D. H., Hearst, J. E., & Sidow, A. (1993). Early evolution of photosynthesis: clues from nitrogenase and chlorophyll iron proteins. *Proceedings of the National Academy of Sciences of the United States of America, 90*, 7134–7138.

Castenholz, R. W. (2001). Phylum BX. Cyanobacteria: oxygenic photosynthetic bacteria. (2nd ed.In D. R. Boone & R. W. Castenholz (Eds.), *Bergey's manual of systematic bacteriology* (Vol. 1, pp. 474–487). New York: Springer.

Chew, A. G., & Bryant, D. A. (2007). Chlorophyll biosynthesis in bacteria: the origins of structural and functional diversity. *Annual Review of Microbiology, 61*, 113–129.

Choudhary, M., & Kaplan, S. (2000). DNA sequence analysis of the photosynthesis region of *Rhodobacter sphaeroides* 2.4.1. *Nucleic Acids Research, 28*(4), 862–867.

Ciccarelli, F. D., Doerks, T., von Mering, C., Creevey, C. J., Snel, B., & Bork, P. (2006). Toward automatic reconstruction of a highly resolved tree of life. *Science, 311*(5765), 1283–1287.

Cole, J. R., Wang, Q., Cardenas, E., Fish, J., Chai, B., Farris, R. J., et al. (2009). The Ribosomal Database Project: improved alignments and new tools for rRNA analysis. *Nucleic Acids Research, 37*, D141–D145. Database issue.

Dismukes, G. C., Klimov, V. V., Baranov, S. V., Kozlov, Y. N., DasGupta, J., & Tyryshkin, A. (2001). The origin of atmospheric oxygen on Earth: the innovation of oxygenic photosynthesis. *Proceedings of the National Academy of Sciences of the United States of America, 98*(5), 2170–2175.

Dutilh, B. E., Snel, B., Ettema, T. J., & Huynen, M. A. (2008). Signature genes as a phylogenomic tool. *Molecular Biology and Evolution, 25*(8), 1659–1667.

Eisen, J. A., Nelson, K. E., Paulsen, I. T., Heidelberg, J. F., Wu, M., Dodson, R. J., et al. (2002). The complete genome sequence of *Chlorobium tepidum* TLS, a photosynthetic, anaerobic, green-sulfur bacterium. *Proceedings of the National Academy of Sciences of the United States of America, 99*(14), 9509–9514.

Euzeby, J. P. (2011). *List of prokaryotic names with standing in nomenclature*. http://www.bacterio.cict.fr/classifphyla.html. Ref Type: Generic.

Fang, G., Rocha, E. P., & Danchin, A. (2008). Persistence drives gene clustering in bacterial genomes. *BMC Genomics, 9*, 4.

Felsenstein, J. (1988). Phylogenies from molecular sequences: inference and reliability. *Annual Review of Genetics, 22*, 521–565.

Felsenstein, J. (2004). *Inferring phylogenies*. Sunderland, MA: Sinauer Associates, Inc.

Frigaard, N. U., Chew, A. G., Li, H., Maresca, J. A., & Bryant, D. A. (2003). *Chlorobium tepidum*: insights into the structure, physiology, and metabolism of a green sulfur bacterium derived from the complete genome sequence. *Photosynthesis Research, 78*(2), 93–117.

Gao, B., & Gupta, R. S. (2012). Microbial systematics in the post-genomics era. *Antonie van Leeuwenhoek, 101*(1), 45–54.

Gao, B., Mohan, R., & Gupta, R. S. (2009). Phylogenomics and protein signatures elucidating the evolutionary relationships among the Gammaproteobacteria. *International Journal of Systematic and Evolutionary Microbiology, 59*(2), 234–247.

Garrity, G. M., & Holt, J. G. (2001a). Phylum BVI, Chloroflexi phy. nov. In D. R. Boone & R. W. Castenholz (Eds.), *Bergey's manual of systematic bacteriology* (2nd ed. *The Archaea and the deeply branching and phototrophic bacteria* (Vol. 1, pp. 427–446). New York: Springer Verlag.

Garrity, G. M., & Holt, J. G. (2001b). Phylum BVIi. Thermomicrobia phy. nov. In D. R. Boone & R. W. Castenholz (Eds.), *Bergey's manual of systematic bacteriology* (2nd ed. *The Archaea and the deeply branching and phototrophic bacteria* (Vol. 1, pp. 447–450). New York: Springer Verlag.

Gest, H., & Blankenship, R. E. (2004). Time line of discoveries: anoxygenic bacterial photosynthesis. *Photosynthesis Research, 80*(1–3), 59–70.

Gest, H., & Favinger, J. (1983). *Heliobacterium chlorum*, an anoxygenic brownish-green photosynthetic bacterium containing a "new" form of bacteriochlorophyll. *Archive Microbiology, 136*, 11–16.

Golbeck, J. H. (1993). Shared thematic elements in photochemical reaction centers. *Proceedings of the National Academy of Sciences of the United States of America, 90*, 1642–1646.

Granick, S. (1965). Evolution of heme and chlorophyll. In V. Bryson & H. J. Vogel (Eds.), *Evolving genes and proteins* (pp. 67–88). New York: Academic Press.

Green, B. R., & Gantt, E. (2000). Is photosynthesis really derived from proteobacteria? *Phycology, 36*, 983–985.

Griffiths, E., & Gupta, R. S. (2001). The use of signature sequences in different proteins to determine the relative branching order of bacterial divisions: evidence that *Fibrobacter* diverged at a similar time to Chlamydia and the Cytophaga-Flavobacterium-Bacteroides division. *Microbiology, 147*(9), 2611–2622.

Griffiths, E., & Gupta, R. S. (2004). Signature sequences in diverse proteins provide evidence for the late divergence of the order Aquificales. *International Microbiology, 7*, 41–52.

Griffiths, E., & Gupta, R. S. (2006). Lateral transfers of serine hydroxymethyl transferase (*glyA*) and UDP-N-acetylglucosamine enolpyruvyl transferase (*murA*) genes from free-living *Actinobacteria* to the parasitic chlamydiae. *Journal of Molecular Evolution, 63*, 283–296.

Griffiths, E., & Gupta, R. S. (2007). Phylogeny and shared conserved inserts in proteins provide evidence that Verrucomicrobia are the closest known free-living relatives of chlamydiae. *Microbiology, 153*(8), 2648–2654.

Gupta, R. S. (1998). Protein phylogenies and signature sequences: a reappraisal of evolutionary relationships among archaebacteria, eubacteria, and eukaryotes. *Microbiology Molecular Biology Reviews, 62*, 1435–1491.

Gupta, R. S. (2000a). The natural evolutionary relationships among prokaryotes. *Critical Reviews in Microbiology, 26*, 111–131.

Gupta, R. S. (2000b). The phylogeny of proteobacteria: relationships to other eubacterial phyla and eukaryotes. *FEMS Microbiology Reviews, 24*, 367–402.

Gupta, R. S. (2001). The branching order and phylogenetic placement of species from completed bacterial genomes, based on conserved indels found in various proteins. *International Microbiology*, *4*, 187–202.

Gupta, R. S. (2003). Evolutionary relationships among photosynthetic bacteria. *Photosynthesis Research*, *76*, 173–183.

Gupta, R. S. (2004). The phylogeny and signature sequences characteristics of Fibrobacteres, Chlorobi and Bacteroidetes. *Critical Reviews in Microbiology*, *30*, 123–143.

Gupta, R. S. (2005a). Molecular sequences and the early history of life. In J. Sapp (Ed.), *Microbial phylogeny and evolution: Concepts and controversies* (pp. 160–183). New York: Oxford University Press.

Gupta, R. S. (2005b). Protein signatures distinctive of alpha proteobacteria and its subgroups and a model for alpha proteobacterial evolution. *Critical Reviews in Microbiology*, *31*(101), 135.

Gupta, R. S. (2009). Protein signatures (molecular synapomorphies) that are distinctive characteristics of the major cyanobacterial clades. *International Journal of Systematic and Evolutionary Microbiology*, *59*, 2510–2526.

Gupta, R. S. (2010a). Applications of conserved indels for understanding microbial phylogeny. In A. Oren & R. T. Papke (Eds.), *Molecular phylogeny of microorganisms* (pp. 135–150). Norfolk, UK: Caister Academic Press.

Gupta, R. S. (2010b). Molecular signatures for the main phyla of photosynthetic bacteria and their subgroups. *Photosynthesis Research*, *104*(2–3), 357–372.

Gupta, R. S. (2011). Origin of diderm (Gram-negative) bacteria: antibiotic selection pressure rather than endosymbiosis likely led to the evolution of bacterial cells with two membranes. *Antonie van Leeuwenhoek*, *100*, 171–182.

Gupta, R. S. (May 24, 2012). Origin and spread of photosynthesis based upon conserved sequence features in key bacteriochlorophyll biosynthesis proteins. *Mol. Biol. Evol.* PMID: 22628531 [Epub ahead of print], vol. *29*, 3397–3412.

Gupta, R. S., Chander, P., & George, S. (2013). Phylogenetic framework and molecular signatures for the class Chloroflexi and its different clades; proposal for division of the class Chloroflexi class. nov. into the suborder Chloroflexineae subord. nov., consisting of the emended family *Oscillochloridaceae* and the family *Chloroflexaceae* fam. nov., and the suborder Roseiflexineae subord. nov., containing the family *Roseiflexaceae* fam. nov. *Antonie van Leeuwenhoek*, vol. *103*, 99–119.

Gupta, R. S., & Griffiths, E. (2002). Critical issues in bacterial phylogeny. *Theoretical Population Biology*, *61*, 423–434.

Gupta, R. S., & Johari, V. (1998). Signature sequences in diverse proteins provide evidence of a close evolutionary relationship between the Deinococcus-thermus group and cyanobacteria. *Journal of Molecular Evolution*, *46*, 716–720.

Gupta, R. S., & Lorenzini, E. (2007). Phylogeny and molecular signatures (conserved proteins and indels) that are specific for the Bacteroidetes and Chlorobi species. *BMC Evolutionary Biology*, *7*, 71.

Gupta, R. S., & Mathews, D. W. (2010). Signature proteins for the major clades of Cyanobacteria. *BMC Evolutionary Biology*, *10*, 24.

Gupta, R. S., & Mok, A. (2007). Phylogenomics and signature proteins for the alpha proteobacteria and its main groups. *BMC Microbiology*, 7(1), 106.

Gupta, R. S., Mukhtar, T., & Singh, B. (1999). Evolutionary relationships among photosynthetic prokaryotes (*Heliobacterium chlorum*, *Chloroflexus aurantiacus*, cyanobacteria, *Chlorobium tepidum* and proteobacteria): implications regarding the origin of photosynthesis. *Molecular Microbiology*, *32*, 893–906.

Gupta, R. S., Pereira, M., Chandrasekera, C., & Johari, V. (2003). Molecular signatures in protein sequences that are characteristic of Cyanobacteria and plastid homologues. *International Journal of Systematic and Evolutionary Microbiology*, *53*, 1833–1842.

Gupta, R. S., & Sneath, P. H. A. (2007). Application of the character compatibility approach to generalized molecular sequence data: branching order of the proteobacterial subdivisions. *Journal of Molecular Evolution, 64*, 90–100.

Hanada, S., & Pierson, B. K. (2006). The family *Chloroflexaceae*. In M. Dworkin, et al. (Ed.), *The prokaryotes: A handbook on the biology of bacteria* (3rd ed.). *Proteobacteria: Delta, epsilon subclass* (Vol. 7, pp. 815–842). New York: Springer. Release 3.12 edn.

Hanada, S., Takaichi, S., Matsuura, K., & Nakamura, K. (2002). *Roseiflexus castenholzii* gen. nov., sp. nov., a thermophilic, filamentous, photosynthetic bacterium that lacks chlorosomes. *International Journal of Systematic and Evolutionary Microbiology, 52*(1), 187–193.

Hartman, H. (1998). Photosynthesis and the origin of life. *Origins of Life and Evolution of Biospheres, 28*, 515–521.

Haselkorn, R. (1986). Organization of the genes for nitrogen fixation in photosynthetic bacteria and cyanobacteria. *Annual Review of Microbiology, 40*, 525–547.

Heinnickel, M., & Golbeck, J. H. (2007). Heliobacterial photosynthesis. *Photosynthesis Research, 92*(1), 35–53.

Hohmann-Marriott, M. F., & Blankenship, R. E. (2011). Evolution of photosynthesis. *Annual Review of Plant Biology, 62*, 515–548.

Hugenholtz, P., & Stackebrandt, E. (2004). Reclassification of *Sphaerobacter thermophilus* from the subclass Sphaerobacteridae in the phylum Actinobacteria to the class Thermomicrobia (emended description) in the phylum Chloroflexi (emended description). *International Journal of Systematic and Evolutionary Microbiology, 54*(6), 2049–2051.

Imhoff, J. F. (2001). The anoxygenic phototrophic purple bacteria. In D. R. Boone & R. W. Castenholz (Eds.), *Bergey's manual of systematic bacteriology* (2nd ed.). *The Archaea and the deeply branching and phototrophic bacteria* (Vol. 1, pp. 631–637). Berlin: Springer-Verlag.

Imhoff, J. F. (2003). Phylogenetic taxonomy of the family *Chlorobiaceae* on the basis of 16S rRNA and fmo (Fenna–Matthews–Olson protein) gene sequences. *International Journal of Systematic and Evolutionary Microbiology, 53*(4), 941–951.

Kainth, P., & Gupta, R. S. (2005). Signature proteins that are distinctive of alpha proteobacteria. *BMC Genomics, 6*, 94.

Kazmierczak, J., & Altermann, W. (2002). Neoarchean biomineralization by benthic cyanobacteria. *Science, 298*(5602), 2351.

Keppen, O. I., Tourova, T. P., Kuznetsov, B. B., Ivanovsky, R. N., & Gorlenko, V. M. (2000). Proposal of *Oscillochloridaceae* fam. nov. on the basis of a phylogenetic analysis of the filamentous anoxygenic phototrophic bacteria, and emended description of *Oscillochloris* and *Oscillochloris trichoides* in comparison with further new isolates. *International Journal of Systematic and Evolutionary Microbiology, 50*(4), 1529–1537.

Kersters, K., Devos, P., Gillis, M., Swings, J., Vandamme, P., & Stackebrandt, E. (2006). Introduction to the Proteobacteria. In M. Dworkin, et al. (Ed.), *The prokaryotes: A handbook on the biology of bacteria* (3rd ed.). *Proteobacteria: Alpha and beta subclasses* (Vol. 5, pp. 3–37). New York: Springer. Release 3.12 edn.

Koonin, E. V., Aravind, L., & Kondrashov, A. S. (2000). The impact of comparative genomics on our understanding of evolution. *Cell, 101*(6), 573–576.

Kuo, C. H., & Ochman, H. (2009). The fate of new bacterial genes. *FEMS Microbiology Reviews, 33*(1), 38–43.

Lerat, E., Daubin, V., Ochman, H., & Moran, N. A. (2005). Evolutionary origins of genomic repertoires in bacteria. *PLoS Biology, 3*(5), e130.

Ludwig, W., & Klenk, H. -P. (2005). Overview: a phylogenetic backbone and taxonomic framework for prokaryotic systematics. In D. J. Brenner, et al. (Ed.), *Bergey's manual of systematic bacteriology* (2nd ed.). *The proteobacteria* (Vol. 2, pp. 49–65). Berlin: Springer-Verlag.

Madigan, M.T. (2006). The family *Heliobacteriaceae*. In M. Dworkin, et al. (Ed.), *The prokaryotes: A handbook on the biology of bacteria* (3rd ed.). *Bacteria: Firmicutes, cyanobacteria* (Vol. 4, pp. 951–964). New York: Springer. Release 3.12 edn.

Margulis, L. (1993). *Symbiosis in cell evolution*. New York: W. H. Freeman and Company.

Mauzerall, D. (1978). Bacteriochlorophyll and photosynthesis evolution. In R. K. Clayton & W. R. Sistrom (Eds.), *The photosynthetic bacteria* (pp. 223–231). New York: Plenum Press.

Moe, W. M., Yan, J., Nobre, M. F., da Costa, M. S., & Rainey, F. A. (2009). *Dehalogenimonas lykanthroporepellens* gen. nov., sp. nov., a reductively dehalogenating bacterium isolated from chlorinated solvent-contaminated groundwater. *International Journal of Systematic and Evolutionary Microbiology*, *59*(11), 2692–2697.

Moore, L. R., Rocap, G., & Chisholm, S. W. (1998). Physiology and molecular phylogeny of coexisting *Prochlorococcus* ecotypes. *Nature*, *393*(6684), 464–467.

Morden, C.W., Delwiche, C. F., Kuhsel, M., & Palmer, J. D. (1992). Gene phylogenies and the endosymbiotic origin of plastids. *Biosystems*, *28*, 75–90.

Moreira, D., & Philippe, H. (2000). Molecular phylogeny: pitfalls and progress. *International Microbiology*, *3*(1), 9–16.

Mulkidjanian, A. Y., Koonin, E. V., Makarova, K. S., Mekhedov, S. L., Sorokin, A., Wolf, Y. I., et al. (2006). The cyanobacterial genome core and the origin of photosynthesis. *Proceedings of the National Academy of Sciences of the United States of America*, *103*(35), 13126–13131.

Muraki, N., Nomata, J., Ebata, K., Mizoguchi, T., Shiba, T., Tamiaki, H., et al. (2010). X-ray crystal structure of the light-independent protochlorophyllide reductase. *Nature*, *465*(7294), 110–114.

Narra, H. P., Cordes, M. H., & Ochman, H. (2008). Structural features and the persistence of acquired proteins. *Proteomics*, *8*(22), 4772–4781.

Nelson, N., & Ben Shem, A. (2005). The structure of photosystem I and evolution of photosynthesis. *Bioessays*, *27*(9), 914–922.

Nitschke, W., & Rutherford, A. W. (1991). Photosynthetic reaction centers: variation on a common structural theme? *Trends Biochemical Sciences*, *16*, 241–245.

Nobrega, M. A., & Pennacchio, L. A. (2004). Comparative genomic analysis as a tool for biological discovery. *Journal of Physiology*, *554*(1), 31–39.

Olsen, G. J., Woese, C. R., & Overbeek, R. (1994). The winds of (evolutionary) change: breathing new life into microbiology. *Journal of Bacteriology*, *176*(1), 1–6.

Olson, J. M. (2006). Photosynthesis in the Archean era. *Photosynthesis Research*, *88*(2), 109–117.

Olson, J. M., & Blankenship, R. E. (2004). Thinking about the evolution of photosynthesis. *Photosynthesis Research*, *80*, 373–386.

Olson, J. M., & Pierson, B. K. (1987). Evolution of reaction centers in photosynthetic prokaryotes. *International Review of Cytology*, *108*, 209–248.

Overmann, J. (2003). The family *Chlorobiaceae*. In *The prokaryotes* (3rd ed.).

Pierson, B. K. (1994). The emergence, diversification, and role of photosynthetic eubacteria. In S. Benston (Ed.), *Early life on earth: Nobel symposium No. 84* (pp. 161–180). New York: Columbia University Press.

Pierson, B. K., & Castenholz, R. W. (1992). The family *Chloroflexaceae*. In A. Balows, et al. (Ed.), *The prokaryotes* (2nd ed., pp. 3754–3775). New York: Springer-Verlag.

Raymond, J. (2008). Coloring in the tree of life. *Trends in Microbiology*, *16*(2), 41–43.

Raymond, J. (2009). The role of horizontal gene transfer in photosynthesis, oxygen production, and oxygen tolerance. *Methods in Molecular Biology*, *532*, 323–338.

Raymond, J., Siefert, J. L., Staples, C. R., & Blankenship, R. E. (2004). The natural history of nitrogen fixation. *Molecular Biology and Evolution*, *21*(3), 541–554.

Raymond, J., Zhaxybayeva, O., Gogarten, J. P., & Blankenship, R. E. (2003). Evolution of photosynthetic prokaryotes: a maximum-likelihood mapping approach. *Philosophical Transactions of the Royal Society of London. Series B, Biological Sciences*, *358*(1429), 223–230.

Raymond, J., Zhaxybayeva, O., Gogarten, J. P., Gerdes, S. Y., & Blankenship, R. E. (2002). Whole-genome analysis of photosynthetic prokaryotes. *Science, 298*(5598), 1616–1620.

Rippka, R., Deruelles, J., Waterbury, J. B., Herdman, M., & Stanier, R. Y. (1979). Generic assignments, strain histories and properties of pure cultures of cyanobacteria. *Journal of General Microbiology, 111*, 1–61.

Rivera, M. C., & Lake, J. A. (1992). Evidence that eukaryotes and eocyte prokaryotes are immediate relatives. *Science, 257*(5066), 74–76.

Rocap, G., Distel, D. L., Waterbury, J. B., & Chisholm, S. W. (2002). Resolution of *Prochlorococcus* and *Synechococcus* ecotypes by using 16S–23S ribosomal DNA internal transcribed spacer sequences. *Applied Environmental Microbiology, 68*(3), 1180–1191.

Rocap, G., Larimer, F. W., Lamerdin, J., Malfatti, S., Chain, P., Ahlgren, N. A., et al. (2003). Genome divergence in two *Prochlorococcus* ecotypes reflects oceanic niche differentiation. *Nature, 424*(6952), 1042–1047.

Rokas, A., & Holland, P. W. (2000). Rare genomic changes as a tool for phylogenetics. *Trends in Ecology and Evolution, 15*(11), 454–459.

Sadekar, S., Raymond, J., & Blankenship, R. E. (2006). Conservation of distantly related membrane proteins: photosynthetic reaction centers share a common structural core. *Molecular Biology and Evolution, 23*(11), 2001–2007.

Sanchez-Baracaldo, P., Hayes, P. K., & Blank, C. E. (2005). Morphological and habitat evolution in the Cyanobacteria using a compartmentalization approach. *Geobiology, 3*, 145–165.

Sarma, R., Barney, B. M., Hamilton, T. L., Jones, A., Seefeldt, L. C., & Peters, J. W. (2008). Crystal structure of the L protein of *Rhodobacter sphaeroides* light-independent protochlorophyllide reductase with MgADP bound: a homologue of the nitrogenase Fe protein. *Biochemistry, 47*(49), 13004–13015.

Sattley, W. M., & Blankenship, R. E. (2010). Insights into heliobacterial photosynthesis and physiology from the genome of *Heliobacterium modesticaldum*. *Photosynthesis Research, 104*(2–3), 113–122.

Sattley, W. M., Madigan, M. T., Swingley, W. D., Cheung, P. C., Clocksin, K. M., Conrad, A. L., et al. (2008). The genome of *Heliobacterium modesticaldum*, a phototrophic representative of the Firmicutes containing the simplest photosynthetic apparatus. *Journal of Bacteriology, 190*(13), 4687–4696.

Schubert, W. D., Klukas, O., Saenger, W., Witt, H. T., Fromme, P., & Krauss, N. (1998). A common ancestor for oxygenic and anoxygenic photosynthetic systems: a comparison based on the structural model of photosystem I. *Journal of Molecular Biology, 280*, 297–314.

Shi, T., & Falkowski, P. G. (2008). Genome evolution in cyanobacteria: the stable core and the variable shell. *Proceedings of the National Academy of Sciences of the United States of America, 105*(7), 2510–2515.

Swingley, W. D., Blankenship, R. E., & Raymond, J. (2008). Integrating Markov clustering and molecular phylogenetics to reconstruct the cyanobacterial species tree from conserved protein families. *Molecular Biology and Evolution, 25*(4), 643–654.

Tice, M. M., & Lowe, D. R. (2004). Photosynthetic microbial mats in the 3,416-Myr-old ocean. *Nature, 431*(7008), 549–552.

Tice, M. M., & Lowe, D. R. (2006). Hydrogen-based carbon fixation in the earliest known photosynthetic organisms. *Geology, 34*, 37–40.

Trost, J. T., & Blankenship, R. E. (1989). Isolation of a photoactive photosynthetic reaction center–core antenna complex from *Heliobacillus mobilis*. *Biochemistry, 28*(26), 9898–9904.

Vassiliev, I. R., Antonkine, M. L., & Golbeck, J. H. (2001). Iron–sulfur clusters in type I reaction centers. *Biochimica Biophysica Acta, 1507*(1–3), 139–160.

Vermaas, W. F.J. (1994). Evolution of heliobacteria: implications for photosynthetic reaction center complexes. *Photosynthesis Research, 41*, 285–294.

Williams, K. P., Sobral, B. W., & Dickerman, A. W. (2007). A robust species tree for the Alphaproteobacteria. *Journal of Bacteriology*, *189*, 4578–4586.

Wilmotte, A., & Golubic, S. (1991). Morphological and genetic criteria in the taxonomy of Cyanophyta/Cyanobacteria. *Arhciv fur Hydrobiologie*, *64*(Suppl. 92), 1–24.

Wilmotte, A., & Herdman, M. (2001). Phylogenetic relationships among the cyanobacteria based on 16S rRNA sequences. (2nd ed.In D. R. Boone & R. W. Castenholz (Eds.), *Bergey's manual of systematic bacteriology* (Vol. 1, pp. 487–493). New York: Springer.

Xiong, J., & Bauer, C. E. (2002). Complex evolution of photosynthesis. *Annual Review of Plant Biology*, *53*, 503–521.

Xiong, J., Fischer, W. M., Inoue, K., Nakahara, M., & Bauer, C. E. (2000). Molecular evidence for the early evolution of photosynthesis. *Science*, *289*(5485), 1724–1730.

Xiong, J., Inoue, K., & Bauer, C. E. (1998). Tracking molecular evolution of photosynthesis by characterization of a major photosynthesis gene cluster from *Heliobacillus mobilis*. *Proceedings of the National Academy of Sciences of the United States of America*, *95*, 14851–14856.

Yabe, S., Aiba, Y., Sakai, Y., Hazaka, M., & Yokota, A. (2010). *Thermosporothrix hazakensis* gen. nov., sp. nov., isolated from compost, description of *Thermosporotrichaceae* fam. nov. within the class Ktedonobacteria Cavaletti et al. 2007 and emended description of the class Ktedonobacteria. *International Journal of Systematic and Evolutionary Microbiology*, *60*(8), 1794–1801.

Yamada, T., Sekiguchi, Y., Hanada, S., Imachi, H., Ohashi, A., Harada, H., et al. (2006). *Anaerolinea thermolimosa* sp. nov., *Levilinea saccharolytica* gen. nov., sp. nov. and *Leptolinea tardivitalis* gen. nov., sp. nov., novel filamentous anaerobes, and description of the new classes Anaerolineae classis nov. and Caldilineae classis nov. in the bacterial phylum Chloroflexi. *International Journal of Systematic and Evolutionary Microbiology*, *56*(6), 1331–1340.

Yarza, P., Ludwig, W., Euzeby, J., Amann, R., Schleifer, K. H., Glockner, F. O., et al. (2010). Update of the all-species living tree project based on 16S and 23S rRNA sequence analyses. *Systematic Applied Microbiology*, *33*(6), 291–299.

Zhaxybayeva, O., Gogarten, J. P., Charlebois, R. L., Doolittle, W. F., & Papke, R. T. (2006). Phylogenetic analyses of cyanobacterial genomes: quantification of horizontal gene transfer events. *Genome Research*, *16*(9), 1099–1108.

CHAPTER THREE

Properties and Evolutionary Implications of the Heliobacterial Genome

W. Matthew Sattley*,[1], Wesley D. Swingley**

*Indiana Wesleyan University, Division of Natural Sciences, Marion, IN, USA
**Northern Illinois University, Department of Biological Sciences, DeKalb, IL, USA
[1]Corresponding author: E-mail: matthew.sattley@indwes.edu

Contents

Abstract

Heliobacteria are strictly anaerobic, anoxygenic phototrophic bacteria belonging to the phylum *Firmicutes*. They are distinct from other anaerobic anoxygenic phototrophs in that they produce unique photosynthetic pigments (bacteriochlorophyll *g* is the major pigment), have no capacity for autotrophic growth, and like their nonphotosynthetic relatives the clostridia, have a Gram-positive cell structure and are capable of producing heat-resistant endospores. Phototrophy in heliobacteria is carried out using an FeS-type (type-I) homodimeric reaction centre that represents the simplest known photosynthetic apparatus. We present herein a summary of the ecological, phylogenetic, photosynthetic, and physiological properties that distinguish heliobacteria from other phototrophs based on an analysis of features of the helio-

Advances in Botanical Research, Volume 66
ISSN 0065-2296, http://dx.doi.org/10.1016/B978-0-12-397923-0.00003-5

bacterial (*Heliobacterium modesticaldum* strain Ice1) genome. Also considered are the implications of pigment biosynthesis elements and components of the heliobacterial photosynthetic apparatus to the complex question of the origin and evolution of photosynthesis.

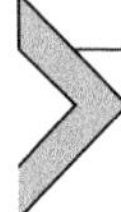

1. INTRODUCTION

1.2. Basic Biology and Ecology of Heliobacteria

The family *Heliobacteriaceae* (Asao & Madigan, 2009; Madigan, Euzéby, & Asao, 2010) comprises a relatively small but important group of Gram-positive phototrophic bacteria. Six bacterial phyla containing phototrophic representatives are known and include the oxygen-evolving (oxygenic) *Cyanobacteria* and five phyla containing anoxygenic phototrophs: (1) *Firmicutes* (heliobacteria), (2) *Chlorobi* (green sulfur bacteria, GSB), (3) *Chloroflexi* (filamentous anoxygenic phototrophs), (4) *Proteobacteria* (purple bacteria, consisting of several classes), and (5) *Acidobacteria* (represented by a single cultured representative, *Candidatus* Chloracidobacterium thermophilum) (Fig. 3.1). The present work reviews the phototrophic *Firmicutes* (the heliobacteria) in terms of their ecology, phototrophic and chemotrophic metabolisms, and evolution, based on features of the *Heliobacterium modesticaldum* strain Ice1^T genome. This bacterium, a moderate thermophile isolated from volcanic soils of Iceland (Kimble, Mandelco, Woese, & Madigan, 1995), has emerged as a model organism for the study of photosynthesis due to the relative simplicity of its photosynthetic apparatus, and it is currently the only heliobacterium for which a complete genome sequence is publically available (Sattley et al., 2008). Table 3.1 provides a summary of the major characteristics of this bacterium.

Several important characteristics distinguish the heliobacteria from all other phototrophic bacteria. The primary defining feature of heliobacteria is their production of bacteriochlorophyll (BChl) *g* as the major pigment for light harvesting and photochemistry (Brockmann & Lipinski, 1983). BChl *g* is unique to the heliobacteria and absorbs maximally in the near infrared at 785–790 nm, specifically between the absorption maxima of major pigments from green bacteria (705–740 nm) and purple bacteria (830–1100 nm) (Madigan, 2006; Asao & Madigan, 2010). This property allows heliobacteria to utilise wavelengths of light that are not absorbed by other phototrophs, thus providing them with a photosynthetic niche that helps to promote their ecological success. A smaller quantity of other unique pigments are synthesised by heliobacteria, including 8^1-hydroxy-chlorophyll *a* and C_{30} carotenoids (either 4-4′-diaponeurosporene or OH-diaponeurosporene glucoside esters) (Fuller, Sprague, Gest, &

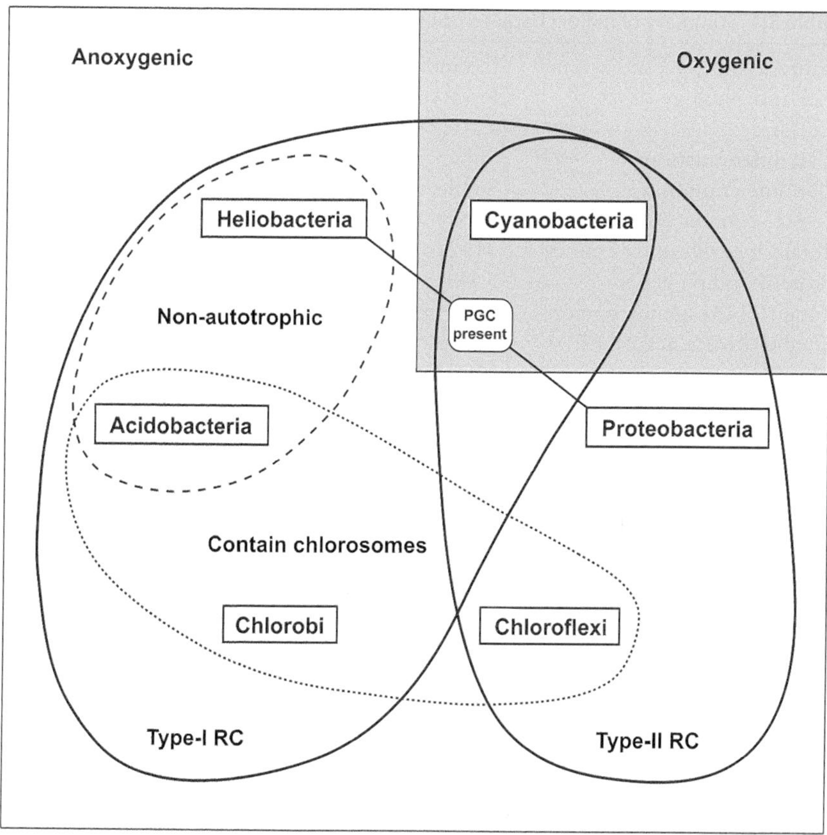

Figure 3.1 ***Relationships among phototrophic bacteria.*** Only one bacterial phylum, the *Cyanobacteria*, contains oxygenic phototrophs. The *cyanobacteria* are the only phototrophic bacteria to contain both iron–sulfur-type and quinone-type RCs (type-I and type-II RCs, respectively). Light-harvesting pigments in *cyanobacteria* and purple bacteria (phototrophic *Proteobacteria*) are typically housed in intracytoplasmic lamellar membranes, whereas pigments in green sulfur bacteria (*Chlorobi*), filamentous anoxygenic phototrophs (*Chloroflexi*), and phototrophic *Acidobacteria* are organised within chlorosomes. Heliobacteria have a much simpler phototrophic apparatus, by comparison, having all light-harvesting pigments contained within plasma membrane-bound, type-I RCs. As indicated in the figure, only purple bacteria and heliobacteria contain PGCs, in which photosynthesis genes are organised into superoperons of 50 kbp or more (Fig. 3.4). Interestingly, autotrophic growth, one of the hallmarks of photosynthesis, has never been observed in heliobacteria or 'Chloracidobacterium thermophilum', the sole cultured representative of phototrophic *Acidobacteria*.

Table 3.1 Summary of major characteristics of *Hbt. modesticaldum* strain Ice1[T]

Source	Icelandic volcanic soil
Cell morphology	Rod/curved rod
Cell size	1 × 2.5–6.5 μm
Cell differentiation	Subterminal endospores produced
Genome composition	Single, circular chromosome of 3,075,407 bp
G + C content (%)	56.98
Total Open Reading Frames	3142
Protein-coding genes	3000
Total rRNAs (genes/operons)	24/8
Total tRNA genes	104
Motility	Flagellar
Growth temperature	Range, 25–56 °C; Optimum, 50–52 °C
Optimal pH	pH 6–7
Phototrophic growth	Photoassimilation of pyruvate, lactate, acetate, or yeast extract
Chemotrophic growth	Fermentation of pyruvate
Diazotrophy	Optimal N_2-fixation at 50 °C via a molybdenum-dependant group I nitrogenase
RC core composition	Type-I PshA homodimer
Pigments per RC	20 BChl g_F, two BChl g'_F, two 8^1-OH-Chl a_F, and one 4,4′ diaponeurosporene

Blankenship, 1985; Takaichi et al., 1997; Takaichi, Oh-Oka, Maoka, Jung, & Madigan, 2003; Trost & Blankenship, 1989; van de Meent et al., 1991). All heliobacterial pigments are contained within a cytoplasmic membrane-bound, type-I reaction centre (RC) (Table 3.1). This is in contrast to other phototrophic bacteria, which organise antenna pigments within specialised internal membrane structures, such as the chlorosomes of green bacteria and phototrophic *acidobacteria*, the chromatophores/membrane lamellae of purple bacteria, and the thylakoid membranes of *cyanobacteria*. Table 3.2 provides a comparison of pigments, pigment-containing membrane systems, and other defining features of phototrophic bacteria from all six phyla.

Like the clostridia, heliobacteria are strict anaerobes, and they are the only phototrophic bacteria to lack an outer membrane and have a Gram-positive cell wall structure (Fig. 3.2.) (Pickett, Weiss, et al., 1994). In addition, unlike all other anaerobic anoxygenic phototrophs, heliobacteria are incapable of autotrophic growth (Table 3.2). While optimal growth of heliobacteria occurs under photoheterotrophic conditions using a limited number of organic carbon sources, most species are also capable of growing

Table 3.2 Summary of the major properties of the six groups of phototrophic bacteria

Property	Heliobacteria	Green Sulfur Bacteria	Purple Bacteria	Filamentous Anoxygenic Phototrophs*	Phototrophic *Acidobacteria*†	*Cyanobacteria*
Phylogenetic group	Phylum *Firmicutes*	Phylum *Chlorobi*	Phylum *Proteobacteria*	Phylum *Chloroflexi*	Phylum *Acidobacteria*	Phylum *Cyanobacteria*
Nature of photosynthesis	Anoxygenic	Anoxygenic	Anoxygenic	Anoxygenic	Anoxygenic	Oxygenic
Habitats	Paddy or volcanic soils, neutral to alkaline hot springs, and soda lakes	Acidic to neutral, high-sulfide (anoxic) hot springs and lakes	Fresh or saline, acidic to alkaline waters, often containing sulfide	Neutral to alkaline hot spring mats, temper-ate fresh and marine waters, and hypersaline mats	Alkaline hot spring mats	Terrestrial, marine, and freshwater habitats, including hot springs and saline lakes
Photosynthetic pigments‡	BChl *g*; 8^1-OH-Chl *a*	BChl *a*, *c*, *d*, *e*; Chl *a*	BChl *a*, *b*; BPhe *a*, *b*	BChl *a*, *c*; BPhe *a*	BChl *a*, *c*	Chl *a*, *b*, *d*; Phycobilins
Photosynthetic membranes	Cytoplasmic membrane	Chlorosomes	Intracellular lamellae	Chlorosomes§	Chlorosomes	Thylakoids

Continued

Table 3.2 Summary of the major properties of the six groups of phototrophic bacteria—cont'd

Property	Heliobacteria	Green Sulfur Bacteria	Purple Bacteria	Filamentous Anoxygenic Phototrophs*	Phototrophic *Acidobacteria*[†]	*Cyanobacteria*
Reaction center(s)	Type-I homodimeric	Type-I homodimeric	Type-II heterodimeric	Type-II heterodimeric	Type-I homodimeric	Type-II heterodimer (PSII) and type-I heterodimer (PSI)
Photoautotrophy	No	Yes, reductive TCA cycle	Yes[‖], Calvin cycle	Yes[¶], 3-OH-propionate pathway or Calvin cycle	No	Yes, Calvin cycle
Oxygen relationship	Strictly anaerobic	Anaerobic	Facultatively aerobic or anaerobic	Facultatively aerobic	Aerobic	Facultatively aerobic

*Also called 'green nonsulfur bacteria'.
†Only one cultured representative from this group is known – *Candidatus* Chloracidobacterium thermophilum.
‡Excludes carotenoids. Some pigments are present in some, but not all, genera of each group of phototrophic bacteria listed.
§Absent in species of the genera *Roseiflexus* and *Heliothrix*, which subsequently also lack BChl *c*.
‖Some aerobic purple phototrophs, which do not carry out photosynthesis under anaerobic conditions, are incapable of autotrophic growth.
¶The capacity for autotrophy in some filamentous anoxygenic phototrophs has not yet been determined.

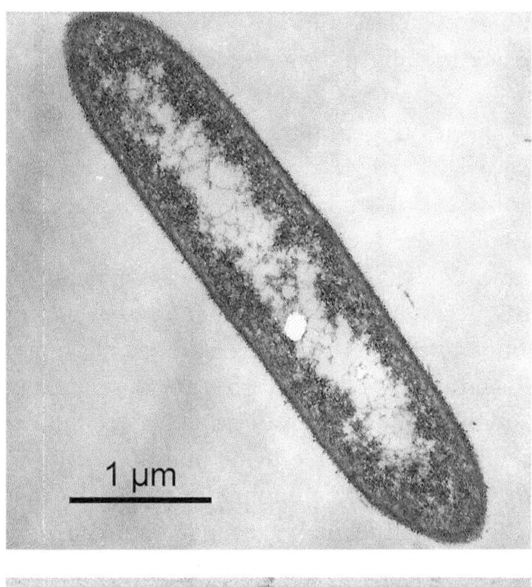

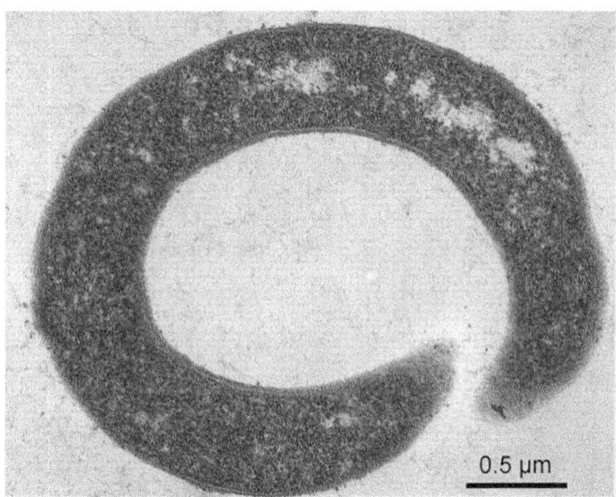

Figure 3.2 ***Transmission electron micrographs of heliobacteria.*** Top: *Heliobacterium modesticaldum* (neutrophilic) *(adapted from Kimble et al., 1995)*. Bottom: *Heliorestis convoluta* (alkaliphilic) *(adapted from Asao et al., 2006)*. Despite the fact that cells of heliobacteria stain Gram-negative, a Gram-positive cell wall structure is evident in ultrathin sections. Note the lack of specialised, intracytoplasmic light-harvesting structures, such as membrane lamellae or chlorosomes. The genome of *Hbt. modesticaldum* has been completely sequenced (Sattley et al., 2008), and genome sequencing for *Hrs. convoluta* is currently underway.

chemotrophically (anoxic/dark) by fermenting pyruvate. The exception to this is the alkaliphilic heliobacteria, which, as a group, appear to be obligate photoheterotrophs (Asao, Jung, Achenbach, & Madigan, 2006; Asao, Takaichi, & Madigan, 2012; Bryantseva, Gorlenko, Kompantseva, Achenbach, & Madigan, 1999; Bryantseva et al., 2000).

Heliobacteria can be placed into two physiologically distinct categories – those that are neutrophilic and those that are alkaliphilic. All neutrophilic heliobacteria, which consist of three genera and comprise the majority of described species (Fig. 3.3), have been isolated from either soils or hot springs. In particular, their consistent presence in paddy soils is well documented from several studies in which heliobacteria were isolated from natural samples (Beer-Romero & Gest, 1987; Ormerod et al., 1996; Stevenson, Kimble, Woese, & Madigan, 1997). A potential explanation for their competitive success in such environments has been proposed based on several observations. First, heliobacteria are strongly diazotrophic (Kimble & Madigan, 1992), and it is possible that they form a mutualistic relationship with rice plants reminiscent of the association between rhizobia and legume plants. While heliobacteria are free living and do not form/colonize root nodules, they may reside in close enough proximity to rice plant roots to provide them with fixed nitrogen in exchange for organic carbon (Madigan, 2006). The flooded paddy environment would presumably facilitate the transfer of nutrients between these organisms during the growing season. In addition, the ability of heliobacteria to form endospores – unique among phototrophs – allows them to survive desiccation and exposure to oxygen during the dry season (Asao & Madigan, 2010; Madigan, 2006). Endospore formation in heliobacteria is discussed in more detail in Section 3.2.

In contrast to the primarily terrestrial neutrophilic heliobacteria, all alkaliphilic heliobacteria have been isolated from soda lake water or sediments (either shoreline or benthic) and have pH optima between 8 and 9.5 (Asao et al., 2012). Currently, all isolated alkaliphilic species of heliobacteria belong to the genus *Heliorestis* (*Hrs.*). However, a phylogenetically distinct strain of alkaliphilic heliobacteria maintained in coculture with a chemotrophic bacterium has recently been described (Asao et al., 2012). This unusual heliobacterium, designated *Candidatus* Heliomonas lunata, was unable to be isolated from the chemotrophic co-inhabitant and is discussed in more detail in the next section.

1.2. Taxonomy and Small-Subunit rRNA Phylogeny of the *Heliobacteriaceae*

The physiologically distinct neutrophilic and alkaliphilic heliobacteria are also distinct phylogenetically, forming two independent clades in 16S ribosomal RNA (rRNA) gene trees (Fig. 3.3). Neutrophilic heliobacteria consist of the genera *Heliobacterium* (five species), *Heliobacillus* (one species), and *Heliophilum* (one species), whereas four alkaliphilic species are

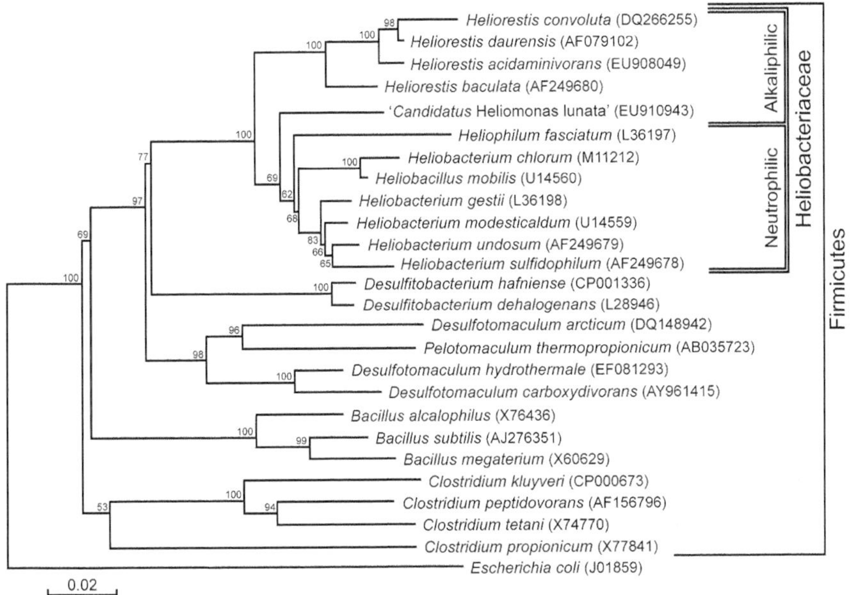

Figure 3.3 ***Phylogenetic tree constructed using the 16S rRNA gene sequences of type strains of currently described heliobacteria and related*** **Firmicutes.** Only members of the *Heliobacteriaceae* are phototrophic. The moderate thermophile *Hbt. modesticaldum* has become a model organism for biochemical and genomic studies of heliobacteria. The tree was generated using the weighted neighbour-joining method (Bruno et al., 2000) in conjunction with the Jukes–Cantor corrected distance model. Bootstrap values (≥50%) based on 100 replicates are indicated at the nodes. The tree was rooted using *Escherichia coli*, a Gammaproteobacterium. All 16S rRNA gene sequences included in the analysis have been deposited into GenBank with the accession numbers indicated in parentheses.

of the genus *Heliorestis* (Fig. 3.3). The curious exception to the otherwise clear bifurcation of alkaliphilic and neutrophilic heliobacteria is the Soap Lake (Washington state, USA) bacterium *Candidatus* Heliomonas lunata. Although only distantly related to neutrophilic heliobacteria based on 16S rRNA gene sequence identity (<94%), this interesting alkaliphile nevertheless clearly groups within the neutrophilic clade rather than the *Heliorestis* clade (Fig. 3.3). It was therefore proposed that *Candidatus* Heliomonas lunata might constitute an extant evolutionarily intermediate or 'transitional' species between the two major groups of heliobacteria (Asao et al., 2012). In this regard, the basal position of *Candidatus* Heliomonas lunata to neutrophilic species suggests that heliobacteria may have an alkaliphilic origin, although it should be noted that in an earlier analysis (Asao et al., 2012), *Candidatus* Heliomonas lunata was positioned between *Heliophilum fasciatum* and a clade containing *Heliobacillus mobilis* and *Heliobacterium chlorum*, which are all neutrophilic species. Unfortunately, the fact that *Candidatus* Heliomonas lunata is not available as a pure culture complicates further characterisation of this unusual heliobacterium (Asao et al., 2012).

A comparison of the percent identity of 16S rRNA gene sequences of heliobacteria suggests a degree of taxonomic uncertainty for *Heliobacillus* (*Hba.*) *mobilis* – the only described species of the genus. *Hba. mobilis* aligns well within the genus *Heliobacterium* (*Hbt.*) in distance calculations. Together with *Hbt. chlorum*, *Hba. mobilis* is positioned basally to a clade containing *Hbt. gestii*, *Hbt. modesticaldum*, *Hbt. undosum*, and *Hbt. sulfidophilum* (Fig. 3.3), and 16S rRNA distance matrices indicate that *Hba. mobilis* is actually more closely related to the four species of this clade than is *Hbt. chlorum* (Asao & Madigan, 2010). In addition, *Hba. mobilis* does not appear to differ significantly from species of *Heliobacterium* based on phenotypic and physiological traits. Therefore, the reclassification of these species is a possibility that should be considered. What is unclear is whether *Hba. mobilis* should be reclassified as a species of *Heliobacterium* or *Hbt. chlorum* should be reassigned to the genus *Heliobacillus*. Sequence data suggest that one or the other of these changes should be made, and based on 16S rRNA distance measurements, the transfer of *Hbt. chlorum* to the genus *Heliobacillus* seems logical. A reasonable way to resolve the question would be to perform a genomic DNA:DNA hybridisation analysis between *Hba. mobilis*, *Hbt. chlorum*, and *Hbt. gestii*, a close relative from the major *Heliobacterium* clade.

2. HELIOBACTERIAL PHOTOSYNTHESIS

2.1. Organisation of Heliobacterial Photosynthesis Genes

Similar to the purple bacteria but unlike all other phototrophs, photosynthesis genes in heliobacteria are organised into a photosynthesis gene cluster (PGC) (Sattley et al., 2008; Xiong, Inoue, & Bauer, 1998). In *Hbt. modesticaldum*, this large (~60 kb) superoperon contains the *pshA* gene, which encodes the core polypeptide for the type-I homodimeric RC, the *hemA* and *bch* genes required for biosynthesis of the primary photosynthetic pigment BChl *g*, and the *pet* genes, which encode components of the cytochrome *bc* complex (PetABCD) and cytochrome c_{553} (PetJ) (Fig. 3.4). By contrast, the *pet* genes in purple bacteria are located outside the PGC (Blankenship, 2002). Also present in the *Hbt. modesticaldum* PGC are genes for cofactor biosynthesis, carotenoid biosynthesis (*crtN*), endospore

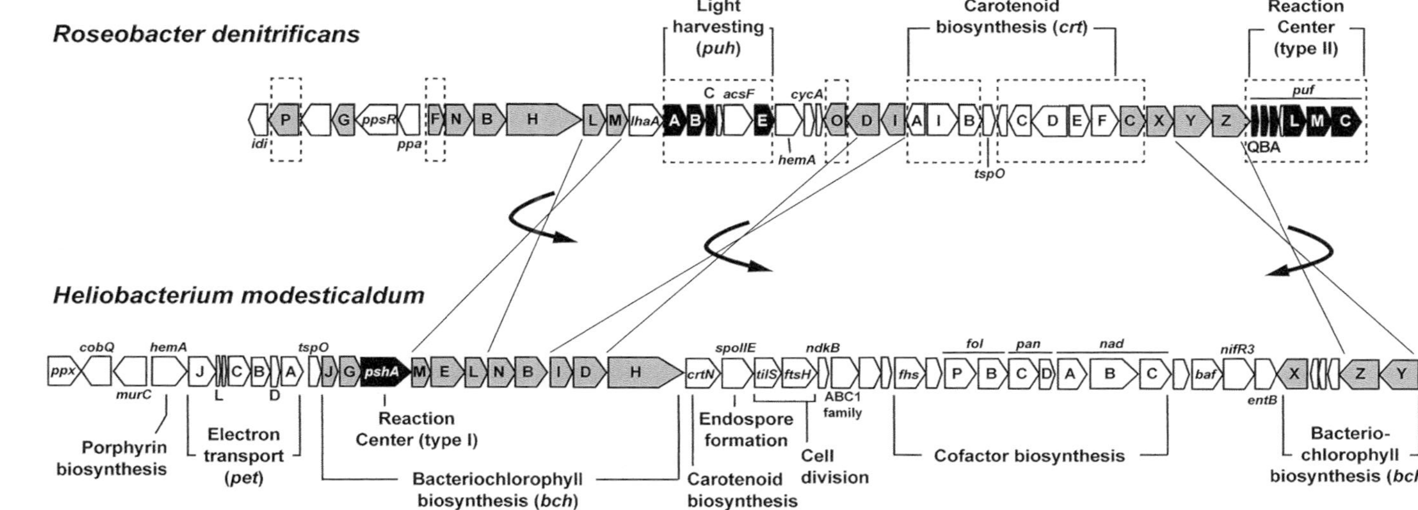

Figure 3.4 ***A comparison of heliobacterial and proteobacterial PGCs.*** A schematic representation of the PGCs from *Hbt. modesticaldum* and the purple bacterium *Roseobacter denitrificans* is shown. Bacteriochlorophyll biosynthesis (*bch*) genes are shaded grey. RC and light-harvesting genes are black. Dashed boxes indicate *R. denitrificans* photosynthesis genes not present in *Hbt. modesticaldum*. Gene rearrangements, including gene inversions and insertions, are highlighted by the connecting lines and curved arrows.

formation (*spoIIE*), and cell division (Fig. 3.4). At least one-half of the genes present in the *Hbt. modesticaldum* PGC do not appear to be directly related to photosynthesis. This is not the case for sequenced purple bacteria, where most of the genes encode proteins for photosynthesis (Fig. 3.4) (Lu et al., 2010; Swingley, Blankenship, & Raymond, 2008). The tighter clustering of photosynthesis genes in the facultatively aerobic purple bacteria may facilitate more efficient control of their expression since these genes are alternatively expressed or repressed depending on prevailing conditions, such as the presence of oxygen (Blankenship, 2002). This is presumably not an issue for heliobacteria, which are strict anaerobes and apparently express photosynthesis genes under all conditions suitable for growth since cultures grown chemotrophically in the dark are fully pigmented (Kimble, Stevenson, & Madigan, 1994; Tang, Yue, et al., 2010).

A comparison of the heliobacterial PGC with a representative purple bacterial PGC shows little gene synteny (Fig. 3.4). Gene rearrangements, including inversions and insertions, are common, and most of the photosynthesis genes present in purple bacteria are not present in heliobacteria (Fig. 3.4). The remarkable simplicity of heliobacterial photosynthesis versus that of purple bacteria becomes clear when comparing the suite of genes that encode RC core and light-harvesting polypeptides in purple bacteria (encoded by the *puf*, *puh*, and *puc* gene sets) to the single gene that encodes the RC core polypeptide in heliobacteria, *pshA*. Considering the number of significant differences in gene synteny and content between the PGCs of heliobacteria and purple bacteria, it seems likely that they developed as a result of convergence rather than from a shared ancestry. However, this would not rule out the transfer of specific genes or even groups of genes through a lateral mechanism.

2.2. Photosynthetic Pigments and Pigment Biosynthesis

Although phototrophy is not the only means by which most cultured heliobacteria can grow, it is more efficient than pyruvate fermentation and supports the most rapid growth of these bacteria (Madigan, 2006). Heliobacteria synthesise unique pigments for photosynthesis that are structurally similar: BChl *g* and 8^1-OH-Chl *a* (Gest & Favinger, 1983; van de Meent et al., 1991). Like BChls *c*, *d*, and *e* of GSB, BChl *g* and 8^1-OH-Chl *a* are both esterified with farnesol (Michalski et al., 1987), which is synthesised through a complete nonmevalonate pathway, according to genes identified in *Hbt. modesticaldum* (Sattley et al., 2008). The differences between the pigments lie in ring B: BChl *g* is a bacteriochlorin (both ring B and D are reduced) with an ethylidene substituent at the C-8 position; 8^1-OH-Chl *a* is a chlorin (containing a double bond on ring B between C-7 and C-8) with a C-8^1-hydroxyethyl group (Fig. 3.5). Because BChl *g* has both a C-3 vinyl and a C-8 ethylidene, it is also structurally similar to both chlorophyll (Chl) *a* and BChl *b* and looks much like a hybrid of the two (although Chl *a* and BChl *b* are both esterified with phytol rather than farnesol; Blankenship, 2002) (Fig. 3.5).

The biosynthetic pathways for BChl *g* and 8^1-OH-Chl *a* have not been determined experimentally, but genes predicted to encode proteins that play a role in these processes are shown in Table 3.3. Putative pathways for biosynthesis of heliobacterial (bacterio)chlorin pigments from a divinyl protochlorophyllide precursor, as well as pathways for Chl *a* and BChl *b*, are shown in Fig. 3.5. The pathway leading to Chl *a* biosynthesis is a three-step process from divinyl protochlorophyllide. Our understanding of BChl *b* biosynthesis is limited due to the fact that the enzyme catalysing the formation of the C-8 ethylidene group to produce bacteriochlorophyllide *b* from bacteriochlorophyllide *a* has not been identified. A similar situation exists in the heliobacteria, where no gene coding for the enzyme that forms the C-8 ethylidene group has yet been identified in the *Hbt. modesticaldum* genome (Sattley et al., 2008). The proposed pathway for BChl *g* biosynthesis shown in Fig. 3.5 incorporates the presence of the *bchXYZ* genes in *Hbt. modesticaldum*, which encode the chlorophyllide reductase enzyme complex (reaction 'iv' in the pathway; Fig. 3.5). Following the synthesis of divinyl bacteriochlorophyllide *a* by chlorophyllide reductase in heliobacteria, an unidentified isomerase could carry out the next reaction (indicated by the '?'), in which the C-8 vinyl group is converted to an ethylidene substituent to produce bacteriochlorophyllide *g* (Sattley et al., 2008). Note that in this scenario, the C-8 ethylidene groups of BChls *b* and *g* would be produced by different mechanisms, suggesting that although the two pigments have an identical ring B structure, they do not appear to be closely related.

The mechanism for biosynthesis of 8^1-OH-Chl *a* is unknown, but the irreversible photoisomerisation of BChl *g* to 8^1-OH-Chl *a* in the presence of light and O_2 has been documented and is well known (Madigan, 2006; Sattley, Asao, Tang, & Collins, in press). Kobayashi et al. (1998) noted that the conversion of BChl g_F to Chl a_F also occurs under oxic conditions in darkness in the presence of a weak acid, which may have been significant in the evolution of Chl *a*. However, because heliobacteria are strict anaerobes that quickly lose viability upon exposure to O_2, the use of this mechanism for 8^1-OH-Chl *a* synthesis seems unlikely to be a means by which

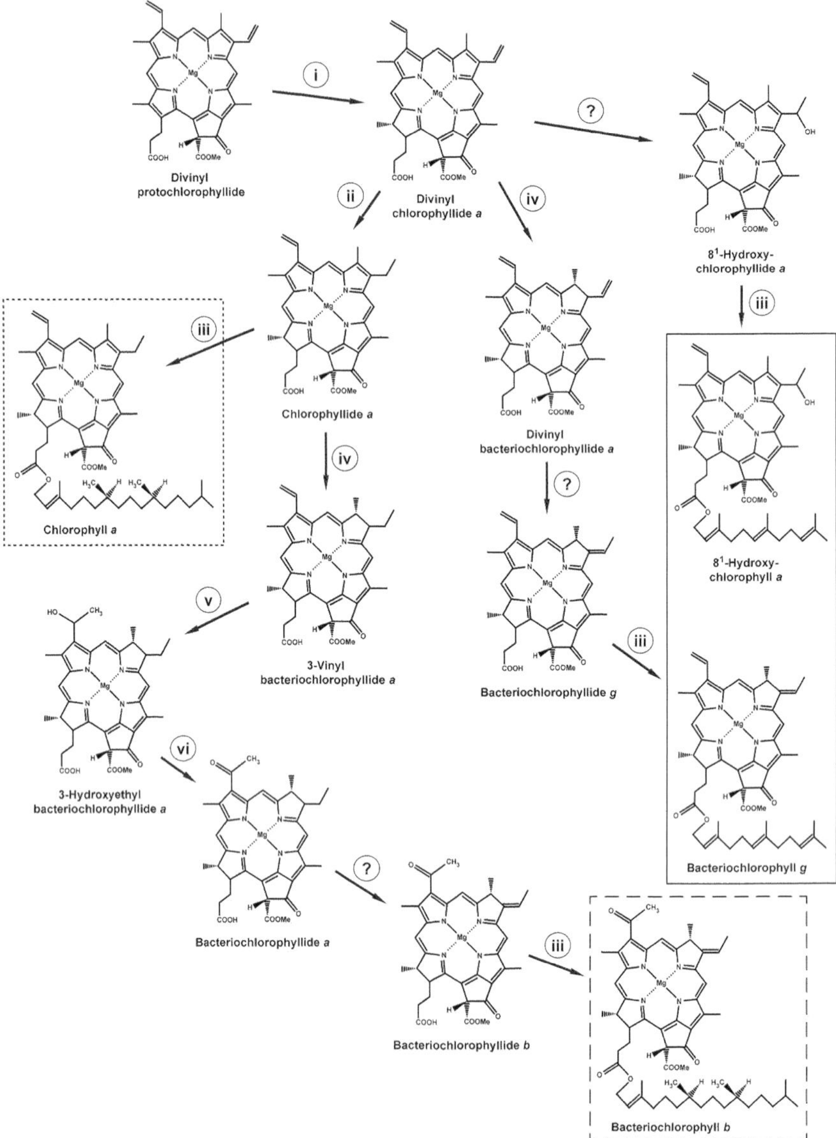

Figure 3.5 ***Proposed biosynthetic pathways of structurally similar photosynthetic pigments from a divinyl protochlorophyllide precursor.*** Pigments shown include chlorophyll *a* (dotted box) of *cyanobacteria* and eukaryotic photosynthetic organisms, bacteriochlorophyll *b* (dashed box), present in some purple bacteria, and 8^1-hydroxy-chlorophyll *a* and bacteriochlorophyll *g* (solid box), found exclusively in heliobacteria. The enzymes that catalyse the numbered reactions are (1) light-independent protochlorophyllide reductase (BchLNB), (2) divinyl chlorophyllide *a* 8-vinyl reductase, (3) chlorophyll/bacteriochlorophyll synthase (BchG), (4) chlorophyllide reductase (BchXYZ),

heliobacteria produce this pigment. Fig. 3.5 includes a putative pathway in which 8^1-OH-chlorophyllide *a* is synthesised directly from divinyl chlorophyllide *a* using an as yet unidentified enzyme. While the model is feasible and straightforward, it is complicated by the fact that no *bchF* homologue that could potentially encode an 8-vinyl chlorophyllide hydratase (similar to 3-vinyl bacteriochlorophyllide hydratase in purple bacteria) was identified in either the *Hbt. modesticaldum* genome (Sattley et al., 2008) or the PGC of *Hba. mobilis* (Mizoguchi, Oh-oka, & Tamiaki, 2005; Xiong et al., 1998). Although there is no evidence to support this pathway from our current understanding of the genomic data, it would circumvent the seemingly inefficient step of reducing ring B only to re-oxidise it, which would be necessary if 8^1-OH-Chl *a* were derived from BChl *g*.

2.3. RC Proteins and Cofactors in *Hbt. modesticaldum*

The heliobacterial RC (HbRC) is the simplest known photosynthetic apparatus (Sattley et al., 2008). Like GSB and phototrophic *acidobacteria*, heliobacteria contain a type-I RC in which the final electron acceptor is an iron–sulfur cluster (Table 3.2). This is in contrast to the type-II RCs of purple bacteria and filamentous anoxygenic phototrophs (the green nonsulfur bacteria), in which the final electron acceptor in the RC is a quinone (Hohmann-Marriott & Blankenship, 2011). Like all type-I RCs of anoxygenic phototrophic bacteria, the HbRC is a homodimer. The homodimeric RCs of GSB and phototrophic *acidobacteria* are composed of the PscA core polypeptide, while heliobacteria contain the homologous PshA (Fig. 3.6). PS I of oxygenic phototrophs contains the only known heterodimeric type-I RC, which is composed of the PsaA and PsaB core polypeptides. While clearly homologous, the type-I RC core polypeptides of these diverse phototrophs are only distantly related based on overall sequence identity (Hohmann-Marriott & Blankenship, 2011; Leibl et al., 1993; Romberger & Golbeck, 2010; Sattley & Blankenship, 2010).

(5) 3-vinyl bacteriochlorophyllide hydratase (BchF), and (6) 3-hydroxyethyl bacteriochlorophyllide *a* dehydrogenase (BchC). Unknown enzymes (?) include those that catalyse the addition of the ethylidene substituent on ring B of BChls *b* and *g* and the enzyme(s) that produce the C-8^1-hydroxyethyl group of 8^1-hydroxy-chlorophyllide *a*. Concerning the latter, it is noteworthy that a *bchF* orthologue was not identified in the *Hbt. modesticaldum* genome. Note that if the proposed pathways are correct, the mechanism of synthesis of the C-8 ethylidene group of Bchl *g* appears to be unrelated to that of Bchl *b*. In this scenario, an as yet unidentified isomerase converts the C-8 vinyl group to an ethylidene group in heliobacteria.

Table 3.3 Summary of protein coding sequences (CDSs) predicted to play major roles in heliobacterial physiology*

Locus	Symbol	Predicted function
Photosynthesis and electron transport		
HM1_0289	*fnr*	Ferredoxin:NADP$^+$ reductase, FNR
HM1_0654, 0655, 0659	*bchXYZ*	Chlorophyllide reductase, subunits Y, Z, and X
HM1_0682, 0683, 0684	*bchDHI*	Magnesium chelatase, subunits H, D, and I
HM1_0685, 0686, 0687	*bchBNL*	Protochlorophyllide reductase, subunits B, N, and L
HM1_0688	*bchE*	Mg-protoporphyrin IX monomethyl ester oxidative cyclase
HM1_0689	*bchM*	Mg-protoporphyrin O-methyltransferase
HM1_0690	*pshA*	RC core polypeptide, PshA
HM1_0692	*bchG*	Bacteriochlorophyll synthetase
HM1_0693	*bchJ*	Bacteriochlorophyll biosynthesis protein, BchJ
HM1_0696, 0697, 0698, 0699	*petADBC*	Cytochrome *bc* complex: diheme cyt *c*, subunit IV, cyt *b*, and Rieske [2Fe–2S] subunit
HM1_0701	*petJ*	18 kDa cytochrome c_{553}, PetJ
HM1_1028, 1029	*nuoEFG*	NADH:quinone oxidoreductase, chain EF (gene fusion) and chain G
HM1_1461	*pshBII*	RC-associated [4Fe–4S] ferredoxin, PshBII
HM1_1462	*pshBI*	RC-associated [4Fe–4S] ferredoxin, PshBI
HM1_2196–2206	*nuoA–D, H–N*	NADH:quinone oxidoreductase, chain A–D and chain H–N
Nitrogen fixation		
HM1_0858	*nifV*	Homocitrate synthase
HM1_0859	*nifB*	Dinitrogenase cofactor biosynthesis protein, NifB
HM1_0860	*fdxB*	Nif-specific type-III ferredoxin
HM1_0861	*nifX*	Dinitrogenase Fe/Mo cofactor
HM1_0862, 0863	*nifNE*	Dinitrogenase Fe/Mo cofactor biosynthesis proteins NifN and NifE

Table 3.3 Summary of protein coding sequences (CDSs) predicted to play major roles in heliobacterial physiology*—cont'd

Locus	Symbol	Predicted function
HM1_0865, 0864	*nifDK*	Dinitrogenase Fe/Mo protein, alpha and beta subunits
HM1_0866	*nifH*	Dinitrogenase iron protein, NifH
HM1_0867, 0868	$nifI_2I_1$	Nitrogen P_{II} regulatory proteins
HM1_0869	*orf1*	*nif* regulatory protein
HM1_0891, 0892 HM1_2617, 2618	*etfAB*	Electron transfer flavoprotein, alpha and beta subunits (two copies)
Central carbon metabolism		
HM1_0076, 1600	*pykA*	Pyruvate kinase (2 copies)
HM1_0080, 0081	*accAD*	Acetyl-coenzyme A carboxylase alpha and beta subunits
HM1_0105	*acn*	Aconitase
HM1_0282	*oadA*	Oxaloacetate decarboxylase
HM1_0371–0373	*sdhCAB*	Fumarate reductase cytochrome b_{558}, flavoprotein, and Fe–S subunits
HM1_0375, 0374	*sucDC*	Succinyl-coenzyme A synthetase, alpha and beta subunits
HM1_0464	*fumB*	Fumarate hydratase (fumarase)
HM1_0807	*porC*	Pyruvate:ferredoxin oxidoreductase, PFOR
HM1_0951	*acsA*	AMP-forming acetyl-coenzyme A synthetase
HM1_1471	*icd*	Isocitrate dehydrogenase, NADP dependant
HM1_1472	*mdh*	Malate dehydrogenase, NAD dependant
HM1_2157	*ackA*	Acetate kinase
HM1_2461	*ppdK*	Pyruvate-phosphate dikinase
HM1_2757	*ldh*	L-lactate dehydrogenase
HM1_2762, 2763, 2766, 2767	*korCBAD*	α-ketoglutarate:ferredoxin oxidoreductase, γ, β, α, and δ subunits, KFOR
HM1_2773	*pckA*	Phosphoenolpyruvate carboxykinase
HM1_2993	*aksA*	Homocitrate synthase; putative (*Re*)-citrate synthase

*Genes of physiological importance identified in the *Hbt. modesticaldum* genome not shown in the table include ATP synthase (*atpABCDEFGH*), [NiFe] hydrogenase (including two copies of structural genes *hupSLC* with one copy of accessory/maturation genes *hupD* and *hypABCDEF*), and a full complement of genes encoding enzymes of the Embden–Meyerhof–Parnas (glycolytic) pathway.

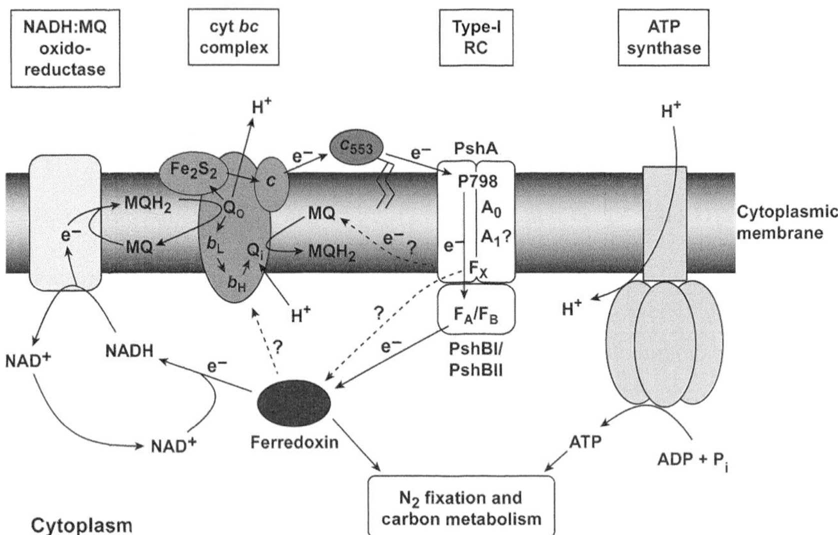

Figure 3.6 ***Proposed pathway of electron transfer and energy conservation in heliobacteria based on genes identified in*** **Hbt. modesticaldum.** Question marks indicate features of the pathway that have not been confirmed experimentally. The F_A/F_B proteins PshBI and PshBII are loosely bound to the P798-F_X core (Heinnickel et al., 2007), and F_X itself may be able to donate electrons to soluble electron acceptors (Romberger & Golbeck, 2012). See Table 3.3 for identification of the genes corresponding to the components shown. Fe_2S_2, Rieske iron–sulfur subunit. *(Adapted from Sattley et al. (2008)).*

However, certain features of these RCs are highly conserved, such as the three-dimensional core structure, which contains 11 transmembrane helices, and the amino acid sequence within the loop region that binds the [4Fe–4S] electron acceptor F_X (Blankenship, Sadekar, & Raymond, 2007; Heinnickel & Golbeck, 2007; Leibl et al., 1993; Oh-oka, 2007; Olson & Blankenship, 2004; Romberger & Golbeck, 2010; Sadekar, Raymond, & Blankenship, 2006; Trost, Brune, & Blankenship, 1992).

The RC of *Hbt. modesticaldum* was recently purified and analysed via liquid chromatography and tandem mass spectrometry (Sarrou et al., 2012). The results reiterated the simple nature of heliobacterial photosynthesis (relative to other phototrophs) and showed that the *Hbt. modesticaldum* RC homodimer binds the following cofactors: 20 BChl *g*, two BChl *g′* (the special pair primary electron donor, P798, a C-13^1 epimer of BChl *g*), two 8^1-OH-Chl *a* (the primary electron acceptor, A_0), one 4-4′-diaponeurosporene, and an average of ~1.6 menaquinone (MQ), most of which is MQ-9 (Kobayashi, Watanabe, Ikegami, van de Meent, & Amesz, 1991; Sarrou et al., 2012; Trost & Blankenship, 1989). The presence of only one carotenoid pigment in the HbRC is unusual, as one would not expect an odd number in a presumably symmetrical homodimeric RC. However, considering the recent development of a procedure to obtain high yields of stable, pure HbRC – even in the presence of O_2 (under low light) (Sarrou et al., 2012) – crystallisation studies leading to a solved HbRC structure may soon be realised, and current questions regarding cofactor geometry would be resolved.

The [4Fe–4S] cluster F_X serves as the terminal bound electron acceptor within the PshA core polypeptide (Heinnickel, Agalarov, Svensen, Krebs, & Golbeck, 2006; Miyamoto et al., 2006; Romberger & Golbeck, 2012). Two additional [4Fe–4S] clusters, F_A and F_B, are contained in each of two highly similar (62% sequence identity) ferredoxins designated PshBI and PshBII (Heinnickel, Shen, & Golbeck, 2007; Romberger, Castro, Sun, & Golbeck, 2010; Romberger & Golbeck, 2010). Adjacent genes located well outside of the PGC in the *Hbt. modesticaldum* genome (HM1_1462 and HM1_1461) encode these two RC-associated polypeptides (Table 3.3). PshBI/PshBII was unexpectedly found to bind loosely to the PshA homodimer, a situation that is in stark contrast to that in PSI, in which the homologous F_A/F_B-containing polypeptide PsaC is tightly bound to the PsaA/PsaB heterodimer (Heinnickel et al., 2007; Parrett, Mehari, Warren, & Golbeck, 1989; Romberger & Golbeck, 2010).

2.4. Electron Transport Pathways and ATP Synthesis

A proposed pathway for heliobacterial electron transfer and photophosphorylation based on genes identified in the *Hbt. modesticaldum* genome is shown in Fig. 3.6. Electron transfer within the HbRC is poorly understood but first proceeds from excited P798, the primary electron donor, to 8^1-OH-Chl *a*, the primary electron acceptor (A_0). There is significant confusion as to whether electrons are then shuttled to MQ, which would function as a secondary electron acceptor (A_1), or are instead donated directly to the [4Fe–4S] cluster F_X (Fig. 3.6).

The presence of MQ within the HbRC has been confirmed, but its role remains a mystery. Despite early studies indicating the presence of two electron acceptors (A_1 and F_X) situated between A_0 and F_A/F_B (Brok et al., 1986; Muhiuddin et al., 1999; Nitschke et al., 1990; Trost et al., 1992), other evidence suggests that MQ does not take part in RC electron transfer (Heinnickel & Golbeck, 2007). This conclusion is based on time-resolved and transient EPR spectroscopic studies (Brettel, Leibl, & Liebl, 1998; Lin et al., 1994; Lin et al., 1995; van der Est, Hager-Braun, Leibl,

Hauska, & Stehlik, 1998) and on the fact that ether extraction of MQ, which does not appear to be covalently bound (Liebl et al., 1993), from *Hbt. chlorum* membranes did not alter electron transfer kinetics (Kleinherenbrink, Ikegami, Hiraishi, Otte, & Amesz, 1993). Also lending support to this argument is the finding that an A_1 cofactor does not appear to participate in electron transfer within the type-I RC of GSB (Hauska, Schoedl, Remigy, & Tsiotis, 2001). Therefore, the role of MQ in the HbRC, if indeed there is one, remains an important unanswered question. If no role can be assigned, then the presence of MQ in the RCs of heliobacteria and GSB (and presumably also the type-I RC of phototrophic *acidobacteria*) may be a vestige of an ancestral photosynthetic apparatus.

Some aspects of electron transfer from F_X in heliobacteria are also not entirely clear. For example, although likely to be the case, cyclic photophosphorylation via the transfer of electrons from F_X back to the MQ pool in heliobacterial membranes has not been unequivocally confirmed (Fig. 3.6) (Kramer, Schoepp, Liebl, & Nitschke, 1997; Oh-oka, 2007). There is also some uncertainty regarding the role of F_A/F_B in heliobacteria. It has recently been shown that F_X can reduce flavodoxin (a soluble redox protein not encoded by the *Hbt. modesticaldum* genome) in the absence of PshBI/PshBII, indicating that F_X functions as the terminal bound electron acceptor rather than F_A/F_B (Romberger & Golbeck, 2012). By contrast, F_X in the PSI core requires F_A/F_B-containing PsaC for electron transfer to flavodoxin. Therefore, in addition to F_A/F_B-containing PshBI/PshBII, F_X in the HbRC may be able to reduce other soluble redox proteins, such as the Nif-specific type-III ferredoxin encoded by *fdxB* (HM1_0860; Table 3.3), which provides reducing power for nitrogen fixation (Fig. 3.6).

Electrons from [4Fe–4S] clusters in the HbRC are used to reduce a cytoplasmic ferredoxin, which is a source of reducing power for nitrogen fixation and carbon metabolism and can also reduce NAD^+ (via ferredoxin:$NADP^+$ reductase; HM1_0289) and possibly cytochrome (cyt) *b*, as well, although the latter has not been demonstrated experimentally (Fig. 3.6). A full complement of 14 NADH:MQ oxidoreductase genes (*nuoA–N*) is present in *Hbt. modesticaldum* (Oh-oka, 2007; Sattley et al., 2008), and therefore electrons from NADH can be transferred to the MQ pool and enter the electron transport chain (Fig. 3.6). MQH_2 is oxidised by a cyt *bc* complex, which is encoded by the *petA–D* genes (Table 3.3) and consists of four subunits: a diheme cyt *c*, cyt *b*, the Rieske [2Fe–2S] subunit, and subunit IV. The presence of a diheme cyt *c* is unusual in the *Firmicutes* but may be a universal feature of the heliobacteria, as this protein is also present in *Hbt. gestii* and

Hba. mobilis with 86.3% and 64.2% sequence identity, respectively, to cyt *c* of *Hbt. modesticaldum*. It was recently proposed that the diheme cyt *c* in heliobacteria originated from a gene duplication and subsequent fusion based on the similarity of N- and C-terminal domains in this protein (Yue, Kang, Zhang, Gao, & Blankenship, 2012). Electron transfer from the cyt *bc* complex to P798 in the HbRC occurs through the membrane-bound cyt c_{553} (Fig. 3.6) (Oh-oka, Iwaki, & Itoh, 2002). Protons pumped through the cyt *bc* complex create a proton motive force for ATP synthesis via ATP synthase, encoded by the *atpA–H* genes, which are located on a single conserved operon (HM1_1098–HM1_1105) (Sattley et al., 2008).

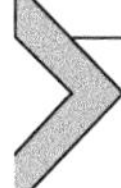

3. HELIOBACTERIA AS 'PHOTOTROPHIC CLOSTRIDIA'

3.1. Chemoorganotrophy and Central Carbon Metabolism in Heliobacteria

In addition to their phylogenetic classification, heliobacteria show a close relationship to clostridia in terms of their carbon metabolism. Unlike all other anaerobic anoxygenic phototrophs, heliobacteria have no genetic capacity for autotrophic CO_2 fixation and are therefore obligate heterotrophs (Asao & Madigan, 2010; Sattley et al., 2008). True to their clostridial roots, most cultured heliobacteria are capable of chemotrophic (dark/anoxic) growth by pyruvate fermentation (Asao & Madigan, 2010; Kimble et al., 1994). The oxidation of pyruvate in *Hbt. modesticaldum* is catalysed by pyruvate:ferredoxin oxidoreductase and is accompanied by the production of acetyl-CoA, CO_2, and reduced ferredoxin (Fig. 3.7) (Pickett, Williamson, et al., 1994; Sattley et al., 2008; Tang, Feng, et al., 2010; Tang, Yue, et al., 2010). Substrate-level phosphorylation generates ATP, as acetyl-CoA is subsequently converted to acetate (Fig. 3.7) (Kimble et al., 1994; Pickett, Williamson, et al., 1994).

Although growth is quite possible through this mechanism, it seems likely that fermentation is not a major mode of growth for heliobacteria in nature for several reasons: (1) because chemotrophic growth yields limited energy, several important energy-demanding processes, most notably nitrogen fixation, are either inactive or down-regulated during fermentative growth, which may severely limit the ability of heliobacteria to compete in complex ecosystems (Tang, Yue, et al., 2010); (2) pyruvate is the only substrate that supports fermentative growth of heliobacteria capable of chemoorganotrophy, and therefore its availability determines whether

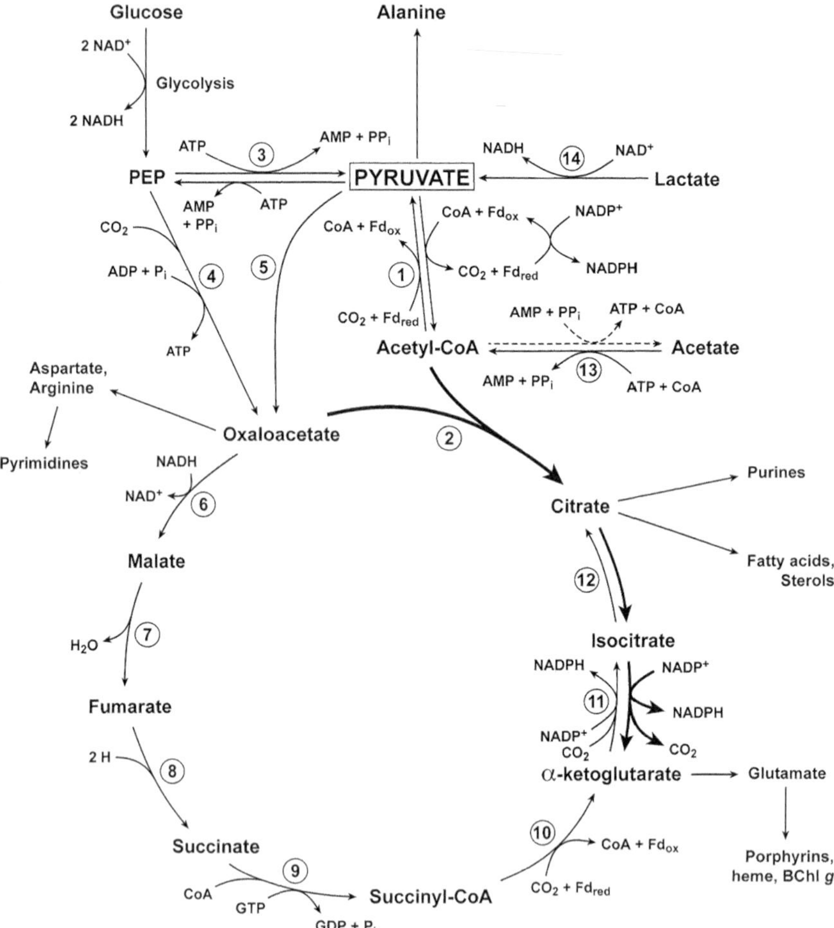

Figure 3.7 ***Schematic overview of carbon metabolism in heliobacteria.*** During phototrophic growth (solid arrows), *Hbt. modesticaldum* employs both a partial reductive TCA cycle and a partial oxidative TCA cycle, with the latter dominating carbon flow. A key enzyme for oxidative carbon flow in heliobacteria is (*Re*)-citrate synthase, which synthesises citrate from acetyl-CoA and oxaloacetate. Unlike the more common (*Si*)-citrate synthase, (*Re*)-citrate synthase is an oxygen-sensitive enzyme found in some anaerobic bacteria. While pyruvate supports the best growth of *Hbt. modesticaldum*, phototrophic growth also occurs with lactate and acetate; weak growth using ribose, fructose, and glucose has also been observed. Chemotrophic growth in *Hbt. modesticaldum* (dashed arrows) occurs through the fermentation of pyruvate to acetate. Numbers correspond to the following enzymes: (1) pyruvate:ferredoxin oxidoreductase; (2) (*Re*)-citrate synthase; (3) pyruvate–phosphate dikinase; (4) phosphoenolpyruvate carboxykinase; (5) oxaloacetate decarboxylase; (6) malate dehydrogenase; (7) fumarase; (8) fumarate reductase; (9) succinyl-CoA synthetase; (10) α-ketoglutarate:ferredoxin oxidoreductase; (11) NADP+-dependant isocitrate dehydrogenase; (12) aconitase; (13) AMP-forming acetyl-CoA synthetase; and (14) lactate dehydrogenase. *(Figure adapted from Tang, Feng, et al. (2010)).*

fermentation is possible as a means of growth at any given time; and (3) although all heliobacteria can grow photoheterotrophically using pyruvate, no cultured alkaliphilic heliobacteria have shown the ability to ferment pyruvate (Asao et al., 2006, 2012; Asao & Madigan, 2010; Bryantseva et al., 1999, 2000), suggesting that these bacteria have lost this ability over time through one or more gene loss events. Regarding this final point, a genome sequence for an alkaliphilic heliobacterium, *Heliorestis convoluta*, should soon be available, and a comparative analysis with the genome of *Hbt. modesticaldum* should reveal the genetic disparities that are responsible for the inability of alkaliphilic heliobacteria to grow chemotrophically by fermentation.

Heliobacteria lack the genetic capacity for autotrophic growth via any of the known pathways (i.e. the Calvin cycle, reductive tricarboxylic acid (RTCA) cycle, reductive acetyl-CoA pathway, 3-hydroxypropionate pathway, and 3-hydroxypropionate/4-hydroxybutyrate pathway). Photoheterotrophy is therefore the major mode of growth for heliobacteria, and pyruvate supports the best phototrophic growth of these organisms (Madigan, 2006). The solid arrows in Fig. 3.7 summarise the central carbon metabolism of phototrophically grown heliobacteria based on genes identified in *Hbt. modesticaldum*. As shown in the figure, heliobacteria possess a nearly complete RTCA cycle. If not for the lack of a gene encoding citrate lyase, an enzyme that splits citrate into acetyl-CoA and oxaloacetate, heliobacteria would presumably exhibit a central carbon metabolism similar to that of GSB, in which the RTCA cycle facilitates autotrophic growth (Hügler & Sievert, 2011; Sirevåg & Ormerod, 1970; Tang & Blankenship, 2010).

It has recently been found that both a partial RTCA cycle and a partial oxidative tricarboxylic acid (OTCA) cycle operate in heliobacteria during phototrophic growth (Tang, Feng, et al., 2010). This became apparent with the measurement of citrate synthase activity, in which acetyl-CoA and oxaloacetate condense to form citrate, in cells of *Hbt. modesticaldum* (Fig. 3.7) (Tang, Feng, et al., 2010). It was determined that this activity was the result of (*Re*)-citrate synthase rather than the more common (*Si*)-citrate synthase (a phylogenetically distinct enzyme encoded by *gltA*, which was not found in the *Hbt. modesticaldum* genome) (Li, Hagemeier, Seedorf, Gottschalk, & Thauer, 2007; Sattley et al., 2008; Tang, Feng, et al., 2010). One of two genes originally annotated as homocitrate synthase in *Hbt. modesticaldum* (HM1_2993) was putatively reassigned as (*Re*)-citrate synthase based on strong sequence similarity to other (*Re*)-citrate synthase genes and on expression levels in metabolic studies (Table 3.3) (Tang, Feng, et al., 2010). The presence of (*Re*)-citrate synthase, an oxygen-sensitive enzyme, has been reported in several species of clostridia and other anaerobic *Bacteria* and *Archaea* (Feng et al., 2009; Jahn, Huber, Eisenreich, Hugler, & Fuchs, 2007; Li et al., 2007; Pingitore, Tang, Kruppa, & Keasling, 2007; Tang et al., 2009), and therefore this important finding provides another interesting link between the heliobacteria and their nonphototrophic, anaerobic relatives.

3.2. Endospore Formation

Similar to species of *Clostridium* and *Bacillus* but unlike all other phototrophic bacteria, heliobacteria have the capacity to differentiate into heat-resistant endospores, and this feature is often used as a selective measure for their enrichment from pasteurised natural samples (Kimble-Long & Madigan, 2001; Madigan, 2006). However, the conditions and regulatory mechanisms that lead to endosporulation of heliobacteria are poorly understood. While endospores have often been observed in primary enrichment cultures containing heliobacteria, highly enriched and pure cultures of these cells almost never yield endospores (Kimble-Long & Madigan, 2001). Unfortunately, genomic sequence analysis alone has not provided the solution to this enigma.

Endosporulation has been most extensively studied in *Bacillus subtilis*, and approximately 100 of the more than 200 sporulation/germination genes in *B. subtilis* have been identified in *Hbt. modesticaldum*. Genes that play major roles in initiating the different stages of sporulation appear to be in place in *Hbt. modesticaldum*, including all genes necessary for the synthesis of the five sigma factors that are thought to be essential for this process: σ^H (initiates the development of the prespore), σ^F (regulates transcription in the forespore), σ^G (regulates maturation of the developing endospore), σ^E (controls transcription in the early mother cell), and σ^K (controls transcription in the late mother cell). One notable omission, however, is that no *spo0M* orthologue was identified. Spo0M regulates stage 0 development of the endospore, and disruption of *spo0M* was found to impair sporulation in *Bacillus subtilis* (Han et al., 1998). The implications of this genetic deficiency have not been explored, and it is likely that extensive research will be necessary to clarify the unusual sporulation patterns observed in heliobacteria.

In consideration of their distinctive carbon metabolism (including the ability to carry out light-independent growth by pyruvate fermentation), the ability to produce endospores, their strictly anaerobic physiology, and their phylogenetic alignment, heliobacteria very much resemble clostridia that have obtained the capacity for photoheterotrophy using a limited set of organic carbon sources and a minimal photosynthetic apparatus.

4. THOUGHTS ON HELIOBACTERIA AND THE EVOLUTION OF PHOTOSYNTHESIS

4.1. The Origin of Pigment–Protein Complexes in the *Firmicutes*

The estimates for the origin of cyanobacterial photosynthesis and oxygenation of Earth's atmosphere place the root bacterial photosystems at no later than ~2.4 billion years ago (Xiong & Bauer, 2002; Bekker et al., 2004). This estimate precedes the presumed differentiation of the order *Clostridiales* by up to 1 billion years (Battistuzzi, Feijao, & Hedges, 2004). This disparity suggests that either current molecular clock estimates are inaccurate or cyanobacterial photosystems predate those in heliobacteria, a fact that is hard to justify given the apparent simplicity of the homodimeric heliobacterial RC. However, recent analyses of whole genome sequences from a diverse array of bacterial phototrophs have begun to clarify the origin of bacterial photosynthesis, indicating more firmly that heliobacteria are key players in the advent of phototrophic metabolism.

Early analysis of 16S rRNA sequences suggested that *Chloroflexi* were the earliest-branching phototrophic phylum (Pace, 1997). However, analyses of HSP60 and HSP70 (Gupta, Mukhtar, & Singh, 1999) and small ribosomal proteins (Ciccarelli et al., 2006) contradict 16S rRNA results, suggesting that the Gram-positive *Firmicutes* were the earliest in the prokaryotic tree of life. It is notable that these analyses focused on nonphotosynthesis-related gene families, and, as has been mentioned elsewhere (Xiong, Fischer, Inoue, Nakahara, & Bauer, 2000), studies of this nature may represent the organismal history without properly describing the complicated evolutionary history of phototrophy.

While type-I and type-II RCs share structural homology (Sadekar et al., 2006) and likely evolutionary history, reconstructing a universal phylogenetic history of the two protein families has proven difficult due to a lack of primary sequence homology (Xiong & Bauer, 2002). Within type-I photosystems, the HbRC is presumed to be the most ancient type-I complex, developed from a simplified type-II complex (Xiong & Bauer, 2002). While direct analysis of the core RC proteins suggests that the *Chlorobi* homodimeric RC may precede that in heliobacteria (Mix, Haig, & Cavanaugh, 2005), the lack of a root for the analysis leaves this longstanding question unanswered.

One major barrier to the analysis of heliobacteria photosystem evolution is the suggestion of gene fusion events prior to the divergence of the heliobacterial lineages (Allen & Vermaas, 2010; Vermaas, 1994). Supporting this hypothesis is evidence that half of the HbRC shares a greater homology with CP43 and CP47 from PSII than its corresponding region in PSI (Vermaas, 1994). However, this apparent homology could also be explained by remnant ancestral sequence features shared in both the HbRC and PSII, derived from the progenitor of type-I and type-II photosystems. This hypothesis would predict an ancestral 'type-1.5' RC that had features resembling both type-I and type-II RCs (Blankenship, 2010).

4.2. The Origin of Pigment Biosynthesis in the *Firmicutes*

The evolution of pigment biosynthesis pathways must certainly have coincided with or preceded the development of pigment-binding protein complexes. In fact, unlike the disparate RC proteins, all phototrophic organisms share a number of pigment biosynthesis proteins. The clustering of genes responsible for photosynthetic RCs and pigment biosynthesis in heliobacteria and purple bacteria (Fig. 3.4) adds to the likelihood that these genes have been laterally transferred into or out of these two phototrophic lineages (a near certainty given the distribution of phototrophy in six bacterial phyla).

Among the nine proteins related to (bacterio)chlorophyll biosynthesis shared by all phototrophs, BchL, BchN, and BchB are unique to all phototrophic organisms (Raymond, Zhaxybayeva, Gerdes, Gogarten, & Blankenship, 2002). These three proteins, which form the light-independent protochlorophyllide oxidoreductase complex, are evolutionarily related to the BchXYZ chlorin reductase complex (Sattley et al., 2008), which is found in all anoxygenic phototrophs. Both of these three-subunit complexes are presumed to originate from the NifHDK nitrogenase complex through two sets of gene duplications (Gupta, 2012).

Recent analysis of conserved signature indels by Gupta (2012) suggests not only that BchX originated prior to BchL, but also that the heliobacterial BchL is more primitive than that of other phototrophs. This is in contrast with an earlier analysis suggesting that *Hba. mobilis* BchL was a newer acquisition, being more closely related to homologues in *Cyanobacteria* (Xiong & Bauer, 2002). An insert in *Chlorobi* and *Chloroflexi* BchX suggests that these two lineages acquired the BchXYZ protein complex more recently than any other phototrophic lineages (Gupta, 2012). Considering the ability of heliobacteria to fix nitrogen, which depends on a protein complex that appears to be the ancestral origin of two large chlorophyll biosynthesis complexes, this recent analysis supports the theory that (bacterio)chlorophyll synthesis may have originated in the heliobacteria.

Although it is still unclear whether heliobacteria represent the earliest lineage of phototrophic bacteria, the fact remains that the homodimeric HbRC, with its lack of an extrinsic antenna system, is the simplest known photosystem. The identification and sequence analysis of additional phototrophic prokaryotes, such as the complete sequencing of the *Heliorestis convoluta* genome currently in progress, should clarify whether the minimal heliobacterial photosystem is due to an ancestral state or a reduction in complexity from a *Chlorobi* or *Cyanobacteria* progenitor. Such information will only serve to improve our understanding of this complex evolutionary story.

ACKNOWLEDGEMENTS

The authors thank Dr Michael T. Madigan for supplying the electron micrographs for this manuscript. We also thank Dr Robert E. Blankenship for helpful discussions during the preparation of the manuscript.

REFERENCES

Allen, J. F., & Vermaas, W. F. J. (2010). Evolution of photosynthesis. In *Encyclopedia of life sciences*. Chichester: John Wiley and Sons, Ltd.

Asao, M., Jung, D. O., Achenbach, L.A., & Madigan, M.T. (2006). *Heliorestis convoluta* sp. nov., a coiled, alkaliphilic heliobacterium from the Wadi El Natroun, Egypt. *Extremophiles, 10*, 403–410.

Asao, M., & Madigan, M.T. (2009). The family Heliobacteriaceae. In W. B. Whitman (Ed.), *Bergey's manual of systematic bacteriology* (2nd ed., pp. 913–920). New York: Springer.

Asao, M., & Madigan, M.T. (2010). Taxonomy, phylogeny, and ecology of the heliobacteria. *Photosynthesis Research, 104*, 103–111.

Asao, M., Takaichi, S., & Madigan, M. T. (2012). Amino acid-assimilating phototrophic heliobacteria from soda lake environments: *Heliorestis acidaminivorans* sp. nov. and '*Candidatus* Heliomonas lunata'. *Extremophiles, 16*, 585–595.

Battistuzzi, F. U., Feijao, A., & Hedges, S. B. (2004). A genomic timescale of prokaryotic evolution: insights into the origin of methanogenesis, phototrophy, and the colonization of land. *BMC Evolutionary Biology, 4*, 44.

Beer-Romero, P., & Gest, H. (1987). *Heliobacillus mobilis*, a peritrichously flagellated anoxyphototroph containing bacteriochlorophyll g. *FEMS Microbiology Letters, 41*, 109–114.

Bekker, A., Holland, H. D., Wang, P. L., Rumble, D., Stein, H. J., Hannah, J. L., et al. (2004). Dating the rise of atmospheric oxygen. *Nature, 427*, 117–120.

Blankenship, R. E. (2002). *Molecular mechanisms of photosynthesis*. Oxford, UK: Blackwell Science.

Blankenship, R. E. (2010). Early evolution of photosynthesis. *Plant Physiology, 154*, 434–438.

Blankenship, R. E., Sadekar, S., & Raymond, J. (2007). The evolutionary transition from anoxygenic to oxygenic photosynthesis. In P. Falkowski & A. N. Knoll (Eds.), *Evolution of aquatic photoautotrophs* (pp. 21–35). New York: Academic Press.

Brettel, K., Leibl, W., & Liebl, U. (1998). Electron transfer in the heliobacterial reaction center: evidence against a quinone-type electron acceptor functioning analogous to A1 in photosystem I. *Biochimica et Biophysica Acta, 1363*, 175–181.

Brok, M., Vasmel, H., Horikx, J. T.G., & Hoff, A. J. (1986). Electron transport components of *Heliobacterium chlorum* investigated by EPR spectroscopy at 9 and 35 GHz. *FEBS Letters, 194*, 322–326.

Brockmann, H., & Lipinski, A. (1983). Bacteriochlorophyll *g*. A new bacteriochlorophyll from *Heliobacterium chlorum*. *Archives of Microbiology, 136*, 17–19.

Bruno, W. J., Nicholas, D. S., & Halpern, A. L. (2000). Weighted neighbor joining: A likelihood-based approach to distance-based phylogeny reconstruction. *Molecular Biology and Evolution, 17*, 189–197.

Bryantseva, I. A., Gorlenko, V. M., Kompantseva, E. I., Achenbach, L. A., & Madigan, M. T. (1999). *Heliorestis daurensis*, gen. nov. sp. nov., an alkaliphilic rod-to-coiled-shaped phototrophic heliobacterium from a siberian soda lake. *Archives of Microbiology, 172*, 167–174.

Bryantseva, I. A., Gorlenko, V. M., Kompantseva, E. I., Tourova, T. P., Kuznetsov, B. B., & Osipov, G. A. (2000). Alkaliphilic heliobacterium *Heliorestis baculata* sp. nov. and emended description of the genus *Heliorestis*. *Archives of Microbiology, 174*, 283–291.

Ciccarelli, F. D., Doerks, T., von Mering, C., Creevey, C. J., Snel, B., & Bork, P. (2006). Toward automatic reconstruction of a highly resolved tree of life. *Science, 311*, 1283–1287.

Feng, X., Mouttaki, H., Lin, L., Huang, R., Wu, B., Hemme, C. L., et al. (2009). Characterization of the central metabolic pathways in *Thermoanaerobacter* sp. Strain X514 via isotopomer-assisted metabolite analysis. *Applied and Environmental Microbiology, 75*, 5001–5008.

Fuller, R. C., Sprague, S. G., Gest, H., & Blankenship, R. E. (1985). A unique photosynthetic reaction center from *Heliobacterium chlorum*. *FEBS Letters, 182*, 345–349.

Gest, H., & Favinger, J. L. (1983). *Heliobacterium chlorum*, and anoxygenic brownish-green photosynthetic bacterium containing a "new" form of bacteriochlorophyll. *Archives of Microbiology, 136*, 11–16.

Gupta, R. S. (2012). Origin and spread of photosynthesis based upon conserved sequence features in key bacteriochlorophyll biosynthesis proteins. *Molecular Biology and Evolution*10.1093/molbev/mss145.

Gupta, R. S., Mukhtar, T., & Singh, B. (1999). Evolutionary relationships among photosynthetic prokaryotes (*Heliobacterium chlorum, Chloroflexus aurantiacus*, cyanobacteria, *Chlorobium tepidum* and proteobacteria): implications regarding the origin of photosynthesis. *Molecular Microbiology, 32*, 893–906.

Han, W. -D., Kawamoto, S., Hosoya, Y., Fujita, M., Sadaie, Y., Suzuki, K., et al. (1998). A novel sporulation-control gene (*spo0M*) of *Bacillus subtilis* with a σ^H-regulated promoter. *Gene, 217*, 31–40.

Hauska, G., Schoedl, T., Remigy, H., & Tsiotis, G. (2001). The reaction center of green sulfur bacteria. *Biochimica et Biophysica Acta, 1507*, 260–277.

Heinnickel, M., Agalarov, R., Svensen, N., Krebs, C., & Golbeck, J. H. (2006). Identification of F_X in the heliobacterial reaction center as a [4Fe-4S] cluster with an S = 3/2 ground spin state. *Biochemistry, 45*, 6756–6764.

Heinnickel, M., & Golbeck, J. H (2007). Heliobacterial photosynthesis. *Photosynthesis Research, 92*, 35–53.

Heinnickel, M., Shen, G., & Golbeck, J. H. (2007). Identification and characterization of PshB, the dicluster ferredoxin that harbors the terminal electron acceptors F_A and F_B in *Heliobacterium modesticaldum*. *Biochemistry, 46*, 2530–2536.

Hohmann-Marriott, M. F., & Blankenship, R. E. (2011). Evolution of photosynthesis. *Annual Review of Plant Biology, 62*, 515–548.

Hügler, M., & Sievert, S. M. (2011). Beyond the Calvin cycle: autotrophic carbon fixation in the ocean. *Annual Review of Marine Science, 3*, 261–289.

Jahn, U., Huber, H., Eisenreich, W., Hugler, M., & Fuchs, G. (2007). Insights into the autotrophic CO_2 fixation pathway of the archaeon *Ignicoccus hospitalis*: comprehensive analysis of the central carbon metabolism. *Journal of Bacteriology, 189*, 4108–4119.

Kimble, L. K., & Madigan, M. T. (1992). Nitrogen fixation and nitrogen metabolism in heliobacteria. *Archives of Microbiology, 158*, 155–161.

Kimble-Long, L. K., & Madigan, M. T. (2001). Molecular evidence that the capacity for endosporulation is universal among phototrophic heliobacteria. *FEMS Microbiology Letters*, *199*, 191–195.

Kimble, L. K., Mandelco, L., Woese, C. R., & Madigan, M. T. (1995). *Heliobacterium modesticaldum*, sp. nov., a thermophilic heliobacterium of hot springs and volcanic soils. *Archives of Microbiology*, *163*, 259–267.

Kimble, L. K., Stevenson, A. K., & Madigan, M. T. (1994). Chemotrophic growth of heliobacteria in darkness. *FEMS Microbiology Letters*, *115*, 51–55.

Kleinherenbrink, F. A.M., Ikegami, I., Hiraishi, A., Otte, S. C.M., & Amesz, J. (1993). Electron transfer in menaquinone-depleted membranes of *Heliobacterium chlorum*. *Biochimica et Biophysica Acta*, *1142*, 69–73.

Kobayashi, M., Hamano, T., Akiyama, M., Watanabe, T., Inoue, K., Oh-oka, H., et al. (1998). Light-independent isomerization of bacteriochlorophyll *g* to chlorophyll *a* catalyzed by weak acid in vitro. *Analytica Chimica Acta*, *365*, 199–203.

Kobayashi, M., Watanabe, T., Ikegami, I., van de Meent, E. J., & Amesz, J. (1991). Enrichment of bacteriochlorophyll *g′* in membranes of *Heliobacterium chlorum* by ether extraction: unequivocal evidence for its existence in vivo. *FEBS Letters*, *284*, 129–131.

Kramer, D. M., Schoepp, B., Liebl, U., & Nitschke, W. (1997). Cyclic electron transfer in *Heliobacillus mobilis* involving a menaquinol-oxidizing cytochrome bc complex and an RCI-type reaction center. *Biochemistry*, *36*, 4203–4211.

Li, F., Hagemeier, C. H., Seedorf, H., Gottschalk, G., & Thauer, R. K. (2007). *Re*-citrate synthase from *Clostridium kluyveri* is phylogenetically related to homocitrate synthase and isopropylmalate synthase rather than to *Si*-citrate synthase. *Journal of Bacteriology*, *189*, 4299–4304.

Liebl, U., Mockensturm-Wilson, M., Trost, J. T., Brune, D. C., Blankenship, R. E., & Vermaas, W. (1993). Single core polypeptide in the reaction center of the photosynthetic bacterium *Heliobacillus mobilis*: structural implications and relations to other photosystems. *Proceedings of the National Academy of Sciences of the USA*, *90*, 7124–7128.

Lin, S., Chiou, H. C., & Blankenship, R. E. (1995). Secondary electron transfer processes in membranes of *Heliobacillus mobilis*. *Biochemistry*, *34*, 12761–12767.

Lin, S., Chiou, H. C., Kleinherenbrink, F. A.M., & Blankenship, R. E. (1994). Time-resolved spectroscopy of energy and electron transfer processes in the photosynthetic bacterium *Heliobacillus mobilis*. *Biophysical Journal*, *66*, 437–445.

Lu, Y.-K., Marden, J., Han, M., Swingley, W. D., Mastrian, S. D., Chowdhury, S. R., et al. (2010). Metabolic flexibility revealed in the genome of the cyst-forming α-1 proteobacterium *Rhodospirillum centenum*. *BMC Genomics*, *11*, 325.

Madigan, M. T. (2006). The family Heliobacteriaceae. (3rd ed.). *The prokaryotes* (Vol. 4, pp. 951–964). New York: Springer.

Madigan, M. T., Euzéby, J. P., & Asao, M. (2010). Proposal of *Heliobacteriaceae* fam. nov. *International Journal of Systematic and Evolutionary Microbiology*, *60*, 1709–1710.

Michalski, T. J., Hunt, J. E., Bowman, M. K., Smith, U., Bardeen, K., Gest, H., et al. (1987). Bacteriopheophytin *g*: properties and some speculations on a possible primary role for bacteriochlorophylls *b* and *g* in the biosynthesis of chlorophylls. *Proceedings for the National Academy of Sciences of the USA*, *84*, 2570–2574.

Mix, L. J., Haig, D., & Cavanaugh, C. M. (2005). Phylogenetic analyses of the core antenna domain: investigating the origin of Photosystem I. *Journal of Molecular Evolution*, *60*, 153–163.

Miyamoto, R., Iwaki, M., Mino, H., Harada, J., Itoh, S., & Oh-oka, H. (2006). ESR signal of the iron–sulfur center F_X and its function in the homodimeric reaction center of *Heliobacterium modesticaldum*. *Biochemistry*, *45*, 6306–6316.

Mizoguchi, T., Oh-oka, H., & Tamiaki, H. (2005). Determination of stereochemistry of bacteriochlorophyll g_F and 8^1-hydroxy-chlorophyll a_F from *Heliobacterium modesticaldum*. *Photochemistry and Photobiology*, *81*, 666–673.

Muhiuddin, I. P., Rigby, S. E.J., Evans, M. C.W., Amesz, J., & Heathcote, P. (1999). ENDOR and special TRIPLE resonance spectroscopy of photoaccumulated semiquinone electron acceptors in the reaction centers of green sulfur bacteria and heliobacteria. *Biochemistry, 38,* 7159–7167.

Nitschke, W., Sétif, P., Liebl, U., Feiler, U., & Rutherford, A. W. (1990). Reaction center photochemistry of *Heliobacterium chlorum*. *Biochemistry, 29,* 11079–11088.

Oh-oka, H. (2007). Type 1 reaction center of photosynthetic heliobacteria. *Photochemistry and Photobiology, 83,* 177–186.

Oh-oka, H., Iwaki, M., & Itoh, S. (2002). Electron donation from membrane-bound cytochrome *c* to the photosynthetic reaction center in whole cells and isolated membranes of *Heliobacterium gestii*. *Photosynthesis Research, 71,* 137–147.

Olson, J. M., & Blankenship, R. E. (2004). Thinking about the evolution of photosynthesis. *Photosynthesis Research, 80,* 373–386.

Ormerod, J. G., Kimble, L. K., Nesbakken, T., Torgersen, Y. A., Woese, C. R., & Madigan, M. T. (1996). *Heliophilum fasciatum* gen. nov. sp. nov. and *Heliobacterium gestii* sp. nov: endospore-forming heliobacteria from rice field soils. *Archives of Microbiology, 165,* 226–234.

Pace, N. R. (1997). A molecular view of microbial diversity and the biosphere. *Science, 276,* 734–740.

Parrett, K. G., Mehari, T., Warren, P. G., & Golbeck, J. H. (1989). Purification and properties of the intact P_{700} and F_X-containing Photosystem I core protein. *Biochimica et Biophysica Acta, 973,* 324–332.

Pickett, M. W., Weiss, N., & Kelly, D. J. (1994). Gram-positive cell wall structure of the A3 gamma type I heliobacteria. *FEMS Microbiology Letters, 122,* 7–12.

Pickett, M. W., Williamson, M. P., & Kelly, D. J. (1994). An enzyme and 13C-NMR study of carbon metabolism in heliobacteria. *Photosynthesis Research, 41,* 75–88.

Pingitore, F., Tang, Y., Kruppa, G. H., & Keasling, J. D. (2007). Analysis of amino acid isotopomers using FT-ICR MS. *Analytical Chemistry, 79,* 2483–2490.

Raymond, J., Zhaxybayeva, O., Gerdes, S., Gogarten, J. P., & Blankenship, R. E. (2002). Whole genome analysis of photosynthetic prokaryotes. *Science, 298,* 1616–1620.

Romberger, S. P., Castro, C., Sun, Y., & Golbeck, J. H. (2010). Identification and characterization of PshBII, a second F_A/F_B-containing polypeptide in the photosynthetic reaction center of *Heliobacterium modesticaldum*. *Photosynthesis Research, 104,* 293–303.

Romberger, S. P., & Golbeck, J. H. (2010). The bound iron–sulfur clusters of Type-I homodimeric reaction centers. *Photosynthesis Research, 104,* 333–346.

Romberger, S. P., & Golbeck, J. H. (2012). The F_X iron–sulfur cluster serves as the terminal bound electron acceptor in heliobacterial reaction centers. *Photosynthesis Research, 111,* 285–290.

Sadekar, S., Raymond, J., & Blankenship, R. E. (2006). Conservation of distantly related membrane proteins: photosynthetic reaction centers share a common structural core. *Molecular Biology and Evolution, 23,* 2001–2007.

Sarrou, I., Khan, Z., Cowgill, J., Lin, S., Brune, D., Romberger, S., et al. (2012). Purification of the photosynthetic reaction center from *Heliobacterium modesticaldum*. *Photosynthesis Research, 111,* 291–302.

Sattley, W.M., Asao, M., Tang, K.H., and Collins, A.C. (in press). Energy conservation in heliobacteria: photosynthesis and central carbon metabolism. In M.F. Hohmann-Marriott, (Ed.), *The structural basis of biological energy generation, section II – physiological integration of biological energy conversion – how physiology shapes structure. Advances in photosynthesis and respiration,* Dordrecht: Springer.

Sattley, W. M., & Blankenship, R. E. (2010). Insights into heliobacterial photosynthesis and physiology from the genome of *Heliobacterium modesticaldum*. *Photosynthesis Research, 104,* 113–122.

Sattley, W. M., Madigan, M. T., Swingley, W. D., Cheung, P. C., Clocksin, K. M., Conrad, A. L., et al. (2008). The genome of *Heliobacterium modesticaldum*, a phototrophic representative of the *Firmicutes* containing the simplest photosynthetic apparatus. *Journal of Bacteriology, 190*, 4687–4696.

Sirevåg, R., & Ormerod, J. G. (1970). Carbon dioxide – fixation in photosynthetic green sulfur bacteria. *Science, 169*, 186–188.

Stevenson, A. K., Kimble, L. K., Woese, C. R., & Madigan, M. T. (1997). Characterization of new phototrophic heliobacteria and their habitats. *Photosynthesis Research, 53*, 1–11.

Swingley, W. D., Blankenship, R. E., & Raymond, J. (2008). Evolutionary relationships among purple photosynthetic bacteria and the origin of proteobacterial photosynthetic systems. In C. N. Hunter, F. Daldal, M. C. Thurnauer & J. T. Beatty (Eds.), *The purple phototrophic bacteria. Advances in photosynthesis and respiration*. Dordrecht: Springer.

Takaichi, S., Inoue, K., Akaike, M., Kobayashi, M., Oh-Oka, H., & Madigan, M. T. (1997). The major carotenoid in all species of heliobacteria is the C_{30} carotenoid 4,4′-diaponeurosporene, not neurosporene. *Archives of Microbiology, 168*, 277–281.

Takaichi, S., Oh-Oka, H., Maoka, T., Jung, D. O., & Madigan, M. T. (2003). Novel carotenoid glucoside esters from alkaliphilic heliobacteria. *Archives of Microbiology, 179*, 95–100.

Tang, Y. J., Yi, S., Zhuang, W. Q., Zinder, S. H., Keasling, J. D., & Alvarez-Cohen, L. (2009). Investigation of carbon metabolism in "*Dehalococcoides ethenogenes*" strain 195 by use of isotopomer and transcriptomic analyses. *Journal of Bacteriology, 191*, 5224–5231.

Tang, K. H., & Blankenship, R. E. (2010). Both forward and reverse TCA cycles operate in green sulfur bacteria. *Journal of Biological Chemistry, 285*, 35848–35854.

Tang, K. H., Feng, X., Zhuang, W. Q., Alvarez-Cohen, L., Blankenship, R. E., & Tang, Y. J. (2010). Carbon flow of heliobacteria is related more to clostridia than to the green sulfur bacteria. *Journal of Biological Chemistry, 285*, 35104–35112.

Tang, K. H., Yue, H., & Blankenship, R. E. (2010). Energy metabolism of *Heliobacterium modesticaldum* during phototrophic and chemotrophic growth. *BMC Microbiology, 10*, 150.

Trost, J. T., & Blankenship, R. E. (1989). Isolation of a photoactive photosynthetic reaction center-core antenna complex from *Heliobacillus mobilis*. *Biochemistry, 28*, 9898–9904.

Trost, J. T., Brune, D. C., & Blankenship, R. E. (1992). Protein sequences and redox titrations indicate that the electron acceptors in reaction centers from heliobacteria are similar to Photosystem I. *Photosynthesis Research, 32*, 11–22.

van de Meent, E. J., Kobayashi, M., Erkelens, C., van Veelen, P. A., Amesz, J., & Watanabe, T. (1991). Identification of 8^1-hydroxychlorophyll *a* as a functional reaction center pigment in heliobacteria. *Biochimica et Biophysica Acta, 1058*, 356–362.

van der Est, A., Hager-Braun, C., Leibl, W., Hauska, G., & Stehlik, D. (1998). Transient electron paramagnetic resonance spectroscopy on green sulfur bacteria and heliobacteria at two microwave frequencies. *Biochimica et Biophysica Acta, 1409*, 87–98.

Vermaas, W. F. J. (1994). Evolution of heliobacteria: implications for photosynthetic reaction center complexes. *Photosynthesis Research, 41*, 285–294.

Xiong, J., & Bauer, C. E. (2002). Complex evolution of photosynthesis. *Annual Review of Plant Biology, 53*, 503–521.

Xiong, J., Fischer, W. M., Inoue, K., Nakahara, M., & Bauer, C. E. (2000). Molecular evidence for the early evolution of photosynthesis. *Science, 289*, 1724–1730.

Xiong, J., Inoue, K., & Bauer, C. E. (1998). Tracking molecular evolution of photosynthesis by characterization of a major photosynthesis gene cluster from *Heliobacillus mobilis*. *Proceedings of the National Academy of Sciences of the USA, 95*, 14851–14856.

Yue, H., Kang, Y., Zhang, H., Gao, X., & Blankenship, R. E. (2012). Expression and characterization of the diheme cytochrome *c* subunit of the cytochrome *bc* complex in *Heliobacterium modesticaldum*. *Archives of Biochemistry and Biophysics, 517*, 131–137.

CHAPTER FOUR

Green Bacteria: Insights into Green Bacterial Evolution through Genomic Analyses

Donald A. Bryant[*,**,1], **Zhenfeng Liu**[*,†]
[*]Department of Biochemistry and Molecular Biology, The Pennsylvania State University, University Park, PA, USA
[**]Department of Chemistry and Biochemistry, Montana State University, Bozeman, MT, USA
[†]Department of Biological Sciences, University of Southern California, Los Angeles, CA, USA
[1]Corresponding author: E-mail: dab14@psu.edu

Contents

Abstract

Green bacteria are physiologically and metabolically heterogeneous assemblage of organisms that belong to three phyla of the Domain Bacteria: Chlorobi, Chloroflex, and Acidobacteria. Green bacteria are defined by their ability to synthesize one of three bacteriochlorophylls (BChls) *c*, *d*, or *e* and to assemble these BChls into large antenna structures known as chlorosomes. Over the past decade, the genome sequences of many green bacteria have been completed and analysed. This chapter mainly focuses on key aspects of the chlorophotoautotrophic physiology of members of the phylum Chlorobi. Until recently, this phylum was generally considered to be synonymous with the green sulfur bacteria (Order Chlorobiales). However, genome and metagenome sequencing has shown that the members of this phylum are more diverse than previously imagined. Surprisingly, early-diverging members of this phylum are aerobic chemoorganoheterotrophs and aerobic anoxygenic photoheterotrophs. The evolution of members of the phylum Chlorobi is discussed. Pathways for the synthesis of (bacterio)chlorophylls ((B)

Advances in Botanical Research, Volume 66
ISSN 0065-2296, http://dx.doi.org/10.1016/B978-0-12-397923-0.00004-7

Chls) are described and discussed in the context of the Granick hypothesis, which states that biosynthetic pathways recapitulate their evolution. Finally, based upon information obtained from genome sequences and the discovery of the first phototrophic member of the phylum Acidobacteria, "*Candidatus* Chloracidobacterium thermophilum" ("*Ca.* Cab. thermophilum"), some thoughts on the evolution of photosynthesis are discussed.

1. INTRODUCTION

Green bacteria are anoxygenic chlorophototrophs: they use (bacterio) chlorophylls ((B)Chls) for light capture and conversion into chemical energy (as adenosine-5'-triphosphate (ATP) and reducing power), they use light as their primary energy source for growth, and unlike cyanobacteria, they do not oxidise water and thus do not produce oxygen as a by-product of photosynthesis. As will be described below, the defining properties of these organisms are that they synthesize bacteriochlorophyll (BChl) *c*, *d*, or *e* and assemble these BChls into chlorosomes for light harvesting (Bryant et al., 2012). Based upon this definition, three phyla, Chlorobi, Chloroflexi, and Acidobacteria, within the Domain Bacteria contain members that are green bacteria. However, these bacteria are strikingly different in many physiological and biochemical properties (Bryant & Frigaard, 2006; Bryant et al., 2012). The most widespread green bacteria, green sulfur bacteria (GSB; Order Chlorobiales) are obligately photolithoautotrophic anaerobes; they oxidise reduced sulfur compounds, ferrous iron or hydrogen and fix carbon dioxide using the enzymes of the reverse tricarboxylic acid (TCA) cycle (Overmann, 2001, 2008). Filamentous anoxygenic phototrophs (FAPs, Chloroflexiales) were formerly called green non-sulfur bacteria (Hanada & Pierson, 2006; Pierson & Castenholz, 1995). Although these organisms are principally facultative photoheterotrophs or photomixotrophs that can usually grow as aerobic chemoorganoheterotrophs, some FAPs (e.g. *Oscillochloris* (*Osc.*) *trichoides*) are photoautotrophs, which can oxidise sulfide and/or hydrogen and fix carbon dioxide using the enzymes of the 3-hydroxypropionate (Fuchs, 2011; Tang, Tang, & Blankenship, 2011; Zarzycki, Brecht, Müller, & Fuchs, 2009) or Calvin–Benson–Bassham cycles (Ivanovsky et al., 1999; Kuznetsov et al., 2011; Turova et al., 2006). "*Candidatus* Chloracidobacterium thermophilum" ("*Ca.* Cab. thermophilum") has only recently been described (Bryant et al., 2007; Garcia Costas, Liu et al., 2012), and this organism is currently the only known chlorophototrophic member of the phylum Acidobacteria. "*Ca.* Cab. thermophilum" has a photosynthetic apparatus that is very similar to that of GSB, but it is an aerobic photo-organoheterotroph and lacks pathways for carbon dioxide fixation (Garcia Costas et al., 2011, Garcia Costas, Liu et al., 2012, Garcia Costas, Tsukatani et al., 2012; Tsukatani, Wen, Blankenship, & Bryant, 2010; Tsukatani, Romberger, Golbeck, & Bryant, 2012; Wen, Tsukatani et al., 2011).

All chlorophototrophs convert light energy into chemical energy by using multi-subunit, pigment–protein complexes known as photochemical reaction centres (RCs). RCs are classified into two types on the basis of their structures and electron transport cofactors (Blankenship, 2010; Golbeck, 1993; Hohmann-Marriott & Blankenship, 2011; Sadekar, Raymond & Blankenship, 2006). All RCs harbour a dimer of (B)Chl molecules, the so-called 'special pair (P)', that acts either as the primary electron donor (or as a very rapid secondary donor in some cases; Müller, Slavov, Luthra, Redding, & Holzwarth, 2010). Type-1 RCs have three [4Fe–4S] clusters that are the terminal electron acceptors, and these RCs characteristically produce a weak oxidant (P^+) and a strong reductant. Type-2 RCs have quinones as the terminal electron acceptors, and these RCs produce a strong oxidant (P^+) and weak reductant. GSB and "*Ca.* Cab. thermophilum" as well as members of the *Heliobacteriaceae* (phylum Firmicutes; not considered here but see Heinnickel & Golbeck, 2007; Oh-Oka, 2007; Romberger & Golbeck, 2010), have homodimeric type-1 RCs that bind about 20 (B)Chl molecules (see section 4.6 below). FAPs have heterodimeric type-2 RCs that are very similar to the type-2 RCs of purple bacteria (phylum Proteobacteria). These RCs only contain six pigment molecules: three (or four) BChl *a* molecules and three (or two) bacteriopheophytin (BPhe) *a* molecules (Pierson & Thornber, 1983). The RCs of cyanobacteria, denoted photosystem (PS) I and PS II, have heterodimeric cores and are much more complex than other RCs but are nevertheless distantly related to them structurally. Cyanobacterial PS I forms trimeric complexes in which each 356-kDa monomer has 11 or 12 polypeptide subunits and binds 96 chlorophyll (Chl) *a* molecules and 22 β-carotene molecules (Grotjohann & Fromme, 2005). Cyanobacterial PS II complexes are dimers in which each 350-kDa monomer has 19 polypeptide subunits and binds 35 Chl *a* molecules and 11 β-carotene molecules (Umena et al., 2011).

As implied by the name, many green bacteria are actually green in colour, but these organisms can also be brown, reddish-brown, or even orange-brown. Thus, the defining property of these organisms is not their colour, but instead, green bacteria are actually defined by two interrelated biochemical properties. All green bacteria synthesize very large quantities of BChl *c*, *d*, or *e*. These BChl molecules are assembled into chlorosomes, which are literally sacs containing self-organized BChls and which function as the principal light-harvesting complexes (Blankenship and Matsuura, 2003; Frigaard & Bryant, 2006; Oostergetel, van Amerongen, & Boekema, 2010). Chlorosomes are very large structures (Ø, 30–60 nm; length, 100–200 nm) (Fig. 4.1), and each chlorosome can contain from 150,000 to 250,000 BChl *c*, *d*, or *e* molecules (Fig. 4.1) (Martinez-Planells et al., 2002; Montaño, Bowen et al., 2003). Because of the extraordinarily large numbers of BChls in their antenna complexes, some green bacteria are able to grow at fantastically low light irradiances (1–10 nmol photons m^{-2} s^{-1}) under which no other types of chlorophototrophs can grow (Beatty et al., 2005; Manske, Glaeser, Kuypers & Overmann, 2005). Chlorosomes are connected to the inner surface of the cytoplasmic membrane through a structure known as the 'baseplate', which is a paracrystalline array of the 7.5-kDa protein, CsmA (Fig. 4.2) (Bryant, Vassilieva, Frigaard, & Li, 2002; Li, Frigaard, & Bryant, 2006; Pedersen, Linnanto, Frigaard, Nielsen, & Miller, 2010; Pšenčík et al., 2009). CsmA binds a single BChl *a* molecule and one or two carotenoid molecules (Bryant et al., 2002; Montaño, Wu et al., 2003). In GSB and "*Ca.* Cab. thermophilum", the baseplate binds to another BChl *a*-binding protein, the Fenna–Matthews–Olson (FMO) protein (Huang, Wen, Blankenship, & Gross, 2012; Larson et al., 2011; Tronrud, Wen, Gay, & Blankenship, 2009; Tsukatani et al., 2010; Wen, Zhang, Gross, & Blankenship, 2009; Wen, Tsukatani, et al., 2011; Wen, Zhang, Gross & Blankenship, 2011) (Fig. 4.2). FMO is a water-soluble, peripheral membrane protein that connects chlorosomes to the type-1 RCs in the cytoplasmic membrane (Hauska, Schödl, Remigy & Tsiotis, 2001; Huang et al., 2012). FMO does not occur in chlorosome-containing FAPs, such as *Chloroflexus* (*Cfx.*) *aurantiacus*, and in these organisms chlorosomes interact directly with antenna and RC proteins embedded in the cytoplasmic membrane. The type-2 RCs in FAPs are surrounded by a ring of membrane-intrinsic, heterodimeric BChl *a*-binding proteins, known as B808-B866, which resembles the light-harvesting-I complexes of purple bacteria (Xin, Lin, Montaño, & Blankenship, 2005).

Because green bacteria are phylogenetically incoherent organisms, this chapter will mostly focus on members of the phyla Chlorobi and Acidobacteria. A genomics-based comparison that includes a description of the properties of members of the Chloroflexi can be found elsewhere

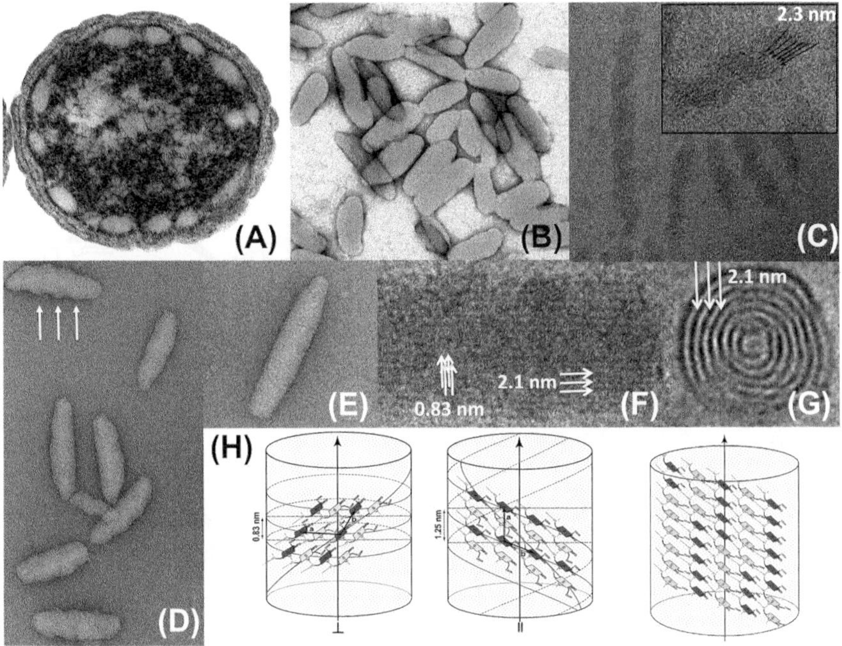

Figure 4.1 ***Structure of chlorosomes from the green sulfur bacterium*** **Cba. tepidum.** **A.**Transmission electron micrograph of a *Cba. tepidum* cell showing chlorosomes, which appear as electron transparent ovoids appressed to the inner surface of the cytoplasmic membrane. **B.** Transmission electron micrograph of isolated, negatively stained chlorosomes from *Cba. tepidum*. **C.** Cryo-electron micrograph showing the structure of chlorosomes from "*Ca.* Cab. thermophilum". The inset shows an enlargement showing the folded or twisted nature of the lamellar sheets and their 2.3-nm spacings (red arrows). **D.** Transmission electron micrograph of isolated, negatively stained chlorosomes from "*Ca.* Cab. thermophilum". The three arrows show irregularly spaced ridges that occur on the cytoplasmic-facing surface of the chlorosomes. **E.** Transmission electron micrograph of a negatively stained chlorosome from the *bchQRU* triple mutant of *Cba. tepidum*. **F.** Cryo-electron microscopic image of a chlorosome from the *bchQRU* triple mutant of *Cba. tepidum*. The three horizontal arrows indicate the 2.1-nm spacings between the surfaces that form coaxial nanotubules. The four vertical arrows indicate the 0.83-nm repeats that arise from the *syn–anti* monomer stacks of BChl d_F molecules. **G.** Cryo-electron micrograph showing an end-on view of a single chlorosome from the *bchQRU* triple mutant of *Cba. tepdium*. The arrows indicate the 2.1-nm lamellar spacings that correspond to those in panel F. **H.** Schemes showing the *syn–anti* monomer stacks of BChl d_F molecules in the *bchQRU* triple mutant of *Cba. tepdium* (far left, ⊥), the syn–anti stacks of BChl c_F molecules in chlorosomes of wild-type *Cba. tepdium* (centre, ||) and mixed all-*syn* and all-*anti*, parallel monomer stacks of [Et, Me]-BChl c_F in the *bchQR* mutant of *Cba. tepidum* (right) (Ganapathy et al., 2009, 2012; Section 5). (For interpretation of the references to colour in this figure legend, the reader is referred to the online version of this book.)

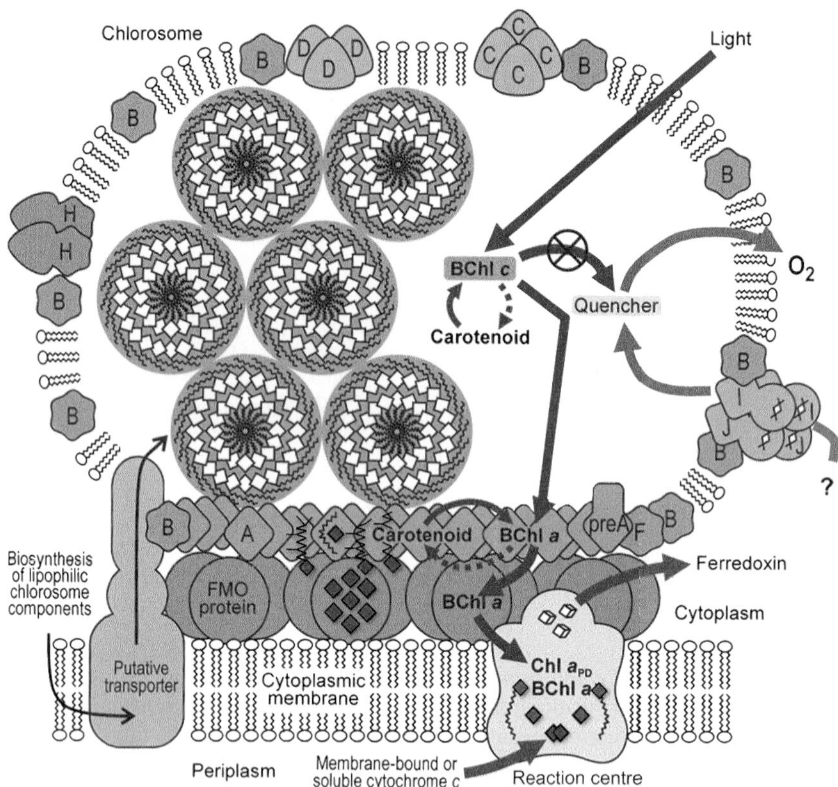

Figure 4.2 ***Scheme showing the structural organization of chlorosomes, FMO protein, and the type-1 reaction centers (RCs) of* Cba. tepidum.** Energy transfer pathways are indicated by red arrows, and electron transfer reactions are indicated by blue arrows. The quencher is mostly oxidised chlorobiumquinone and menaquinone-7. Ten proteins occur in the chlorosome envelope, and the paracrystalline array of CsmA is shown in the baseplate. Blue diamonds indicate BChl a_P, and green diamonds in the RC indicate four Chl a_{PD} molecules. In *Cba. tepidum*, the BChl c_F (white diamonds with green coloured halos) form *syn–anti* monomer stacks that self-organize into coaxial nanotubes. In "*Ca.* Cab. thermophilum", the BChl c_S molecules appear to form folded or twisted sheets (Section 5). (For interpretation of the references to colour in this figure legend, the reader is referred to the online version of this book.)

(Bryant et al., 2012). Furthermore, this chapter will focus on only a few key aspects of green bacterial metabolism, principally relating to chlorophototrophy in green bacteria: bacteriochlorophyll biosynthesis, type-1 RCs, carbon fixation, and sulfur metabolism. Finally, based upon our analyses, we will provide some perspectives about the evolution of green bacteria, (B) Chl biosynthesis, and photosynthesis.

2. SULFUR METABOLISM

A defining characteristic of GSB, reflected in their name, is the ability to oxidise reduced sulfur compounds for photoautotrophic growth. This ability is nearly universal among GSB strains. *Chlorobium* (*Chl.*) *ferrooxidans*, which oxidises Fe^{2+} and actually performs assimilatory sulfate reduction, is currently the only well-characterized exception (Heising, Richter, Ludwig, & Schink, 1999). Genomic and mechanistic aspects of sulfur oxidation in GSB have been extensively described elsewhere (Frigaard & Bryant, 2008; Frigaard & Dahl, 2009; Gregersen, Bryant, & Frigaard, 2011), and only a few salient points will be presented here.

All sulfur-oxidizing GSB strains can oxidise sulfide, and most strains oxidise sulfide completely to sulfate. Some strains can additionally oxidise thiosulfate to sulfate. Nearly all GSB have FccA, a large flavoprotein, and FccB, a small *c*-type cytochrome, but the precise role of these proteins in sulfide oxidation is still uncertain. Sulfide-quinone reductases (SQRs) usually perform the first of several oxidation reactions of sulfur compounds. SQRs are membrane-associated proteins that oxidise sulfide to polysulfide using menaquinone as the electron acceptor. SQRs comprise a large protein family, and six types are currently recognized, four of which are found in GSB (Gregersen et al., 2011). Among these four, type III and type IV SQRs are found in most strains, and type IV (SqrD) is responsible for most of the SQR activity in *Chlorobaculum* (*Cba.*) *tepidum* (Chan, Morgan-Kiss, & Hanson, 2009). The wide distribution of SQRs in GSB suggests that SQR was present in the common ancestor(s) of GSB (Gregersen et al., 2011), and this notion is supported by the fact that the earliest diverging member of the phylum Chlorobi, *Ignavibacterium* (*Ign.*) *album*, also has SqrD and SqrE.

Polysulfide, a reduced, linear form of elemental sulfur, which forms zero-valence S_8 rings, is subsequently oxidised to sulfite in most GSB strains. This reaction is catalysed by dissimilatory sulfite reductase (DSR, Holkenbrink, Barbas, Mellerup, Otaki & Frigaard, 2011). The DSR complex contains 17 different genes, *dsrNCABLUEFHTMKJOPVW*, often encoded in a single gene cluster (Gregersen et al., 2011). A few exceptions are known. In addition to *Chl. ferrooxidans*, which lacks sulfur oxidation genes except *sqrD* and *sqrF*, *dsr* genes are also missing in *Chloroherpeton* (*Chp.*) *thalassium*. This strongly implies that the *dsr* genes and the ability to oxidise elemental sulfur to sulfite were possibly acquired after the divergence of Chlorobiaceae from Chloroherpetonaceae (see Section 7). *dsrEFH* genes are missing in *Chlorobaculum parvum*, which is unable to oxidise sulfide to sulfate (Kelly, 2008).

The oxidation of sulfite to sulfate occurs by two pathways in GSB. Adenosine-5′-phosphosulfate (APS) reductase oxidises sulfite to APS, and ATP sulfurylase cleaves APS forming ATP and sulfate. The clustering of the genes encoding the membrane-bound oxidoreductase, QmoABC, and APS reductase suggests that these proteins might be coupled to mediate electron transfer to menaquinone in GSB (Frigaard & Dahl, 2009). Another possible sulfite oxidation enzyme is the polysulfide reductase-like complex 3. The gene encoding this protein is always clustered with the *dsr* genes in GSB, which suggests it may play a role in sulfur oxidation (Frigaard & Bryant, 2008; Frigaard & Dahl, 2009; Gregersen et al., 2011).

Some GSB strains possess the Sox enzyme system and thus can also oxidise thiosulfate to sulfate. All strains with this ability have an operon of eight genes, *soxJXYZAKBW*, which differs from the *sox* gene clusters found in other thiosulfate-oxidizing bacteria (Frigaard & Bryant, 2008; Gregersen et al., 2011). Among eight GSB strains that have *sox* gene clusters, four are *Cba.* spp., including all *Cba.* spp. strains whose genomes have been sequenced, and *Chlorobaculum thiosulfatiphilum* Deutsche Sammlung von Mikroorganismen (DSM 249) (Verté et al., 2002). Detailed comparisons of the phylogenetic relationships and gene structures strongly suggest that the *sox* gene cluster might represent an ecologically important and readily selectable mobile genetic element. A virus might have horizontally transferred this gene cassette from a *Cba.* spp. strain to *Chlorobium phaeovibrioides* (Gregersen et al., 2011). It seems likely that the *sox* gene cluster was first acquired by ancestral *Cba.* spp. and that the capacity for thiosulfate oxidation subsequently spread to other GSB.

3. REVERSE TCA CYCLE

Unlike all other chlorophototrophic bacteria, GSB fix CO_2 by the reverse TCA cycle. The reverse TCA cycle, also known as the reductive TCA cycle, operates oppositely to its usual oxidative direction, which leads to the production of acetyl-coenzyme A (CoA) from CO_2 using electrons derived from highly reducing ferredoxins and NAD(P)H (Buchanan & Arnon, 1990; Wahlund & Tabita, 1997). Most of the reactions of the oxidative TCA cycle are reversible, and the enzymes catalysing these steps are shared with the reductive TCA cycle. One key difference between the two is noteworthy: the reverse TCA cycle includes ATP-dependent citrate lyase, which catalyses the cleavage of citrate to form acetyl-CoA and oxaloacetate, instead of citrate synthase of the TCA cycle.

ATP-dependent citrate lyase in GSB is a heterodimer that is dependent on ATP, CoA, and Mg^{2+} (Kanao, Fukui, Atomi & Imanaka, 2001). The mechanism of the reaction has not been fully elucidated, but it is probably similar to that of homologous eukaryotic enzymes, in which autophosphorylation of a histidine residue occurs at the active site, followed by formation of citryl-phosphate, which is then cleaved to form the products, oxaloacetate and acetyl-CoA (Kim and Tabita, 2006; Wahlund & Tabita, 1997). The two genes encoding citrate lyase are extremely conserved among all GSB strains but are missing in a recently described aerobic photoheterotrophic member of the phylum Chlorobi, "*Candidatus* Thermochlorobacter aerophilum" ("*Ca.* Tcb. aerophilum") (Liu, Klatt et al., 2012). Two alternative enzymes, citryl-CoA synthetase and citryl-CoA lyase, catalyse the key citrate cleavage step of the reverse TCA cycle in other bacteria, such as *Aquifex* and *Hydrogenobacter* spp. (Aoshima, Ishii, & Igarashi, 2004a, 2004b; Hügler, Huber, Molyneaux, Vetriani, & Sievert, 2007). In contrast to ATP-dependent citrate lyase, these two enzymes carry out the reaction in two independent steps with citryl-CoA as the intermediate. Hügler and Sievert (2011) also postulated the existence of a third type of enzyme, which could also catalyse the same reaction. This enzyme is similar to ATP-dependent citrate lyase but has one additional domain with unknown function. These different types of citrate-cleaving enzymes are related to each other in sequence, the domains of these proteins that are responsible for citryl-phosphate or citryl-CoA cleavage are also related to citrate synthase. It has been speculated that all of these proteins share a common ancestor and evolved differently to favour either synthesis or cleavage of citrate in various organisms (Aoshima, Ishii, & Igarashi, 2004b; Hügler et al., 2007).

ATP-independent citrate lyase is a fourth enzyme that can cleave citrate. This enzyme is completely unrelated to the three types of enzyme mentioned above and is not dependent on ATP or CoA. It is involved in citrate fermentation in some organisms but has not yet been shown to participate in the reverse TCA cycle (Meyer, Dimroth, & Bott, 2001). However, this enzyme occurs in a recently described, non-chlorophototrophic member of the Chlorobi, *Ign. album* (Iino et al., 2010), and Liu, Frigaard et al. (2012) proposed this potentially new function for this enzyme. *Ign. album* contains all components necessary for CO_2 fixation by the reverse TCA cycle, including ferredoxins and 2-oxoglutarate:ferredoxin oxidoreductase. Therefore, ATP-independent citrate lyase could potentially complete the reverse TCA cycle in that organism (Liu, Frigaard et al., 2012). Interestingly, homologues of *citE*, which encode one subunit of this enzyme, are found in four GSB strains: *Chlorobium limicola*, *Chlorobium phaeobacteroides* DSM 266, *Prosthecochloris* (*Ptc.*) *phaeobacteroides* BS1, and *Prosthecochloris aestuarii*. This distribution suggests that ATP-independent citrate lyase may have been present in ancestors of extant members of the phylum Chlorobi.

In addition to the key citrate-cleavage enzyme, 2-oxoglutarate:ferredoxin oxidoreductase (KOR) and pyruvate:ferredoxin oxidoreductase (POR) also play important roles in CO_2 fixation in GSB. These two related enzymes use reduced ferredoxin generated in a light-dependent manner from RCs as the source of electrons for CO_2 reduction. Although the reaction catalysed by KOR is reversible, it favours the reductive/carboxylative direction and thus replaces 2-oxoglutarate dehydrogenase in the reverse TCA cycle. All GSB strains have KOR, but interestingly, three GSB strains, *Chl. limicola, Chlorobium clathratiforme*, and *Chp. thalassium,* also have 2-oxoglutarate dehydrogenase. POR catalyses the carboxylation of acetyl-CoA, obtained from either the reverse TCA cycle or from assimilated acetate, to form pyruvate, which is the growth-rate controlling reaction of GSB during mixotrophic growth (Feng, Tang, Blankenship, & Tang, 2010). POR is also reversible and favours carboxylation when ferredoxin acts as the electron donor and favours decarboxylation when rubredoxin acts as the electron acceptor (Yoon, Hille, & Tabita, 1999, 2001). Interestingly, POR and KOR are also found in two aerobic photoheterophic members of the phylum Chlorobi, "*Ca.* Tcb. aerophilum", and *Ign. album*; these enzymes might play important roles during mixotrophic growth, either via an incomplete or new variant of the reverse TCA cycle (Liu, Frigaard et al., 2012, Liu, Klatt et al., 2012). The presence of these enzymes in early diverging and aerobic members of the phylum Chlorobi suggests that the evolutionary acquisition of these two CO_2-fixing enzymes might have been independent of, and have occurred prior to, the acquisition of an enzyme for citrate cleavage.

4. BACTERIOCHLOROPHYLL BIOSYNTHESIS IN GREEN BACTERIA

All characterized members of the Chlorobiales synthesize three types of Chls: BChl a_P, Chl a_{PD}, and either BChl *c*, *d*, or *e* (the subscripts indicate the esterifying alcohol: P, phytol; PD, Δ2,6-phytadienol; F, farnesol; S, stearol (octadecanol)). The structure of BChl c_F, which is actually a family of compounds with different numbers of methyl groups at the C-8^2 and

C-12[1] positions, is shown in Fig. 4.3 (Gomez Maqueo Chew & Bryant, 2007b; Liu & Bryant, 2012). "*Ca.* Cab. thermophilum" synthesizes BChl a_P, Chl a_{PD}, and BChl c_S, and additionally synthesizes Zn-BChl a_P', which occurs in its RCs (Bryant et al., 2007; Garcia Costas, Tsukatani et al., 2012; Tsukatani et al., 2012; see Section 6). Chlorosome-producing Chloroflexi, e.g. *Cfx. aurantiacus*, synthesise BChl a_P and BChl c_S but do not synthesize Chl *a* (Gloe & Risch, 1978).

The biosynthetic pathway for these (B)Chls is shown in Fig. 4.4 and 4.5 (Gomez Maqueo Chew and Bryant, 2007b; Liu & Bryant, 2012). The synthesis of (B)Chls proceeds through a series of core reactions that are found in all chlorophototrophs, which convert protoporphyrin (Proto) IX (compound **[1]**; the numbers in brackets in bold font indicate the numbered compounds in Figs 4.4 and 4.5) into chlorophyllide (Chlide) *a* **[6]**, the immediate precursor of Chl *a* **[7]** (Figs 4.4 and 4.5). The first committed step in (B)Chl biosynthesis is the ATP-dependent insertion of Mg^{2+} into Proto IX to produce Mg-Proto IX **[2]**, a reaction that is catalysed by Mg:Proto IX chelatase (BchHID/ChlHID). In vitro studies of the closely related enzyme from cyanobacteria suggest that ~15 ATP molecules are hydrolysed during the synthesis of one Mg-Proto IX (Reid and Hunter, 2004). GSB typically contain

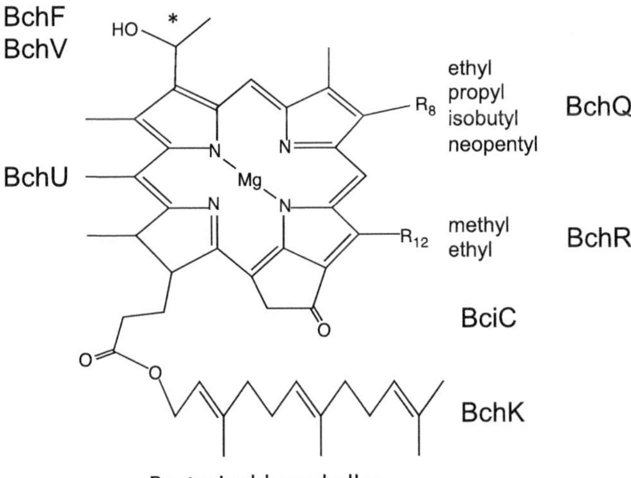

Figure 4.3 ***Structure of BChl c_F esterified with farnesol.*** The side chains at R8 may be ethyl, propyl, isobutyl, or neopentyl. The side chains at R12 may be methyl or ethyl. The asterisk near C-3[1] indicates a chiral carbon. The protein names around the periphery indicate the gene products responsible for modifications that occur after the Chlide *a* intermediate (see Section 4).

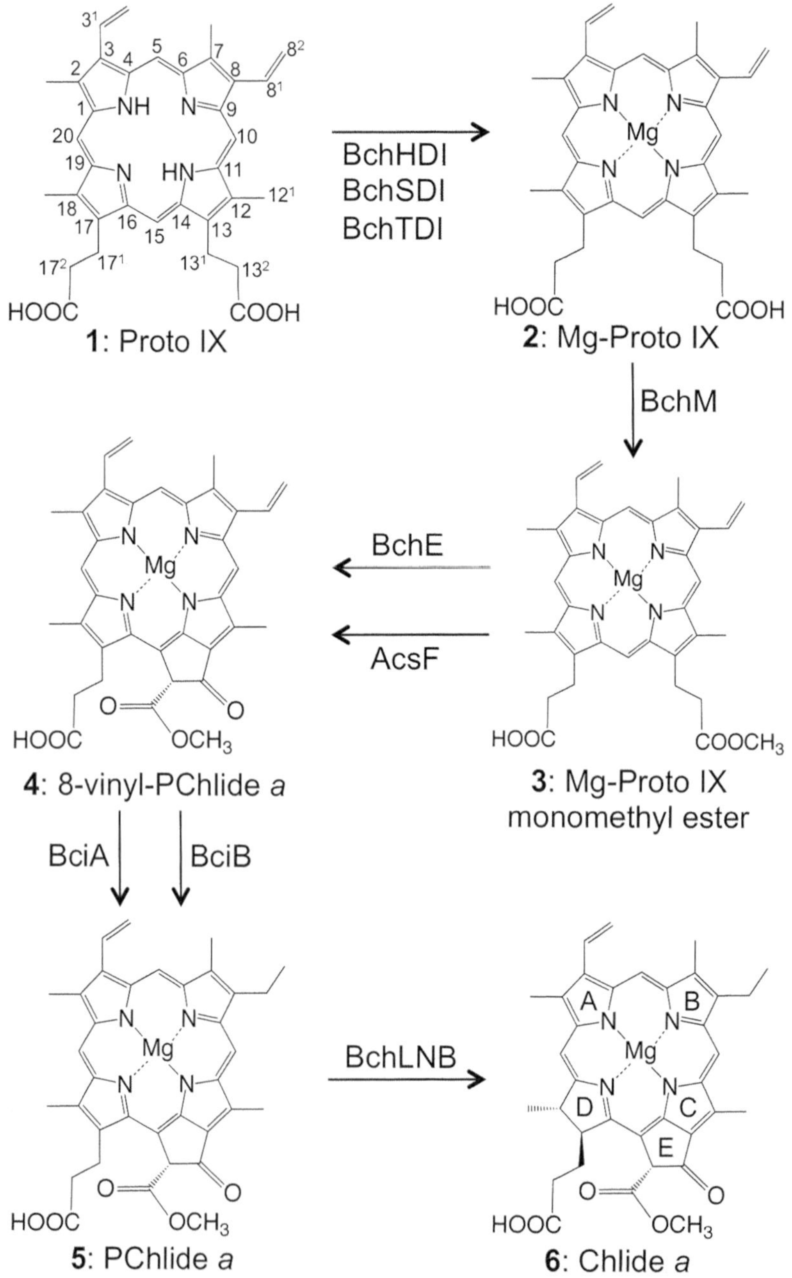

Figure 4.4 ***Scheme showing the biosynthesis of Chlide*** **a** ***from Proto IX.*** The numbered compounds correspond to compounds mentioned in the text in brackets and in bold font (see Section 4).

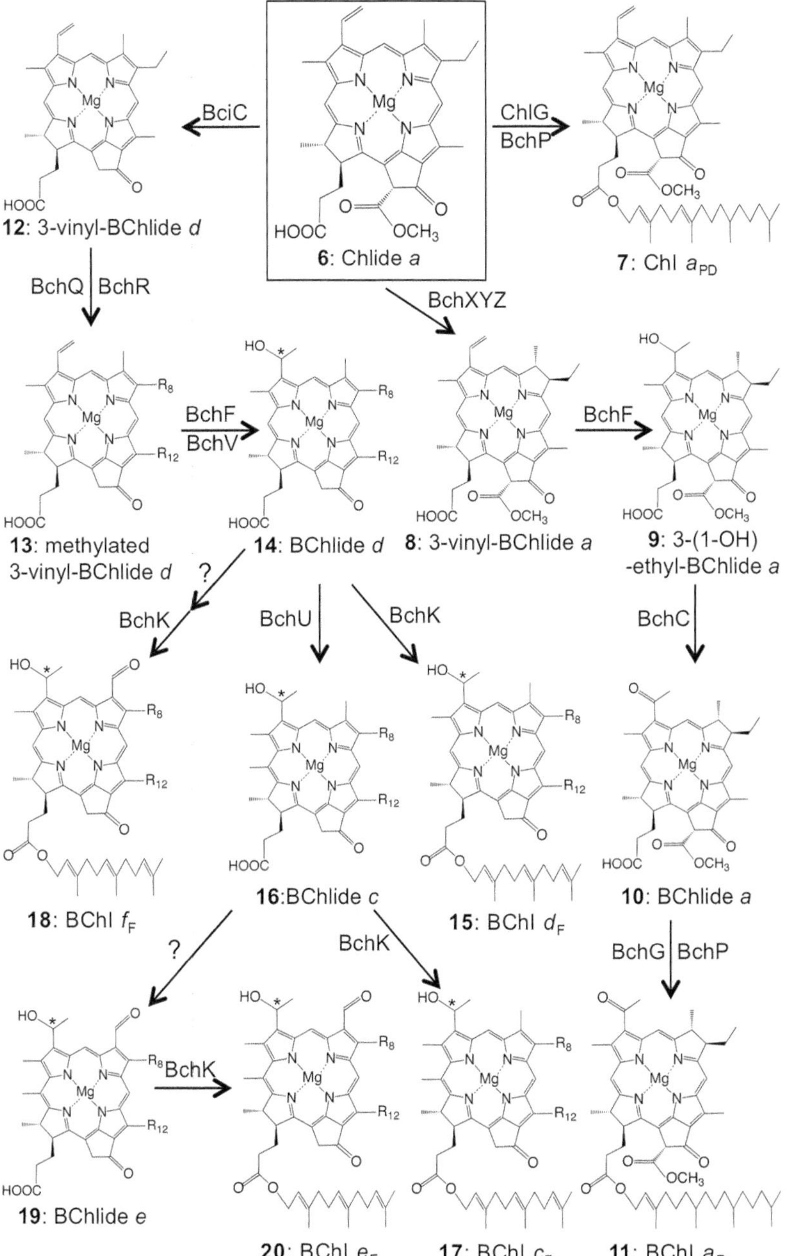

Figure 4.5 ***Scheme showing the biosynthesis of BChl* a, c, d, *and* e *from Chlide* a *(boxed).*** The numbered compounds correspond to compounds mentioned the text in brackets and in bold font (see Section 4).

two or three isoforms of this enzyme (Eisen et al., 2002; Frigaard, Li, Gomez Maqueo Chew, Maresca, & Bryant, 2003; Gomez Maqueo Chew, Frigaard, & Bryant, 2009; Liu, Klatt et al., 2012), and enzymological studies in vitro as well as mutagenesis studies in vivo showed that the paralogous enzymes have different biochemical properties and regulation (Gomez Maqueo Chew et al., 2009; Johnson and Schmidt-Dannert, 2008). For example, inactivation of the *bchS* gene in *Cba. tepidum*, which encodes one of three members of the BchH gene family, caused a severe reduction in the cellular content of BChl *c* and led to the excretion of Proto IX from cells (Gomez Maqueo Chew et al., 2009). The mutagenesis studies further showed that the enzymatic activity of BchT was insufficient to support the BChl a_P and Chl a_{PD} requirements of cells, but that enzymes produced with BchS or BchH subunits were sufficient to provide sufficient (B)Chl to support photoautotrophic growth under laboratory conditions (Gomez Maqueo Chew et al., 2009). The origin of Zn–BChl a_P' in the RCs of "*Ca*. Cab. thermophilum" is currently unknown. The "*Ca*. Cab. thermophilum" genome does not contain any paralogs of the single-copy *bchH* gene. In a *bchI* mutant of *Rhodobacter capsulatus*, Zn–BChl a_P was produced and incorporated into components of the photosynthetic apparatus (Jaschke, Hardjasa, Digby, Hunter & Beatty, 2011). However, in this case it appears that Zn-Proto IX was synthesized by ferrochelatase and subsequently converted by the usual pathway into Zn-BChl a_P.

The subsequent reaction in the synthesis of Chlide *a* is the methylation of the C-13^2 carboxyl group by BchM/ChlM, which yields Mg-Proto IX monomethyl ester **[3]** (Johnson and Schmidt-Dannert, 2008). Interestingly, this is the only reaction in the pathway leading from Proto IX to Chlide *a* that is catalysed by only one enzyme. The subsequent conversion of Mg-Proto IX monomethyl ester [**3**] into 8-vinyl-protochlorophyllide *a* (8-vinyl-PChlide) *a* **[4]**, is catalysed by two different enzymes: BchE in anaerobes and AcsF in aerobes. BchE is an oxygen-sensitive member of the radical-SAM superfamily (Booker, 2009; Gough, Petersen & Duus, 2000), while AcsF is an oxygen-dependent di-iron oxidase (Pinta, Picaud, Reiss-Husson & Astier, 2002). Both introduce the C-13^1 keto group and isocylic ring while performing a six-electron oxidation of the substrate. All members of the Chlorobiaceae and Chloroherpetonaceae that have been characterized by genome sequencing have *bchE*, which is consistent with their obligately anaerobic lifestyle. However, *bchE* was not detected in the metagenome of "*Ca*. Tcb. aerophilum", and furthermore, no evidence for *bchE* was found in the metatranscriptome as well (Liu, Klatt et al., 2012). Because AcsF was not found in the scaffolds assigned to "*Ca.* Tcb. aerophilum", and because AcsF is not as highly conserved as BchE, it was not possible to determine whether "*Ca.* Tcb. aerophilum" has an *acsF* gene as expected for an aerobe. However, the genome of "*Ca.* Cab. thermophilum" includes *acsF*, and this organism produces (B)Chls under fully oxic conditions (Bryant et al., 2007; Garcia Costas, Liu et al., 2012; Garcia Costas, Tsukatani et al., 2012). "*Ca.* Cab. thermophilum" also has a homologue of *bchE*, but it is not yet known if this gene functions in (B)Chl biosynthesis (Garcia Costas, Liu et al., 2012).

Green bacteria have at least three different enzymes, BciA, BciB, and a still unidentified enzyme, that can reduce the C-8 vinyl group of 8-vinyl-PChlide *a* **[4]** to produce PChlide *a* **[5]** (Gomez Maqueo Chew & Bryant, 2007a; Liu & Bryant, 2011b). BciA of GSB is very similar to the NADPH-dependent enzyme of plants (Gomez Maqueo Chew & Bryant, 2007a; Nagata, Tanaka, & Tanaka, 2007; Tanaka & Tanaka, 2007), and a homologue *bciA* is found in the genomes of purple bacteria. BciB is similar to the 8-vinyl reductase of cyanobacteria (Ito, Yokono, Tanaka & Tanaka, 2008; Liu & Bryant, 2011b). BciB, which utilises ferredoxin as source of electrons (A. M. Saunders and D. A. Bryant, unpublished results), is also found in most GSB, Cyanobacteria, "*Ca.* Cab. thermophilum" and some members of the Chloroflexi. *Roseiflexus* spp., which synthesizes BChl *a* and must therefore reduce the C-8 vinyl group, lacks homologues of both BciA and BciB, and therefore *Roseiflexus* spp. and perhaps other organisms have a third type of 8-vinyl reductase.

Unlike Cyanobacteria, algae and plants, green bacteria do not have a light- and NADPH-dependent PChlide oxidoreductase (LPOR) (Sytina et al., 2009). Instead, in all anoxygenic bacteria the reduction of the C-17,18 double bond of PChlide *a* **[5]** to produce Chlide *a* **[6]** is catalysed by a light-independent (dark-active) enzyme (DPOR) that is structurally related to nitrogenase (Bröcker et al., 2010; Muraki et al., 2010; Sarma et al., 2008). This enzyme is encoded by the *bchB*, *bchN*, and *bchL* genes, and like the reduction of dinitrogen, catalysis requires ATP hydrolysis. Chl *a* **[7]** is synthesized by Chl *a* synthase (ChlG) and BchP (geranylgeranyl reductase) from Chlide *a* **[6]**, which is one of the two 'hub' compounds of (B)Chl biosynthesis (see Section 8). Chlide *a* can be converted into 9 of the 14 major BChl and Chl derivatives currently known to occur in chlorophototrophic bacteria (Figs 4.5 and 4.7; Sections 8 and 9) (Liu & Bryant, 2012).

Aside from the BChl *a* synthase (BchG) and geranylgeranyl reductase (BchP), the synthesis of BChl *a* **[11]** requires three enzymes. The first committed step in the synthesis of BChl *a* from Chlide *a* is the reduction of the C-7,8 double bond of the B-ring by Chlide reductase (chlorophyllide oxidoreductase (COR)) to produce 3-vinyl-bacteriochlorophyllide a (3-vinyl-BChlide *a*) **[8]**. Chlide reductase is encoded by the *bchX*, *bchY*, and *bchZ* genes and is closely related to PChlide reductase by a gene duplication event (and by extension is also homologous to nitrogenase). The remaining steps in BChl *a* synthesis are the hydration of the C-3 vinyl group by BchF to produce 3^1-OH-ethyl-BChlide *a* **[9]** and the subsequent oxidation of the resulting hydroxyl group by BchC, yielding BChlide *a* **[10]**. These three reactions also probably occur during the synthesis of BChl *b*, but as will be discussed in Section 8, BChlide *g*, and not BChlide *a* **[10]**, is possibly the precursor of BChlide *b*.

The signature Chls of green bacteria are those that form the self-assembling, major antenna pigments in chlorosomes, namely BChls *d* **[15]**, *c* **[17]**, and *e* **[20]**, all of which are synthesized in a pathway leading from Chlide *a* **[6]** (Liu & Bryant, 2011a). The removal the C-13^2 methylcarboxyl group, which produces 3-vinyl-BChlide *d* (**12**), requires BciC, but the mechanism of this first committed step is not known. Two paralogous methyltransferases, BchQ and BchR, are members of the radical-SAM protein superfamily, and these enzymes add methyl groups from S-adenosyl-L-methionine to the C-8^2 and C-12^1 positions **[13]**. Thus, BChl *c*, *d*, and *e* are actually families of compounds differing in their degree of methylation at these two positions (Fig. 4.3). BchR can add a single methyl group at C-12^1, and in the GSB *Cba. tepidum*, ~90–95% of the BChl *c* molecules have an ethyl group at this position. In *Chlorobaculum limnaeum*, however, nearly all molecules of BChl *e* have an ethyl side chain at C-12 (Vogl, Tank, Orf, Blankenship, & Bryant, 2012). BchQ can add one, two, or three methyl groups at the C-8^2 position, producing propyl, isobutyl, or neopentyl side chains at this position. BchQ and BchR are not found in *Cfx. aurantiacus*, which synthesizes only [8-Et, 12-Me]-BChl *c*. However, the BchQ and BchR methyltransferases occur in some other green FAPs (e.g. *Osc. trichoides*; Kuznetsov et al., 2011), and both are also found in "*Ca.* Cab. thermophilum" (Garcia Costas, Liu et al., 2012, Garcia Costas, Tsukatani et al., 2012).

Hydration of the C-3 vinyl group by BchF or BchV must occur after the methylation reactions at C-8^2 and C-12^1, because the extent of methylation at those positions influences the chirality created at the C-3^1 position (Huster and Smith, 1990; A. Gomez Maqueo Chew and D. A.

Bryant, unpublished results). Highly methylated BChl *c* molecules have predominantly *S*-chirality at this position (Huster and Smith, 1990); however, [8-Et, 12-Me]-BChl *c* has almost exclusively *R*-chirality (Ganapathy et al., 2009). Both hydratases can introduce water in such a way that molecules with both *R*- and *S*-chirality are products (A. Gomez Maqueo Chew and D. A. Bryant, unpublished results); however, BchF is inferred to have highest activity with molecules that are not extensively methylated at C8 and C12. Moreover, because BchF is essential for viability in *Cba. tepidum*, it probably also functions in the synthesis of BChl *a*. Conversely, BchV is not essential for viability and appears to be selectively active with more extensively methylated substrates (Gomez Maqueo Chew, 2007).

BchU is an S-adenosyl-methionine-dependent methyl transferase that methylates BChlide *d* [**14**] at C-20 to produce BChlide *c* [**16**]. The X-ray structure of BchU was solved with substrate bound (Wada et al., 2006), and this showed that the C-3^1 hydroxyl group is important for binding to the active site. This observation strongly suggests that the C-20 methylation reaction probably occurs immediately before the esterification reaction catalysed by BchK, the BChl *c*/*d*/*e* synthase (Frigaard, Voigt, & Bryant, 2002). BChl *c*, *d*, and *e* are produced when BchK esterifies the corresponding BChlides with alcohols such as farnesol, geranylgeraniol, hexadecanol, and octadecanol (stearol). BchU is missing or mutated by a frameshift mutation (Maresca et al., 2004) in organisms that synthesize BChl *d*, and BchU also occurs in organisms that synthesize BChl *e* [**20**]. Insertional inactivation of the *bchU* gene in the brown-coloured, BChl *e*-producing organism, *Cba. limnaeum*, resulted in a mutant that synthesizes BChl *f* [**18**], a BChl that had previously never been observed in nature (Harada, Mizoguchi, Tsukatani, Noguchi, & Tamiaki, 2012; Tamiaki et al., 2011; Vogl et al., 2012). In depth characterization of the chlorosomes from this *bchU* mutant suggested why BChl *f* may not occur naturally (Orf et al., 2012; Vogl et al., 2012). Chlorosomes containing BChl *f* transfer energy much less efficiently from BChl *f* to the BChl *a* of the baseplate complex than chlorosomes containing BChl *e*. This is reflected by a much lower growth rate at low irradiance for the mutant than for the wild type (Vogl et al., 2012). Organisms that produce BChl *e* grow at the lowest irradiance levels of all green bacteria (Manske et al., 2005). Organisms producing BChl *f* would not be able to compete effectively at low irradiance with organisms producing BChl *e* [**20**] because of inefficient energy transfer.

BChlide *e* [**19**] differs from BChlide *c* [**16**] by having a formyl side chain like Chl *b* at C7 rather than a methyl group. Phylogenetic profiling analyses of GSB strains synthesizing BChl *c* or BChl *e* had identified a small cluster of genes near *cruB*, which encodes γ-carotene cyclase (Maresca, Romberger & Bryant, 2008), that seemed likely to be involved in BChl *e* biosynthesis. In particular, an apparent operon of three genes seemed most likely to be involved. One of these genes encodes a paralog of BchF and BchV, and another is a member of the radical-SAM superfamily. Inactivation of the latter caused BChl *c* to accumulate in cells. This result suggests that this gene is involved in BChl *e* biosynthesis and probably catalyses the first step in the conversion of BChl *c* into BChl *e* (J. Thweatt and D. A. Bryant, unpublished results). This small cluster of genes appears in different genomic contexts in various brown-coloured GSB strains. Like the eight-gene cluster of genes required for thiosulfate utilisation (see Section 2), it is possible that the genes that convert BChl *c* into BChl *e* form a cassette that can be laterally transferred among GSB strains and that immediately confer the ability to occupy a new light niche when this occurs. The close association between BChl *e* and the carotenoid isorenieratene suggests that the *cruB* gene may also be transferred as a component of the same gene cassette.

5. CHLOROSOME STRUCTURE AND ASSEMBLY

BChls *c*, *d*, and *e* assemble into suprastructures in chlorosomes (Figs 4.1 and 4.2), which can contain from 150,000 to 250,000 BChl molecules, without the direct participation of proteins. Cells of the model GSB, *Cba. tepidum*, contain about 200 chlorosomes per cell (Fig. 4.1A), and thus each *Cba. tepidum* cell can contain up to 50 million BChl *c* molecules. The chlorosomes of *Cba. tepidum* are large and heterogeneous in size, with lengths of 100–200 nm and diameters of 30–60 nm (Fig. 4.1B, D) (Frigaard & Bryant, 2006; Oostergetel et al., 2010). *Chlorobaculum tepidum* chlorosomes also contain about 2500 BChl *a* molecules; 5000 protein molecules (all are located in the envelope; half is CsmA); 20,000 carotenoids; 18,000 chlorobiumquinone, 1′-OH-chlorobiumquinone, and menaquinone-7 molecules; and 20,000 lipid molecules (glycolipids, phospholipids, and wax esters) (Frigaard & Bryant, 2006).

The chlorosome envelope is an asymmetric bilayer membrane, in which glycolipids and proteins form the outer leaflet and the esterifying alcohol tails of BChls form the inner leaflet (Fig. 4.2). The envelope is stabilized and mostly comprised of 10 different proteins, which belong to four structure motif families: CsmA/CsmE, CsmB/CsmF/CsmH, CsmC/CsmD/CsmH, and CsmI/CsmJ/CsmX (Frigaard, Li et al., 2004; Vassilieva et al., 2002).

Mutational studies in *Cba. tepidum* showed that CsmA is the only essential envelope protein (Frigaard, Li et al., 2004; Li & Bryant, 2009; Li, Frigaard, & Bryant, 2012). CsmA forms a paracrystalline array, which is known as the baseplate and which attaches the chlorosomes to the FMO protein in the cytoplasmic membrane (Bryant et al., 2002; Li et al., 2006; Pedersen et al., 2010). Cross-linking studies have identified a number of protein–protein interactions in the chlorosome envelope (Li et al., 2006). For example, CsmI and CsmJ form homodimers, heterodimers, and heterotetramers. Each of these proteins has an adrenodoxin-like domain that harbours a [2Fe-2S] cluster (Li et al., 2006, 2012; Johnson, Li, Frigaard, Golbeck & Bryant, 2012). Other chlorosome proteins, for example CsmB, CsmF, CsmC, CsmD, and CsmH, apparently are functionally redundant and modify the size and shape of chlorosomes. The observed changes in chlorosome shape and dimensions may arise because these envelope proteins influence the orientation and organization of the BChl suprastructure that forms inside the chlorosome (Li, Jubilerer, Costas, Frigaard, & Bryant, 2009).

Chlorosomes are not only heterogeneous in size, but as revealed by cryo-electron microscopy, the organization of the BChl *c* molecules inside each chlorosome is unique – i.e. no two chlorosomes are structurally identical (Oostergetel et al., 2007). When chlorosomes are viewed end-on, it can be observed that the BChl *c* molecules form coaxial, concentric nanotubular structures. As originally reported by Pšenčík and co-workers (Pšenčík et al., 2004), these lamellae have a spacing of ~2.1 nm (vertical arrows in Fig. 4.1F, G). It is now clear that these features have arisen from the surfaces formed by the BChl tetrapyrrole head groups and that the lamellar spacings are determined by the hydrophobic interactions of esterifying alcohols (Ganapathy et al., 2009, 2012). The BChl molecules in two adjacent layers interact through their esterifying alcohols in a manner analogous to the hydrophobic interactions of fatty acids of lipids in a bilayer membrane. The glycolipids of the envelope interact with the esterifying alcohols with the outer most BChl layer.

Solid-state NMR and cryo-electron microscopy further showed that the BChls in a *bchQ bchR bchU* mutant and the wild-type strain of *Cba. tepidum* form *syn–anti* dimer stacks, in which the esterifying alcohol of one molecule points above and the other points below the surface formed by the tetrapyrrole rings (Ganapathy et al., 2009). The *syn–anti* stacks are actually arranged as shallow helices, and the 0.83-nm repeats observed in the cryo-electron micrographs (Fig. 4.1F, vertical arrows) result from these *syn–anti* stacked BChls (Fig. 4.1F, H) (Ganapathy et al., 2009). These stacks run parallel to the long axis of the chlorosome in wild-type *Cba. tepidum,* which synthesizes a complex mixture of BChl *c* homologues (Fig. 4.1H). However, these stacks are approximately perpendicular to the long axis of the chlorosome in a *bchQ bchR bchU* mutant, which synthesizes almost exclusively *R*-[8-Et, 12-Me]-BChl d_F (Fig. 4.1H and Ganapathy et al., 2009).

It should be noted that not all chlorosomes have the same suprastructure as that just described. In a *bchQ bchR* mutant of *Cba. tepidum* that synthesizes almost exclusively *R*-[8-Et, 12-Me]-BChl c_F, the BChls form surfaces made up of parallel stacks of all-*syn* and all-*anti* coordinated BChl molecules (Fig. 4.1H) (Ganapathy et al., 2012). Furthermore, the chlorosomes of "*Ca.* Cab. thermophilum" are structurally distinct from those of *Cba. tepidum* (Garcia Costas et al., 2011). Cryo-electron microscopy showed that, similar to other chlorosomes, the BChl *c* suprastructure has 2.3-nm lamellar spacings (Fig. 4.1C, D). However, the BChls form twisted or folded sheets rather than concentric nanotubes, and the underlying structural organization of the BChl *c* molecules responsible for this structure is not yet known. Furthermore, the chlorosomes of *Cfx. aurantiacus* are much smaller than those of *Cba. tepidum* (140–220 nm in length, 30–60 nm in width, and 10–20 nm in height) (Pšenčík et al., 2009; Staehelin, Golecki, & Drews, 1980; Staehelin, Golecki, Fuller, & Drews, 1978). Their baseplate has a very distinctive paracrystalline lattice with a 3.3-nm spacing and that is probably formed from CsmA dimers. This baseplate complex has spectroscopic properties similar to other BChl *a*–CsmA complexes, however (Montaño, Wu et al., 2003; Pšenčík et al., 2009; Sakuragi, Frigaard, Shimada, & Matsuura, 1999). The suprastructural organization of the BChl c_S molecules in chlorosomes of *Cfx. aurantiacus* has not yet been determined. The proteins of the chlorosome envelope of *Cfx. aurantiacus* are very different from those of *Cba. tepidum,* but the same four structure motif families appear to be present (Bryant et al., 2012).

The self-assembly of BChls *c, d,* or *e* into suprastructures causes a significant redshift in the Q_y absorption spectrum. For BChl *c*, the absorption shifts from ~670 to 740–750 nm, and for BChl *d* the shift is from about 654to 729 nm (Orf et al., 2012). Although there is a similar shift for BChl *e* in the Q_y absorption region from 656 to 721 nm, aggregation of the BChls causes a splitting in the Soret region that leads to an intense absorption band at about 528 nm (Harada et al., 2012; Orf et al., 2012; Vogl et al., 2012). Although BChl *f* does not naturally occur in GSB, studies with *bchU* mutants of *Cba. limnaeum* showed a behaviour similar to that of BChl *e*: the Q_y absorption shifts from 641 to 705 nm and the Soret splits to produce an intense absorption at about 508 nm (Harada et al., 2012; Orf et al., 2012; Vogl et al., 2012).

The degree of methylation at C-8[2] and C-12[1] causes the bandwidth of the Q_y absorption band to increase due to inhomogenous broadening (Gomez Maqueo Chew, Frigaard, & Bryant, 2007). These methylations, together with methylation at C20 (Maresca et al., 2004), fine-tune the absorption maximum by shifting it still further to longer wavelength and they determine how much BChl is produced in response to irradiance (Gomez Maqueo Chew et al., 2007). These effects combine to increase the absorption cross-section of chlorosomes and reduce overlap with Chl *a*, which is found in plants, algae, and cyanobacteria – all of which filter the light reaching these anaerobes because they occur in the upper, oxic layers of stratified lakes or microbial mats (Gomez Maqueo Chew et al., 2007; Maresca et al., 2004). The very high density of BChl molecules in chlorosomes produces very short inter-BChl distances, which leads to very strong excitonic coupling (Ganapathy et al., 2009, 2012). Absorption events produce excited states that are delocalized over many BChls (Prokhorenko, Steensgaard, & Holzwarth, 2000) and that migrate very rapidly with energy-transfer times on the sub-100 fs timescale (Dostál et al., 2012).

As noted in Section 6, upon illumination, the type-1 RCs of GSB and "*Ca.* Cab. thermophilum" produce strong (i.e. low potential) reductants in the form of reduced ferredoxins. When cells are exposed to oxygen in the light, highly toxic, reactive oxygen species are produced (Li et al., 2009). However, oxygen also rapidly quenches energy transfer in chlorosomes, probably by oxidation of some of the numerous quinones that are found in the chlorosomes (Blankenship et al., 1993; Frigaard & Matsuura, 1999; Frigaard, Takaichi, Hirota, Shimada, & Matsuura, 1997, Frigaard, Matsuura, Hirota, Miller, & Cox, 1998; Hohmann-Marriott & Blankenship, 2007; Wang, Brune, & 1990). Energy transfer through the FMO protein is also inhibited after exposure to oxygen (Zhou LoBrutto, Lin, & Blankenship, 1994). These oxidised quenching compounds in chlorosomes must be reduced again to reactivate energy transfer. The chlorosome envelopes of *Cba. tepidum* contain heterotetrameric complexes of CsmI and CsmJ, which participate in this process. These proteins have adrenodoxin-like domains, each of which ligates one [2Fe–2S] cluster (Johnson et al., 2012; Li et al., 2012). Mutants lacking CsmI and CsmJ are defective in a light-dependent process that reactivates energy transfer, but still have an undefined pathway that can reduce the quenching quinones in the dark. The chlorosomes of "*Ca.* Cab. thermophilum" are also highly susceptible to quenching under oxidizing conditions, and these chlorosomes contain large amounts of menaquinone-7. Although an Fe/S protein related to CsmI/CsmJ/CsmX is present in the chlorosome envelopes of "*Ca.* Cab. thermophilum", the chlorosomes of this bacterium additionally include a type-2 NADH dehydrogenase, which could transfer electrons from NADH to oxidised menaquinone-7 to reactivate energy transfer. Chlorosomes of *Cfx. aurantiacus* contain menaquinone-10, but energy transfer in these chlorosomes is not as sensitive to oxygen as in the cases described above (Frigaard, Tokita, & Matsuura, 1999; Tokita, Frigaard, Hirota, Shimada, & Matsuura, 2000; Garcia Costas et al., 2011). Nevertheless, the chlorosome envelope of *Cfx. aurantiacus* contains a protein (CsmP) that is homologous to CsmI and CsmJ and is predicted to harbour a [2Fe–2S] cluster. A gene for a type-II NADH dehydrogenase occurs in an operon encoding other chlorosome envelope proteins and may also participate in reduction of the quencher.

6. TYPE-1 RCs

The type-1 RCs of GSB are composed of four protein subunits, designated PscA, PscB, PscC, and PscD; they harbour ~16 BChl *a*, four Chl *a*, two menaquinone-7, one or more carotenoids, and three [4Fe–4S] clusters (Hauska et al., 2001). The RC core is a homodimer of PscA subunits, which are 82-kDa membrane-intrinsic polypeptides with 11 transmembrane α-helices (Sadekar et al., 2006). PscA binds all of the pigments and electron transfer cofactors except for the two terminal Fe/S clusters, denoted F_A and F_B. PscB resembles bacterial ferredoxins and harbours the two terminal oxygen-sensitive [4Fe–4S] clusters (Jagannathan & Golbeck, 2008, 2009a, 2009b). Water-soluble and freely diffusing bacterial ferredoxins accept electrons from these very strongly reducing Fe/S clusters ($E_0' = \sim -600$ mV).

The RCs of GSB have two additional subunits, PscC and PscD, which do not have homologues in the RCs of heliobacteria or "*Ca.* Cab. thermophilum". PscC is a membrane-intrinsic *c*-type cytochrome, also known as cytochrome (Cyt) c_Z or Cyt c_{551}, and it is the immediate electron donor to the oxidised special pair P840$^+$, but PscC/Cyt c_Z can also receive electrons derived from thiosulfate oxidation and transfer them to P840$^+$ (Okkels et al., 1992; Higuchi et al., 2009; Azai, Tsukatani, Harada & Oh-oka, 2009). Each RC complex probably contains two Cyt c_Z subunits (Oh-Oka, Kamei, Matsubara, Iwaki &Itoh, 1995, Oh-oka, Iwaki & Itoh, 1997, 1998). Another small membrane-associated, *c*-type Cyt, Cyt c_{556}, is probably the product of ORF CT0073 in *Cba. tepidum*. This cytochrome apparently shuttles electrons derived from the oxidation of menaquinol between the Cyt *b*/Rieske complex and PscC/Cyt c_Z. When *Cba. tepidum* cells are oxidising thiosulfate, an alternative electron transfer pathway, which utilizes a soluble, periplasmic *c*-cytochrome, Cyt $c_{554/555}$, appears to participate in the electron transfer chain. Cyt $c_{554/555}$ is the product of ORF CT0075 in *Cba. tepidum*, and it donates electrons directly to Cyt c_Z and does not act as a soluble shuttle between Cyt c_{556} and Cyt c_Z (Azai et al., 2009; Tsukatani, Azai, Kondo, Itoh, & Oh-oka, 2008). The predicted monoheme-binding domains of CT0073 and CT0075 are very similar in amino acid sequence, and both genes are found in all GSB strains except *Chp. thalassium*, even though only a few of these strains can oxidise thiosulfate. In *Chp. thalassium* ORF Ctha_1874 encodes a protein with strong sequence similarity to both PscC and Cyt $c_{554/555}$. This protein appears to be the result of a gene fusion event, in which the N-terminus is similar to PscC (Cyt c_Z) and the C-terminal domain is similar to Cyt $c_{554/555}$.

PscD is a nonessential 16-kDa subunit, which is bound to the cytoplasmic surface of the RC but sometimes missing from preparations of RCs (Hager-Braun et al., 1995; Tsukatani, Miyamoto, Itoh, & Oh-oka, 2004). PscD may interact with PscB, and/or it might participate in interactions of soluble electron carriers with the RC. An extremely high-resolution, 1.3-Å X-ray structure of the FMO protein of *Ptc. aestuarii* 2K has recently been determined, and this structure suggested the sub-stoichiometric presence of an eighth BChl *a* molecule (Tronrud et al., 2009), which was recently confirmed by mass spectrometry (Wen, Zhang, et al., 2011). Two FMO trimers are tightly associated with each RC complex (Hauska et al., 2001).

When the special pair directly absorbs a photon, or when it receives excitation energy initially absorbed by chlorosomes or FMO, an excited state, denoted as P840★, is formed. The formation of this excited state causes photochemical charge separation to occur within a few picoseconds, producing $P840^+$ A_0^-, which results in the absorption decrease at 840 nm due to the oxidation of the special pair (Hauska et al., 2001). The primary electron acceptors, denoted spectroscopically as A_0, are two of the four Chl *a* molecules found in the RC. P840 has a weakly oxidizing midpoint potential, $E_0' = +240$ mV, and $P840^+$ is rapidly reduced by electron transfer from Cyt c_Z ($E_0' = +180$ mV). The rapid reduction of $P840^+$ prevents charge recombination from the reduced primary acceptor, A_0^-. Although two menaquinone-7 molecules occur in the RCs of GSB, these quinones have not been conclusively demonstrated to participate in the electron transfer chain. The terminal electron acceptor of the core complex is the intersubunit F_X cluster, which is a [4Fe-4S] cluster with a midpoint potential lower than −650 mV; two cysteine residues associated with each PscA subunit of the core homodimer ligate the F_X [4Fe-4S] cluster (Jagannathan & Golbeck, 2008).

The type-1 RCs of "*Ca.* Cab. thermophilum" have recently been isolated and initially characterized (Tsukatani et al., 2012). As noted above, the "*Ca.* Cab. thermophilum" genome does not contain homologues of *pscC* and *pscD* genes, and thus the RCs were expected to contain only PscA and PscB. This simple subunit structure should resemble that observed in heliobacteria; however, it is not yet known whether PscB is permanently or transiently bound to the RC core as it is in heliobacteria (Romberger & Golbeck, 2010, 2012). Strong chaotropic agents were required to remove chlorosomes from cytoplasmic membranes during RC purification, and PscB was probably lost during isolation (Tsukatani et al., 2012). "*Ca.* Cab. thermophilum" PscA (865 amino acids) is larger than the apoproteins of most other type-1 RCs because of the insertion of ~165 amino acids between *trans*-membrane helices six and 7 (Bryant et al., 2007). A carotenoid-binding protein (CBP), which is related to the subunits of type-IV pili, copurified with the PscA homodimers in the RC complex (Tsukatani et al., 2012). This RC–CBP complex had an apparent mass of ~480 kDa and probably contained 12 CBP monomers, each of which may bind three to four carotenoid molecules. The RC core complex had absorption maxima at 812 and 670 nm, and HPLC and LC–MS analyses showed the presence of 12 BChl a_P molecules, eight Chl a_{PD} molecules, and 2 Zn-BChl a'_P molecules per P840 (Tsukatani et al., 2012). Surprisingly, no quinones were detected in this RC preparation. Whole cells, cell membranes, and isolated RCs showed very similar light-induced difference spectra with maximal bleaching at 840 nm, and the observed difference spectra were similar to that of RC preparations from *Cba. tepidum*. However, it is presently unknown whether the two Zn-BChl a'_P molecules are the primary donor or the primary acceptor (or neither) in these RCs (Tsukatani et al., 2012).

"*Ca.* Tcb. aerophilum" is the first described example of a newly discovered family of aerobic chlorophototrophic organisms within the phylum Chlorobi (Liu et al., 2011, Liu, Klatt et al., 2012) (see Section 7). Although this organism has not yet been cultured, metagenomic and metatranscriptomic analyses allowed a detailed description of the organism to be made (Liu et al., 2011, Liu, Klatt et al., 2012). Transcriptional analyses and distribution studies in mats show that this bacterium is an aerobic photoheterotroph and that it is probably physiologically very similar to "*Ca.* Cab.

thermophilum", which is known to be an aerobe. Interestingly, the RCs in "*Ca.* Tcb. aerophilum" have the same subunit structure as those of other GSB strains rather than that of "*Ca.* Cab. thermophilum".

7. EVOLUTION OF GREEN BACTERIA

Compared to other phototrophic bacteria or (bacterial) photosynthesis in general, the evolution of GSB, or more inclusively the phylum Chlorobi, has not received much attention until recently. This was mainly because of the absence of diversity among cultured organisms belonging to this phylum, both in terms of physiology and genomic content (Bryant et al., 2012). Until very recently, almost all cultured strains of the phylum Chlorobi had the typical characteristics of GSB with only a few exceptions. Thus, it was not surprising that the terms 'GSB' and 'Chlorobi' were considered to be synonymous. A logical extention of this thinking was that the ancestors of modern organisms belonging to the Chlorobi must have had similar if not identical properties as extant GSB, and that relatively few changes were adopted since their emergence. However, the recent discoveries and genomic descriptions of two different members of the Chlorobi, "*Ca.* Tcb. aerophilum" (Liu, Klatt et al., 2012) and *Ign. album* (Iino et al., 2010; Liu, Frigaard et al., 2012), require a re-evaluation of these perceptions.

"*Ca.* Tcb. aerophilum" is a photoheterotroph that has a photosynthetic apparatus that is nearly identical to those of GSB, and it has an incomplete reverse TCA cycle because of the absence of ATP-dependent citrate lyase (Liu, Klatt et al., 2012). *Ign. album* is a heterotroph that completely lacks genes for Chl biosynthesis and photosynthetic RCs and furthermore lacks the reverse TCA cycle, or at least has an unconventional cycle (Liu, Frigaard et al., 2012). Both organisms appear to be unable to oxidise sulfide, grow under oxic conditions, and apparently lack the ability to synthesize some amino acids. Finally, unlike any other member of the phylum Chlorobi, *Ign. album* has a complete set of genes to produce flagella, which suggest that it is capable (or was recently capable) of swimming motility.

Phylogenetic analyses of 16S rRNA genes of cultured and environmental organisms belonging to phylum Chlorobi indicate that the collection of cultured GSB only represents a late-diverging, monophyletic group, in addition to several class-level lineages with few or no cultured representatives (Iino et al., 2010). "*Ca.* Tcb. aerophilum" and *Ign. album* represent early-diverging new lineages of the phylum Chlorobi at the family and class levels, respectively (Liu, Frigaard et al., 2012; Liu, Klatt et al., 2012). Furthermore, *Chp. thalassium* and its uncultured relatives should actually be regarded as a family-level lineage, *Chloroherpetonaceae,* that is distinct from the family, *Chlorobiaceae,* which houses the other three genera of GSB (Liu, Frigaard et al., 2012; Fig. 4.6).

Combined with their phylogenetic relationships, comparisons of the genomes of these newly described organisms carry several implications for the evolution of GSB and the phylum Chlorobi. Only very minor differences are observed for genes encoding the photosynthetic RCs, FMO, the (B)Chl biosynthesis pathway, and chlorosomes. The genes for the type-1 RCs (*pscA, pscB, pscC,* and *pscD*) are universally present in all chlorophototrophic members of the phylum. Furthermore, most of the genes for enzymes of (B)Chl biosynthesis and chlorosome envelope proteins are conserved as well, although only 4 (CsmA, CsmC, CsmI, and CsmX) of 10 chlorosome envelope proteins found in *Cba. tepidum* (Frigaard, Li et al., 2004), were found in "*Ca.* Tcb. aerophilum" CsmX might be able to substitute for CsmJ (Li et al., 2012), and mutants lacking CsmB, CsmF, and CsmH, or lacking CsmC, CsmD, CsmE and CsmH produce functional chlorosomes (Li & Bryant, 2009). Therefore, these four proteins might be sufficient to produce functional chlorosomes. These observations suggest that the genes encoding these proteins and enzymes were probably vertically transmitted for organisms belonging to the phylum Chlorobi.

Dramatic changes have occurred for electron transfer complexes for members of this phylum. GSB that live in anoxic conditions use Cyt *b*-Rieske complex to mediate electron transfer between menaquinol and Cyt *c*. However, *Ign. album* uses alternative complex III to accomplish the same function (Liu, Frigaard et al., 2012), while "*Ca.* Tcb. aerophilum" has both complexes (Liu, Klatt et al., 2012). The presence of alternative complex III is probably linked to the need for aerobic respiration. Phylogenetic analyses of the Cyt *b*-Rieske complex and alternative complex III suggested that they were probably vertically inherited as well (Liu, Frigaard et al., 2012; Liu, Klatt et al., 2012). Interestingly, phototrophic Chloroflexi also have alternative complex III. This is another evolutionary connection between members of these two phyla in addition to other phylogenetic evidence from (B) Chl biosynthesis (see Section 9). Terminal oxidases also differed between these organisms according to their life styles. The anaerobic GSB strains have high-affinity Cyt *bd*-quinol oxidases and/or bb_3-type Cyt *c* oxidases as protective mechanisms against O_2. The aerobic "*Ca.* Tcb. aerophilum" has an aa_3-type Cyt *c* oxidase for respiration. However, the metabolic versatile

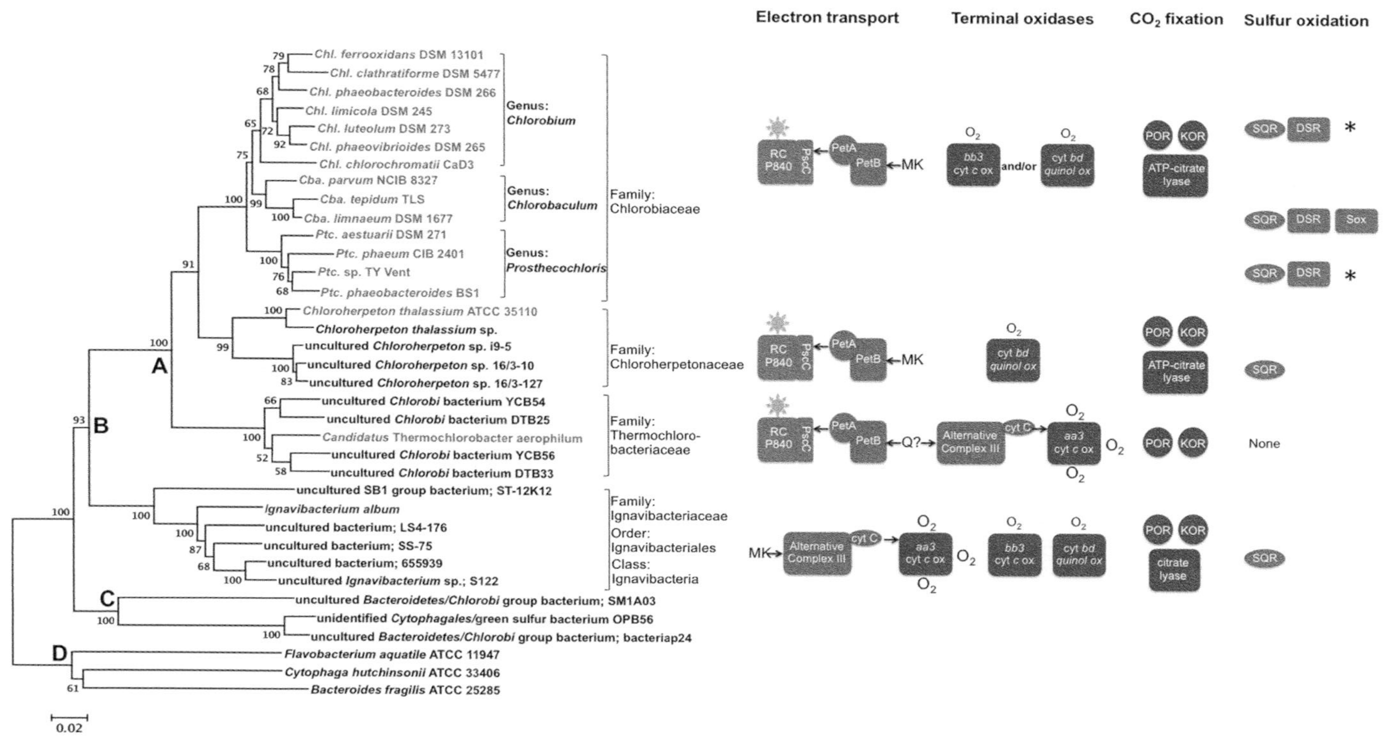

Figure 4.6 ***Metabolic differences among organisms of the phylum* Chlorobi *in the context of phylogeny.*** The phylogeny of organisms belonging to the phylum *Chlorobi* is based on a neighbour-joining tree of 16S rRNA sequences with Jukes–Cantor correction. Statistical support among 1000 bootstrap-sampled trees is shown. The bar denotes 0.02 changes per nucleotide. This phylogeny is also generally supported by whole-genome analyses and trees derived from concatenated, shared proteins. The tree is adapted from Liu, Frigaard et al. (2012). Organisms whose (meta)genomes are sequenced are labelled in colour. Green, phototrophic organisms; blue, non-phototrophic organism. Branch point **A** denotes the class Chlorobia, order Chlorobiales; **B**, phylum Chlorobi, including the class Ignavibacteria; **C**, Bacteroidetes/Chlorobi, a group of uncertain affiliation; **D**, phylum Bacteroidetes. MK, menaquinone; *Q?*, an undefined quinone; cyt, cytochrome; ox, oxidase; PetAB, cytochrome *b*-Rieske complex. Star (*) indicate the presence of the Sox enzyme system in some but not all strains. (For interpretation of the references to colour in this figure legend, the reader is referred to the online version of this book.)

Ign. album has two Cyt *bd*-type quinol oxidases and both bb_3- and aa_3-type Cyt oxidases (Liu, Frigaard et al., 2012; Liu, Klatt et al., 2012). This is another example that illustrates how members of the Chlorobi have adapted to different environments by acquiring different electron transfer complexes.

Extant GSB have adapted to live in stable, anoxic environments, while "*Ca.* Tcb. aerophilum" and *Ign. album* apparently can survive in both oxic and anoxic conditions. GSB may have lost genes related to environmental adaptation for this reason, although in general they have retained some genes for protection from oxygen (Li et al., 2009). *Ign. album* has a nearly complete set of genes for the production of flagella and for chemotaxis, and some of these chemotaxis genes also occur in "*Ca.* Tcb. aerophilum" and *Chp. thalassium*. *Ign. album* also has many more regulatory genes compared to GSB (Liu, Frigaard et al., 2012).

GSB have presumably acquired some functions that enabled autotrophic growth. One of the key acquisitions was ATP-dependent citrate lyase. It seems that other key reverse TCA cycle enzymes, such as KOR and POR, preceded the acquisition of ATP-dependent citrate lyase in Chlorobi (Section 3); ancestors of modern Chlorobi could probably perform some mixotrophic CO_2 fixation like "*Ca.* Tcb. aerophilum" and *Ign. album* using these enzymes. ATP-dependent citrate lyase and a complete reverse TCA cycle would have enabled GSB to become independent of organic carbon sources, which could have become limiting in the environments where these organisms now occur. An equally important gain of function would have been the ability to oxidise sulfur to sulfite. Only two electrons are available for CO_2 reduction when sulfide is oxidised to polysulfide (sulfur), but complete oxidation of sulfide to sulfate provides eight electrons. These extra electrons are critical for both CO_2 fixation and protonmotive force generation for GSB that live in energy-limited environments.

8. GRANICK HYPOTHESIS

Several hypotheses to explain the evolution of complex biochemical pathways have been proposed. The retrograde hypothesis (Horowitz, 1945) and the patchwork hypothesis (Jensen, 1976; Ycas, 1974) are just two examples. Specifically relating to photosynthesis, the Granick hypothesis, which states that biosynthetic pathways recapitulate their evolution, has probably been discussed most frequently. This might be because Granick (1957), (1965) used haems and chlorophylls as his primary example to explain forward biochemical evolution: how simpler biological molecules such as haem appeared prior to more complex molecules like Chls. However, with a few exceptions (Mauzerall 1973, 1978, 1992; Olson, 1970, 2000), the Granick hypothesis as it specifically relates to photosynthesis has gained surprisingly little acceptance and has been 'proven' incorrect several times (e.g. Xiong, 2006; Xiong and Bauer, 2002a; Xiong, Fischer, Inoue, Nakahara, & Bauer, 2000). However, the pathway leading from Chlide *a* to BChlides *d, c,* and *e* strongly supports the Granick hypothesis (Gomez Maqueo Chew & Bryant, 2007b; Liu & Bryant, 2012). Intermediates in this pathway produce functional chlorosomes (Gomez Maqueo Chew et al., 2007;Vogl et al., 2012), yet nature clearly continued to extend this pathway by adding methylation reactions as well as the conversion of the C7 methyl group to the formyl group of BChl *f* and *e*, to achieve novel functional outcomes that helped to define new light niches and reduce the effects of competition with Chl *a* for light. If the Granick hypothesis is true for one arm of the branching pathway of (B)Chl biosynthesis, it seems logical that it should be true for all branches.

A strict interpretation of the Granick hypothesis argues that Chl *a* predated most if not all other (B)Chls. Some have argued that this is tantamount to saying that oxygen-evolving photosynthesis predated anoxygenic photosynthesis, but this is clearly not the case. There is no information available that can inform one about the type(s) of (B)Chls present in the RCs and antenna complexes of ancestral chlorophototrophs, and even among extant organisms this statement is clearly incorrect. Further, it is entirely possible – even highly likely – that ancestral organisms had RCs that used pigment complements that are different from those of extant chlorophototrophs. Experiments have shown that Chls can often replace other Chls with only modest effects on functionality, and organisms with Chls *b*, *d* and *f* are very similar to other cyanobacteria (Chen et al., 2010; Chen & Blankenship, 2010; Paulsen, 2006; Xu, Vavilin, & Vermaas, 2001). Many writers have overlooked the fact that all known chlorophototrophic members of the phyla Chlorobi and Acidobacteria synthesize Chl *a* (Sections 4 and 5); furthermore, heliobacteria synthesize the closely related compound 8-OH-Chl *a* (van de Meent et al., 1991). In fact, the only chlorophototrophs that do *not* synthesize Chl *a* are chlorophototrophic members of the Chloroflexi and Proteobacteria; however, even these organisms synthesize Chlide *a* as an intermediate in the biosynthesis of nearly all, if not all, BChls. Ancestral homodimeric type-1 RCs could easily have had electron transport chains with only Chl *a*, and just as Chl *d* and Chl *f* have appeared later in some cyanobacterial lineages, BChl *a* could have been incorporated into ancestral

RCs to take advantage of near-infrared light. Additionally, the related argument that 'less specific' enzymes may have acted on more symmetric substrates and only later gained additional specificity through gene duplication and divergence is also problematic. Often-cited examples for this idea are the light-independent protochlorophyllide (BchNBL/ChlNBL) and chlorophyllide reductases (BchXYZ). However, the substrate is not symmetric for binding or catalysis. The carboxyl group of the propionic acid chain at C17 is proposed to play a specific role in protonation of the substrate during catalysis by protochlorophyllide reductase (Reinbothe et al., 2010). Because no equivalent carboxyl group occurs on the C7 and C8 side chains, the same mechanism cannot apply to the reduction of the B ring.

In a previous review of (B)Chl biosynthesis in bacteria, it was argued that Chlide *a* is the central 'hub' compound of (B)Chl biosynthesis, analogous to the position of pyruvate in the pathways of fermentation (Gomez Maqueo Chew & Bryant, 2007b). An implication of this argument was that the Granick hypothesis was correct and that all Chls and BChls can be synthesized by pathways leading from Chlide *a*. The driving force for such forward evolution would have been competition for light, which clearly is responsible for defining communities in which chlorophototrophs are prominent members today (Stomp, Huisman, Stal, & Matthijs, 2007). If one assumes that photosynthesis became highly successful with the invention of Chl *a* by some ancestral chlorophototroph, there would have been competition for light as soon as there were two cells using Chl *a* as the primary photopigment. Arguments for 'why Chl *a*' have been discussed by others (e.g. Björn, Papageorgiou, Blankenship, & Govindjee, 2009; Mauzerall, 1973), but one of the most important is that Chl *a* is actually 'black' rather than 'green' – it absorbs all visible wavelengths of light. It would have been necessary for organisms to produce pigments that do not directly compete with Chl *a* for light absorption. Competition for a limiting energy source is a very powerful selection pressure, but the invention of oxygenic photosynthesis by cyanobacteria would have eventually made this competition even more demanding. Anoxygenic chlorophototrophs would have necessarily retreated to anoxic refugia, where irradiance levels were probably even more limiting, and the selection pressure to diversify light harvesting correspondingly even greater. Diversification of Chls and BChls would have resulted to cope with these selection pressures. With the exception of Chl *b*, divinyl-Chl *a* and divinyl-Chl *b*, and perhaps BChl *e* and *f*, all of which have enhanced absorption in the blue, modifications to produce pigments that absorb light wavelengths longer than the Q_y absorption of Chl *a* in the far red and near infrared were the solutions most often adopted by nature (Chl *d*, Chl *f*, BChl *d*, BChl *c*, BChl *e*, BChl *a*, BChl *g*, and BChl *b*),

Although the 'single hub' concept was useful in providing a framework for discussion of (B)Chl biosynthesis, we now propose that there are in fact two hub compounds: Chlide *a* and 3,8-divinyl-Chlide *a*. Note that these two hubs are connected by one enzyme that is not essential (Gomez Maqueo Chew & Bryant, 2007b) and for which multiple enzymes exist (Liu & Bryant, 2012). Figure 4.7 shows a unified scheme for (B)Chl biosynthesis that explains the biosynthetic origins of all known Chls and BChls to date. Several of the branches share the same or functionally similar enzymes. As suggested by Gomez Maqueo Chew and Bryant (2007b), the majority of Chls and BChls can be synthesized from Chlide *a*: Chls *a*, *b*, *d*, and *f*, and BChls *a*, *c*, *d*, *e* and *f*. We propose that five additional Chls and BChls can be synthesized from 3,8-divinyl-Chlide *a*: divinyl-Chl *a*, divinyl-Chl *b*, 8^1-OH-Chl *a*, BChl *g*, and BChl *b*. An interesting and testable feature of

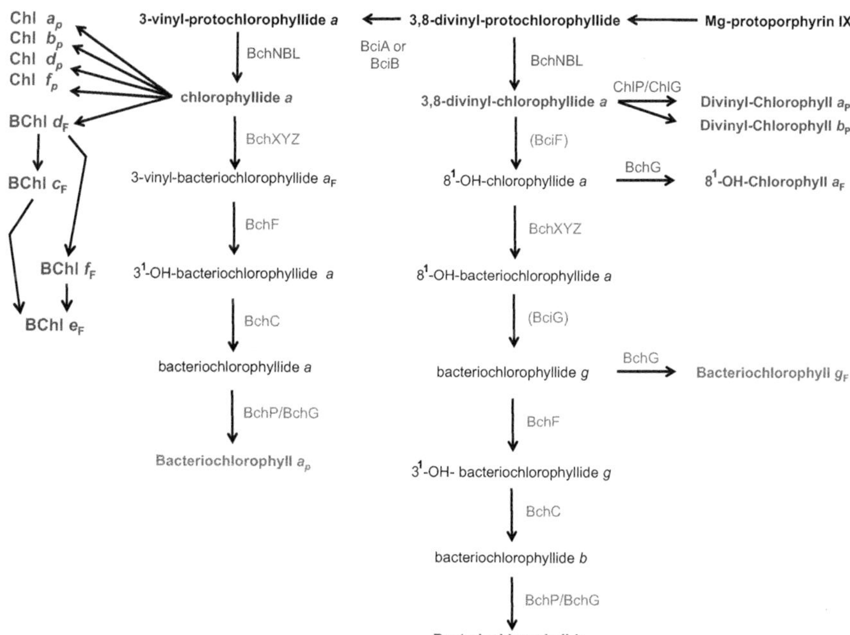

Figure 4.7 ***Scheme showing the pathways leading to the 14 major (B)Chls currently known to occur in chlorophototrophic bacteria.*** The compounds in red, Chlide *a* and 3,8-divinyl-Chlide *a*, are proposed to be the two central, 'hub' compounds that serve as the precursors for the 14 end-product (B)Chls. The pathway leading from 3,8-divinyl-Chlide to BChl *b* is proposed in the text; this pathway and the enzymes 'BciF' and 'BciG' have not yet been demonstrated (see Section 8).

this proposal is that BChlide *g*, and not BChlide *a*, is an intermediate in the synthesis of BChl *b*. This scheme proposes a specific route for the synthesis of 8^1-OH-Chl *a*, an essential Chl that is the primary acceptor of electrons in the heliobacterial RC (van de Meent et al., 1991). Note that BchXYZ, BchF, BchC, BchP, and BchG are shared along with BchNBL in the pathways leading to BChl *a* and BChl *b*. We propose that only two new enzymes are required: a hydratase for the C8 vinyl group (BciF) and a dehydratase (BciG) that acts subsequent to reduction of the B ring by Chlide reductase to produce the ethylidene side chain that is common to BChl *g* and BChl *b* at the C8 carbon. This scheme further provides a rationale for why the *bchXYZ* genes for Chlide reductase are found in heliobacterial genomes. The B ring must first be reduced after hydration of the 8-vinyl group to provide the proton that can be eliminated by dehydration to produce the ethylidene side chain at C8. Finally, hydration of the 8-vinyl group is a reaction that is analogous to the hydration reaction catalysed by BchF or BchV that occurs with the C3 vinyl group in other arms of the pathway.

A number of reactions in the scheme in Fig. 4.7 remain to be validated by identification of the necessary genes and enzymes. In addition to the proposed enzymes, BciF and BciG, mentioned above, the enzymes responsible for BChl *e* biosynthesis from BChl *c* have not yet identified (but see Section 4), and the enzymes responsible for the synthesis of Chlide *d* and Chlide *f* from Chlide *a* are likewise currently unknown. Furthermore, it is known that *Roseiflexus* spp. and perhaps other organisms lack BciA and BciA and thus must have an alternative 8-vinyl reductase. However, the broad implications of the relationships shown in Fig. 4.7 already provide strong support for the Granick hypothesis.

9. THOUGHTS ON EVOLUTION OF PHOTOSYNTHESIS

The ability to use light as an energy source, i.e. phototrophy, has evolved twice and occurs by using distinctively different mechanisms. Organisms that synthesize bacteriorhodopsin or proteorhodopsin, i.e. retinalophototrophs (Bryant and Frigaard, 2006), can directly couple light absorption and proton translocation through light-induced conformation changes of the protein that are driven by photo-isomerization and dark relaxation of the retinal chromophore. Although there appears to be no a priori reason to preclude coupling of this mechanism of ATP production to carbon dioxide fixation, no example of an organism that can do so has been reported (Bryant and Frigaard, 2006). Possible reasons of why such organisms may not exist include the low absorption cross-section of retinal-based systems or the inability of retinal-based systems to use light to generate low-potential reductants. Chlorophototrophs, organisms that produce energy using (B)Chl-dependent RCs, have evolved to overcome both of these limitations (Bryant and Frigaard, 2006). The invention of photosynthesis was one of the most important events in biology because it coupled a biological mode of energy production to an inexhaustible energy source. Energy and nutritional independence resulted from coupling solar energy transduction to carbon dioxide and nitrogen reduction, and the eventual development of oxygenic photosynthesis dramatically altered the trajectory of life and evolution on Earth.

Because photochemical RCs are the essential engines of (B)Chl-based phototrophy, the evolutionary origins of Chls and RCs are tightly linked. Many suggestions have been proposed to explain how and why RCs might have evolved. One plausible possibility is that primitive RCs were porphyrin- or haem-containing membrane proteins that evolved to perform a light-induced electron transfer from some abundant donor (e.g. Fe^{+2}) to a suitable biological acceptor (e.g. an Fe/S cluster) across a biological membrane (Olson & Pierson, 1987; Olson, 1999). Examples of photoferrotrophs, organisms that can photooxidise ferrous iron, still exist on Earth (Heising et al., 1999; Widdel et al., 1993). Alternatively, RCs might have evolved from proteins used for sensing near-infrared radiation from thermal vents (Nisbet, Cann, & van Dover, 1995). Biliverdin-containing bacteriophytochromes exist and are used to sense near-infrared light (Giraud and Verméglio, 2008). Although there is presently no experimental evidence to support this proposal, still another possibility is that ancestral RCs provided photoprotection from ultraviolet radiation by absorption and dissipation of the energy (Mulkidjanian & Junge, 1997). Finally, others have suggested that RCs are descended from a respiratory protein, e.g. cytochrome *b*; presumably, light might have facilitated an electron transfer reaction that could have been improved by selection (Meyer, 1994; Xiong and Bauer, 2002a, 2002b).

As noted above, two families of RCs, designated type-1 and type-2, are currently known (Golbeck, 1993). All members of both RC families contain a bifurcating chain of six (B)Chls and/or (B)Phe. These pigments are arranged as three pairs of pigment molecules and constitute a donor–acceptor cassette. It has recently become less certain which specific (B)Chl(s) are primary versus secondary donors (Müller et al., 2010), and some authors have referred to the 'special set' of four pigments as the primary donor rather than the 'special pair' (Hohmann-Marriott & Blankenship, 2011). These electron transport cofactors are bound at the interface of two polypeptide chains that can be homodimers or heterodimers. Structural similarities of the cofactor and polypeptide arrangements suggest that all photochemical RCs are descended from a common ancestor and that this family expanded by gene duplication and divergence (Mix, Haig, & Cavanaugh, 2005; Sadekar et al., 2006; Schubert et al., 1998; Fig. 4.10 and text below). Although there is insufficient sequence similarity between type-1 and type-2 RCs to draw any reliable conclusions about phylogeny, a combination of structural comparisons and more limited sequence comparisons within families suggest that several gene duplication and divergence events have occurred to account for the four extant families of RCs: (1) homodimeric type-1 RCs (PscA or PshA); (2) heterodimeric type-1 RCs (i.e. PS I, PsaA and PsaB); (3) heterodimeric bacterial type-2 RCs (PufL and PufM); (4) heterodimeric, oxygen-evolving type-2 RCs (i.e. PS II; PsbA and PsbD).

Phylogenetic inference methods have been employed in attempts to gain insights into the origins of photosynthesis. However, because there is negligible sequence similarity between the electron transfer domains of type-1 and type-2 RCs, quantitative measures of structural similarities and differences among RCs were used to build phylogenetic trees (Sadekar et al., 2006). This method led to the conclusion that two independent gene duplication events produced the two types of type-2 RCs: the heterodimeric Proteobacteria/Chloroflexi-type RC, and the heterodimeric PS II-type RC. A third gene duplication event led to the heterodimeric type-1 RC of cyanobacteria (Hohmann-Marriott & Blankenship, 2011; Sadekar et al., 2006). At least a fourth gene duplication might also have occurred. This duplication would have ultimately led to the creation of two different RC types in one cellular compartment. The innovation–amplification–divergence model (Näsvall, Sun, Roth, & Andersson, 2012), as well as the idea that partial duplications of a gene can promote the emergence of new functions (Katju, 2012), suggest that it may not have been difficult for cells producing different RCs to evolve, provided these RCs performed specialized functions (Allen and Martin, 2007). Although lateral gene transfer could perhaps have accounted for the acquisition of a second type of RC (the fusion hypothesis; Olson & Blankenship, 2004), fusion of cells harbouring different RC types is highly unlikely because cell fusion and endocytosis are uniquely eukaryotic capabilities.

Whole-genome phylogenetic methods have met with very limited success in defining the origins of photosynthesis, because the number of genes that are shared across all chlorophototrophs is small, and this method therefore tends to reflect the phylogenetic relationships of the housekeeping properties of the organism rather than photosynthesis (Mulkidjanian et al., 2006; Raymond, Zhaxybayeva, Gogarten, Gerdes, & Blankenship, 2002). Genes for carotenoid biosynthesis apparently can be readily exchanged by horizontal gene transfer (Phadwal, 2005) and are nonessential components of RCs under anoxic conditions (Frigaard, Maresca, Yunker, Jones, & Bryant, 2004). Thus, the enzymes of (B)Chl biosynthesis may be the only reliable markers for the origins of photosynthesis, and even these must be considered very carefully in the context of the RCs and phylogenetic/taxonomic affiliations of the organisms in which these genes occur.

Figure 4.8 shows a phylogenetic tree for the concatenated sequences of the ATP-dependent Mg-Proto IX chelatase, ChlHID/BchHID, the enzyme that catalyses the first committed step in the synthesis of (B)Chls. Three pairs of chlorophototrophic taxa are strongly supported by this tree, but the branching order of these three groups cannot be determined from these data. These three pairs of sister taxa are: (1) Chloroflexi and Chlorobi; (2) Acidobacteria and Proteobacteria; and (3) Firmicutes (heliobacteria) and Cyanobacteria. Interestingly, within each of these three subgroups, organisms having homodimeric type-1 RCs diverge early from clades of organisms that have heterodimeric RCs in the other phylum of the pair. This is in agreement with suggestions that homodimeric RCs were the ancestral type of RCs (Mix et al., 2005; Sadekar et al., 2006). It is also interesting to note that these early-diverging organisms (Chlorobi, green FAPs (e.g. *Chloroflexus* spp.), and "*Ca.* Cab. thermophilum") have chlorosomes for light harvesting. Heliobacteria (Firmicutes), which do not have light-harvesting complexes, are the exception. Considering that members of the Chlorobi have homodimeric type-1 RCs and Chloroflexi have heterodimeric type-2 RCs, the sister relationship of these two clades of organisms is surprising. This observation suggests that gene gain by horizontal gene transfer or gene loss (Fig. 4.10) probably occurred in one of these lineages. The early divergence of "*Ca.* Cab. thermophilum" in the Acidobacteria–Proteobacteria pair is also consistent with the 'type-1 homodimeric RCs early' hypothesis. Moreover, this relationship of these two phyla reflects their sister-group relationship of Acidobacteria and Proteobacteria observed in a global phylogenetic analysis of 31 orthologous, concatenated proteins (Ciccarelli et al., 2006).

Figure 4.9 shows a similar phylogenetic analysis of concatenated sequences encoding light-independent (Dark-operative) PChlide reductase (BchlL/ChlL, BchN/ChlN, and BchB/ChlB = DPOR) and Chlide reductase (BchX, BchY, and BchZ = COR) (Section 4). This analysis used

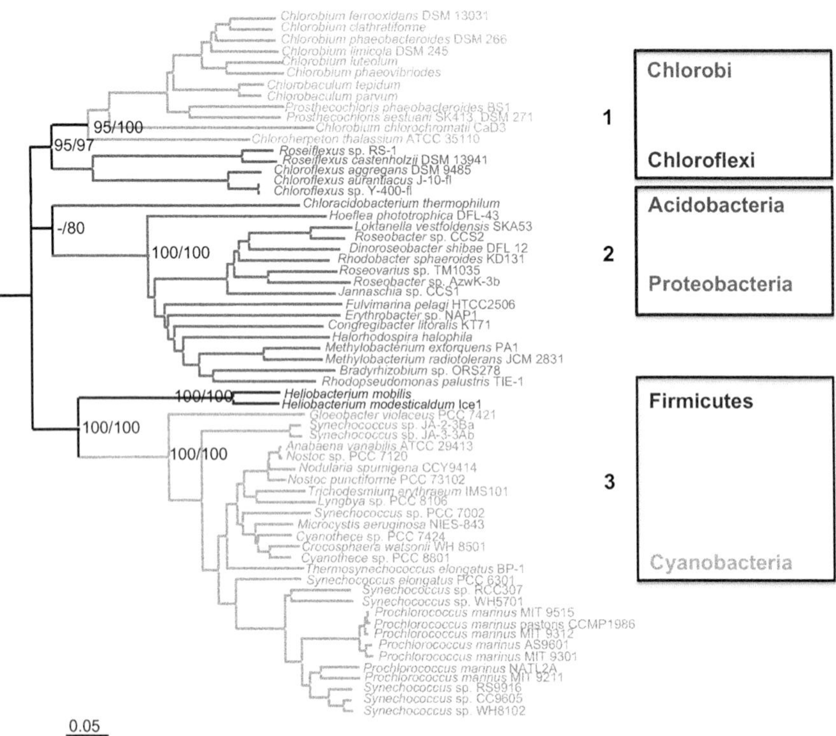

Figure 4.8 ***Phylogenetic relationships among photosynthetic bacteria based on the concatenated BchIDH/ChlIDH protein sequences.*** Mg-chelatase family proteins (NP_614634, NP_613479) from *Methanopyrus kandleri* AV19 (NC_003551) were used as the outgroup. Sets of homologous protein sequences were aligned using MUSCLE (Edgar 2004) and conserved regions of the resulting multiple sequence alignments were identified with GBlocks (Castresana 2000). The neighbour-joining (NJ) method implemented in MEGA 4.0 (Tamura, Dudley, Nei, & Kumar, 2007) and the maximum likelihood (ML) method implemented in PhyML 2.4.4 (Guindon & Gascuel, 2003) with 500 replicates were used to construct phylogenetic tree. Numbers at nodes indicate the NJ and ML bootstrap values, respectively; a dash indicates a branch with less than 50% support. The six phyla/kingdoms containing chlorophototrophic members are indicated at the right and the names are colour coded. Three pairs of phyla are boxed and numbered (see Section 9). (For interpretation of the references to colour in this figure legend, the reader is referred to the online version of this book.)

parologous NifH, NifD, and NifK sequences of nitrogenase as outgroup. The inclusion of *Prochlorococcus* spp. and marine *Synechococcus* spp. sequences for PChlide reductase in the clade that includes all proteobacterial sequences is a very clear evidence for a horizontal gene transfer event. However, other than this one anomaly, the same pairs of taxa are observed as in Fig. 4.8, but this analysis provides at least some hints about the relationships among these three pairs. The upper part of the tree shows the phylogenetic

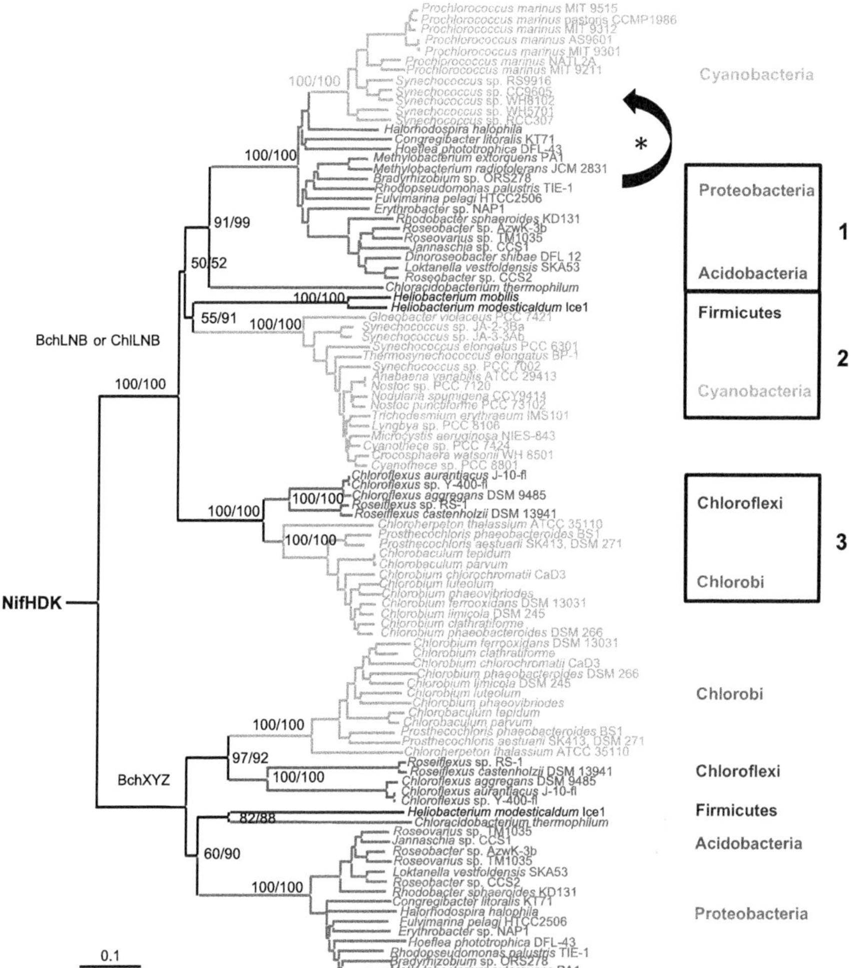

Figure 4.9 ***Phylogenetic relationships among photosynthetic bacteria based on the concatenated BchLNB/ChlLNB/BchXYZ protein sequences.*** Nitrogenase proteins (NifHDK) from *Methanococcus maripaludis* C5 (NC_009135) were used as outgroup. Sets of homologous protein sequences were aligned using MUSCLE (Edgar 2004) and conserved regions of the resulting multiple sequence alignments were identified with GBlocks (Castresana 2000). The neighbour-joining (NJ) method implemented in MEGA 4.0 (Tamura et al., 2007) and the maximum likelihood (ML) method implemented in PhyML 2.4.4 (Guindon & Gascuel, 2003) with 500 replicates were used to construct phylogenetic tree. Numbers at nodes indicate the NJ and ML bootstrap values, respectively. The six phyla/kingdoms containing chlorophototrophic members are indicated at the right and the names are colour coded. Three pairs of phyla are boxed and numbered. The arrow and asterisk indicate a probable lateral gene transfer event between a proteobacterium and an ancestor of the *Prochlorococcus*/marine *Synechococcus* clade (see Section 9). (For interpretation of the references to colour in this figure legend, the reader is referred to the online version of this book.)

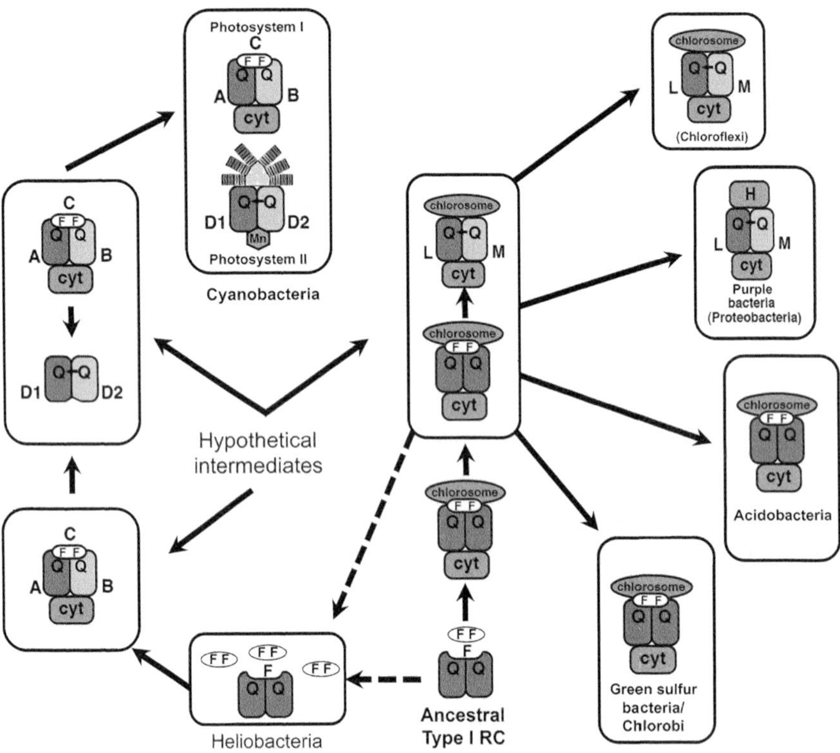

Figure 4.10 ***Scheme showing evolution of chlorophototrophs based on phylogeny of photochemical RCs.*** The scheme is an adaptation of a proposal by Cavalier-Smith (2006) but is based on homodimeric type-1 RCs as the ancestral type. This scheme includes four gene duplication events to account for the different structure classes of RCs currently recognized (Sadekar et al., 2006), and includes several organisms as hypothetical intermediates (see Section 9).

relationships among PChlide reductase sequences and contains three subgroups. The Chloroflexi and Chlorobi are the earliest diverging lineage, and the Firmicutes (heliobacteria) and Cyanobacteria diverge next. "*Ca.* Cab. thermophilum" (Acidobacteria) and Proteobacteria diverge last, and as seen previously for Mg-chelatase, the acidobacterium "*Ca.* Cab. thermophilum" diverges earliest within this clade. The lower portion of the tree for Chlide reductase is quite similar to the upper part for PChlide reductase, except for the position of *Heliobacterium modesticaldum*. In the absence of cyanobacteria sequences (cyanobacteria do not synthesize BChl *a* and thus lack BchX, BchY, and BchZ), the heliobacterial sequence groups with "*Ca.* Cab. thermophilum" probably due to long-branch attraction. When one considers the distribution of RCs among these taxa, it is again clear that each pair of phyla includes earliest diverging groups that have homodimeric type-1

RCs, and as suggested before, chlorosomes occur in the earliest diverging members of two of the pairs of taxa. Only the Firmicutes–Cyanobacteria pair lacks any members with chlorosomes, so chlorosomes may have evolved as one of the earliest types of antenna systems.

Figure 4.10 shows a speculative scheme incorporating results from some of the ideas discussed above: homodimeric type-1 RCs first, chlorosomes early, gene duplications and divergence, and importantly hypothetical intermediates with duplicated RCs combined with lineage-specific gene loss. The model is an adaptation of a figure that appeared in a paper by Cavalier-Smith (2006). In this model, a simple, homodimeric type-1 RC, similar to that found now in extant heliobacteria, was the ancestor of all RCs, and it gained the ability to interact with chlorosomes at an early stage. A hypothetical intermediate organism is shown that contains a homodimeric RC, which might have been specialized for the production of strong reductants, as well as a heterodimeric, bacterial type-2 RC, similar to that found today in members of the Chloroflexi and which might have been specialized for reducing quinones for cyclic electron transport. Lineage-specific retention of one or the other of these RC types would explain the close relationship in (B)Chl biosynthesis between members of the Chloroflexi and Chlorobi, and would also explain the relationship among the PufL and PufM polypeptide sequences of Chloroflexi (earlier diverging) and Proteobacteria (later diverging). Phylogenetic trees of both RCs and enzymes of Chl biosynthesis suggest that heliobacteria and cyanobacteria are sister clades (see above), although heliobacteria have very simple homodimeric RCs. Multiple gene duplication events are required to transform homodimeric type-1 RCs into two types of heterodimeric RCs. Gene duplication to produce a heterodimeric core could have been an important innovation in allowing a bacterial 8Fe-8S protein to bind to remain permanently associated with the complex. This might have provided improved functionality and protection from oxygen and other inhibitors (Jagannathan & Golbeck, 2009a, 2009b). In a sense, the recently discovered marine 'cyanobacterium' which lacks PS II and RuBisCO but has PS I and nitrogenase, at least superficially resembles this hypothetical intermediate (Tripp et al., 2010; Zehr et al., 2008), although it probably arose by genome reduction and gene loss. A second gene duplication and divergence could have produced a type-2 RC, which would have been the ancestor of PS II and which eventually diverged and became functionally specialized for water oxidation. An organism like this would have been the ancestor of cyanobacteria, an organism that has been called a 'protocyanobacterium' (Mulkidjanian et al., 2006). It should be noted that this scheme makes no predictions about the specific types of (B)Chls that were associated with any ancestral organism, although it can certainly be argued that all or nearly of them could have employed Chl *a* in the beginning.

Allen and Martin (2007) have suggested that gene regulation might have allowed the use of different RCs under different environmental conditions for specialised purposes, and this of course might have been the case. However, there are some obvious features of type-1 and type-2 RCs that naturally allow their use for specialized purposes. No matter which CO_2 fixation pathway is used, this process requires a reducing agent (reduced ferredoxin or NADPH) and energy in the form of ATP. Type-2 RCs are ideal for producing protonmotive force via cyclic electron transport because their product, a reduced quinone, accepts two electrons and two protons. In the context of a quinone: Cyt *c* oxidoreductase complex, vectorial translocation of protons associated with return of electrons to the oxidised special pair creates light-driven protonmotive force for ATP synthesis. On the other hand, Type-1 RCs are ideal for producing strong reductants from weaker reductants using the energy of a photon. However, the soluble electron donors and electron acceptors of type-1 RCs are one-electron carriers that do also carry protons. Although it is possible for Type-1 RCs to drive cyclic electron transport, the process requires the participation of additional proteins that can transduce electrical potential into chemical potential energy in the form of protonmotive force. If one further considers that different substrates could be preferentially oxidised by different types of RCs, although this is not a requirement, there would be a clear possibility for RC specialization, just as there is today in the RCs of GSB and purple sulfur bacteria.

A final comment is that there is currently no 'grand unification theory' for the evolution of modern anoxygenic and oxygenic phototrophs. The unfathomably long times involved, and the complexity of the processes and mechanisms that likely contributed to this evolution, currently allow far too many possibilities to produce a single plausible explanation. However, as the discovery of "*Ca.* Cab. thermophilum" illustrates, important new phototrophs are likely still being discovered, and the 'Rosetta Stone' to explain the evolution of photosynthesis may still be awaiting discovery.

ABBREVIATIONS

APS adenosine-5′-phosphosulfate
BChl bacteriochlorophyll
(B)Chl bacteriochlorophyll and/or chlorophyll
BChlide bacteriochlorophyllide

BPhe bacteriopheophytin
Ca. *Candidatus*
Cab. Chloracidobacterium
Cba. *Chlorobaculum*
Cfx. *Chloroflexus*
CBP carotenoid-binding protein
Chl chlorophyll
Chl. *Chlorobium*
Chlide chlorophyllide
COR chlorophyllide oxidoreductase
Cyt cytochrome
DPOR light-independent protochlorophyllide oxidoreductase
DSM Deutsche Sammlung von Mikroorganismen
DSR dissimilatory sulfite reductase
FAP filamentous anoxygenic phototroph
FMO Fenna–Matthews–Olson BChl *a*-binding protein
GSB green sulfur bacteria
Ign. *Ignavibacterium*
KOR 2-oxoglutarate:ferredoxin oxidoreductase
LPOR light-dependent protochlorophyllide oxidoreductase
Osc. *Oscillochloris*
P special pair
PChlide protochlorophyllide
POR pyruvate:ferredoxin oxidoreductase
Proto IX protoporphyrin IX
PS photosystem
Ptc. *Prosthecochloris*
RC reaction centre
SQR sulfide-quinone reductases
Tcb. Thermochlorobacter

ACKNOWLEDGEMENTS

The US Department of Energy Office of Energy Biosciences (DE-FG02-94ER20137), the National Science Foundation (MCB-0523100 and MCB-1021725), and the National Aeronautics and Space Administration Exobiology Program (NX09AM87G) supported research described in this article from the laboratory of D.A.B. The community sequencing program of the Joint Genome Institute of the Department of Energy provided support to D.A.B. for sequencing the genomes of many of the phototrophs mentioned in this article. The Office of Science of the U. S. Department of Energy supports the work conducted by the DOE-Joint Genome Institute under Contract No. DE-AC02-05CH11231. The authors thank their many collaborators and colleagues for helpful discussions and suggestions.

REFERENCES

Allen, J. F., & Martin, W. (2007). Out of thin air. *Nature*, *445*, 610–612.
Aoshima, M., Ishii, M., & Igarashi, Y. (2004a). A novel enzyme, citryl-CoA synthetase, catalysing the first step of the citrate cleavage reaction in *Hydrogenobacter thermophilus* TK-6. *Molecular Microbiology*, *52*, 751–761.

Aoshima, M., Ishii, M., & Igarashi, Y. (2004b). A novel enzyme, cirtyl-CoA lyase, catalysing the second step of the citrate cleavage reaction in *Hydrogenobacter thermophilus* TK-6. *Molecular Microbiology, 52*, 763–770.

Azai, C., Tsukatani, Y., Harada, J., & Oh-oka, H. (2009). Sulfur oxidation in the mutants of the photosynthetic green sulfur bacterium *Chlorobium tepidum* devoid of cytochrome *c*-554 and SoxB. *Photosynthesis Research, 100*, 57–65.

Beatty, J. T., Overmann, J., Lince, M. T., Mankske, A. K., Lang, A. S., Blankenship, R. E., et al. (2005). An obligately photosynthetic bacterial anaerobe from a deep-sea hydrothermal vent. *Proceedings of the National Academy of Sciences, USA, 102*, 9306–9310.

Björn, L. O., Papageorgiou, G. C., Blankenship, R. E., & Govindjee (2009). A viewpoint: why chlorophyll *a? Photosynthesis Research, 99*, 85–98.

Blankenship, R. E. (2010). Early evolution of photosynthesis. *Plant Physiology, 154*, 434–438.

Blankenship, R. E., Cheng, P., Causgrove, T. P., Brune, D. C., Wang, S.-H., Choh, J.-U., et al. (1993). Redox regulation of energy transfer efficiency in antennas of green photosynthetic bacteria. *Photochemistry and Photobiology, 57*, 103–107.

Blankenship, R. E., & Matsuura, K. (2003). Antenna complexes from green photosynthetic bacteria. In B. R. Green & W. W. Parson (Eds.), *Advances in photosynthesis and respiration Light-harvesting antennas* (Volume 13, pp. 195–217). Dordrecht, The Netherlands: Kluwer.

Booker, S. J. (2009). Anaerobic functionalization of unactivated C–H bonds. *Current Opinion in Chemical Biology, 13*, 58–73.

Bröcker, M. J., Schomburg, S., Heinz, D. W., Jahn, D., Schubert, W. -D., & Moser, J. (2010). Crystal structure of the nitrogenase-like dark operative protochlorophyllide oxidoreductase catalytic complex (Chln/ChlB)2. *Journal of Biological Chemistry, 285*, 27336–27345.

Bryant, D. A., Garcia Costas, A. M., Maresca, J. A., Gomez Maqueo Chew, A., Klatt, C. G., Bateson, M. M., et al. (2007). "*Ca*. Cab. thermophilum" an aerobic phototrophic acidobacterium. *Science, 317*, 523–526.

Bryant, D. A., & Frigaard, N.-U. (2006). Prokaryotic photosynthesis and phototrophy illuminated. *Trends in Microbiology, 14*, 488–496.

Bryant, D. A., Liu, Z., Li, T., Zhao, F., Garcia Costas, A. M., Klatt, C. G., et al. (2012). Comparative and functional genomics of anoxygenic green bacteria from the taxa Chlorobi, Chloroflexi, and Acidobacteria. In R. L. Burnap & W. Vermaas (Eds.), *Advances in photosynthesis and respiration Functional genomics and evolution of photosynthetic systems* (Vol. 35, pp. 47–102). New York: Springer.

Bryant, D. A., Vassilieva, E. V., Frigaard, N.-U., & Li, H. (2002). Selective protein extraction from *Chlorobium tepidum* chlorosomes using detergents. Evidence that CsmA forms multimers and binds bacteriochlorophyll *a*. *Biochemistry, 41*, 14403–14411.

Buchanan, B. B., & Arnon, D. I. (1990). A reverse KREBS cycle in photosynthesis: consensus at last. *Photosynthesis Research, 24*, 47–53.

Castresana, J. (2000). Selection of conserved blocks from multiple alignments for their use in phylogenetic analysis. *Molecular Biology and Evolution, 17*, 540–552.

Cavalier-Smith, T. (2006). Rooting the tree of life by transition analyses. *Biology Direct, 1*, 19.

Chan, L. K., Morgan-Kiss, R. M., & Hanson, T. M. (2009). Functional analysis of three sulfide:quinone oxidoreductase homologs in *Chlorobaculum tepidum*. *Journal of Bacteriology, 191*, 1026–1034.

Chen, M., & Blankenship, R. E. (2010). Expanding the solar spectrum used by photosynthesis. *Trends in Plant Sciences, 16*, 427–431.

Chen, M., Schliep, M., Willows, R. D., Cai, Z.-L., Neilan, B. A., & Scheer, H. (2010). A red-shifted chlorophyll. *Science, 329*, 1318–1319.

Ciccarelli, F. D., Doerks, T., von Mering, C., Creevey, C. J., Snel, B., & Bork, P. (2006). Toward automatic reconstruction of a highly resolved tree of life. *Science, 311*, 1283–1287.

Dostál, J., Mančal, T., Augulis, R., Vácha, F., Pšenčík, J., & Zigmantas, D. (2012). Two-dimensional electronic spectroscopy reveals ultrafast energy diffusion in chlorosomes. *Journal of the American Chemical Society, 134*, 11611–11617.

Edgar, R. C. (2004). MUSCLE: multiple sequence alignment with high accuracy and high throughput. *Nucleic Acids Research, 32*, 1792–1797.

Eisen, J. A., Nelson, K. E., Paulsen, I. T., Heidelberg, J. F., Wu, M., Dodson, R. J., et al. (2002). The complete genome sequence of the green sulfur bacterium *Chlorobium tepidum*. *Proceedings of the National Academy of Sciences, USA, 99*, 9509–9514.

Feng, X., Tang, K. H., Blankenship, R. E., & Tang, Y. J. (2010). Metabolic flux analysis of the mixotrophic metabolisms in the green sulfur bacterium *Chlorobaculum tepidum*. *Journal of Biological Chemistry, 285*, 39544–39550.

Frigaard, N.-U., & Bryant, D. A. (2006). Chlorosomes: antenna organelles in green photosynthetic bacteria. In J. M. Shively (Ed.), *Microbiology monographs Complex intracellular structures in prokaryotes* (Vol. 2, pp. 79–114). Berlin: Springer.

Frigaard, N.-U., & Bryant, D. A. (2008). Genomic insights into the sulfur metabolism of phototrophic sulfur bacteria. In R. Hell, C. Dahl, D. B. Knaff & T. Leustek (Eds.), *Advances in photosynthesis and respiration Sulfur metabolism in phototrophic organisms* (Vol. 27, pp. 343–361). Dordrecht, The Netherlands: Springer.

Frigaard, N.-U., & Dahl, C. (2009). Sulfur metabolism in phototrophic sulfur bacteria. *Advances in Microbial Physiology, 54*, 103–200.

Frigaard, N.-U., Li, H., Gomez Maqueo Chew, A., Maresca, J. A., & Bryant, D. A. (2003). *Chlorobium tepidum*: insights into the physiology and biochemistry of green sulfur bacteria from the complete genome sequence. *Photosynthesis Research, 78*, 93–117.

Frigaard, N.-U., Li, H., Milks, K. J., & Bryant, D. A. (2004). Nine mutants of *Chlorobium tepidum* each unable to synthesize a different chlorosome protein still assemble functional chlorosomes. *Journal of Bacteriology, 186*, 646–653.

Frigaard, N.-U., Maresca, J. A., Yunker, C. E., Jones, A. D., & Bryant, D. A. (2004). Genetic manipulation of carotenoid biosynthesis in the green sulfur bacterium *Chlorobium tepidum*. *Journal of Bacteriology, 186*, 5210–5220.

Frigaard, N.-U., & Matsuura, K. (1999). Oxygen uncouples light absorption by the chlorosome antenna and photosynthetic electron transfer in the green sulfur bacterium *Chlorobium tepidum*. *Biochimica Biophysica Acta, 1412*, 108–117.

Frigaard, N.-U., Matsuura, K., Hirota, M., Miller, M., & Cox, R. P. (1998). Studies of the location and function of isoprenoid quinones in chlorosome from green sulfur bacteria. *Photosynthesis Research, 58*, 81–90.

Frigaard, N.-U., Takaichi, S., Hirota, M., Shimada, K., & Matsuura, K. (1997). Quinones in chlorosomes of green sulfur bacteria and their role in the redox-dependent fluorescence studied in chlorosome-like bacteriochlorophyll *c* aggregates. *Archives of Microbiology, 167*, 343–349.

Frigaard, N.-U., Tokita, S., & Matsuura, K. (1999). Exogenous quinones inhibit photosynthetic electron transfer in *Chloroflexus aurantiacus* by specific quenching of the excited bacteriochlorophyll *c* antenna. *Biochimica Biophysica Acta, 1413*, 108–116.

Frigaard, N.-U., Voigt, G. D., & Bryant, D. A. (2002). *Chlorobium tepidum* mutant lacking bacteriochlorophyll *c* made by inactivation of the *bchK* gene, encoding bacteriochlorophyll *c* synthase. *Journal of Bacteriology, 184*, 3368–3376.

Fuchs, G. (2011). Alternative pathways of carbon dioxide fixation: insights into the early evolution of life. *Annual Review of Microbiology, 65*, 631–658.

Ganapathy, S., Oostergetel, G. T., Wawrzyniak, P. K., Reus, M., Gomez Maqueo Chew, A., Buda, F., et al. (2009). Alternating *syn–anti* bacteriochlorophylls form concentric helical nanotubes in chlorosomes. *Proceedings of the National Academy of Sciences, USA, 106*, 8525–8530.

Ganapathy, S., Reus, M., Oostergetel, G., Wawrzyniak, P. K., Tsukatani, Y., Gomez Maqueo Chew, A., et al. (2012). Self-assembly of BChl *c* in chlorosomes of the green sulfur bacterium, *Chlorobaculum tepidum*: a comparison of the *bchQR* mutant and the wild type. *Biochemistry, 51*, 4488–4498.

Garcia Costas, A. M., Liu, Z., Tomsho, L. P., Schuster, S. C., Ward, D. M., & Bryant, D. A. (2012). Complete genome of "*Ca.* Cab. thermophilum" a chlorophyll-based photoheterotroph belonging to the phylum Acidobacteria. *Environmental Microbiology, 14*, 177–190.

Garcia Costas, A. M., Tsukatani, Y., Rijpstra, W. I.C., Schouten, S., Welander, P. V., Summons, R. E., et al. (2012). Identification of the bacteriochlorophylls, carotenoids, quinones, lipids, and hopanoids of "*Ca.* Cab. thermophilum" *Journal of Bacteriology, 194*, 1158–1168.

Garcia Costas, A. M., Tsukatani, Y., Romberger, S. P., Oostergetel, G., Boekema, E., Golbeck, J. H., et al. (2011). Ultrastructural analysis and identification of envelope proteins of "*Ca.* Cab. thermophilum" chlorosomes. *Journal of Bacteriology, 193*, 6701–6711.

Giraud, E., & Verméglio, A. (2008). Bacteriophytochromes in anoxygenic photosynthetic bacteria. *Photosynthesis Research, 97*, 141–153.

Gloe, A., & Risch, N. (1978). Bacteriochlorophyll c_S, a new bacteriochlorophyll from *Chloroflexus aurantiacus*. *Archives of Microbiology, 118*, 153–156.

Golbeck, J. H. (1993). Shared thematic elements in photochemical reaction centers. *Proceedings of the National Academy of Sciences, USA, 90*, 1642–1646.

Gomez Maqueo Chew, A. (2007). *Elucidation of the bacteriochlorophyll c biosynthesis pathway in green sulfur bacterium Chlorobium tepidum.* PhD thesis, The Pennsylvania State University: University Park, PA.

Gomez Maqueo Chew, A., & Bryant, D. A. (2007a). Characterization of a plant-like protochlorophyllide *a* divinyl reductase in green sulfur bacteria. *Journal of Biological Chemistry, 282*, 2967–2975.

Gomez Maqueo Chew, A., & Bryant, D. A. (2007b). Chlorophyll biosynthesis in bacteria: the origins of structural and functional diversity. *Annual Reviews of Microbiology, 61*, 113–129.

Gomez Maqueo Chew, A., Frigaard, N.-U., & Bryant, D. A. (2007). Bacteriochlorophyllide c C-82 and C-121 methyltransferases are essential for adaptation to low light in *Chlorobaculum tepidum*. *Journal of Bacteriology, 189*, 6176–6184.

Gomez Maqueo Chew, A., Frigaard, N.-U., & Bryant, D. A. (2009). Mutational analysis of three *bchH* paralogs in (bacterio-)chlorophyll biosynthesis in *Chlorobaculum tepidum*. *Photosynthesis Research, 101*, 21–34.

Gough, S. P., Petersen, B. O., & Duus, J. Ø. (2000). Anaerobic chlorophyll isocyclic ring formation in *Rhodobacter capsulatus* requires a cobalamin cofactor. *Proceedings of the National Academy of Sciences, USA, 97*, 6908–6913.

Granick, S. (1957). Speculations on the origins and evolution of photosynthesis. *Annals of the New York Academy of Sciences, 69*, 292–308.

Granick, S. (1965). Evolution of heme and chlorophyll. In V. Bryson & H. G. Vogel (Eds.), *Evolving genes and proteins* (pp. 67–68). New York, NY: Academic Press.

Gregersen, L. H., Bryant, D. A., & Frigaard, N.-U. (2011). Mechanisms and evolution of oxidative sulfur metabolism in green sulfur bacteria. *Frontiers in Microbiology, 2*, 116.

Grotjohann, I., & Fromme, P. (2005). Structure of cyanobacterial photosystem I. *Photosynthesis Research, 85*, 51–72.

Guindon, S., & Gascuel, O. (2003). A simple, fast, and accurate algorithm to estimate large phylogenies by maximum likelihood. *System Biology, 52*, 696–704.

Hager-Braun, C., Xie, D. L., Jarosch, U., Herold, E., Büttner, M., Zimmermann, R., et al. (1995). Stable photobleaching of P840 in *Chlorobium* reaction center preparations: presence of the 42-kDa bacteriochlorophyll *a* protein and a 17-kDa polypeptide. *Biochemistry, 34*, 9617–9624.

Hanada, S., & Pierson, B. K. (2006). The family *Chloroflexaceae*. In S. Falkow, E. Rosenberg, K.-H. Schleifer & E. Stackebrandt (Eds.), *The prokaryotes* (Vol. 7, pp. 815–842). Berlin: Springer.

Harada, J., Mizoguchi, T., Tsukatani, Y., Noguchi, M., & Tamiaki, H. (2012). A seventh bacterial chlorophyll driving a large light-harvesting antenna. *Scientific Reports, 2*, 671.

Hauska, G., Schödl, T., Remigy, H., & Tsiotis, G. (2001). The reaction center of green sulfur bacteria. *Biochimica Biophysica Acta, 1507*, 260–277.

Heinnickel, M., & Golbeck, J. H. (2007). Heliobacterial photosynthesis. *Photosynthesis Research, 92*, 35–53.

Heising, S., Richter, L., Ludwig, W., & Schink, B. (1999). *Chlorobium ferrooxidans* sp. nov., a phototrophic green sulfur bacterium that oxidizes ferrous iron in coculture with a "*Geospirillum*" sp. strain. *Archives of Microbiology, 172*, 116–124.

Higuchi, M., Hirano, Y., Kimura, Y., Oh-oka, H., Miki, K., & Wang, Z.-Y. (2009). Overexpression, characterization, and crystallization of the functional domain of cytochrome c_z from *Chlorobium tepidum*. *Photosynthesis Research, 102*, 77–84.

Hohmann-Marriott, M. F., & Blankenship, R. E. (2007). Variable fluorescence in green sulfur bacteria. *Biochimica Biophysica Acta, 1767*, 106–113.

Hohmann-Marriott, M. F., & Blankenship, R. E. (2011). Evolution of photosynthesis. *Annual Review of Plant Biology, 62*, 515–548.

Holkenbrink, C., Barbas, S. O., Mellerup, A., Otaki, H., & Frigaard, N.-U. (2011). Sulfur globule oxidation in green sulfur bacteria is dependent on the dissimilatory sulfite reductase system. *Microbiology, 157*, 1229–1239.

Horowitz, N. H. (1945). On the evolution of biochemical syntheses. *Proceedings of the National Academy of Sciences, USA, 31*, 153–157.

Huang, R. Y.-C., Wen, J., Blankenship, R. E., & Gross, M. L. (2012). Hydrogen-deuterium exchange mass spectrometry reveals the interaction of Fenna–Matthews–Olson protein and chlorosome CsmA protein. *Biochemistry, 51*, 187–193.

Hügler, M., Huber, H., Molyneaux, S. J., Vetriani, C., & Sievert, S. M. (2007). Autotrophic CO_2 fixation via the reductive tricarboxylic acid cycle in different lineages within the phylum Aquificae: evidence for two ways of citrate cleavage. *Environmental Microbiology, 9*, 81–92.

Hügler, M., & Sievert, S. M. (2011). Beyond the Calvin cycle: autotrophic carbon fixation in the ocean. *Annual Review of Marine Science, 3*, 261–289.

Huster, M. S., & Smith, K. M. (1990). Biosynthetic studies of substituent homologation in bacteriochlorophylls *c* and *d*. *Biochemistry, 29*, 4348–4355.

Iino, T., Mori, K., Uchino, Y., Nakagawa, T., Harayama, S., & Suzuki, K. (2010). *Ignavibacterium album* gen. nov., sp. nov., a moderately thermophilic anaerobic bacterium isolated from microbial mats at a terrestrial hot spring and proposal of *Ignavibacteria* classis nov., for a novel lineage at the periphery of green sulfur bacteria. *International Journal of Systematic and Evolutionary Microbiology, 60*, 1376–1382.

Ito, H., Yokono, M., Tanaka, R., & Tanaka, A. (2008). Identification of a novel vinyl reductase gene essential for the biosynthesis of monovinyl chlorophyll in *Synechocystis* sp. PCC6803. *Journal of Biological Chemistry, 283*, 9002–9011.

Ivanovsky, R. N., Fal, Y. I., Berg, I. A., Ugolkova, N. V., Karsilnikova, E. N., Keppen, O. I., et al. (1999). Evidence for the presence of the reductive pentose phosphate cycle in a filamentous anoxygenic photosynthetic bacterium *Oscillochloris triochoides* strain DG-6. *Microbiology, 145*, 1743–1748.

Jagannathan, B., & Golbeck, J. H. (2008). Unifying principles in homodimeric type I photosynthetic reaction centers: properties of PscB and the F_A, F_B and F_x iron clusters in green sulfur bacteria. *Biochimica Biophysica Acta, 1777*, 1371–1380.

Jagannathan, B., & Golbeck, J. H. (2009a). Understanding of the binding interface between PsaC and the PsaA/PsaB heterodimer in photosystem I. *Biochemistry, 48*, 5405–5416.

Jagannathan, B., & Golbeck, J. H. (2009b). Breaking biological symmetry in membrane proteins: the asymmetrical orientation of PsaC on the pseudo-C_2-symmetric photosystem I core. *Cellular and Molecular Life Sciences, 66*, 1257–1270.

Jaschke, P. R., Hardjasa, A., Digby, E. L., Hunter, C. N., & Beatty, J. T. (2011). A BchD (magnesium chelatase) mutant of *Rhodobacter sphaeroides* synthesizes zinc bacteriochlorophyll through novel zinc-containing intermediates. *Journal of Biological Chemistry, 286,* 20313–20322.

Jensen, R. A. (1976). Enzyme recruitment in evolution of new function. *Annual Review of Microbiology, 30,* 409–425.

Johnson, T. W., Li, H., Frigaard, N-U., Golbeck, J. H. and Bryant, D. A. (2013). [2Fe–2S] proteins in chlorosomes: redox properties of CsmI, CsmJ and CsmX of the chlorosome envelope of Chlorobaculum Tepidum. Biochemistry, in press.

Johnson, E. T., & Schmidt-Dannert, C. (2008). Characterization of three homologs of the large subunit of the magnesium chelatase from *Chlorobaculum tepidum* and interaction with the magnesium protoporphyrin IX methyltransferase. *Journal of Biological Chemistry, 283,* 27776–27784.

Kanao, T., Fukui, T., Atomi, H., & Imanaka, T. (2001). ATP-citrate lyase from the green sulfur bacterium *Chlorobium limicola* is a heterodimeric enzyme composed of two distinct gene products. *European Journal of Biochemistry, 288,* 1670–1678.

Katju, V. (2012). In with the old, in with the new: the promiscuity of the duplication process engenders diverse pathways for novel gene creation. *International Journal of Evolutionary Biology, 2012,* 341932.

Kelly, D. P. (2008). Stable sulfur isotope fractionation by the green bacterium *Chlorobaculum parvum* during photolithoautotrophic growth on sulfide. *Polish Journal of Microbiology, 57,* 275–279.

Kim, W., & Tabita, F. R. (2006). Both subunits of ATP-citrate lyase from *Chlorobium tepidum* contribute to catalytic activity. *Journal of Bacteriology, 188,* 6544–6552.

Kuznetsov, B. B., Ivanovsky, R. N., Keppen, O. I., Sukhacheva, M. V., Bumazhkin, B. K., Patutina, E. O., et al. (2011). Draft genome sequence of the anoxygenic filamentous phototrophic bacterium *Oscillochloris trichoides* subsp. DG-6. *Journal of Bacteriology, 193,* 321–322.

Larson, C. R., Seng, C. O., Lauman, L., Matthies, H. J., Wen, J., Blankenship, R. E., et al. (2011). The three-dimensional structure of the FMO protein from *Pelodictyon phaeum* and the implications for energy transfer. *Photosynthesis Research, 107,* 139–150.

Li, H., & Bryant, D. A. (2009). Envelope proteins of the CsmB/CsmF and CsmC/CsmD motif families influence the size, shape, and composition of chlorosomes in *Chlorobaculum tepidum*. *Journal of Bacteriology, 191,* 7109–7120.

Li, H., Frigaard, N.-U., & Bryant, D. A. (2006). Molecular contacts for chlorosome envelope proteins revealed by cross-linking studies with chlorosomes from *Chlorobium tepidum*. *Biochemistry, 45,* 9095–9103.

Li, H., Frigaard, N-U., and Bryant, D. A. (2013). [2Fe–2S]-proteins in chlorosomes: CsmI and CsmJ participate in oxidizing and reducing the quinone quenchers that control energy transfer in chlorosomes of Chlorobaculum tepidum. Biochemistry, in press.

Li, H., Jubilerer, S., Garcia Costas, A. M., Frigaard, N.-U., & Bryant, D. A. (2009). Multiple antioxidant proteins protect *Chlorobaculum tepidum* against oxygen and reactive oxygen species. *Archives of Microbiology, 191,* 853–867.

Liu, Z., & Bryant, D. A. (2011a). Identification of a gene essential for the first committed step in the synthesis of bacteriochlorophyll c. *Journal of Biological Chemistry, 286,* 22393–22402.

Liu, Z., & Bryant, D. A. (2011b). Multiple types of 8-vinyl reductases for (bacterio) chlorophyll biosynthesis occur in some green sulfur bacteria. *Journal of Bacteriology, 193,* 4996–4998.

Liu, Z., & Bryant, D. A. (2012). Biosynthesis and assembly of bacteriochlorophyll *c* in green bacteria: theme and variations. In K. M. Kadish, K. M. Smith & R. Guilard (Eds.), *Handbook of porphyrin science* (Vol. 20, pp. 108–142). Hackensack, NJ: World Scientific Publishing.

Liu, Z., Frigaard, N.-U., Vogl, K., Iino, T., Ohkuma, M., Overmann, J., et al. (2012). Complete genome sequence of *Ignavibacterium album*, a non-phototrophic member of the phylum Chlorobi. *Frontiers of Microbiology, 3*, 185.

Liu, Z., Klatt, C. G., Ludwig, M., Rusch, D. B., Jensen, S. I., Kühl, M., et al. (2012). "*Ca.* Tcb. aerophilum" an aerobic chlorophotoheterotrophic member of the phylum Chlorobi defined by metagenomics and metratranscriptomics. *ISME Journal, 6*, 1869–1882.

Liu, Z., Klatt, C. G., Wood, J. M., Rusch, D. B., Wittekindt, N., Tomsho, L. P., et al. (2011). Metatranscriptomic analyses of chlorophototrophs of a hot-spring microbial mat. *ISME Journal, 5*, 1279–1290.

Manske, A. K., Glaeser, J., Kuypers, M. M., & Overmann, J. (2005). Physiology and phylogeny of green sulfur bacteria forming a monospecific phototrophic assemblage at a depth of 100 meters in the Black Sea. *Applied and Environmental Microbiology, 71*, 8049–8060.

Maresca, J. A., Gomez Maqueo Chew, A., Ros Ponsatí, M., Frigaard, N.-U., Ormerod, J. G., Jones, A. D., et al. (2004). The *bchU* gene of *Chlorobium tepidum* encodes the bacteriochlorophyll C-20 methyltransferase. *Journal of Bacteriology, 186*, 2558–2566.

Maresca, J. A., Romberger, S. P., & Bryant, D. A. (2008). Isorenieratene biosynthesis in green sulfur bacteria requires the cooperative actions of two carotenoid cyclases. *Journal of Bacteriology, 190*, 6384–6391.

Martinez-Planells, A., Arellano, J. B., Borrego, C. M., López-Iglesias, C., Gich, F., & Garcia-Gil, J. (2002). Determination of the topography and biometry of chlorosomes by atomic force microscopy. *Photosynthesis Research, 71*, 83–90.

Mauzerall, D. (1973). Why chlorophyll? *Annals of the New York Academy of Science, 206*, 483–494.

Mauzerall, D. (1978). Porphyrins, chlorophyll and photosynthesis. In A. A. Trebst & M. Avron (Eds.), *Encyclopedia of plant physiology* (Vol. 5, pp. 117–124). New York: Springer.

Mauzerall, D. (1992). Light, iron, Sam Granick, and the origin of life. *Photosynthesis Research, 33*, 163–170.

Meyer, T. E. (1994). Evolution of photosynthetic reaction centers and light harvesting chlorophyll proteins. *BioSystems, 33*, 165–175.

Meyer, M., Dimroth, P., & Bott, M. (2001). Catabolite repression of the citrate fermentation genes in *Klebsiella pneumoniae*: evidence for involvement of the cyclic AMP receptor protein. *Journal of Bacteriology, 183*, 5248–5256.

Mix, L. J., Haig, D., & Cavanaugh, C. M. (2005). Phylogenetic analyses of the core antenna domain: investigating the origin of photosystem I. *Journal of Molecular Evolution, 60*, 153–163.

Montaño, G. A., Bowen, B. P., LaBelle, J. T., Woodbury, N. W., Pizziconi, V. B., & Blankenship, R. E. (2003). Characterization of *Chlorobium tepidum* chlorosomes: a calculation of bacteriochlorophyll *c* per chlorosome and oligomer modeling. *Biophysical Journal, 85*, 2560–2565.

Montaño, G. A., Wu, H. M., Lin, S., Brune, D. C., & Blankenship, R. E. (2003). Isolation and characterization of the B798 light-harvesting baseplate from the chlorosome of *Chloroflexus aurantiacus*. *Biochemistry, 42*, 10246–10251.

Mulkidjanian, A. Y., & Junge, W. (1997). On the origin of photosynthesis as inferred from sequence analysis. A primordial UV-protector as common ancestor of reaction centers and antenna proteins. *Photosynthesis Research, 51*, 27–42.

Mulkidjanian, A. Y., Koonin, E. V., Makarova, K. S., Mekhedov, S. L., Sorokin, A., Wolf, Y. I., et al. (2006). The cyanobacterial genome core and the origin of photosynthesis. *Proceedings of the National Academy of Sciences, USA, 103*, 13126–13131.

Müller, M. G., Slavov, C., Luthra, R., Redding, K. E., & Holzwarth, A. R. (2010). Independent initiation of primary electron transfer in the two branches of the photosystem I reaction center. *Proceedings of the National Academy of Sciences, USA, 107*, 4123–4128.

Muraki, N., Nomata, J., Ebata, K., Mizoguchi, T., Shiba, T., Tamiaki, H., et al. (2010). X-ray crystal structure of the light-independent protochlorophyllide reductase. *Nature, 465*, 110–114.

Nagata, N., Tanaka, R., & Tanaka, A. (2007). The major route for chlorophyll synthesis includes [3,8-divinyl]-chlorophyllide *a* reduction in *Arabidopsis thaliana*. *Plant Cell Physiology, 48*, 439–445.

Näsvall, J., Sun, L., Roth, J. R., & Andersson, D. I. (2012). Real-time evolution of new genes by innovation, amplification and divergence. *Science, 338*, 384–387.

Nisbet, E. G., Cann, J. R., & van Dover, C. L. (1995). Did photosynthesis begin from thermotaxis? *Nature, 373*, 479–480.

Oh-Oka, H. (2007). Type 1 reaction center of photosynthetic heliobacteria. *Photochemistry and Photobiology, 83*, 177–186.

Oh-oka, H., Iwaki, M., & Itoh, S. (1997). Vicosity dependence of the electron transfer rate from bound cytochrome *c* to P840 in the photosynthetic reaction center of the green sulfur bacterium *Chlorobium tepidum*. *Biochemistry, 36*, 9267–9272.

Oh-oka, H., Iwaki, M., & Itoh, S. (1998). Membrane-bound cytochrome cz couples quinol oxidoreductase to the P840 reaction center complex in isolated membranes of the green sulfur bacterium *Chlorobium tepidum*. *Biochemistry, 37*, 12293–12300.

Oh-oka, H., Kamei, S., Matsubara, H., Iwaki, M., & Itoh, S. (1995). Two molecules of cytochrome *c* function as the electron donors to P840 in the reaction center complex isolated from a green sulfur bacterium *Chlorobium tepidum*. *FEBS Letters, 365*, 24–30.

Okkels, J. S., Kjær, B., Hansson, O., Svendsen, I., Møller, B. L., & Scheller, H. V. (1992). A membrane-bound monoheme cytochrome c_{551} of a novel type is the immediate electron donor to P840 of the *Chlorobium vibrioforme* photosynthetic reaction center complex. *Journal of Biological Chemistry, 267*, 21139–21145.

Olson, J. M. (1970). Evolution of photosynthesis. *Science, 168*, 438–487.

Olson, J. M. (1999). Early evolution of chlorophyll-based photosystems. *Biochemistry and Molecular Biology, 12*, 468–482.

Olson, J. M. (2000). Evolution of photosynthesis (1970). Re-examined thirty years later. *Photosynthesis Research, 68*, 95–117.

Olson, J. M., & Blankenship, R. E. (2004). Thinking about the evolution of photosynthesis. *Photosynthesis Research, 80*, 373–386.

Olson, J. M., & Pierson, B. K. (1987). Evolution of reaction centers in photosynthetic prokaryotes. *International Review of Cytology, 108*, 209–248.

Oostergetel, G. T., Reus, M., Gomez Maqueo Chew, A., Bryant, D. A., Boekema, E. J., et al. (2007). Long-range organization of bacteriochlorophyll in chlorosomes of *Chlorobium tepidum* investigated by cryo-electron microscopy. *FEBS Letters, 581*, 5435–5439.

Oostergetel, G.T., van Amerongen, H., & Boekema, E. J. (2010). The chlorosome: a prototype for efficient light harvesting in photosynthesis. *Photosynthesis Research, 104*, 245–255.

Orf, G. S., Tank, M., Vogl, K., Niedzwiedzki, D. M., Bryant, D. A. and Blankenship, R. E. (2013). Spectroscopic insights into the decreased efficiency of chlorosomes containing bacteriochlorophyll *f*. Biochimica Biophysica Acta, in press.

Overmann, J. (2001). Green sulfur bacteria. (2nd ed.). In D. R. Boone & R. W. Castenholz (Eds.), *Bergey's manual of systematic bacteriology* (Vol. 1, pp. 601–605). Berlin: Springer.

Overmann, J. (2008). Ecology of phototrophic sulfur bacteria. In R. Hell, C. Dahl, D. B. Knaff & T. Leustek (Eds.), *Advances in photosynthesis and respiration Sulfur metabolism in phototrophic organisms* (Vol. 27, pp. 375–396). New York: Springer.

Paulsen, H. (2006). Reconstitution and pigment exchange. In B. Grimm, R. J. Porra, W. Rüdiger & H. Scheer (Eds.), *Advances in photosynthesis and respiration Chlorophylls and bacteriochlorophylls: Biochemistry, biophysics, functions and applications* (Vol. 25, pp. 375–385). New York: Springer.

Pedersen, M. Ø., Linnanto, J., Frigaard, N.-U., Nielsen, N. C., & Miller, M. (2010). A model of the protein–pigment baseplate complex in chlorosomes. *Photosynthesis Research, 104*, 233–243.

Phadwal, K. (2005). Carotenoid biosynthetic pathway: molecular phylogenies and evolutionary behavior of *crt* genes in eubacteria. *Gene, 345*, 35–43.

Pierson, B. K., & Castenholz, R. W. (1995). Taxonomy and physiology of filamentous anoxygenic phototrophs. In R. E. Blankenship, M. T. Madigan & C. E. Bauer (Eds.), *Advances in photosynthesis and respiration Anoxygenic photosynthetic bacteria* (Vol. 2, pp. 31–47). Dordrecht, The Netherlands: Kluwer.

Pierson, B. K., & Thornber, J. P. (1983). Isolation and spectral characterization of photochemical reaction centers from the thermophilic green bacterium *Chloroflexus aurantiacus* strain j-10-fl. *Proceedings of the National Academy of Sciences, USA, 80*, 80–84.

Pinta, V., Picaud, M., Reiss-Husson, F., & Astier, C. (2002). *Rubrivivax gelatinosus acsF* (previously orf358) codes for a conserved, putative binuclear-iron-cluster-containing protein involved in aerobic oxidative cyclization of Mg-protoporphyrin IX monomethylester. *Journal of Bacteriology, 184*, 746–753.

Prokhorenko, V. I., Steensgaard, D. B., & Holzwarth, A. R. (2000). Exciton dynamics in the chlorosomal antennae of the green bacteria *Chloroflexus aurantiacus* and *Chlorobium tepidum*. *Biophysical Journal, 79*, 2105–2120.

Pšenčík, J., Collins, A. M., Lijeroos, L., Torkkeli, M., Laurinmäki, P., Ansink, H. M., et al. (2009). Structure of chlorosomes from the green filamentous bacterium *Chloroflexus aurantiacus*. *Journal of Bacteriology, 191*, 6701–6708.

Pšenčík, J., Ikonen, T. P., Laurinmäki, P., Merckel, M. C., Butcher, S. J., Serimaa, R. E., et al. (2004). Lamellar organization of pigments in chlorosomes, the light harvesting complexes of green photosynthetic bacteria. *Biophysical Journal, 87*, 1165–1172.

Raymond, J., Zhaxybayeva, O., Gogarten, J. P., Gerdes, S. Y., & Blankenship, R. E. (2002). Whole-genome analysis of photosynthetic prokaryotes. *Science, 298*, 1616–1620.

Reid, J. D., & Hunter, C. N. (2004). Magnesium-dependent ATPase activity and cooperativity of magnesium chelatase from *Synechocystis* sp. PCC6803. *Journal of Biological Chemistry, 279*, 26893–26899.

Reinbothe, C., El Bakkoiuri, M., Buhr, F., Muraki, N., Nomata, J., Kurisu, G., et al. (2010). Chlorophyll biosynthesis: spotlight on protochlorophyllide reduction. *Trends in Plant Sciences, 15*, 614–624.

Romberger, S. P., & Golbeck, J. H. (2010). The bound iron-sulfur clusters of type-I homodimeric reaction centers. *Photosynthesis Research, 104*, 333–346.

Romberger, S. P., & Golbeck, J. H. (2012). The F_X iron-sulfur cluster serves as the terminal bound electron acceptor in heliobacterial reaction centers. *Photosynthesis Research, 111*, 285–290.

Sadekar, S., Raymond, J., & Blankenship, R. E. (2006). Conservation of distantly related membrane proteins: photosynthetic reaction centers share a common structural core. *Molecular Biology and Evolution, 23*, 2001–2007.

Sakuragi, Y., Frigaard, N.-U., Shimada, K., & Matsuura, K. (1999). Association of bacteriochlorophyll *a* with the CsmA protein in chlorosomes of the photosynthetic green filamentous bacterium *Chloroflexus aurantiacus*. *Biochimica Biophysica Acta, 413*, 172–180.

Sarma, R., Barney, B. M., Hamilton, T. L., Jones, A., Seefeldt, L. C., & Peters, J. W. (2008). Crystal structure of the L protein of *Rhodobacter sphaeroides* light-independent protochlorophyllide reductase with MgADP bound: a homolog of the nitrogenase Fe protein. *Biochemistry, 47*, 13004–13015.

Schubert, W.-D., Klukas, O., Saenger, W., Witt, H. T., Fromme, P., & Krauß, N. (1998). A common ancestor for oxygenic and anoxygenic photosynthetic systems: a comparison based on the structural model of photosystem I. *Journal of Molecular Biology, 280*, 297–314.

Staehelin, L. A., Golecki, J. R., & Drews, G. (1980). Supramolecular organization of chlorosomes (*Chlorobium* vesicles) and of their membrane attachments sites in *Chlorobium limicola*. *Biochimica Biophysica Acta, 589*, 30–45.

Staehelin, L. A., Golecki, J. R., Fuller, R. C., & Drews, G. (1978). Visualization of the supramolecular architecture of chlorosomes (*Chlorobium* type vesicles) in freeze-fractured cells of *Chloroflexus aurantiacus*. *Archives of Microbiology, 119*, 269–277.

Stomp, M., Huisman, J., Stal, L. J., & Matthijs, H. C. (2007). Colorful niches of phototrophic microorganisms shaped by vibrations of the water molecule. *ISME Journal, 1*, 271–282.

Sytina, O. A., Heyes, D. J., Hunter, C. N., & Groot, M. L. (2009). Ultrafast catalytic processes and conformational changes in the light-driven enzyme protochlorophyllide oxidoreductase (POR). *Biochemical Society Transactions, 37*, 387–391.

Tamiaki, H., Komada, J., Kunieda, M., Fukai, K., Yoshitomi, T., Harada, J., et al. (2011). In vitro synthesis and characterization of bacteriochlorophyll-*f* and its absence in bacteriochlorophyll-*e* producing organisms. *Photosynthesis Research, 107*, 133–138.

Tamura, K., Dudley, J., Nei, M., & Kumar, S. (2007). MEGA4: molecular evolutionary genetics analysis (MEGA) software version 4.0. *Molecular Biology and Evolution, 24*, 1596–1599.

Tanaka, R., & Tanaka, A. (2007). Tetrapyrrole biosynthesis in higher plants. *Annual Review of Plant Biology, 58*, 321–346.

Tang, K. H., Tang, Y. J., & Blankenship, R. E. (2011). Carbon metabolic pathways in phototrophic bacteria and their broader evolutionary implications. *Frontiers of Microbiology, 2*, 165.

Tokita, S., Frigaard, N.-U., Hirota, M., Shimada, K., & Matsuura, K. (2000). Quenching of bacteriochlorophyll fluorescence in chlorosomes from *Chloroflexus aurantiacus* by exogenous quinones. *Photochemistry and Photobiology, 72*, 345–350.

Tripp, H. J., Bench, S. R., Turk, K. A., Foster, R. A., Desany, B. A., Niazi, F., et al. (2010). Metabolic streamlining in an open-ocean nitrogen-fixing cyanobacterium. *Nature, 464*, 90–94.

Tronrud, D., Wen, J., Gay, L., & Blankenship, R. E. (2009). The structural basis for the difference in absorbance spectra for the FMO antenna protein from various green sulfur bacteria. *Photosynthesis Research, 100*, 79–87.

Tsukatani, Y., Azai, C., Kondo, T., Itoh, S., & Oh-oka, H. (2008). Parallel electron donation pathways to cytochrome c_z in the type I homodimeric photosynthetic reaction center complex of *Chlorobium tepidum*. *Biochimica Biophysica Acta, 1777*, 1211–1217.

Tsukatani, Y., Miyamoto, R., Itoh, S., & Oh-oka, H. (2004). Function of a PscD subunit in the homodimeric reaction center complex of the photosynthetic green sulfur bacterium *Chlorobium tepidum* studied by insertional gene inactivation. Regulation of energy transfer and ferredoxin-mediated $NADP^+$ reduction on the cytoplasmic side. *Journal of Biological Chemistry, 279*, 51122–51130.

Tsukatani, Y., Romberger, S. P., Golbeck, J. H., & Bryant, D. A. (2012). Characterization of the oxygen-tolerant, homodimeric type-1 reaction centers of "*Ca.* Cab. thermophilum". *Journal of Biological Chemistry, 287*, 5720–5732.

Tsukatani, Y., Wen, J., Blankenship, R. E., & Bryant, D. A. (2010). Characterization of the bacteriochlorophyll *a*-binding, Fenna–Matthews–Olson protein from "*Ca.* Cab. thermophilum". *Photosynthesis Research, 104*, 201–209.

Turova, T. P., Spiridonova, E. M., Slobodova, N. V., Bulygina, E. S., Keppen, O. I., Kuznetsov, B. B., et al. (2006). Phylogeny of anoxygenic filamentous phototrophic bacteria of the family Oscillochloridaceae as inferred from comparative analyses of the *rrs, cbbL*, and *nifH* genes. *Mikrobiologia, 75*, 235–244.

Umena, Y., Kawakami, K., Shen, J. R., & Kamiya, N. (2011). Crystal structure of oxygen-evolving photosystem II at a resolution of 1.9 Å. *Nature, 473*, 55–60.

van de Meent, E. J., Kobayashi, M., Erkelens, C., van Veelen, P. A., Amesz, J., & Watanabe, T. (1991). Identification of 8^1-hydroxy-chlorophyll *a* as a functional reaction center pigment in heliobacteria. *Biochimica Biophysica Acta, 1058*, 356–362.

Vassilieva, E. V., Stirewalt, V. L., Jakobs, C. U., Frigaard, N. -U., Inoue-Sakamoto, K., Baker, M. A., et al. (2002). Subcellular localization of chlorosome proteins in *Chlorobium tepidum* and characterization of three new chlorosome proteins: CsmF, CsmH, and CsmX. *Biochemistry, 41*, 4358–4370.

Verté, F., Kostanjevecki, V., De Smet, L., Meyer, T. E., Cusanovich, M. A., & Van Beeumen, J. J. (2002). Identification of a thiosulfate utilization gene cluster from the green phototrophic bacterium *Chlorobium limicola*. *Biochemistry, 41*, 2932–2945.

Vogl, K., Tank, M., Orf, G. S., Blankenship, R. E., & Bryant, D. A. (2012). Bacteriochlorophyll *f*: properties of chlorosomes containing the "forbidden chlorophyll." *Frontiers in Microbiology, 3*, 298.

Wada, K., Yamaguchi, H., Harada, J., Niimi, K., Osumi, S., Saga, Y., et al. (2006). Crystal structures of BchU, a methyltransferase involved in bacteriochlorophyll *c* biosynthesis, and its complex with S-adenosylhomosysteine: implications for reaction mechanism. *Journal of Molecular Biology, 360*, 839–849.

Wahlund, T. M., & Tabita, F. R. (1997). The reductive tricarboxylic acid cycle of carbon dioxide assimilation: initial studies and purification of ATP-citrate lyase from the green sulfur bacterium *Chlorobium tepidum*. *Journal of Bacteriology, 179*, 4859–4867.

Wang, J., Brune, D. C., & Blankenship, R. E. (1990). Effects of oxidants and reductants on the efficiency of excitation transfer in green photosynthetic bacteria. *Biochimica Biophysica Acta, 1015*, 457–463.

Wen, J., Tsukatani, Y., Cui, W., Zhang, H., Gross, M. L., Bryant, D. A., et al. (2011a). Structural and spectroscopic insights of the FMO antenna protein of the aerobic chlorophototroph "*Ca.* Cab. thermophilum". *Biochimica Biophysica Acta, 1807*, 157–164.

Wen, J., Zhang, H., Gross, M. L., & Blankenship, R. E. (2009). Membrane orientation of the FMO antenna protein from *Chlorobaculum tepidum* as determined by mass spectrometry-based footprinting. *Proceedings of the National Academy of Sciences, USA, 106*, 6134–6139.

Wen, J., Zhang, H., Gross, M. L., & Blankenship, R. E. (2011). Native electrospray mass spectrometry reveals the nature and stoichiometry of pigments in the FMO photosynthetic antenna protein. *Biochemistry, 50*, 3502–3511.

Widdel, F., Schnell, S., Heising, S., Ehrenreich, A., Assmus, B., & Schink, B. (1993). Ferrous iron oxidation by anoxygenic phototrophic bacteria. *Nature, 362*, 834–836.

Xin, Y., Lin, S., Montaño, G. A., & Blankenship, R. E. (2005). Purification and characterization of the B808-866 light-harvesting complex from green filamentous bacterium *Chloroflexus aurantiacus*. *Photosynthesis Research, 86*, 155–163.

Xiong, J. (2006). Photosynthesis: what color was its origin? *Genome Biology, 7*, 245.

Xiong, J., & Bauer, C. E. (2002a). Complex evolution of photosynthesis. *Annual Review of Plant Biology, 53*, 503–521.

Xiong, J., & Bauer, C. E. (2002b). A cytochrome *b* origin of photosynthetic reaction centers: an evolutionary link between respiration and photosynthesis. *Journal of Molecular Biology, 322*, 1025–1037.

Xiong, J., Fischer, W. M., Inoue, K., Nakahara, M., & Bauer, C. E. (2000). Molecular evidence for the early evolution of photosynthesis. *Science, 289*, 1724–1730.

Xu, H., Vavilin, D., & Vermaas, W. (2001). Chlorophyll *b* can serve as the major pigment in functional photosystem II complexes of cyanobacteria. *Proceedings of the National Academy of Sciences, USA, 98*, 14168–14173.

Ycas, M. (1974). On earlier states of the biochemical system. *Journal of Theoretical Biology, 44*, 145–160.

Yoon, K. S., Bobst, C., Hemann, C. F., Hille, R., & Tabita, F. R. (2001). Spectroscopic and functional properties of novel 2[4Fe-4S] cluster-containing ferredoxins from the green sulfur bacterium *Chlorobium tepidum*. *Journal of Biological Chemistry*, *276*, 44027–44036.

Yoon, K. S., Hille, R., & Tabita, F. R. (1999). Rubredoxin from the green sulfur bacterium *Chlorobium tepidum* functions as an electron acceptor for pyruvate ferredoxin oxidoreductase. *Journal of Biological Chemistry*, *274*, 29772–29778.

Zarzycki, J., Brecht, V., Müller, M., & Fuchs, G. (2009). Final steps of an autotrophic pathway: closing the 3-hydroxypropionate bicycle in *Chloroflexus aurantiacus*. *Proceedings of the National Academy of Sciences, USA*, *106*, 21317–21322.

Zehr, J. P., Bench, S. R., Carter, B. J., Hewson, I., Niazi, F., Shi, T., et al. (2008). Globally distributed uncultivated oceanic N_2-fixing cyanobacteria lack oxygenic photosystem II. *Science*, *322*, 1110–1112.

Zhou, W., LoBrutto, R., Lin, S., & Blankenship, R. E. (1994). Redox effects on the bacteriochlorophyll *a*-containing Fenna–Matthews–Olson protein from *Chlorobium tepidum*. *Photosynthesis Research*, *41*, 89–96.

CHAPTER FIVE

Comparison of Photosynthesis Gene Clusters Retrieved from Total Genome Sequences of Purple Bacteria

Sakiko Nagashima[*,**], **Kenji V. P. Nagashima**[*,†,1]

[*]Research Institute for Photosynthetic Hydrogen Production, Kanagawa University, Hiratsuka, Kanagawa, Japan

[**]Department of Biological Science, Tokyo Metropolitan University, Hachioji, Tokyo, Japan

[†]Precursory Research for Embryonic Science and Technology (PRESTO), Japan Science and Technology Agency (JST), Kawaguchi, Saitama, Japan

[1]Corresponding author: E-mail: wt503649bw@kanagawa-u.ac.jp

Contents

Abstract

Studies on some purple photosynthetic bacteria have shown that the genes required for photosynthesis form a large cluster. This is often called a photosynthesis gene cluster (PGC) and has been described as a unique feature of purple photosynthetic bacteria. Here, photosynthesis genes were retrieved from the rapidly increasing genomic data of purple bacteria. Arrangements of these genes on the genomes were compared among 24 species and strains were selected from the wide distribution of alpha, beta, and gamma classes of purple bacteria. The presence of PGC was confirmed in the all of these, although it is divided into two or several islands in some species. Gene arrangements in several subclusters possibly forming an operon or a superoperon are well conserved while the directions and locations of these subclusters in the PGC vary according to the strains. Phylogenetic trees based on the sequences of the suspected products of the genes in the PGCs consistently place the species of the beta and gamma classes in a cluster within the branch of the alpha class, showing

Advances in Botanical Research, Volume 66
ISSN 0065-2296, http://dx.doi.org/10.1016/B978-0-12-397923-0.00005-9

that a horizontal transfer of PGC had occurred in the process of evolution. Information obtained from the increasing genome projects on bacteria is very useful for the study on evolution of photosynthesis and even for its application in metabolic engineering.

1. INTRODUCTION

The reaction process of photosynthesis starts with the absorption of light energy and is then followed by electron transfer reactions driven by light-excited (bacterio)chlorophylls with quite a low reduction potential. These electron transfer reactions are coupled with proton transfer reactions across the membrane (generation of proton motive force) to produce adenosine triphosphate (ATP). In purple photosynthetic bacteria, the main pathway of the photosynthetic electron transfer is a cyclic process consisting of four components, the photochemical reaction centre complex, the pool of quinones, the cytochrome bc_1 complex, and soluble electron carrier proteins such as cytochrome c_2 and the high potential iron–sulphur protein (Lavergne, Verméglio, & Joliot, 2009). Among these components, the quinone pool, the cytochrome bc_1 complex, and the soluble electron carrier are the essential members in both aerobic and anaerobic respiration pathways. Thus, a component specific to the photosynthetic electron transfer is only the reaction centre (RC) complex, although it is often accompanied by light-harvesting (LH) complexes. Photosynthesis may be only one of the various energy metabolisms found in purple bacteria. However, there is little doubt that the photosynthetic ability has contributed significantly to the dispersal and survival of this group of bacteria.

The phylum Proteobacteria (purple bacteria) is a well-studied large group in the Bacteria kingdom and includes more than 1500 isolated and described species. There are more than 200 photosynthetic species (including 'aerobic' anoxygenic photosynthetic species) among them. Phylogenetic trees based on 16S ribosomal RNA (rRNA) sequences of purple bacteria have shown that the photosynthetic species are not monophyletic but, rather, have a scattered distribution in the alpha, beta, and gamma classes (Fig. 5.1). Therefore, progenitors of purple bacteria should have photosynthetic ability, but this ability has been lost in many lineages. Such an explanation was first given by Woese (1987) and has been widely accepted. Purple bacterial species show a variety of energy metabolisms, i.e. oxygen respiration, fermentation, denitrification, metal oxidation, and oxidation/reduction of sulphur compounds. Therefore, on one hand, it is not surprising that species adapting to the environments where light energy is unavailable have given

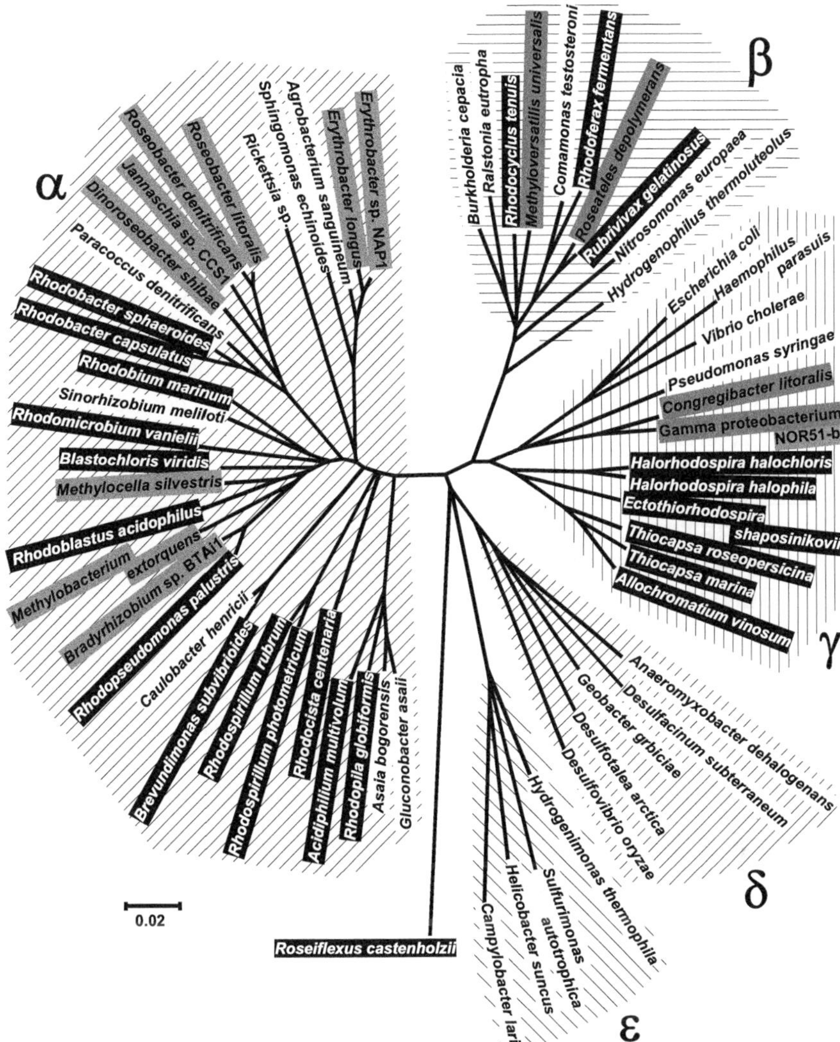

Figure 5.1 ***Phylogenetic tree of purple bacteria (Proteobacteria) based on nucleotide sequences of 16S rRNA.*** Phylogenetic analysis was performed using the programs ClustalX (Thompson, Gibson, Plewniak, Jeanmougin, & Higgins, 1997) and MEGA (Kumar, Tamura, & Nei, 2004). All gaps in the sequence alignment were omitted. The tree was generated by the neighbour-joining method, applying the Kimura 2-parameter distance as a distance estimator. The names of phototrophic species are shown by reverse contrast. Aerobic photosynthetic species are marked with grey background. The sequence of the filamentous anoxygenic phototroph (green filamentous bacteria), *Roseiflexus castenholzii*, was used as an outgroup. See the color plate.

up the photosynthetic ability in the course of evolution. On the other hand, consideration should be given to the possibility that the photosynthetic ability had been renewed in some lineages. Indeed, the horizontal transfer of the genes coding the photochemical reaction centre subunits among purple bacteria has been suggested (Nagashima, Hiraishi, Shimada, & Matsuura, 1997). The genes encoding biosynthetic enzymes for bacteriochlorophylls and carotenoids form a photosynthesis gene cluster (PGC) with the genes coding the subunits of the RC and LH complexes. Such clustering should be convenient to lose and/or gain the photosynthetic ability. In this chapter, PGCs of purple bacteria found in genome project information are compared and analysed. It is hoped that such analyses will be useful for studying the costs and benefits of photosynthesis in bacteria and the course of evolution of bacterial photosynthesis.

2. RETRIEVAL OF PHOTOSYNTHESIS GENE SEQUENCES OF PURPLE BACTERIA FROM GENOMIC DNA DATABASES

The presence of a PGC was first reported in the Alphaproteobacterium, *Rhodobacter capsulatus* (Taylor, Cohen, Clark, & Marrs, 1983). Later, similar gene clustering was also found in *Rhodobacter sphaeroides* (Coomber, Chaudhri, Connor, Britton, & Hunter, 1990). Precise gene mapping and biochemical/physiological identification of the gene products included in the PGC have been performed mainly in these two *Rhodobacter* species through transposon mutagenesis and analyses of the accumulated products in the mutants (Armstrong, Schmidt, Sandmann, & Hearst, 1990; Bollivar, Suzuki, Beatty, Dobrowolski, & Bauer, 1994; Coomber et al., 1990). A PGC contains a complete set of genes required for the synthesis of photochemical reaction centre complex and LH complexes. Genes coding for enzymes in the biosynthetic processes of carotenoids (*crt*) and bacteriochlorophylls (*bch*), in which isoprenoids and protoporphyrin IX are used as substrates, respectively, are also included in the cluster. Later, PGCs of some other purple bacterial species, i.e. *Rubrivivax gelatinosus* (Igarashi et al., 2001) and *Thiocapsa roseopersicina* (Kovács, Rákhely, & Kovács, 2003), were cloned and sequenced.

Recent advances in DNA sequencing technologies have made it remarkably easy to determine the whole genome sequences of bacteria, in which many photosynthetic species are included. Now, we can retrieve genomic information of over 70 strains of purple photosynthetic bacteria.

Some open databases on the World Wide Web, i.e. Genomes OnLine Database (GOLD: http://www.genomesonline.org/), frequently update the list of the species names for which genomic information has become available. It is noteworthy that the PGC is sometimes found in species that have been assumed to be non-photosynthetic, i.e. some nodule bacteria. The Basic Local Alignment Search Tool (BLAST) is a very powerful tool for the identification of stealthy photosynthesis genes and is easily accessible on the websites of National Center for Biotechnology Information (http://blast.ncbi.nlm.nih.gov/Blast.cgi) and DNA Data Bank of Japan (DDBJ)/European Molecular Biology Laboratory (EMBL)/GenBank DNA databases. When a DNA sequence of a known photosynthesis gene, such as *bchX*, is input to the BLAST program, over 100 orthologue sequences will soon be returned with the names of the species that they are derived from.

At the time of this review, completed genome projects for 29 species (43 strains) of photosynthetic proteobacteria could be found on the GOLD website (out of 1192 completed projects for proteobacteria strains). In addition, data from 24 other genomic sequence of photosynthetic proteobacteria (including incomplete projects) were found in DDBJ/EMBL/GenBank DNA databases when using BchX and BchL amino acid sequences of *R. capsulatus* as query sequences to the BLAST search. These are summarised in Table 5.1. Many of the species found in the GOLD database are the Alphaproteobacteria species, of which about half are aerobic photosynthetic bacteria. Some key species, such as *R. sphaeroides*, *Rhodopseudomonas palustris* and *Methylobacterium extorquens*, are well examined over a range of several strains. However, the data from the species of Betaproteobacteria and Gammaproteobacteria are restricted in those from *R. gelatinosus* (β), *Allochromatium vinosum* (γ), and *Thiocapsa marina* (γ). Genome projects for the photosynthetic species belonging to these classes seem to be in progress now. Among the incomplete projects, new isolates and uncultured strains from various environmental samples are also included. In the present chapter, some such isolates and strains are used for sequence comparison and phylogenetic analyses.

The genome sizes of most purple photosynthetic bacteria are in the range of 3–5 Mb in length (average 4.7 Mb, Table 5.1). Strains of *R. palustris* and its relatives, *Bradyrhizobium* and *Methylobacterium* species and strains, have considerably large genomes exceeding 5 Mb, possibly reflecting their versatile metabolisms and/or unique phenotypes, i.e. degradation of aromatic compounds and nodulation in legume plants. The number of genes (or open reading frames, ORFs) on the genome is roughly proportional to

Table 5.1 Purple photosynthetic bacteria species and strains of which total genomic information is available

Organism	Strain	Size of genome (Kb)	Number of ORFs	GC %	Accession number	References
Acidiphilium cryptum	JF-5	3963	3701	67	CP000697	Magnuson et al., 2010
Acidiphilium multivorum	AIU301	4215	4004	67	AP012035	Unpublished
Allochromatium vinosum	DSM180	3669	3366	64	CP001896	Weissgerber et al., 2011
Bradyrhizobium sp.	BTAi1	8494	7819	65	CP000494	Giraud et al., 2007
	ORS278	7457	6825	66	CU234118	Giraud et al., 2007
	S23321	7232	6898	64	AP012279	Okubo et al., 2012
Brevundimonas subvibrioides	ATCC15264	3445	3327	68	CP002102	Brown, Kysela, Buechlein, Hemmerich, & Brun, 2011
Dinoroseobacter shibae	DFL-12	4418	4271	66	CP000830	Wagner-Döbler et al., 2010
Erythrobacter litoralis	HTCC2594	3052	3068	63	CP000157	Oh, Giovannoni, Ferriera, Johnson, & Cho, 2009
Erythrobacter sp.	NAP-1	3265	3226	61	AAMW00000000	Koblížek et al., 2011
Halorhodospira halophila	SL1	2678	2470	68	CP000544	Unpublished
Jannaschia sp.	CCS1	4404	4339	62	CP000264	Moran et al., 2007
Methylobacterium chloromethanicum	CM4	6181	5847	68	CP001298	Unpublished
Methylobacterium extorquens	AM1	6880	6294	68	CP001510	Vuilleumier et al., 2009
	DM4	6124	5829	68	FP103042	Vuilleumier et al., 2009
	PA1	5471	4939	68	CP000908	Unpublished
Methylobacterium populi	BJ001	5849	5538	69	CP001029	Unpublished
Methylobacterium radiotolerans	JCM 2831	6899	6510	71	CP001001	Unpublished

Methylobacterium sp.	4–46	7737	7125	72	CP000943	Unpublished
Methylocella silvestris	BL2	4305	3971	63	CP001280	Chen et al., 2010
Rhodobacter capsulatus	SB1003	3872	3708	67	CP001312	Strnad et al., 2010
Rhodobacter sp.	SW2	3514	3453	65	ACYY00000000	Unpublished
Rhodobacter sphaeroides	2.4.1	4603	4383	69	CP000143	Choudhary, Fu, Mackenzie, & Kaplan, 2004
	ATCC17025	4557	4475	68	CP000661	Choudhary, Zanhua, Fu, & Kaplan, 2007
	ATCC17029	4489	4268	69	CP000577	Choudhary et al., 2007
	KD131	4711	4635	69	CP001150	Lim et al., 2009
Rhodobacterales sp.	HTCC2150	3583	3713	49	AAXZ00000000	Kang, Oh, & Vergin, et al., 2010
Rhodocista centenaria	SW	4356	4065	70	CP000613	Lu et al., 2010
Rhodomicrobium vannielii	ATCC17100	4014	3565	62	CP002292	Brown et al., 2011
Rhodopseudomonas palustris	BisA53	5505	4996	64	CP000463	Oda et al., 2008
	BisB18	5514	5028	65	CP000301	Oda et al., 2008
	BisB5	4893	4501	65	CP000283	Oda et al., 2008
	CGA009	5468	4918	65	BX571963	Larimer et al., 2004
	DX-1	5404	5081	65	CP002418	Unpublished
	HaA2	5332	4788	66	CP000250	Oda et al., 2008
	TIE-1	5744	5377	65	CP001096	Oda et al., 2008
Rhodospirillum photometricum	DSM 122	3876	3287	65	HE663493	Duquesne & Sturgis, 2012
Rhodospirillum rubrum	F11	4353	3945	65	CP003046	Lonjers et al., 2012
	S1	4407	3933	65	CP000230	Munk et al., 2011

Continued

Table 5.1 Purple photosynthetic bacteria species and strains of which total genomic information is available—cont'd

Organism	Strain	Size of genome (Kb)	Number of ORFs	GC %	Accession number	References
Roseobacter denitrificans	OCh 114	4331	4201	59	CP000362	Swingley et al., 2007
Roseobacter litoralis	OCh 149	4745	4577	57	CP002623	Kalhoefer et al., 2011
Rubrivivax gelatinosus	IL144	5043	4768	71	AP012320	Nagashima et al., 2012
*Thiocapsa marina**	5811	5667	5059	64	AFWV00000000	Unpublished
Acidiphilium sp.[†]	PM	3929	3908		AFPR01000000	Martin-Uriz, Gomez, Arcas, Bargiela, & Amils, 2011
Ahrensia sp.	R2A130	3731	3669		AEEB01000000	Unpublished
Congregibacter litoralis	KT71	4328	3988	58	AAOA01000000	Fuchs et al., 2007
Ectothiorhodospira sp.	PHS-1	2943	2841		AGBG01000000	Zargar et al., 2012
Fulvimarina pelagi	HTCC2506	3706	3812	61	AATP01000000	Kang, Oh, & Lim, et al., 2010a
Gammaproteobacterium	NOR5-3	4204	3716	56	ACCX01000012	Unpublished
	NOR51-B	3248	2969	57	ACCY01000000	Unpublished
Hoeflea phototrophica	DFL-43	4458	4407	60	ABIA02000000	Unpublished
Labrenzia alexandrii	DFL-11	5502	5436	56	ACCU01000000	Unpublished
Loktanella vestfoldensis	SKA53	3064	3109	60	AAMS01000000	Unpublished
Marichromatium purpuratum	984	3761	3332		AFWU01000000	Unpublished
Marine Gammaproteobacterium	HTCC2080	3576	3224	52	AAVV01000000	Thrash et al., 2010

Methyloversatilis universalis	FAM5	4235	3981		AFHG01000000	Kittichotirat et al., 2011
Rhodobacter sphaeroides	WS8N	4418	4266		AFER01000000	Porter et al., 2011
Roseobacter sp.	CCS2	3497	3696	56	AAYB00000000	Unpublished
	AzwK-3b	4179	4186	62	ABCR01000000	Unpublished
Roseovarius sp.	TM1035	4210	4150	61	ABCL01000000	Unpublished
	217	4763	4817	61	AAMV01000000	Baldock, Denger, Smits, & Cook, 2007
Rubrivivax benzoatilyticus	JA2	4130	3947		AEWG01000000	Mohammed et al., 2011
Thalassiobium sp.	R2A62	3488	3698	55	ACOA00000000	Unpublished
Thiocystis violascens	DSM 198	5017	4627	63	CP003154	Unpublished
Thiorhodococcus drewsii	AZ1	5524	4785		AFWT01000000	Unpublished
Thiorhodospira sibirica	A12	3189	3004		AGFD00000000	Unpublished
Thiorhodovibrio sp.	970	5338		60	AFWS02000000	Unpublished

*Upper; genomic data listed as completed projects in GOLD.

†Lower; incompleted projects but available on the databases.

the genome size, 3000–5000, in many species and strains and 4500–8000 in *Rhodopseudomonas* and the related species and strains. Annotation of genes (a procedure to attach biological functions and names to ORFs) on the scale of the whole genome is a tremendous task, and its validity depends on the data contributor. In fact, many photosynthesis genes are hypothetically annotated or unknown and sometimes misassigned. Therefore, we have reviewed the gene annotation and revised it when appropriate on the basis of amino acid sequence comparisons with the products of the photosynthesis genes reported in the well-studied *Rhodobacter* species. Table 5.2 is a summary of the averaged sequence identity among species (24 strains) calculated for each of the photosynthesis gene products, which reflects evolutionary conservation. The arrangements of the photosynthesis genes in the species used for the above calculations are shown in Fig. 5.2. It is noteworthy that the gene and product names are, in principle, given in reference to those of *R. capsulatus* or *R. sphaeroides*, so that sometimes the gene and the product names differ from those given by the data contributors.

3. SEQUENCE COMPARISON OF PHOTOSYNTHETIC GENE PRODUCTS

Among 18 Bch, 7 Crt, 6 Puf, 5 Puh and 5 PS-related proteins, BchX, a subunit of chlorin reductase, is the most conserved product (Table 5.2). Two other subunits of the chlorin reductase, BchY and BchZ, also show considerably high sequence conservativeness. BchX, BchY, and BchZ are closely related with the subunits of protochlorophyllide reductase, BchL, BchN, and BchB, respectively. For example, the sequence identity between BchX and BchL is 31% in *R. capsulatus*. It is obvious that these two enzyme complexes have a common origin, although the former specifically recognises the B-ring of chlorophyllide and the latter recognises the D-ring of protochlorophyllide. Each of the latter subunits, BchL, BchN, and BchB, also show a high degree of conservation but at a slightly lower level than that shown in BchX, BchY, or BchZ. Assuming that the rates of amino acid substitutions are equal between the two enzyme complexes, it is concluded that BchXYZ diverged at a later time than BchLNB. This is somewhat inconsistent with the results described by Burke, Hearst, and Sidow (1993). Amino acid sequences of most of the other Bch enzymes are also highly conserved (over 60% in average). Against such a tendency, BchD, BchJ, BchO, and Bch2 show rather low conservativeness. BchD is a subunit of the magnesium chelatase consisting of BchI, BchD and BchH and

Table 5.2 Sequence conservativeness of products of genes found in PGC of 24 species and strains of purple bacteria

Gene product	Number of orthologues found	Average of sequence identities	Standard deviation	Minimum identity	Functions
BchB	24	68.0	4.6	58.1	Protochlorophyllide reductase, subunit B
BchC	24	59.8	5.9	48.1	3-Hydroxyethyl bacteriochlorophyllide a dehydrogenase
BchD	24	48.3	6.2	38.4	Mg-chelatase, subunit D
BchE	16	66.4	6.2	54.9	Mg-protoporphyrin IX monomethyl ester oxidative cyclase
BchF	24	62.5	5.4	50.8	3-Vinyl bacteriochlorophyllide a hydroxylase
BchG	24	61.3	5.6	50.3	Bacteriochlorophyll synthase
BchH	24	61.9	5.4	53.7	Mg-chelatase, subunit H
BchI	24	68.1	4.6	59.6	Mg-chelatase, subunit I
BchJ	16	42.2	10.0	28.6	8-Vinyl reductase
BchL	24	74.4	4.6	64.9	Protochlorophyllide reductase, subunit L
BchM	24	57.5	7.0	43.6	Mg-protoporphyrin IX SAM O-methyltransferase
BchN	24	65.9	5.4	57.2	Protochlorophyllide reductase, subunit N
BchO	24	44.6	7.4	31.6	Unknown
BchP	24	61.9	5.4	50.5	Geranylgeranyl reductase
BchX	24	76.4	3.7	68.9	Chlorin reductase, subunit X
BchY	24	70.5	4.0	62.1	Chlorin reductase, subunit Y
BchZ	24	72.0	5.0	61.6	Chlorin reductase, subunit Z
Bch2	21	52.4	6.4	38.5	Unknown
AcsF	18	59.5	7.5	47.8	Mg-protoporphyrin IX monomethyl ester oxidative cyclase
CrtB	24	46.7	9.1	28.6	Phytoene synthase
CrtC	24	45.1	7.1	31.8	C-1,2 hydratase (Hydroxyneurosporene synthase)

Continued

Table 5.2 Sequence conservativeness of products of genes found in PGC of 24 species and strains of purple bacteria—cont'd

Gene product	Number of orthologues found	Average of sequence identities	Standard deviation	Minimum identity	Functions
CrtD	25	49.3	5.8	38.5	C-3,4 desaturase (Methoxyneurosporene dehydrogenase)
CrtE	24	50.7	12.1	25.3	Geranylgeranyl pyrophospate synthase
CrtF	24	47.0	6.6	36.4	Hydroxynuerosporene-O-methyltransferase
CrtI	24	54.1	9.2	38.2	Phytoene desaturase
PufA	28	53.5	10.6	27.0	LH 1 subunit alpha
PufB	28	52.2	12.0	32.0	LH 1 subunit beta
PufC	16	45.1	9.7	31.6	Photosynthetic reaction centre cytochrome c subunit
PufL	24	73.9	6.0	60.6	Photosynthetic reaction centre L subunit
PufM	24	68.4	6.4	56.0	Photosynthetic reaction centre M subunit
PuhA	24	46.7	7.2	33.6	Photosynthetic reaction centre H subunit
PuhB	24	38.6	6.3	26.3	Possible assembly factor of RC
PuhC	24	29.9	7.2	18.9	Possible reorganisation factor of RC/LH1/PufX
PuhD	14	47.4	12.9	29.2	Unknown
PuhE	24	40.0	7.8	27.9	Possible assembly factor of RC/LH1
LhaA	24	57.5	5.9	43.1	Possible assembly factor of LH1
PpsR (CrtJ)	26	37.2	9.0	24.0	Photosynthesis-specific repressor
PpaA (AerR)	19	27.1	8.5	12.3	Transcriptional regulator

Products of photosynthetic genes found in more than half of the 24 strains selected are listed. The number of the compared sequences can be over 24 when some species have multiple copies. Functions of *bch* and *crt* gene products were assigned in reference to Willows & Kriegel, 2009 and Takaichi 2009, respectively.

forms a stable complex with BchI, which provides ATP hydrolysis activity (Lucien et al., 1999). The amino acid sequence of BchD is similar to that of the N-terminal part of BchI but lacks the motif binding to ATP (Fodje et al., 2001). A structural constraint to maintain the function might be low in the evolutionary course of BchD. BchJ has been identified as 4-vinyl reductase (8-vinyl reductase, according to International Union of Pure and Applied Chemistry tetrapyrrole numbering) to reduce divinyl protochlorophyllide to monovinyl protochlorophyllide, but its deletion mutants are capable of synthesising bacteriochlorophylls in *Rhodobacter* species (Bollivar et al., 1994). Instead, it was recently proposed that BchJ should be a carrier protein of Mg-protoporphyrin IX, similar to the porphyrin carrier protein, Gun4, in oxygenic phototrophs (Sawicki & Willows, 2010). The mechanisms for these aspects of the bacteriochlorophyll biosynthetic pathway are still unclear. At least, BchJ does not seem to be essential to the pathway since one-third of the species addressed here lack the gene that possibly codes this protein and, in addition, recognisable homologues have weak sequence similarity. Another mystery is that species lacking the *bchJ* gene always lack the *bchE* gene, so far as we have found. Possible products of two other *bch* genes with low sequence conservation, BchO and Bch2, have not been assigned enzymatic activities in the bacteriochlorophyll biosynthetic pathway. These genes are often situated in specific loci in the PGCs of most purple photosynthetic bacteria, although their physiological functions remain ambiguous.

Carotenoids have two important functions in photosynthetic bacteria: (1) absorption of light energy in the wavelength region 400–550 nm, which is not covered by bacteriochlorophylls and (2) photoprotection against harmful free radicals (Tuveson, Larson, & Kagan, 1988). Many non-photosynthetic species also produce various carotenoids, possibly because of the latter function. Spheroidene and spirilloxanthin are the major carotenoids and final products of the biosynthetic pathways in many purple photosynthetic bacteria, but intermediary products such as lycopene and neurosporene are also accumulated. The carotenoid composition varies according to the species. The genes coding for the carotenoid biosynthetic enzymes, *crtB*, *crtC*, *crtD*, *crtE*, *crtF*, and *crtI*, are included in the PGC in almost all photosynthetic purple bacteria, although the sequence conservation of each enzyme is not very high. The gene coding for the ketorase specifically recognising the C-2 position of spheroidene and/or spirilloxanthin to produce spheroidenone and/or 2-ketospirillozanthin, *crtA* (Gerjets, Steiger, & Sandmann, 2009), is not detected in many species. Genes coding for proteins other than bacteriochlorophyll and carotenoid biosynthesis enzymes found in the PGC of *Rhodobacter* species are not very highly conserved except for *pufB* and *pufA*, encoding LH proteins, and *pufL* and *pufM* coding for the core proteins of the photochemical reaction centre complex. Details of these genes are discussed below.

4. COMPARISON OF GENE ARRANGEMENTS IN PGCs OF PURPLE BACTERIA

Among purple photosynthetic bacteria, about 40 genes required for photosynthesis show a strong tendency to form a large cluster, as has been shown in *Rhodobacter* species. Such PGCs may be split into several subclusters in some species. Figure 5.2 shows the arrangements of genes in the PGCs of the selected 24 species of purple photosynthetic bacteria for which the total genomic information is available. The arrangements seem to be irregular at first glance, and the difference seems to be pronounced when compared between less closely related species. However, it is noteworthy that the arrangements in several subclusters are well conserved and correspond to operonal units that have been studied so far in *Rhodobacter* and other species.

The genes encoding the L and M core subunits of the RC complex are tandemly arranged and form an operon called *puf* together with the genes coding the alpha and beta subunits of the LH1 complex. The *puf* operon has a transcriptional promoter controlled by the RegA/RegB two-component system (Bauer, Setterdahl, Wu, & Robinson, 2009) but has been known to be co-transcribed with upstream genes responsible for a part of the carotenoid and bacteriochlorophyll biosynthetic processes, *crtE*, *crtF*, *bchC*, *bchX*, *bchY*, and *bchZ*, in *R. capsulatus* (Wellington, Taggart, & Beatty, 1991). In the regions directly upstream of *crtE* and *bchC*, transcription initiation sites presumably regulated by PpsR/CrtJ are located (Penfold & Pemberton, 1994), and, thus, long transcripts covering these *crt*, *bch*, and *puf* genes are generated. Such superoperonal *crtEF–bchCXYZ–pufBALM* gene arrangement has been found in all the species of purple bacteria except for *Rhodomicrobium vannielii*, *Brevundimonas subvibrioides*, *Erythrobacter* sp. NAP-1, and *T. marina* in which neither *crtE* nor *crtF* is included. Some species have additional ORFs between *bch* and *puf* in this subcluster. For example, in *A. vinosum*, four small ORFs with unknown functions are present. In *Rhodobacter* and *Roseobacter* species, *pufQ*, the product that is involved in bacteriochlorophyll *a* biosynthesis (Bauer & Marrs, 1988; Fidai, Dahl, & Richards, 1995) is found in the region directly upstream of *pufB*. Species possessing a multiple haem cytochrome subunit bound to the RC have the gene *pufC* at the region downstream of *pufM*, but this location is replaced by *pufX* encoding a membrane protein associated with the LH1 complex in *Rhodobacter* species, which has no RC-bound cytochrome subunits.

The gene coding for the H subunit of RC, *puhA*, has been known to form a well-conserved gene arrangement with bacteriochlorophyll biosynthesis genes such as *bchFNBHLM–lhaA–puhA*. Exceptions are those of *A. vinosum* and *T. marina*, in which the *lhaA* gene is absent. The gene called *lhaA*, flanked by *bchM* and *puhA*, has been shown to contribute to the assembly of the LH1 complex in *R. capsulatus* (Young, Reyes, & Beatty, 1998). In addition, recent studies have shown that disruptions of genes located in the region downstream of *puhA* in *R. capsulatus* cause the reduction of bacteriochlorophyll biosynthesis or decrease in stability of the RC-LH1 complex, and may impair growth under the photosynthetic conditions (Aklujkar, Prince, & Beatty, 2005a, 2005b, 2006). These genes have been named *puhB*, *puhC*, and *puhE* and are possibly co-transcribed with *puhA*. All the species shown in Fig. 5.2 have ORFs showing high sequence identities (more than 29%) to these *R. capsulatus puh* genes at the corresponding locations. Many of the Alphaproteobacteria species have an other small ORF at the direct downstream of *puhC*. Here, this ORF is tentatively named *puhD*, although its function has not been examined. Possibly, *puhBCDE* genes (often includes *acsF* directly upstream of *puhE*) are members of the *bchFNBHLM–lhaA–puh* superoperonal gene subcluster, and their products are not essential but advantageous for photosynthesis.

In many species, *acsF*, a gene encoding Mg-protoporphyrin IX monomethyl ester cyclase, is found among *puh* genes, especially between *puhC/D* and *puhE*. The enzymatic function assigned to AcsF is the same as that to BchE, but AcsF activity exists only under aerobic growth conditions in a Betaproteobacterium, *R. gelatinosus* (Pinta, Picaud, Reiss-Husson, & Astier, 2002). Some species, including *R. capsulatus* and *A. vinosum*, lack *acsF* in their genomes. On the other hand, many of the aerobic photosynthetic species lack *bchE* but possess *acsF*. The primary structures of these two gene products are completely different. The reason that only this step in the bacteriochlorophyll biosynthetic pathway is redundant is unclear, but may relate to whether bacteriochlorophyll is produced mainly under aerobic or anaerobic conditions. The locations of *bchE* and *bchJ* in PGCs do not seem to be fixed and, in more than a few species, are completely separate from the PGCs. Nevertheless, these genes tend to be tandemly arranged as *bchEJ*.

One of the other well-conserved gene arrangements in the PGC is the *bchIDO* subcluster. The gene products of *bchI* and *bchD* form an enzyme complex of Mg-chelatase with the large subunit coded by *bchH* (Gibson, Willows, Kannangara, von Wettstein, & Hunter, 1995; Gibson, Jensen, & Hunter, 1999) found in the superoperonal gene subcluster, *bchFNBHLM–lhaA–puh*. The function of *bchO* has not been well characterised and has often been described as hypothetical or an ORF with unknown functions; here, however, it is referred to as *bchO* because of its well-conserved structure (44.6% sequence identities) and broad distribution in purple photosynthetic bacteria, as seen in Fig. 5.2. In some species, the *bchIDO* subcluster is separated by insertions of other genes, as seen in *Bradyrhizobium* species and *R. palustris*. *bchI* and *bchD* are usually tandem but completely detached only in *M. extorquens*.

bchG and *bchP*, the products of which catalyse the final steps of bacteriochlorophyll biosynthesis, often form a cluster with an ORF showing high sequence similarity with *pucC* as *bchG*–ORF–*bchP*. This ORF is well conserved on the basis of the amino acid sequence comparison of the products and is tentatively called *bch2*. However, this ORF is missing in *Acidiphilium multivorum*, *A. vinosum*, and *T. marina*. In 19 out of 24 species shown in Fig. 5.2, *ppsR*, a gene encoding a repressor protein of transcription of various photosynthesis genes, is found directly upstream of *bchG*. Some species, i.e. *R. palustris* and *Bradyrhizobium* species, have two *ppsR* genes in their PGCs. The region upstream of *ppsR* is usually occupied by *ppaA*, which codes an oxygen-dependent transcription factor (Gomelsky et al., 2003). However, the *ppaA* gene is not included in the PGCs of Gammaproteobacteria species, and not even *ppsR* could be found in *Halorhodospira halophila*.

Genes encoding enzymes for carotenoid biosynthesis often form a cluster, although the directions of transcription are not uniform. As already stated, *crtE* and *crtF* are tandemly arranged and included in the superoperonal subcluster *crtEF–bchCXYZ–pufBALM* in many purple bacteria. *crtD* and *crtC* also have a strong tendency to be tandemly arranged as *crtDC* and are often found in the region directly upstream of the *crtEF–bchCXYZ–pufBALM* subcluster with the opposite transcriptional direction. *crtI* and *crtB* code for enzymes showing some sequence identities to the products of *crtD* and *crtC*, respectively, and also have the tendency to be tandemly arranged as *crtIB*. When the *crtIB* genes are clustered with other *crt* genes, they are usually located in the region directly downstream of *crtDC* in the opposite transcriptional direction.

5. PHYLOGENETIC ANALYSES OF GENES IN PGCs AND HORIZONTAL GENE TRANSFER

As noted above, photosynthetic species are not a monophyletic group but are phylogenetically distributed with many non-photosynthetic species. Such a distribution has been ascribed to the loss of photosynthesis genes in different lineages during the course of evolution of Proteobacteria (Woese, 1987). On the other hand, horizontal (interspecies) transfer of photosynthesis genes has also been suggested to explain the inconsistencies between the branching topologies of the phylogenetic trees based on photosynthetic genes; for example, genes coding for the photochemical reaction centre (*puf* genes) and phylogenetic marker genes, i.e. the 16S rRNA gene (Nagashima et al., 1997). Here, phylogenetic analysis using the information of photosynthetic genes that has become available through genome projects is performed to review the suspected course of evolution of photosynthesis within the purple bacteria (Proteobacteria). The phylogenetic tree based on the 16S rRNA sequences of the selected 24 species of purple photosynthetic bacteria (used in Table 5.2 and Fig. 5.2) show clear branching to divide the species into three groups, alpha, beta, and gamma, corresponding to the Alphaproteobacteria, Betaproteobacteria, and Gammaproteobacteria, as shown in Fig. 5.3. The alpha group can be further divided into four subgroups, alpha-1 to -4, although the alpha-1 subgroup does not show a clear clustering, as has been presented in various publications. When the amino acid sequences of the subunits of the cytochrome bc_1 complex (PetB/PetC) were used in the calculation, the tree topology is nearly consistent with that of 16S rRNA tree, as shown in Fig. 5.3. It is quite natural to think that these phylogenetic trees reflect the species evolution and that respiration mechanisms including the cytochrome bc_1 complex have evolved in parallel with the species.

However, phylogenetic trees based on the sequences of photosynthesis gene products show different branching shapes. The most pronounced feature found in the phylogenetic tree based on the amino acid sequences of the L and M subunits of the photochemical reaction centre complex is that the species of Betaproteobacteria and Gammaproteobacteria are positioned among Alphaproteobacteria, especially among the alpha-1 and alpha-2 clusters, in contrast to the branching patterns in the 16S rRNA tree (Fig. 5.3). The branching pattern of the Alphaproteobacteria cluster is basically consistent with that in the 16S rRNA tree in this tree, although some nesting distributions are observed between alpha-1 and alpha-2, and between

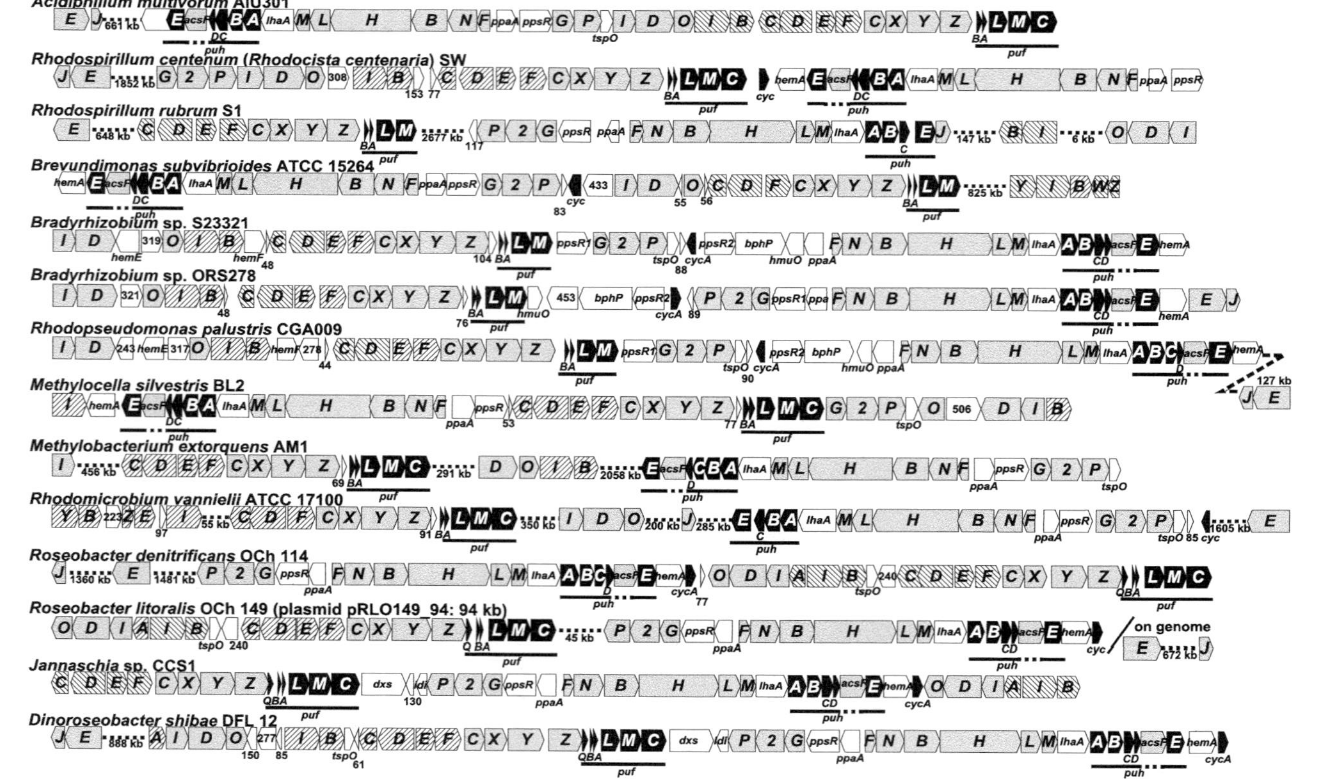
Acidiphilium multivorum AIU301
Rhodospirillum centenum (Rhodocista centenaria) SW
Rhodospirillum rubrum S1
Brevundimonas subvibrioides ATCC 15264
Bradyrhizobium sp. S23321
Bradyrhizobium sp. ORS278
Rhodopseudomonas palustris CGA009
Methylocella silvestris BL2
Methylobacterium extorquens AM1
Rhodomicrobium vannielii ATCC 17100
Roseobacter denitrificans OCh 114
Roseobacter litoralis OCh 149 (plasmid pRLO149_94: 94 kb)
Jannaschia sp. CCS1
Dinoroseobacter shibae DFL 12

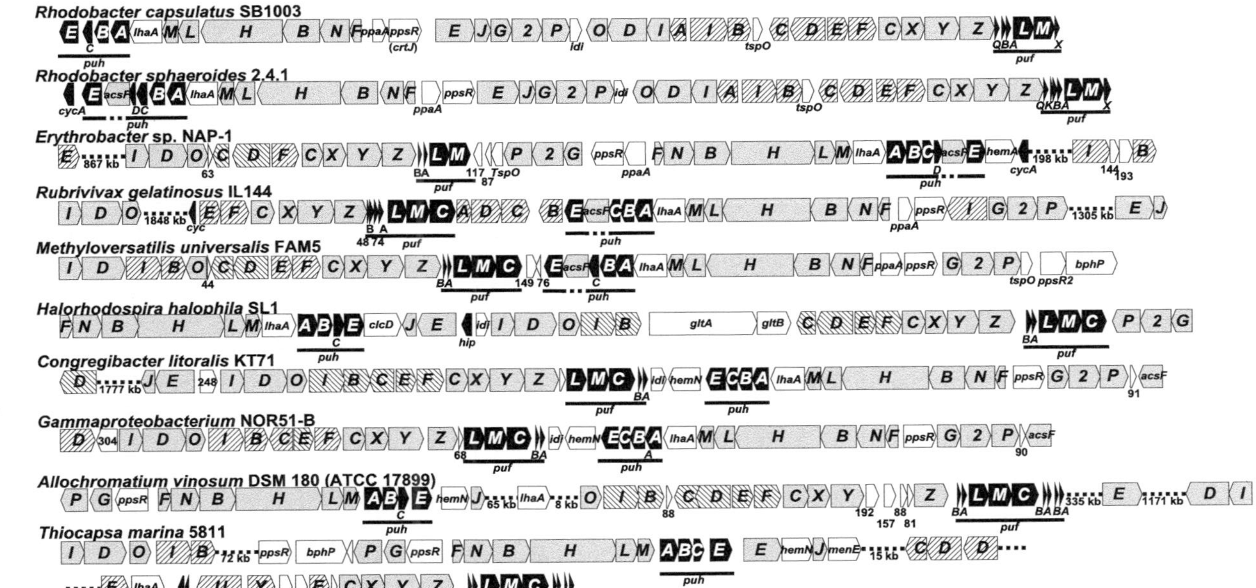

Figure 5.2 ***Arrangements of genes in the PGCs found in the total genomic sequences of 24 species of purple bacteria.*** The genes are presented as arrows pointing in the direction of their transcriptions. Genes coding for the LH and the RC apoproteins and the related products are indicated by solid arrows (*puf* and *puh*). Genes assigned to bacteriochlorophyll (*bch* and *acsF*) and carotenoid (*crt*) biosynthesis genes are shown by green and orange arrows, respectively. ORFs without assigned functions are marked by the lengths of the amino acid sequences of their predicted products. The PGCs are ordered corresponding to the layout of the species in the 16S rRNA tree shown in Fig. 5.3. Each of the PGCs is displayed to direct the *puf* gene transcription at right. See the color plate.

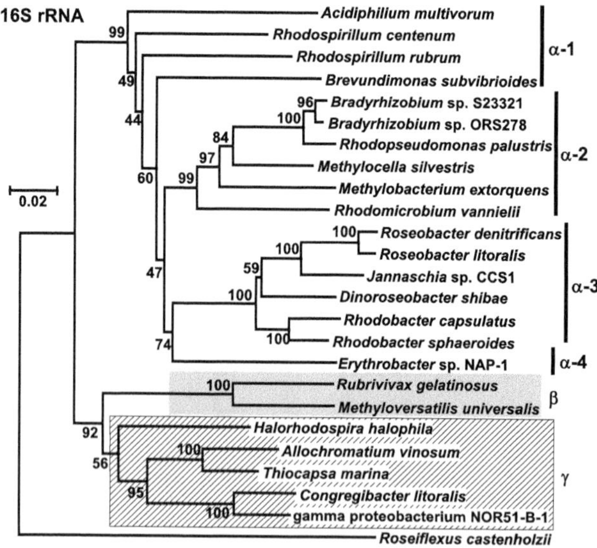

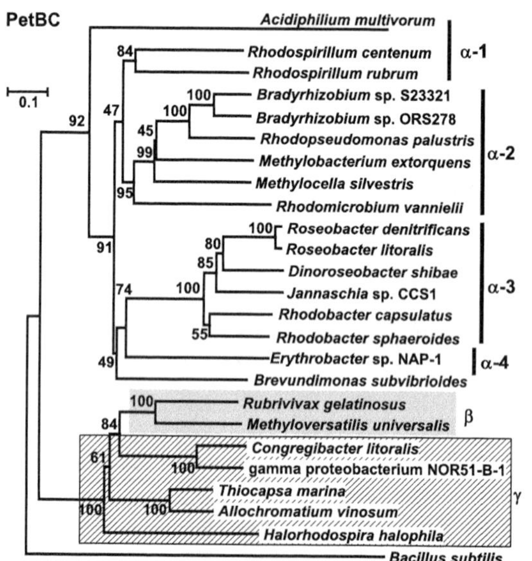

Figure 5.3 *Phylogenetic trees based on the nucleotide sequences of 16S rRNA and suspected amino acid sequences of subunits of the cytochrome* bc_1 *complex, PetB and PetC, L and M subunits of the RC complex, and bacteriochlorophyll biosynthesis enzymes BchLNB and BchXYZ.* The sequence from a gram-positive bacterium, *Bacillus subtilis*, was used as an outgroup in the phylogenetic tree of PetBC. Phylogenetic trees were drawn using the programs ClustalX (Thompson et al., 1997) and MEGA (Kumar et al., 2004, http://www.megasoftware.net/). All gaps in the sequence alignment were omitted in a pairwise manner. Construction of the trees was performed by the neighbour-joining method, applying the Kimura 2-parameter distance estimator to the 16S rRNA tree and the Poisson correction estimator to the others. The obtained bootstrap values are presented at the corresponding nodes. (For colour version of this figure, the reader is referred to the online version of this book.)

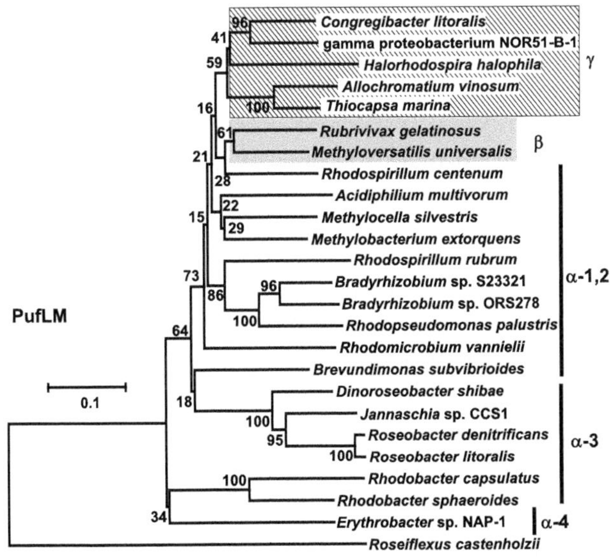

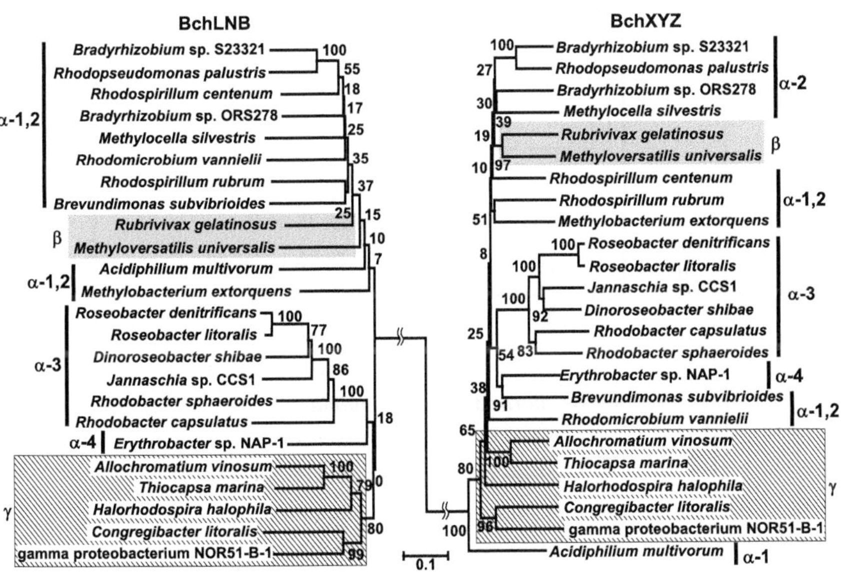

Figure 5.3 Cont'd

alpha-3 and alpha-4, possibly because of a bias resulting from functional evolutionary pressures, i.e. presence or absence of the cytochrome subunit. Such a discrepancy between the tree topologies of PufL/M and 16S rRNA, as has been proposed in studies using a restricted number of species (Nagashima et al., 1997), can be ascribed to the horizontal transfer of the photosynthesis genes between ancestral species of the Alphaproteobacteria and the Beta/Gammaproteobacteria. Analyses using the sequence information of other photosynthesis genes also produce the same conclusion. Figure 5.3 includes a phylogenetic tree based on the amino acid sequence comparison of BchLNB and BchXYZ, both of which show significantly conserved primary structures. In this tree, the species of Betaproteobacteria and

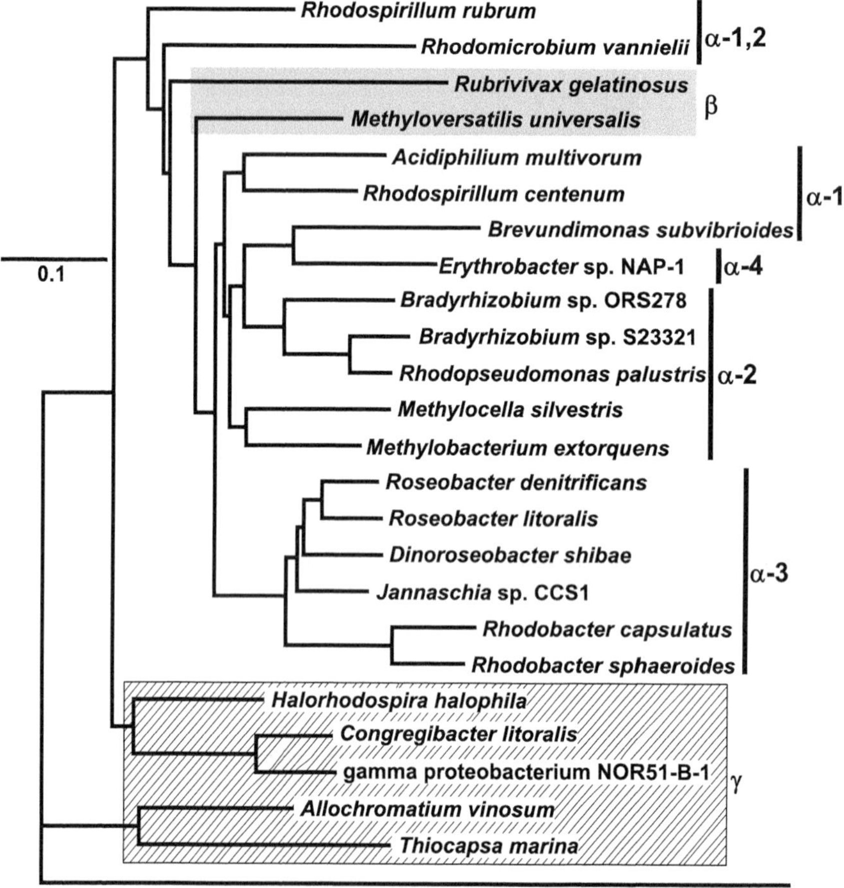

Figure 5.4 ***Phylogenetic tree based on difference in the arrangement of photosynthesis genes.*** Genes directly upstream and downstream of each of the common photosynthesis genes are compared between every two strains to obtain the distance factor represented by the ratios of the numbers of the consistent genes to the numbers of the compared loci. A 25 × 25 strains distance matrix was constructed and analysed by the programs PHYLIP (Felsenstein, 2004) and TreeView (Page, 1996). (For colour version of this figure, the reader is referred to the online version of this book.)

Gammaproteobacteria are again positioned in Alphaproteobacteria, especially among the alpha-1 and alpha-2 groups, while basically maintaining the branching patterns. This feature is commonly observed in the phylogenetic trees calculated from the products of other *bch* genes and *crt*, *puf*, and *puh* and the related genes included in PGCs.

Inconsistencies between gene arrangements in PGCs may result from successive events of gene relocations, i.e. insertions, deletions, inversions, and transpositions. Assuming that the frequency of such events is constant, the difference in gene arrangements among distinct species reflects the time of the divergence and can be used in construction of phylogenetic trees. Indeed, such a method has been used to estimate phylogenetic relationships (Sankoff et al., 1992). Here, a phylogenetic tree was constructed on the basis of the difference in the arrangements of the photosynthetic genes, as shown in Fig. 5.4. The relationships of the species of Alphaproteobacteria in this tree are consistent with those shown in the 16S rRNA tree. Species of Betaproteobacteria are positioned within the cluster consisting the species of alpha-1 and alpha-2 groups, being consistent with that shown in the phylogenetic trees based on the difference in the sequences of the photosynthetic gene products, and supporting the idea of horizontal gene transfer. On the other hand, species of Gammaproteobacteria are separately positioned from others in the tree based on the PGC rearrangement, as seen in the 16S rRNA tree. However, the branches connecting these species are not well congruent at the root. Possibly, selection pressure to maintain the photosynthesis gene arrangement is relatively low in Gammaproteobacteria.

6. CONCLUSION AND PERSPECTIVES

Genes specifically required for the photosynthetic phenotype, genes coding for the photochemical reaction centre and LH complexes (*puf*, *puhA*), bacteriochlorophyll biosynthesis genes (*bch*, *acsF*), carotenoid biosynthesis genes (*crt*), and genes encoding proteins functioning in the assembly, maintenance, and expression control of these products are basically assembled to form homologous PGCs in purple bacteria (Proteobacteria). This assembly seems to consist of subclusters with recognisable patterns, i.e. *crtEF–bchCXYZ–pufBALM*, *bchFNBHLM–lhaA–puhA*, *bchIDO*, and *bchG2P*. The genes included in these subclusters are presumably co-transcribed as operons or superoperons, as has in *Rhodobacter* species. Such an assembly is advantageous in the concerted expression of photosynthesis genes according to environmental changes. One concern is how such an arrangement has been established.

On the basis of phylogenetic analyses of the genes in the PGC, it is almost certain that the present photosynthetic apparatuses in species belonging to Beta/Gammaproteobacteria are derived from those of the ancestral species of Alphaproteobacteria through horizontal gene transfer. Such an event might be mediated by a broad-host-range plasmid or a bacteriophage-related element such as gene transfer agent (GTA), known in *R. capsulatus* (Fogg, Westbye, & Beatty, 2012). A plasmid recently found in the aerobic photosynthetic species *Roseobacter litoralis*, which contains a nearly entire PGC (Kalhoefer et al., 2011), may be a living candidate of such a mediator. Such a form of the PGC should facilitate transfer.

The question is whether the newly introduced PGC is successfully expressed, i.e. whether the RC complex newly obtained is easily incorporated into the existing electron transfer system in the host. If this coordination fails, unwanted energy brought by the new RC may damage the host cells. A thorough understanding of the biochemical and physiological factors required for the successful incorporation of foreign photosynthesis systems is a worthy pursuit. It is currently understood that a simple introduction of a plasmid containing the *R. capsulatus* PGC to the cells of *Escherichia coli* cause no photosynthetic phenotypes (Marrs, 1981). However, recent interest in renewable resources has promoted the attempt to achieve the functional incorporation of the photosynthetic apparatuses into the cells of *E. coli* (Johnson & Schmidt-Dannert, 2008). Such an attempt will be also useful for studies on the horizontal transfer of PGC and its evolutionary course.

ACKNOWLEDGEMENTS

This work was supported by PRESTO of the Japan Science and Technology Agency.

REFERENCES

Aklujkar, M., Prince, R. C., & Beatty, J. T. (2005a). The *puhE* gene of *Rhodobacter capsulatus* is needed for optimal transition from aerobic to photosynthetic growth and encodes a putative negative modulator of bacteriochlorophyll production. *Archives of Biochemistry and Biophysics*, *437*, 186–198.

Aklujkar, M., Prince, R. C., & Beatty, J. T. (2005b). The PuhB protein of *Rhodobacter capsulatus* functions in photosynthetic reaction center assembly with a secondary effect on light-harvesting complex 1. *Journal of Bacteriology*, *187*, 1334–1343.

Aklujkar, M., Prince, R. C., & Beatty, J. T. (2006). The photosynthetic deficiency due to *puhC* gene deletion in *Rhodobacter capsulatus* suggests a PuhC protein-dependent process of RC/LH1/PufX complex reorganisation. *Archives of Biochemistry and Biophysics*, *454*, 59–71.

Armstrong, G. A., Schmidt, A., Sandmann, G., & Hearst, J. E. (1990). Genetic and biochemical characterization of carotenoid biosynthesis mutants of *Rhodobacter capsulatus*. *Journal of Biological Chemistry*, *265*, 8329–8338.

Baldock, M. I., Denger, K., Smits, T. H., & Cook, A. M. (2007). *Roseovarius* sp. strain 217: aerobic taurine dissimilation via acetate kinase and acetate-CoA ligase. *FEMS Microbiology Letters, 271*, 202–206.

Bauer, C. E., & Marrs, B. L. (1988). *Rhodobacter capsulatus puf* operon encodes a regulatory protein (PufQ) for bacteriochlorophyll biosynthesis. *Proceedings of National Academy of Science of the United States of America, 85*, 7074–7078.

Bauer, C. E., Setterdahl, A., Wu, J., & Robinson, B. R. (2009). Reguration of gene expression in response to oxygen tension. In C. N. Hunter, F. Daldal, M. C. Thurnauer & J. T. Beatty (Eds.), *The purple phototrophic bacteria* (pp. 707–725). Dordrecht, The Netherlands: Springer.

Bollivar, D. W., Suzuki, J. Y., Beatty, J. T., Dobrowolski, J. M., & Bauer, C. E. (1994). Directed mutational analysis of bacteriochlorophyll *a* biosynthesis in *Rhodobacter capsulatus*. *Journal of Molecular Biology, 237*, 622–640.

Brown, P. J., Kysela, D. T., Buechlein, A., Hemmerich, C., & Brun, Y. V. (2011). Genome sequences of eight morphologically diverse Alphaproteobacteria. *Journal of Bacteriology, 193*, 4567–4568.

Burke, D. H., Hearst, J. E., & Sidow, A. (1993). Early evolution of photosynthesis: clues from nitrogenase and chlorophyll iron proteins. *Proceedings of National Academy of Science of the United States of America, 90*, 7134–7138.

Chen, Y., Crombie, A., Rahman, M. T., Dedysh, S. N., Liesack, W., Stott, M. B., et al. (2010). Complete genome sequence of the aerobic facultative methanotroph *Methylocella silvestris* BL2. *Journal of Bacteriology, 192*, 3840–3841.

Choudhary, M., Fu, Y. X., Mackenzie, C., & Kaplan, S. (2004). DNA sequence duplication in *Rhodobacter sphaeroides* 2.4.1: evidence of an ancient partnership between chromosomes I and II. *Journal of Bacteriology, 186*, 2019–2027.

Choudhary, M., Zanhua, X., Fu, Y. X., & Kaplan, S. (2007). Genome analyses of three strains of *Rhodobacter sphaeroides*: evidence of rapid evolution of chromosome II. *Journal of Bacteriology, 189*, 1914–1921.

Coomber, S. A., Chaudhri, M., Connor, A., Britton, G., & Hunter, C. N. (1990). Localized transposon Tn5 mutagenesis of the photosynthetic gene cluster of *Rhodobacter sphaeroides*. *Molecular Microbiology, 4*, 977–989.

Duquesne, K., & Sturgis, J. N. (2012). Shotgun genome sequence of the large purple photosynthetic bacterium *Rhodospirillum photometricum* DSM122. *Journal of Bacteriology, 194*, 2380.

Felsenstein, J. (2004). *PHYLIP (Phylogeny Inference Package) version 3.6.* Distributed by the author, Seattle: Department of Genome Sciences, University of Washington.

Fidai, S., Dahl, J. A., & Richards, W. R. (1995). Effect of the PufQ protein on early steps in the pathway of bacteriochlorophyll biosynthesis in *Rhodobacter capsulatus*. *FEBS Letters, 372*, 264–268.

Fodje, M. N., Hansson, A., Hansson, M., Olsen, J. G., Gough, S., Willows, R. D., et al. (2001). Interplay between an AAA module and an integrin I domain may regulate the function of magnesium chelatase. *Journal of Molecular Biology, 311*, 111–122.

Fogg, P. C., Westbye, A. B., & Beatty, J. T. (2012). One for all or all for one: heterogeneous expression and host cell lysis are key to gene transfer agent activity in *Rhodobacter capsulatus*. *PLoS One*, 7. e43772.

Fuchs, B. M., Spring, S., Teeling, H., Quast, C., Wulf, J., Schattenhofer, M., et al. (2007). Characterization of a marine gammaproteobacterium capable of aerobic anoxygenic photosynthesis. *Proceedings of National Academy of Science of the United States of America, 104*, 2891–2896.

Gerjets, T., Steiger, S., & Sandmann, G. (2009). Catalytic properties of the expressed acyclic carotenoid 2-ketolases from *Rhodobacter capsulatus* and *Rubrivivax gelatinosus*. *Biochimica et Biophysica Acta, 1791*, 125–131.

Gibson, L. C. D., Jensen, P. E., & Hunter, C. N. (1999). Magnesium chelatase from *Rhodobacter sphaeroides*: initial characterization of the enzyme using purified subunits and evidence for a BchI–BchD complex. *Biochemical Journal, 337*, 243–251.

Gibson, L. C. D., Willows, R. D., Kannangara, C. G., von Wettstein, D., & Hunter, C. N. (1995). Magnesium-protoporphyrin chelatase of *Rhodobacter sphaeroides*: reconstitution of activity by combining the products of the *bchH, -I,* and *-D* genes expressed in *Escherichia coli*. *Proceedings of National Academy of Science of the United States of America, 92*, 1941–1944.

Giraud, E., Moulin, L., Vallenet, D., Barbe, V., Cytryn, E., Avarre, J. C., et al. (2007). Legumes symbioses: absence of *Nod* genes in photosynthetic bradyrhizobia. *Science, 316*, 1307–1312.

Gomelsky, L., Sram, J., Moskvin, O.V., Horne, I. M., Dodd, H. N., Pemberton, J. M., et al. (2003). Identification and in vivo characterization of PpaA, a regulator of photosystem formation in *Rhodobacter sphaeroides*. *Microbiology, 149*, 377–388.

Igarashi, N., Harada, J., Nagashima, S., Matsuura, K., Shimada, K., & Nagashima, K. V. P. (2001). Horizontal transfer of the photosynthesis gene cluster and operon rearrangement in purple bacteria. *Journal of Molecular Evolution, 52*, 333–341.

Johnson, E. T., & Schmidt-Dannert, C. (2008). Light-energy conversion in engineered microorganisms. *Trends in Biotechnology, 26*, 682–689.

Kalhoefer, D., Thole, S., Voget, S., Lehmann, R., Liesegang, H., Wollher, A., et al. (2011). Comparative genome analysis and genome-guided physiological analysis of *Roseobacter litoralis*. *BMC Genomics, 12*, 324.

Kang, I., Oh, H. M., Lim, S. I., Ferriera, S., Giovannoni, S. J., & Cho, J. C. (2010). Genome sequence of *Fulvimarina pelagi* HTCC2506T, a Mn(II)-oxidizing alphaproteobacterium possessing an aerobic anoxygenic photosynthetic gene cluster and Xanthorhodopsin. *Journal of Bacteriology, 192*, 4798–4799.

Kang, I., Oh, H. M., Vergin, K. L., Giovannoni, S. J., & Cho, J. C. (2010). Genome sequence of the marine alphaproteobacterium HTCC2150, assigned to the *Roseobacter* clade. *Journal of Bacteriology, 192*, 6315–6316.

Kittichotirat, W., Good, N. M., Hall, R., Bringel, F., Lajus, A., Medigue, C., et al. (2011). Genome sequence of *Methyloversatilis universalis* FAM5T, a methylotrophic representative of the order *Rhodocyclales*. *Journal of Bacteriology, 193*, 4541–4542.

Koblížek, M., Janouškovec, J., Oborník, M., Johnson, J. H., Ferriera, S., & Falkowski, P. G. (2011). Genome sequence of the marine photoheterotrophic bacterium *Erythrobacter* sp. strain NAP1. *Journal of Bacteriology, 193*, 5881–5882.

Kovács, A. T., Rákhely, G., & Kovács, K. L. (2003). Genes involved in the biosynthesis of photosynthetic pigments in the purple sulfur photosynthetic bacterium *Thiocapsa roseopersicina*. *Applied Environmental Microbiology, 69*, 3093–3102.

Kumar, S., Tamura, K., & Nei, M. (2004). MEGA3: integrated software for molecular evolutionary genetics analysis and sequence alignment. *Briefings in Bioinformatics, 5*, 150–163.

Larimer, F. W., Chain, P., Hauser, L., Lamerdin, J., Malfatti, S., Do, L., et al. (2004). Complete genome sequence of the metabolically versatile photosynthetic bacterium *Rhodopseudomonas palustris*. *Nature Biotechnology, 22*, 55–61.

Lavergne, J., Verméglio, A., & Joliot, P. (2009). Functional coupling between reaction centers and cytochrome *bc1* complexes. In C. N. Hunter, F. Daldal, M. C. Thurnauer & J. T. Beatty (Eds.), *The purple phototrophic bacteria* (pp. 509–536). Dordrecht, The Netherlands: Springer.

Lim, S. K., Kim, S. J., Cha, S. H., Oh, Y. K., Rhee, H. J., Kim, M. S., et al. (2009). Complete genome sequence of *Rhodobacter sphaeroides* KD131. *Journal of Bacteriology, 191*, 1118–1119.

Lonjers, Z. T., Dickson, E. L., Chu, T. P., Kreutz, J. E., Neacsu, F. A., Anders, K. R., et al. (2012). Identification of a new gene required for the biosynthesis of rhodoquinone in *Rhodospirillum rubrum*. *Journal of Bacteriology, 194*, 965–971.

Lu, Y. K., Marden, J., Han, M., Swingley, W. D., Mastrian, S. D., Chowdhury, S. R., et al. (2010). Metabolic flexibility revealed in the genome of the cyst-forming alpha-1 proteobacterium *Rhodospirillum centenum*. *BMC Genomics, 11*, 325.

Lucien, C. D. G., Jensen, P. E., & Hunter, C. N. (1999). Magnesium chelatase from *Rhodobacter sphaeroides*: initial characterization of the enzyme using purified subunits and evidence for a BchI–BchD complex. *Biochemical Journal, 337*, 243–251.

Magnuson, T. S., Swenson, M. W., Paszczynski, A. J., Deobald, L. A., Kerk, D., & Cummings, D. E. (2010). Proteogenomic and functional analysis of chromate reduction in *Acidiphilium cryptum* JF-5, an Fe(III)-respiring acidophile. *Biometals, 23*, 1129–1138.

Marrs, B. (1981). Mobilization of the genes for photosynthesis from *Rhodopseudomonas capsulata* by a promiscuous plasmid. *Journal of Bacteriology, 146*, 1003–1012.

Martin-Uriz, P. S., Gomez, M. J., Arcas, A., Bargiela, R., & Amils, R. (2011). Draft genome sequence of the electricigen *Acidiphilium* sp. strain PM (DSM 24941). *Journal of Bacteriology, 193*, 5585–5586.

Mohammed, M., Isukapatla, A., Mekala, L. P., Eedara Veera Venkata, R. P., Chintalapati, S., & Chintalapati, V. R. (2011). Genome sequence of the phototrophic betaproteobacterium *Rubrivivax benzoatilyticus* strain JA2T. *Journal of Bacteriology, 193*, 2898–2899.

Moran, M. A., Belas, R., Schell, M. A., Gonzalez, J. M., Sun, F., Sun, S., et al. (2007). Ecological genomics of marine Roseobacters. *Applied and Environmental Microbiology, 73*, 4559–4569.

Munk, A. C., Copeland, A., Lucas, S., Lapidus, A., Del Rio, T. G., Barry, K., et al. (2011). Complete genome sequence of *Rhodospirillum rubrum* type strain (S1). *Standards in Genomic Sciences, 4*, 293–302.

Nagashima, K. V. P., Hiraishi, A., Shimada, K., & Matsuura, K. (1997). Horizontal transfer of genes coding for the photosynthetic reaction centers of purple bacteria. *Journal of Molecular Evolution, 45*, 131–136.

Nagashima, S., Kamimura, A., Shimizu, T., Nakamura-Isaki, S., Aono, E., Sakamoto, K., et al. (2012). Complete genome sequence of phototrophic betaproteobacterium *Rubrivivax gelatinosus* IL144. *Journal of Bacteriology, 194*, 3541–3542.

Oda, Y., Larimer, F. W., Chain, P. S., Malfatti, S., Shin, M. V., Vergez, L. M., et al. (2008). Multiple genome sequences reveal adaptations of a phototrophic bacterium to sediment microenvironments. *Proceedings of National Academy of Science of the United States of America, 105*, 18543–18548.

Oh, H. M., Giovannoni, S. J., Ferriera, S., Johnson, J., & Cho, J. C. (2009). Complete genome sequence of *Erythrobacter litoralis* HTCC2594. *Journal of Bacteriology, 191*, 2419–2420.

Okubo, T., Tsukui, T., Maita, H., Okamoto, S., Oshima, K., Fujisawa, T., et al. (2012). Complete genome sequence of *Bradyrhizobium* sp. S23321: insights into symbiosis evolution in soil oligotrophs. *Microbes and Environments, 27*, 306–315.

Page, R. D. M. (1996). TREEVIEW: an application to display phylogenetic trees on personal computers. *Computer Applications in the Biosciences, 12*, 357–358.

Penfold, R. J., & Pemberton, J. M. (1994). Sequencing, chromosomal inactivation, and functional expression in *Escherichia coli* of *ppsR*, a gene which represses carotenoid and bacteriochlorophyll synthesis in *Rhodobacter sphaeroides*. *Journal of Bacteriology, 176*, 2869–2876.

Pinta, V., Picaud, M., Reiss-Husson, F., & Astier, C. (2002). *Rubrivivax gelatinosus acsF* (previously orf358) codes for a conserved, putative binuclear-iron-cluster-containing protein involved in aerobic oxidative cyclization of Mg-protoporphyrin IX monomethylester. *Journal of Bacteriology, 184*, 746–753.

Porter, S. L., Wilkinson, D. A., Byles, E. D., Wadhams, G. H., Taylor, S., Saunders, N. J., et al. (2011). Genome sequence of *Rhodobacter sphaeroides* strain WS8N. *Journal of Bacteriology, 193*, 4027–4028.

Sankoff, D., Leduc, G., Antoine, N., Paquin, B., Lang, B. F., & Cedergren, R. (1992). Gene order comparisons for phylogenetic inference: evolution of the mitochondrial genome. *Proceedings of National Academy of Science of the United States of America, 89*, 6575–6579.

Sawicki, A., & Willows, R. D. (2010). BchJ and BchM interact in a 1: 1 ratio with the magnesium chelatase BchH subunit of *Rhodobacter capsulatus*. *FEBS Journal, 277*, 4709–4721.

Strnad, H., Lapidus, A., Paces, J., Ulbrich, P., Vlcek, C., Paces, V., et al. (2010). Complete genome sequence of the photosynthetic purple nonsulfur bacterium *Rhodobacter capsulatus* SB 1003. *Journal of Bacteriology, 192*, 3545–3546.

Swingley, W. D., Sadekar, S., Mastrian, S. D., Matthies, H. J., Hao, J., Ramos, H., et al. (2007). The complete genome sequence of *Roseobacter denitrificans* reveals a mixotrophic rather than photosynthetic metabolism. *Journal of Bacteriology, 189*, 683–690.

Takaichi, S. (2009). Distribution and biosynthesis of carotenoids. In C. N. Hunter, F. Daldal, M. C. Thurnauer & J. T. Beatty (Eds.), *The purple phototrophic bacteria* (pp. 97–117). Dordrecht, The Netherlands: Springer.

Taylor, D. P., Cohen, S. N., Clark, W. G., & Marrs, B. L. (1983). Alignment of genetic and restriction maps of the photosynthesis region of the *Rhodopseudomonas capsulata* chromosome by a conjugation-mediated marker rescue technique. *Journal of Bacteriology, 154*, 580–590.

Thompson, J. D., Gibson, T. J., Plewniak, F., Jeanmougin, F., & Higgins, D. G. (1997). The CLUSTAL_X windows interface: flexible strategies for multiple sequence alignment aided by quality analysis tools. *Nucleic Acids Research, 25*, 4876–4882.

Thrash, J. C., Cho, J. C., Ferriera, S., Johnson, J., Vergin, K. L., & Giovannoni, S. J. (2010). Genome sequences of strains HTCC2148 and HTCC2080, belonging to the OM60/NOR5 clade of the Gammaproteobacteria. *Journal of Bacteriology, 192*, 3842–3843.

Tuveson, R. W., Larson, R. A., & Kagan, J. (1988). Role of cloned carotenoid genes expressed in *Escherichia coli* in protecting against inactivation by near-UV light and specific phototoxic molecules. *Journal of Bacteriology, 170*, 4675–4680.

Vuilleumier, S., Chistoserdova, L., Lee, M. C., Bringel, F., Lajus, A., Zhou, Y., et al. (2009). *Methylobacterium* genome sequences: a reference blueprint to investigate microbial metabolism of C1 compounds from natural and industrial sources. *PLoS One, 4*. e5584.

Wagner-Döbler, I., Ballhausen, B., Berger, M., Brinkhoff, T., Buchholz, I., Bunk, B., et al. (2010). The complete genome sequence of the algal symbiont *Dinoroseobacter shibae*: a hitchhiker's guide to life in the sea. *The ISME Journal, 4*, 61–77.

Weissgerber, T., Zigann, R., Bruce, D., Chang, Y. J., Detter, J. C., Han, C., et al. (2011). Complete genome sequence of *Allochromatium vinosum* DSM 180(T). *Standards in Genomic Sciences, 5*, 311–330.

Wellington, C. L., Taggart, A. K.P., & Beatty, J. T. (1991). Functional significance of overlapping transcripts of *crtEF*, *bchCA*, and *puf* photosynthesis gene operons in *Rhodobacter capsulatus*. *Journal of Bacteriology, 173*, 2954–2961.

Willows, R. D., & Kriegel, A. M. (2009). Biosynthesis of bacteriochlorophylls in purple bacteria. In C. N. Hunter, F. Daldal, M. C. Thurnauer & J. T. Beatty (Eds.), *The purple phototrophic bacteria* (pp. 57–79). Dordrecht, The Netherlands: Springer.

Woese, C. R. (1987). Bacterial evolution. *Microbiological Reviews, 51*, 221–271.

Young, C. S., Reyes, R. C., & Beatty, J. T. (1998). Genetic complementation and kinetic analyses of *Rhodobacter capsulatus* ORF1696 mutants indicate that the ORF1696 protein enhances assembly of the light-harvesting I complex. *Journal of Bacteriology, 180*, 1759–1765.

Zargar, K., Conrad, A., Bernick, D. L., Lowe, T. M., Stolc, V., Hoeft, S., et al. (2012). ArxA, a new clade of arsenite oxidase within the DMSO reductase family of molybdenum oxidoreductases. *Environmental Microbiology, 14*, 1635–1645.

CHAPTER SIX

The Living Genome of a Purple Nonsulfur Photosynthetic Bacterium: Overview of the *Rhodobacter sphaeroides* Transcriptome Landscapes

Mark Gomelsky*[,1], Jill H. Zeilstra-Ryalls[,1]**

*Department of Molecular Biology, University of Wyoming, Laramie, WY, USA
**Department of Biological Sciences, Bowling Green State University, Bowling Green, OH, USA
[1]Corresponding authors: E-mail: gomelsky@uwyo.edu; jzeilst@bgsu.edu

Contents

Abstract

Oxygen (O_2) and light are key environmental factors that control metabolic choices in the purple nonsulfur anoxygenic phototrophic bacteria. Whole-genome transcription profiling has proved to be remarkably informative in assessing metabolic changes in this group of bacteria, and in uncovering the regulatory networks underlying metabolic changes. Here we review transcriptome profiling studies of *Rhodobacter sphaeroides*, one of the best-understood purple nonsulfur bacteria. Our focus is on the responses of *R. sphaeroides* to changes in O_2 concentration and light intensity, and the description of the key regulons and transcription factors controlling these changes. Transcriptomic studies have helped bring the genome of *R. sphaeroides* to life by showing massive restructuring of gene expression upon changes in O_2 and light levels as well as the transient nature of many transcription changes. Genome-wide transcription analyses

Advances in Botanical Research, Volume 66
ISSN 0065-2296, http://dx.doi.org/10.1016/B978-0-12-397923-0.00006-0

have helped us understand metabolic strategies employed by *R. sphaeroides* when changing its energy generation mode from aerobic respiration to anaerobic respiration or photosynthesis. It also uncovered multiple grey areas in our understanding of adaptations of this bacterium to changing environments.

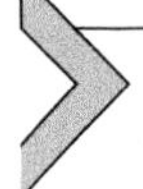

1. TRANSCRIPTOMICS AS A TOOL TO ASSESS METABOLIC CHANGES AND UNDERSTAND REGULATORY NETWORKS

Phototrophic purple nonsulfur bacteria are renowned for their metabolic versatility. First and foremost they have a rich repertoire of energy generation modes, from aerobic respiration to anoxygenic photosynthesis (PS) in the absence of O_2 in the light, to anaerobic respiration in the absence of O_2 in the dark. A decrease in O_2 tension (in the light or dark) brings about a transformation from nonpigmented or lightly pigmented *Escherichia coli*-like cells to pigmented cells filled with intracytoplasmic membranes (ICMs) that house protein–pigment photosynthetic complexes. In addition to diverse energy generation modes, purple nonsulfur bacteria can use both organic and inorganic (CO_2) carbon, fix (di)nitrogen, and degrade various toxic compounds. Key gene expression regulators controlling these metabolic pathways have been identified via genetic and biochemical approaches. But the true scope of gene expression fluctuations in response to environmental changes was revealed through the recent application of genome-wide transcriptome profiling. Combined with bioinformatic analysis transcriptomic studies also helped uncover stimulons (sets of genes that respond to a given environmental factor) and regulons (sets of genes controlled by individual regulatory factors), decipher putative binding sites of transcription factors, gain insights into the hierarchy and connectivity of the regulatory networks, and link genes of unknown function with genes of known function based on expression profile similarity.

We note that in bacteria transcription changes generally qualitatively correlate with changes in protein abundance (which is not the case in higher eukaryotes; Tebaldi et al. (2012)). The overall validity of this notion for purple nonsulfur bacteria was experimentally demonstrated by comparing transcriptome and proteome data (Callister et al., 2006; Eraso et al., 2008; Woronowicz, Olubanjo, Sung, Lamptey, & Niederman, 2012). Since quantitative transcriptomics is more accessible and reliable than quantitative proteomics at present, transcriptomics is a better tool for predicting physiological and behavioural responses of bacteria to new conditions. Whole-genome DNA microarrays have been used to study transcriptome changes in several purple bacteria including *Rhodobacter sphaeroides* (Pappas et al., 2004), *Rhodopseudomonas palustris* (Braatsch et al., 2006) and *Rhodobacter capsulatus* (Hynes, Mercer, Watton, Buckley, & Lang, 2012). Now RNA sequencing has emerged as an additional global transcriptome analysis tool (Berghoff, Glaeser, Sharma, Vogel, & Klug, 2009).

In this chapter, we provide an overview of genome-wide gene expression studies in *R. sphaeroides*, the organism where the largest number of such studies have been performed. In combination with investigations of gene regulation (Mackenzie et al., 2007; Zeilstra-Ryalls & Kaplan, 2004), proteomics (Zeng et al., 2007), metabolic modelling (Golomysova, Gomelsky, & Ivanov, 2010; Imam et al., 2011; Tao et al., 2012), and regulatory network reconstruction (Moskvin, Bolotin, Wang, Ivanov, & Gomelsky, 2011), the transcriptomic studies make *R. sphaeroides* one of the best-understood purple bacteria. We discuss gene expression alterations in this organism in response to changes in O_2 and light, which are critical factors for metabolic and physiological decisions in all purple bacteria. While our focus on a single species may seem narrow, it is clear that the findings reviewed here are broadly applicable to the purple nonsulfur bacteria, although the regulatory details in each species may vary. We suggest that the *R. sphaeroides*-centric description may function as a *lingua franca* for key regulatory pathways common to the purple bacteria, and to a certain extent beyond this group. As the focus here is on gene regulation, we will omit detailed description of the metabolic processes involved, which are described in other chapters of this book and elsewhere (Hunter, Daldal, Thurnauer, & Beatty, 2008).

2. METABOLIC VERSATILITY REQUIRES TRANSCRIPTOME FLEXIBILITY: *R. SPHAEROIDES* TRANSCRIPTOME LANDSCAPES OF DIVERSE GROWTH MODES

The Gomelsky laboratory, in collaboration with the Kaplan laboratory and Affymetrix, developed a DNA microarray, GeneChip, of *R. sphaeroides* 2.4.1 (Pappas et al., 2004). The sequences of 4292 Open reading frames, 47 tRNA and rRNA genes, and 394 intergenic regions that may contain small ORFs or small regulatory RNA genes were manually curated based on the primary genome annotation by Oak Ridge National Laboratory. Most of the transcriptomic studies to date have been performed using this GeneChip, which generates highly reproducible data.

Pappas et al. (2004) profiled transcriptomes of *R. sphaeroides* captured in each of its major energy generation modes under heterotrophic growth conditions: (1) aerobic respiration at saturating O_2 levels, (2) anaerobic PS, and (3) anaerobic respiration in the dark with dimethyl sulfoxide (DMSO) as a terminal electron acceptor. These experiments revealed genes whose products are associated with each metabolism type and uncovered the scope of the influence of O_2 and light on global gene expression. The transcript abundance of approximately 20% of all ORF-encoding genes was significantly different in cells growing under anaerobic photosynthetic compared to aerobic conditions (Fig. 6.1). This study revealed for the first time the extent of flexibility of the *R. sphaeroides* transcriptome required for shifting between energy generation modes.

The largest cluster of co-regulated genes was comprised of genes responsive to O_2 (approximately 600 ORFs), irrespective of the presence or absence of light (Fig. 6.1). The large size of this cluster confirms the notion that O_2 plays the dominant role in controlling *R. sphaeroides* metabolism and physiology. The expression patterns of key gene categories associated with the energy-generating pathways operational under the three conditions are diagrammed in Fig. 6.2. Many of the O_2-responsive genes are required for harvesting of light energy, including those that code for structural proteins of photosynthetic complexes and enzymes involved in the biosynthesis of carotenoid and bacteriochlorophyll, which are clustered together in the so-called PS gene cluster (genes RSP0255–0292) (Fig. 6.3). Details about these genes and their products can be found in earlier reviews (Igarashi et al., 2001; Naylor, Addlesee, Gibson, & Hunter, 1999; Zeilstra-Ryalls & Kaplan, 2004). All genes in the PS gene cluster, with the exception of *ppsR* (RSP0282) encoding a PS gene repressor, were up-regulated under anoxic conditions regardless of the presence or absence of light. The extent of up-regulation varied – from the highest up-regulation of 39- to 49-fold observed for the *puc1BAC* operon (RSP0314–0315) encoding the light-harvesting complex LH2, to more moderate up-regulation of reaction centre (RC) genes *pufLMX* (RSP0255–0257, 13- to 18-fold), *puhA* (RSP0291, 7-fold) and the LH1 complex genes *pufBA* (RSP0258, 15- to 18-fold), and genes encoding bacteriochlorophyll (e.g. *bchFNB* (RSP0284–0286, 8- to 19-fold)) and carotenoid (e.g. *crtA* (RSP0272, 10- to 13-fold)) synthesis enzymes. This up-regulation trend extends to genes located outside of the PS cluster, such as an operon that codes for a second set of LH2 antenna proteins (*puc2BA*, RSP1556–1557

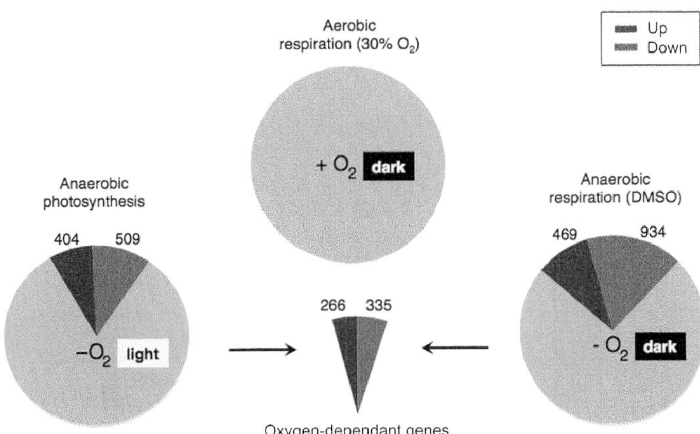

Figure 6.1 ***Transcriptome profiles of* R. sphaeroides *using different energy generation modes.*** The growth conditions used are aerobic respiration (30% O_2), anaerobic PS (10 W/m^2), and anaerobic respiration (using DMSO as a terminal electron acceptor) (Pappas et al., 2004). mRNA levels for all genes under aerobic respiration conditions are assigned a value of one. The numbers of genes with altered (red, up-regulated; blue, down-regulated) expression are shown. O_2-dependant genes are defined as those whose expression has changed under both anoxic conditions, compared to aerobic respiration. (For interpretation of the references to colour in this figure legend, the reader is referred to the online version of this book.)

(Zeng, Choundhary, & Kaplan, 2003)), as well as certain *hem* genes whose products catalyse early steps of tetrapyrrole biosynthesis that are common to bacteriochlorophyll and haem synthesis.

Aerobic respiration in *R. sphaeroides* involves a branched respiratory chain (Fig. 6.2) that includes two NADH–ubiquinone oxidoreductases, succinate dehydrogenase, the bc_1 complex, two major cytochrome *c* species, cyt c_2 and cyt c_y, involved in electron transport to three cyt *c* oxidases, and a quinol *bd* oxidase. The transcriptome profiles performed by Pappas et al. (2004) and Roh, Smith, and Kaplan (2004) provide insights into how expression of genes encoding these components is modulated in response to O_2 availability.

One of the unexpected findings from transcriptome analysis was differential expression of the NADH–ubiquinone oxidoreductases, *nuo1* (RSP2512–2530) and *nuo2* (RSP0100–0112) (Fig. 6.2). Expression of the *nuo1* genes was decreased under anaerobic conditions, while expression of the *nuo2* genes was increased. The physiological significance of this observation is yet to be determined. However, our preliminary data (Ryu and Gomelsky, unpublished) suggest that Nuo1 is essential and Nuo2 is

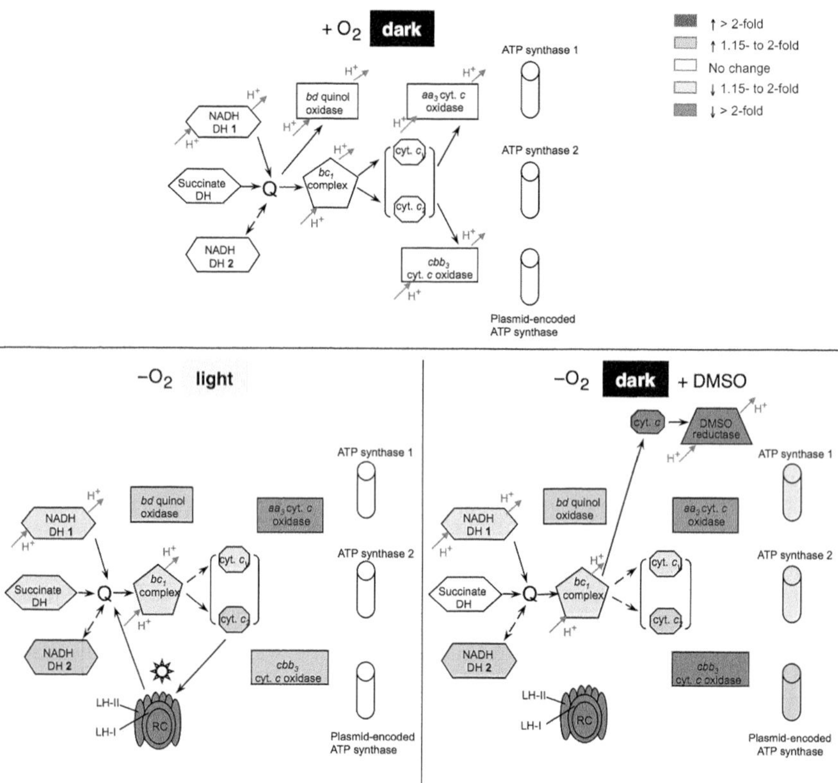

Figure 6.2 R. sphaeroides *electron transport chain complexes and ATP synthases involved in energy generation under oxic, anoxic-light, and anoxic-dark-DMSO conditions.* Arrows indicate electron flow; dashed line arrows indicate anticipated direction of electron flow. The known sites for generation of proton motive force are shown. The expression of genes under oxic conditions is set as the background for comparisons (not coloured). Blue corresponds to decreased gene expression of the corresponding proteins under anoxic compared to oxic conditions as follows: light blue, less than 2-fold decrease; dark blue, at least 2-fold decrease. Pink corresponds to increased gene expression compared to the expression under oxic conditions as follows: light pink, less than 2-fold increase; dark pink, at least 2-fold increase. No colour corresponds to anoxic expression that is not significantly different from expression under oxic conditions. Proteins whose genes are expressed below reliable detection are not shown. DH, dehydrogenase; Q, quinone–quinol pool; LH, light-harvesting complex; RC, reaction centre complex. The RC and light-harvesting complexes are not shown under oxic conditions because several *bch* genes involved in bacteriochlorophyll biosynthesis are not expressed under these conditions; hence, no photosynthetic apparatus is made. *(Modified from Pappas et al. (2004))*. See the color plate.

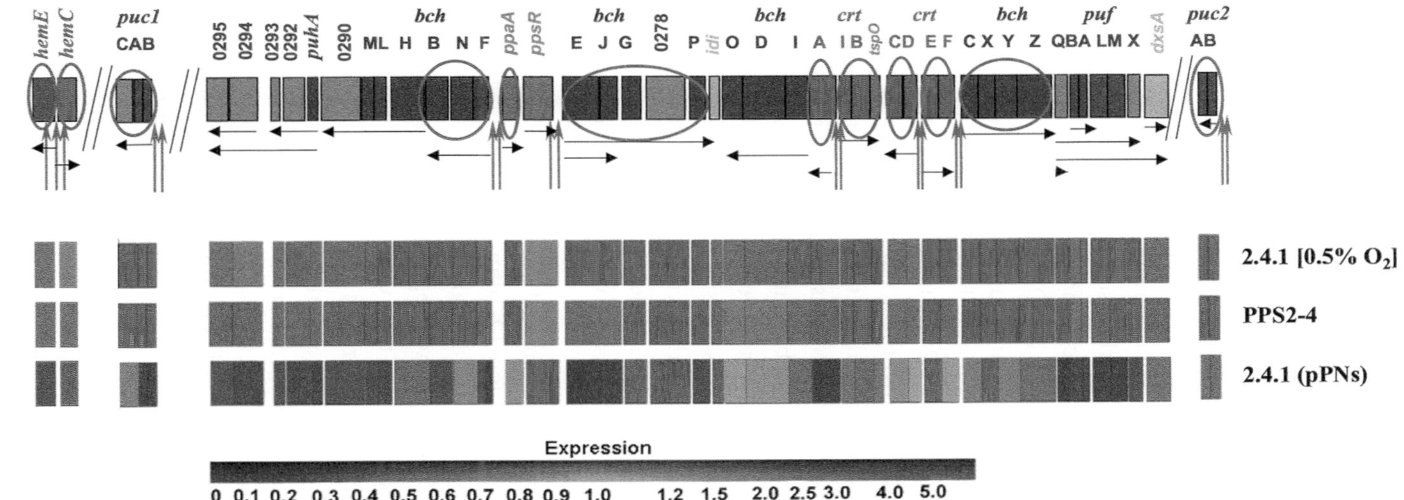

Figure 6.3 ***Composition (upper panel) and expression (lower panel) of the*** **R. sphaeroides** ***PpsR regulon.*** Each gene is represented by a box, coloured according to its function. Green, *bch* genes; red, *crt* genes; blue, genes encoding structural polypeptides of photocomplexes; grey, genes encoding assembly factors or proteins of unknown function; orange, genes encoding regulatory factors; pink, genes encoding enzymes common to bacteriochlorophyll and ubiquinone biosynthesis; magenta, protoporphyrin IX biosynthesis genes. PpsR-binding sites are shown as red vertical arrows. Putative transcripts are shown as black horizontal arrows. Circled genes are repressed by PpsR directly. Relative expression of PpsR-dependant genes, indicated according to the included expression colour scheme, reflect comparisons to the wild type strain grown with 20% O_2. 2.4.1, wild type strain; PPS2-4, *ppsR* point mutant; pPNs, plasmid overexpressing the *ppsR* gene. *Copyright © American Society for Microbiology,* Journal of Bacteriology *(2005),* 187, *2148–2156; doi: 10.1128/JB.187.6.2148-2156.2005.* See the color plate.

dispensable for aerobic respiration, while the opposite is true for anaerobic PS. This example demonstrates how transcriptomic insights are contributing to a better understanding of energy generation pathways in *R. sphaeroides*.

Expression of the succinate dehydrogenase genes (*sdh*, RSP0974–0979) was down-regulated under photosynthetic conditions, but was essentially unchanged between aerobic and anaerobic respiratory conditions. Decreased *sdh* gene expression under photosynthetic conditions may indicate a lower utilisation of succinate, which is the main carbon source in the medium used by Pappas et al. (2004). Two factors likely contribute to this: (1) PS-based energy generation decreases the need for NADH produced from succinate oxidation and (2) CO_2 fixation operating during PS provides an additional source of carbon. Consistent with this possibility, two clusters of genes involved in CO_2 fixation (*cbbI* (RSP1280–1285) and *cbbII* (RSP3266–3271)) were strongly up-regulated under photosynthetic conditions.

Of the two major cyt *c* genes whose products transfer electrons from the bc_1 complex, the cyt c_2 gene, *cycA* (RSP0296), was up-regulated in response to the lack of O_2 (in line with earlier observations (Brander, McEwan, Kaplan, & Donohue, 1989)), while the cyt c_y gene, *cycY* (RSP0705), was down-regulated. This expression pattern is consistent with the ability of both cytochromes to transfer electrons to the cyt *c* oxidases (Daldal et al., 2001; Myllykallio, Zannoni, & Daldal, 1999), while only cyt c_2 transfers electrons to the photosynthetic RC (Donohue, McEwan, Van Doren, Crofts, & Kaplan, 1988; Rott et al., 1990).

The microarray data also revealed differential regulation of the terminal O_2 reductases in *R. sphaeroides*. The low O_2 affinity aa_3-type cyt *c* oxidase encoded by the *cta/cox* (RSP1826–1829 and RSP1877) genes was transcribed maximally at high O_2 and down-regulated during anaerobic growth, whereas the high O_2 affinity cbb_3-type cyt *c* oxidase encoded by the *cco* (RSP0692–0696) operon was expressed at elevated levels under anoxic conditions. This transcription profile conforms with the view that the aa_3-type oxidase operates primarily under highly oxic conditions, whereas the cbb_3-type oxidase functions in cells grown under low and no O_2 (Shapleigh, Hill, Alben, & Gennis, 1992). The third cyt *c* oxidase of the caa_3-type (RSP0115–0118) is not expressed under standard laboratory conditions, and neither is the putative quinol oxidase encoded by the *qoxAB* operon (RSP3097–3098). However, the second quinol oxidase QxtAB (RSP3211–3212) is expressed (Moncey et al., 2000) and up-regulated under anoxic conditions (Fig. 6.2).

Under anaerobic-dark-DMSO conditions, PS genes are highly expressed despite the absence of light, reinforcing the dominant role of O_2 in PS gene expression. Also highly up-regulated are genes involved in DMSO reduction, including those coding for the DMSO reductase, *dorCBA* (RSP3046–3048), and for biosynthesis of its requisite molybdopterin cofactor *moaAB* (RSP3049–3050). The expression pattern of DMSO reduction genes suggests that two factors, the lack of O_2 and the presence of DMSO, are required, which is consistent with earlier studies (Mouncey & Kaplan, 1998a,b; Zeilstra-Ryalls, Gabbert, Mouncey, Kranz, & Kaplan, 1997).

Comparative transcriptomic studies also helped uncover new genes whose products are required for anaerobic photosynthetic growth. Tavano, Podevels, and Donohue (2005) identified RSP4157–4164 as genes whose expression profiles were similar to the profiles of PS genes. A polar insertion mutant of RSP4157 was unable to grow under the heterotrophic anaerobic photosynthetic conditions, unless an exogenous electron acceptor was provided. Although the precise functions of the genes have not yet been determined, it appears that they contribute to redox homoeostasis in photosynthetic cells.

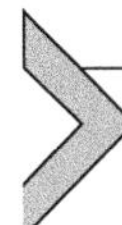

3. THE EFFECTS OF O_2 AND LIGHT: A TRANSCRIPTOME PERSPECTIVE

The effect of O_2 on PS gene expression has been examined in more detail by Moskvin, Kaplan, Gilles-Gonzalez, and Gomelsky (2007), who profiled the transcriptome of *R. sphaeroides* cultures bubbled with gas mixtures containing from 30 to 0.5% O_2. The expression of PS genes was highly sensitive to O_2 levels and increased gradually as O_2 tension decreased. This expression pattern was unexpected because the cytoplasmic membrane invaginations that give rise to the ICM system form only at low O_2 concentrations (Huang & Kaplan, 1973). From the absence of an O_2 threshold in the expression pattern follows that the rate-limiting step(s) in the formation of the PS apparatus is the abundance or activity of requisite proteins, but not the abundance of PS gene transcripts.

Arai, Roh, and Kaplan (2008) explored the effect of O_2 on genome-wide gene expression by exposing photosynthetically grown bacteria to O_2. They observed a remarkably good inverse correlation in transcriptome responses compared to the responses to lower O_2 levels (Moskvin, Gomelsky, & Gomelsky, 2005; Moskvin, Kaplan, et al., 2007; Pappas et al., 2004).

Expression responses to the introduction of O_2 to anaerobic photosynthetic cultures can be readily rationalised for only a fraction of the genes. For example, genes whose products are involved in energy generation respond in a predictable manner, whereas the oxidative stress response may account for the immediate drastic increase in expression of the genes encoding chaperones and proteases (RSP1016; RSP1572; RSP1172–1173; RSP1531–1532; RSP2806), and *suf* genes (RSP0434–0443) involved in FeS cluster assembly. However, expression patterns of many other genes, including genes involved in fatty acid and carbohydrate metabolism, remain unexplained, which illustrates our limited grasp of these important processes. The lack of understanding of lipid biosynthesis genes is a particularly glaring omission given that ICM formation requires massive lipid production.

Interestingly, major changes in gene expression (affecting approximately 20% of all genes) occurred very quickly, within 15 min, after the introduction of O_2. After 4 h, expression of over one-third of the genes had changed significantly. The speed at which gene expression changes occurred demonstrates that, while analysis of the transcriptomes of steady-state cultures is clearly useful, transcriptome dynamics studies capture the temporal nature of the adaptive changes.

To examine the role of light in gene expression in anaerobic photosynthetic cultures, Roh et al. (2004) compared the transcriptome profiles of *R. sphaeroides* cultured at different light intensities. Under the low light intensity of 3 W/m^2, the cell growth rate was several-fold lower, and the abundance of photosynthetic complexes was several-fold higher compared to the high light intensity of 100 W/m^2. However, mRNA abundances of the majority of the PS genes differed only modestly, i.e. within the 1.5- to 2-fold range, between the high- and low light intensity cultures. This seemingly paradoxical expression pattern suggests that post-transcriptional events, but not mRNA levels, play more prominent roles in controlling the light-dependant abundance of photosynthetic complexes in the absence of O_2.

While the role of light in gene regulation in *R. sphaeroides* under anoxic conditions is relatively limited, light plays a very important role in the presence of O_2, particularly during the transition from aerobic respiration to anaerobic PS, i.e. conditions when the photosynthetic apparatus is being actively assembled but is not yet functional. Exposure of abundant photosynthetic pigments to sunlight in the presence of O_2 provides ample opportunities for formation of singlet O_2 and other reactive O_2 species. Transcriptomic and mechanistic studies on photooxidative damage in

R. sphaeroides have been recently reviewed by Ziegelhoffer and Donohue (2009) and Glaeser et al. (2011). We will not discuss these studies here for the lack of space, except for the original study (Braatsch, Moskvin, Klug, & Gomelsky, 2004) that uncovered transcriptome responses to irradiation under low O_2 growth conditions.

Braatsch et al. (2004), who investigated genome-wide expression changes following an exposure of semi-aerobic-dark cultures to blue light, revealed two kinds of responses. One response closely resembled the transcriptomic response from low to increased O_2, as typified by down-regulation of all PS genes. The second type of transcriptome changes involved temporal up-regulation of two alternative sigma factor genes, *rpoE* (RSP1092) and *rpoH2* (RSP0601), and regulons transcribed by these sigma factors. Among the σ^E- and σ^{H2}-dependant genes were those specifically involved in photooxidative stress defence, such as deoxypyrimidine photolyase *phrA* (RSP2143) (Hendrischk, Braatsch, Glaeser, & Klug, 2007), glutathione peroxidase (RSP2389), and quinol oxidoreductase *qxtAB*. That elevated expression of the σ^E and σ^{H2} photooxidative stress regulons was temporary, suggested that in approximately 2 h the cells were able to adjust to the stress imposed by irradiation (Braatsch et al., 2004).

4. THE KEY O_2- AND LIGHT-RESPONSIVE REGULONS

Multiple mechanisms have evolved to allow cells to respond to changes in O_2 levels. Three O_2-sensing regulatory systems play a central role in expression of *R. sphaeroides* genes: (1) the O_2-sensing transcription factor FnrL, a homologue of *E. coli* Fnr (Kiley & Beinert, 1998; Körner, Sofia, & Zumft, 2003); (2) the redox-sensitive PrrB–PrrA two-component regulatory system, known as RegB–RegA in *R. capsulatus* (Wu & Bauer, 2008); and (3) the O_2-sensing AppA–PpsR antirepressor–repressor system (Elsen, Jaubert, Pignol, & Giraud, 2005). Each of these systems was identified via genetic analyses prior to the advent of genome-wide transcriptional tools. Transcriptome studies helped define the scope of each of these systems (i.e. individual regulons), an unsuspected interconnectivity between them, and exposed grey areas in our understanding of O_2-dependant gene regulation in *R. sphaeroides*. These studies also helped better define the binding sites for the transcription regulators of these systems, FnrL, PrrA and PpsR. Two of the above-mentioned systems, AppA–PpsR and PrrB–PrrA, have also been found to play central roles in light-dependant regulation.

4.1. The FnrL Regulon

FnrL is a DNA-binding transcription activator and repressor that plays a central role in the anaerobic metabolisms of *R. sphaeroides*. As is true of *E. coli* Fnr (Kiley & Beinert, 1998), the ability of FnrL to bind DNA is believed to depend upon an O_2-labile 4Fe–4S cluster, and so it is most active under anaerobic conditions. The importance of FnrL in *R. sphaeroides* is apparent from the fact that an *fnrL* null mutant is unable to grow by any anaerobic growth mode (Zeilstra-Ryalls & Kaplan, 1995). Fortunately, it grows under low O_2 conditions, where DNA binding is not maximal but detectable (Roh & Kaplan, 2002). To identify the genes belonging to the FnrL regulon, Moskvin, Lar, Suwansaard, Gomelsky, and Zeilstra-Ryalls (in preparation) compared the transcription profiles of the wild type and an *fnrL* null mutant grown under low O_2 conditions. Separately, Dufour, Kiley, and Donohue (2010) used a ChIP–chip analysis of wild type cells grown under anaerobic photosynthetic conditions to identify FnrL DNA-binding sites. Each method has its limitations, such as the requirement for growth of the *fnrL* mutant in the presence of some O_2 which limited the magnitude of FnrL-dependant expression changes in the transcriptomic study, while in the ChIP–chip study, detection of FnrL binding depended upon its accessibility to target sequences and the strength of binding. Thus, FnrL binding to the *hemA* (RSP2984) upstream sequence was detected, although the transcript levels were not appreciably different between the *fnrL* mutant and the wild type. The opposite situation was observed for the known FnrL target, *rdxBH* (RSP0692–0693) (Roh & Kaplan, 2002), whose transcripts were significantly reduced in the *fnrL* mutant although FnrL binding was not detected. A consensus FnrL-binding sequence deduced from these studies is TTGA(T/C)-N_4-(A/G/C)TCAA.

The answer to the question as to why *R. sphaeroides* FnrL is required for anaerobic growth is unknown. It is possible that FnrL-dependant genes are essential for anaerobic respiration and anaerobic PS. The transcriptomic data supported earlier reports that FnrL directly activates expression of the *dorS* (RSP3044) gene encoding the DMSO sensor kinase. In the absence of DorS, its cognate response regulator DorR does not turn on expression of the DMSO reductase operon *dorCBA* (Mouncey & Kaplan, 1998a,b). While this regulatory cascade may explain the inability of the *fnrL* mutant to grow using anaerobic respiration with DMSO (Zeilstra-Ryalls & Kaplan, 1995; Zeilstra-Ryalls et al., 1997), we cannot explain why FnrL is essential for anaerobic PS, although certain FnrL-regulated genes have emerged as potential targets. One important function of FnrL is activation of transcription of the *hemN* (RSP0317) and *hemZ* (RSP0699) genes whose products catalyse the anoxic decarboxylation of coproporphyrinogen III to protoporphyrinogen IX, the common precursor of haem and bacteriochlorophyll synthesis (reviewed in Zeilstra-Ryalls, 2008). Under oxic conditions, this reaction is catalysed by HemF (RSP0682) that uses molecular O_2 as substrate (Zeilstra-Ryalls & Schornberg, 2006). Interestingly, although FnrL is essential for anaerobic growth in *R. sphaeroides* and *Rubrivivax gelatinosus*, an *R. capsulatus fnrL* mutant is capable of photosynthetic growth (Zeilstra-Ryalls et al., 1997).

Among other important FnrL targets are the aa_3-type and cbb_3-type cyt *c* oxidase genes. In response to lower O_2 tension, FnrL represses transcription of the former and activates transcription of the latter genes. Therefore, FnrL contributes to transcriptome restructuring in response to anaerobiosis and also during changes in O_2 availability. This is thought to be possible by virtue of the retention of some DNA-binding activity by FnrL in a low O_2 environment.

4.2. The PrrA Regulon

A two-component system involving a sensor kinase PrrB and its response regulator PrrA (RegBA in *R. capsulatus*) is a global redox-dependant regulatory system. Originally, RegA (Sganga & Bauer, 1992) and PrrA (Eraso & Kaplan, 1994; Lee & Kaplan, 1992; Phillips-Jones & Hunter, 1994) were identified as activators required for expression of PS genes. Subsequently, it was realised that the PrrBA/RegBA system also controls transcription of genes involved in aerobic and anaerobic respiration, CO_2 assimilation, nitrogen fixation and hydrogen uptake (for reviews see (Dubbs & Tabita, 2004; Mackenzie et al., 2007; Wu & Bauer, 2008)). Transcriptomic studies involving a *prrA* null mutant (Eraso et al., 2008) uncovered the full scope of the PrrA regulon that involves as many as 20% of all expressed genes. One easily discernable theme among the PrrA-regulated genes is that their products participate in processes that affect or are dependant upon the redox poise of the electron transport chain.

The nature of the signal sensed by PrrB/RegB is a matter of some controversy. A compelling case has been presented that RegB of *R. capsulatus* senses the ubiquinol/ubiquinone ratio in the cytoplasmic membrane (Wu & Bauer, 2010), similar to the redox-sensing ArcB kinase of *E. coli* (Bekker et al., 2010), whereas *R. sphaeroides* PrrB has been proposed to monitor electron flow to the cbb_3-type cyt *c* oxidase (Kim et al., 2007;

Oh and Kaplan, 2004). A second sensory mechanism involves a redox-sensitive cysteine within the cytoplasmic portion of PrrB/RegB (Potter, Jeong, Williamson, Henderson, & Phillips-Jones, 2006; Swem et al., 2003) that can be oxidised either by O_2 or by an as yet unidentified redox-active intermediate.

The consensus PrrA binding sequence (C/T)(G/C)CGG(C/G)-N_{0-10}-G(T/A)C(G/A)(C/A) is based upon a combination of DNase protection assays and transcriptomics studies (Mao et al., 2005). But its low conservation and high GC content make genome-wide predictions of PrrA binding sites difficult in the bacterium whose genome GC content is 69%. PrrA most likely binds DNA as a dimer (Laguri, Stenzel, Donohue, Phillips-Jones, & Williamson, 2006; Ranson-Olson & Zeilstra-Ryalls, 2008). It was thought that the phosphorylated form, PrrA~P, binds the majority of its DNA targets with greater affinity to alter transcription. However, both transcriptomic and further studies have revealed that unphosphorylated PrrA also binds DNA (Eraso et al., 2008), and in some cases the affinity of the unphosphorylated PrrA is higher than that of PrrA~P (Comolli, Carl, Hall, & Donohue, 2002; Ranson-Olson & Zeilstra-Ryalls, 2008).

The kinase activity of PrrB increases in response to over-reduction of electron transport chain components, which can occur under various scenarios; e.g. during aerobic respiration when the concentration of O_2 drops, during anaerobic respiration when the concentration of alternative electron acceptors is decreased, or during heterotrophic anaerobic photosynthetic growth when no electron sinks are available. This situation can be avoided by supplying terminal electron acceptors of the electron transport chain, or oxidised compounds that function as electron sinks by other processes such as CO_2 fixation, nitrogen fixation or nitrate reduction; the latter is a process that does not contribute to energy generation in *R. sphaeroides* 2.4.1 (Hartsock & Shapleigh, 2011). This may explain why PrrA up-regulates transcription of genes for the *Calvin-Benson-Bassham* cycle (*cbbI* and *cbbII* operons), nitrogen fixation (*nifHDK* RSP0541-0539) and nitrate reductase (RSP4113-4118), as also noted in earlier studies, etc. (González, Correia, Moura, Brondino, & Moura, 2006; Joshi & Tabita, 1996; Zhu & Kaplan, 1985).

4.3. The PpsR Regulon

In addition to the global transcription regulators, FnrL and PrrBA, *R. sphaeroides* has a PS-specific regulatory system responsive to O_2 and the redox state of the cytoplasm. The system is composed of the transcription repressor PpsR, originally described as a repressor of the *bch*, *crt* and the *puc1* operons (Gomelsky & Kaplan, 1995a; Penfold & Pemberton, 1994), and the antirepressor AppA (Gomelsky & Kaplan, 1995b, 1997, 1998; Masuda & Bauer, 2002). PpsR (CrtJ in *R. capsulatus* (Ponnampalam, Buggy, & Bauer, 1995)) is common to photosynthetic purple nonsulfur bacteria (reviewed in Elsen et al., (2005)), whereas AppA is unique to the *R. sphaeroides* species (Moskvin et al., 2007). Over-expression of PpsR impairs photosynthetic growth, while *ppsR* inactivation results in uncontrolled production of photosynthetic pigments even in the presence of O_2 (Gomelsky & Kaplan, 1995a, 1997).

PpsR repressor activity is regulated by oxidation of a cysteine residue in its DNA-binding domain that results in the formation of an intramolecular disulfide bond (Cheng et al., 2012), and by binding haem that enables it to sense haem levels (Yin, Dragnea, & Bauer, 2012). However, the most potent regulator of the repressor activity of *R. sphaeroides* PpsR is the fascinating antirepressor protein AppA. In an *appA* null mutant, PS genes remain repressed even in the absence of O_2, and photosynthetic growth is impaired (Gomelsky & Kaplan, 1995b). The AppA protein binds haem (Han, Meyer, Keusgen, & Klug, 2007; Moskvin et al., 2007) via a novel haem-binding domain, SCHIC. O_2 binding to haem iron causes discoordination of the haem, resulting in a conformation change that is believed to promote dissociation of AppA-PpsR antirepressor-repressor complexes and consequently initiation of PS gene repression (Moskvin et al., 2007). AppA responds to a wide range of O_2 concentrations in vitro, which allows it to control PpsR repressor activity, and so can adjust expression levels of PS genes proportionally to O_2 tension. Therefore, the AppA–PpsR system functions as an O_2-dependant transcriptional rheostat (Moskvin et al., 2007). In addition to sensing O_2, AppA senses blue light (Braatsch, Gomelsky, Kuphal, & Klug, 2002; Masuda & Bauer, 2002) via its N-terminal BLUF domain (Gomelsky and Klug, 2002; Losi & Gartner, 2011). Because the conformation and/or the stability of the AppA–PpsR complexes is affected by both O_2 and light, expression of the PpsR-regulon genes is responsive to both of these stimuli.

The transcriptome profiling by Moskvin et al. (2005) confirmed that PpsR primarily regulates genes in the PS gene cluster, and also the *puc2* operon. Expression profiles of two genes outside the PS gene cluster, *hemC* (RSP0679) and *hemE* (RSP0680), were similar to the profiles of the known PpsR-regulated genes, and these *hem* genes have been found to be directly regulated by PpsR. Their products are involved in protoporphyrin IX biosynthesis; hence, PpsR derepression results in up-regulation of both haem and bacteriochlorophyll synthesis. From the transcriptomic data, the consensus PpsR-binding sequence has been defined as TGTc-N_{10}-gACA. Although 240 genes in the *R. sphaeroides* genome contain this consensus, the majority, approximately 200 genes, are not directly affected by PpsR because two binding sites were shown to be mandatory for repression in vivo. Interestingly, for the divergently transcribed *hemC* and *hemE*, genes, the requirement for two PpsR-binding sites is met by combinations of one common site positioned within the intergenic region and a second site located within the coding sequences (Moskvin et al., 2005).

Quite unexpectedly, comparisons of the transcriptomes of a *ppsR* mutant, a PpsR over-expression strain and an *appA* null mutant showed that PpsR regulates expression of all PS genes, including those encoding structural proteins and also assembly factors of the RC and LH1 antenna complexes (Fig. 6.3). Regardless of the mechanism of the indirect regulation evoked by this finding, which remains to be determined, because derepression of the PpsR regulon is needed for photosynthetic growth, and the PpsR regulon encompasses all PS genes, PpsR must be regarded as a master regulator of PS gene expression in *R. sphaeroides*.

To parse the roles of PrrBA and AppA–PpsR in PS gene regulation, Moskvin et al. (2005) constructed a *prrA ppsR* double mutant. They showed that *ppsR* inactivation restores anaerobic photosynthetic growth to a *prrA* null mutant, an unanticipated development that prompted an investigation of a possible hierarchical relationship between the two regulatory systems. It was found that PrrA strongly regulates *appA* gene expression, and that the level of *appA* mRNA in a *prrA* mutant is approximately 13% of the level of the wild type, which likely contributes to the inability of *prrA* mutants to grow phototrophically (Gomelsky et al., 2008). Interestingly, simple restoration of *appA* mRNA to wild type levels proved insufficient to restore photosynthetic growth of the *prrA* mutant. Therefore, in addition to affecting *appA* expression, PrrA affects the activity of the AppA-PpsR system on a post-transcriptional level in a manner that remains unknown. Furthermore, *prrA* mRNA abundance was found to depend upon the levels of the PpsR repressor via an enigmatic mechanism (Fig. 6.4; Gomelsky et al., 2008). This unanticipated finding may be responsible, at least in part, for the observation by Bruscella, Eraso, Roh, and Kaplan (2008) of the 'extended' PpsR regulon that was reported to contain numerous genes lacking the PpsR consensus sites.

The inclusion of all PS genes in the PpsR regulon revealed by transcriptome profiling, combined with the understanding that AppA senses blue light (Braatsch et al., 2002; Masuda & Bauer, 2002), provides an explanation as to why all PS genes were down-regulated in response to blue light irradiation in semi-aerobic *R. sphaeroides* cultures (Braatsch et al., 2004). Since O_2 and light affect the AppA–PpsR complex via AppA, transcriptome responses to increased O_2 or to blue light appear qualitatively similar.

It is worth noting that blue light sensing via the BLUF domain of AppA explains part but not all responses of the PS genes to blue light (Braatsch et al., 2004). For example, the so-called onset of the blue light-induced gene down-regulation is independent of the BLUF domain. These observations could be reconciled by the presence of additional blue light sensors. A second blue light receptor, CryB (RSP3077), has been identified (Hendrischk et al., 2009a), and both in vitro and in vivo studies have established that CryB interacts with AppA in a light-dependant manner, making AppA less available for interacting with and antagonising PpsR (Metz et al., 2012a). Thus blue light sensing by CryB may contribute to the rapid down-regulation of PS gene expression following irradiation.

The genome-wide transcriptomic data for a *cryB* null mutant revealed that expression levels of genes unrelated to PS genes are also mildly altered

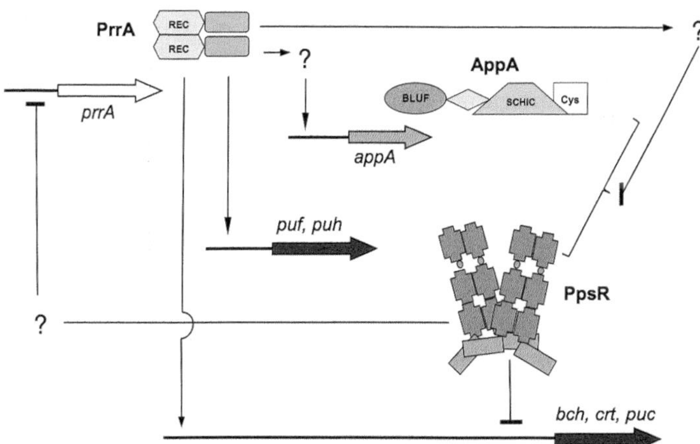

Figure 6.4 Complex hierarchical relationships between the PrrBA and AppA-PpsR pathways controlling O_2-dependant PS gene expression. Thick arrows indicate regulatory genes (*prrA, appA, ppsR*) or categories of PS genes (*bch, crt, puc, puf, puh*). Regulatory proteins (PrrA, AppA, PpsR) are represented in their oligomeric functional forms, i.e. PrrA is a dimer and PpsR is a tetramer. '?', proposed unknown regulator. *(Modified from Gomelsky et al., 2008).* (For colour version of this figure, the reader is referred to the online version of this book.)

compared to the wild type, both in the dark and in the light (Frühwirth, Teich, & Klug, 2012). Among the affected genes were global regulators including *prrA*, the RNA chaperone *hfq* (RSP2843), genes encoding sigma factors and RNA processing enzymes, pyruvate dehydrogenase *pdh* (RSP4047–4050), and some *cco* genes. Peculiarly, the genes identified by Tavano et al. (2005) as involved in redox regulation, RSP4157–4158, were among those most affected by the lack of *cryB*. Therefore, it appears that CryB acts through AppA–PpsR and also through additional regulatory networks that remain to be identified. The recently solved crystal structure of CryB revealed the complexity of this protein. It contains, in addition to FAD that undergoes blue light-dependant photoreduction, two other cofactors: 6,7-dimethyl-8-ribityl-lumazine, that serves as a second chromophore and a redox-sensitive 4Fe–4S cluster (Geisselbreicht, Frühwirth, Schroeder, Pierik, & Klug, 2012). Interestingly, transcriptome profiling suggests that yet another blue light-sensing protein RsLOV (RSP2228) (Hendrischk, Moldt, Frühwirth, & Klug, 2009b; Metz, Jäger, & Klug, 2012b), in the LOV (light–oxygen–voltage) domain family (Losi & Gartner, 2011), may be involved in the RpoE-dependant response to photooxidative stress.

5. CONCLUSIONS AND PERSPECTIVES

Transcriptomic approaches have been fruitful for studying phototrophic purple nonsulfur bacteria, and *R. sphaeroides* in particular. Transcriptome profiling under diverse growth conditions revealed major classes of genes associated with three energy generation modes, and the extent of transcriptome restructuring (20–30% of all genes) required for adopting a different energy generation mode. The observation of these relatively large expression changes have solidified the notion that modulation of transcription is the main means by which purple nonsulfur bacteria adapt to changes in environmental conditions, and that metabolic versatility requires transcriptome flexibility. The genome-wide transcriptome profiling has also uncovered new and previously unsuspected metabolic links.

Transcriptome analyses of the effects of O_2 and light confirmed the dominant role of O_2 in governing energy generation processes in phototrophic purple nonsulfur bacteria, and a lesser role of light in regulating gene expression under anaerobic photosynthetic conditions. A combination of low O_2 and light was found to down-regulate PS genes and induce a photooxidative stress response. Although shutting down PS in response to light in a phototrophic bacterium seems counter-intuitive, the avoidance of photooxidative stress provides a plausible explanation – safety comes first.

The major regulons associated with individual regulatory factors that mediate O_2 and light control have been inventoried, and some of these yielded surprises including hierarchy and interconnectivity between regulons that were not apparent from earlier genetic studies. While many of these connections are now better understood, some remain elusive. At this point, not even all known major transcription factors mediating O_2 and light responses have been investigated (e.g. a mutant lacking the SCHIC domain-containing O_2-sensitive transcription factor PpaA (Moskvin, Gilles-Gonzalez, & Gomelsky, 2010) has yet to be subjected to transcriptome analysis).

A number of *R. sphaeroides* transcriptome profiling studies have not been discussed here because of our focused approach. These studies include H_2O_2 response (Zeller, Moskvin, Li, Klug, & Gomelsky, 2005; 2007), photooxidative stress (Glaeser, Nuss, Berghoff, & Klug, 2011; Ziegelhoffer & Donohue, 2009), heat shock response regulons (Dufour, Imam, Koo, Green, & Donohue, 2012; Nuss, Glaeser, Berghoff, & Klug, 2010), iron-dependant regulation (Peuser, Remes, & Klug, 2012), identification of small regulatory RNAs and their chaperone Hfq that play important roles in photooxidative stress responses (Berghoff et al., 2009; 2011), studies of electron flow during photoheterotrophic hydrogen production (Kontur et al., 2011), and salt stress responses (Tsuzuki et al., 2011). More broadly, only a small fraction of the approximately 200 transcription factors, 16 sigma factors, and dozen regulatory sRNA encoded in the *R. sphaeroides* genome have been analysed by transcriptomic profiling. Future genome-wide transcript analyses along with new computational tools for meta-analysis of regulatory networks, e.g. Rhodobase (Moskvin et al., 2011), and other investigations of gene regulation and function, will help to bring the genome of *R. sphaeroides* to life.

ACKNOWLEDGEMENTS

This work was supported by the US Department of Agriculture Cooperative State Research, Education, Extension and Service and Center for Photoconversion and Catalysis, School of Energy Resources, University of Wyoming (MG), and funds from the US National Science Foundation (MCB-0921449 and other support while JHZR was working at the Foundation). Any opinions, findings, and conclusions or recommendations expressed in this material are those of the authors and do not necessarily reflect the views of the supporting agencies. We are grateful to Kurt W. Miller for manuscript proofreading.

REFERENCES

Arai, H., Roh, J. H., & Kaplan, S. (2008). Transcriptome dynamics during the transition from anaerobic photosynthesis to aerobic respiration in *Rhodobacter sphaeroides* 2.4.1. *Journal of Bacteriology, 190*, 286–299.

Bekker, M., Alexeeva, S., Laan, W., Sawers, G., Teixeira de Mattos, J., & Hellingwerf, K. (2010). The ArcBA two-component system of *Escherichia coli* is regulated by the redox state of both the ubiquinone and the menaquinone pool. *Journal of Bacteriology, 192*, 745–754.

Berghoff, B. A., Glaeser, J., Sharma, C. M., Vogel, J., & Klug, G. (2009). Photooxidative stress-induced and abundant small RNAs in *Rhodobacter sphaeroides*. *Molecular Microbiology, 74*, 1497–1512.

Berghoff, B. A., Glaeser, J., Sharma, C. M., Zobawa, M., Lottspeich, F., Vogel, J., et al. (2011). Contribution of Hfq to photooxidative stress resistance and global regulation in *Rhodobacter sphaeroides*. *Molecular Microbiology, 80*, 1479–1495.

Braatsch, S., Bernstein, J. R., Lessner, F., Morgan, J., Liao, J. C., Harwood, C. S., et al. (2006). *Rhodopseudomonas palustris* CGA009 has two functional *ppsR* genes, each of which encodes a repressor of photosynthesis gene expression. *Biochemistry, 45*, 14441–14451.

Braatsch, S., Gomelsky, M., Kuphal, S., & Klug, G. (2002). A single flavoprotein, AppA, integrates both redox and light signals in *Rhodobacter sphaeroides*. *Molecular Microbiology, 45*, 827–836.

Braatsch, S., Moskvin, O.V., Klug, G., & Gomelsky, M. (2004). Responses of the *Rhodobacter sphaeroides* transcriptome to blue light under semiaerobic conditions. *Journal of Bacteriology, 186*, 7726–7735.

Brander, J. P., McEwan, A. G., Kaplan, S., & Donohue, T. J. (1989). Expression of the *Rhodobacter sphaeroides* cytochrome c_2 structural gene. *Journal of Bacteriology, 171*, 360–368.

Bruscella, P., Eraso, J. M., Roh, J. H., & Kaplan, S. (2008). The use of chromatin immunoprecipitation to define PpsR binding activity in *Rhodobacter sphaeroides* 2.4.1. *Journal of Bacteriology, 190*, 6817–6828.

Callister, S. J., Dominguez, M. A., Nicora, C. D., Zeng, X., Tavano, C. L., Kaplan, S., et al. (2006). Application of the accurate mass and time tag approach to the proteome analysis of sub-cellular fractions obtained from *Rhodobacter sphaeroides* 2.4.1 aerobic and photosynthetic cell cultures. *Journal of Proteome Research, 5*, 1940–1947.

Cheng, Z., Wu, J., Setterdahl, A., Reddie, K., Carroll, K., Hammad, L. A., et al. (2012). Activity of the tetrapyrrole regulator CrtJ is controlled by oxidation of a redox active cysteine located in the DNA binding domain. *Molecular Microbiology, 85*, 734–746.

Comolli, J. C., Carl, A. J., Hall, C., & Donohue, T. (2002). Transcriptional activation of the *Rhodobacter sphaeroides* cytochrome c_2 gene P2 promoter by the response regulator PrrA. *Journal of Bacteriology, 184*, 390–399.

Daldal, F., Mandaci, S., Winterstein, C., Myllykallio, H., Duyck, K., & Zannoni, D. (2001). Mobile cytochrome *c2* and membrane-anchored cytochrome c_y are both efficient electron donors to the *cbb*3- and *aa*3-type cytochrome *c* oxidases during respiratory growth of *Rhodobacter sphaeroides*. *Journal of Bacteriology, 183*, 2013–2024.

Donohue, T. J., McEwan, A. G., Van Doren, S., Crofts, A. R., & Kaplan, S. (1988). Phenotypic and genetic characterization of cytochrome c_2 deficient mutants of *Rhodobacter sphaeroides*. *Biochemistry, 27*, 1918–1925.

Dubbs, J. M., & Tabita, R. F. (2004). Regulators of nonsulfur purple phototrophic bacteria and the interactive control of CO_2 assimilation, nitrogen fixation, hydrogen metabolism, and energy generation. *FEMS Microbiology Reviews, 28*, 353–376.

Dufour, Y. S., Imam, S., Koo, B. M., Green, H. A., & Donohue, T. J. (2012). Convergence of the transcriptional responses to heat shock and singlet oxygen stresses. *Public Library of Science Genetics, 8*, e1002929.

Dufour, Y. S., Kiley, P. J., & Donohue, T. J. (2010). Reconstruction of the core and extended regulons of global transcription factors. *Public Library of Science Genetics*, *6*, e1001027.

Elsen, S., Jaubert, M., Pignol, D., & Giraud, E. (2005). PpsR: a multifaceted regulator of photosynthesis gene expression in purple bacteria. *Molecular Microbiology*, *57*, 17–26.

Eraso, J. M., & Kaplan, S. (1994). *prrA*, a putative response regulator involved in oxygen regulation in photosynthesis gene expression in *Rhodobacter sphaeroides*. *Journal of Bacteriology*, *176*, 32–43.

Eraso, J. M., Roh, J. H., Zeng, X., Callister, S. J., Lipton, M. S., & Kaplan, S. (2008). Role of the global transcriptional regulator PrrA in *Rhodobacter sphaeroides* 2.4.1: combined transcriptome and proteome analysis. *Journal of Bacteriology*, *190*, 4831–4848.

Frühwirth, S., Teich, K., & Klug, G. (2012). Effects of the cryptochrome CryB from *Rhodobacter sphaeroides* on global gene expression in the dark or blue light or in the presence of singlet oxygen. *Public Library of Science One*, 7, e33791.

Geisselbreicht, Y., Frühwirth, S., Schroeder, C., Pierik, A., & Klug, G. (2012). CryB from *Rhodobacter sphaeroides*: a unique class of cryptochromes with new cofactors. *EMBO Reports*, *13*, 223–229.

Glaeser, J., Nuss, A. M., Berghoff, B. A., & Klug, G. (2011). Singlet oxygen stress in microorganisms, chapter 4. *Advances in Microbial Physiology*, *58*, 141–173.

Golomysova, A. N., Gomelsky, M., & Ivanov, P. S. (2010). Flux balance analysis of the photoheterotrophic growth of *Rhodobacter sphaeroides* relevant to biohydrogen production. *International Journal of Hydrogen Energy*, *35*, 12751–12760.

Gomelsky, M., & Kaplan, S. (1995a). Genetic evidence that PpsR from *Rhodobacter sphaeroides* 2.4.1 functions as a repressor of *puc* and *bchF* expression. *Journal of Bacteriology*, *177*, 1634–1637.

Gomelsky, M., & Kaplan, S. (1995b). *appA*, a novel gene encoding a *trans*-acting factor involved in the regulation of photosynthesis gene expression in *Rhodobacter sphaeroides* 2.4.1. *Journal of Bacteriology*, *177*, 4609–4618.

Gomelsky, M., & Kaplan, S. (1997). Molecular genetic evidence suggesting interactions between AppA and PpsR in regulation of photosynthesis gene expression in *Rhodobacter sphaeroides* 2.4.1. *Journal of Bacteriology*, *179*, 128–134.

Gomelsky, M., & Kaplan, S. (1998). AppA, a redox regulator of photosystem formation in *Rhodobacter sphaeroides* 2.4.1 is a flavoprotein. Identification of a novel FAD binding domain. *Journal of Biological Chemistry*, *273*, 35319–35325.

Gomelsky, M., & Klug, G. (2002). BLUF: a novel FAD-binding domain involved in sensory transduction in microorganisms. *Trends in Biochemical Sciences*, *27*, 497–500.

Gomelsky, L., Moskvin, O. V., Stenzel, R. A., Jones, D. F., Donohue, T. J., & Gomelsky, M. (2008). Hierarchical regulation of photosynthesis gene expression by the oxygen-responsive PrrBA and AppA-PpsR systems of *Rhodobacter sphaeroides*. *Journal of Bacteriology*, *190*, 8106–8114.

González, P. J., Correia, C., Moura, I., Brondino, C. D., & Moura, J. J. (2006). Bacterial nitrate reductases: molecular and biological aspects of nitrate reduction. *Journal of Inorganic Biochemistry*, *100*, 1015–1023.

Han, Y., Meyer, M. H., Keusgen, M., & Klug, G. (2007). A haem cofactor is required for redox and light signaling by the AppA protein of *Rhodobacter sphaeroides*. *Molecular Microbiology*, *64*, 1090–1104.

Hartsock, A., & Shapleigh, J. P. (2011). Physiological roles for two periplasmic nitrate reductases in *Rhodobacter sphaeroides* 2.4.3 (ATCC17025). *Journal of Bacteriology*, *193*, 6483–6489.

Hendrischk, A. K., Braatsch, S., Glaeser, J., & Klug, G. (2007). The *phrA* gene of *Rhodobacter sphaeroides* encodes a photolyase and is regulated by singlet oxygen and peroxide in a sigma(E)-dependent manner. *Microbiology*, *153*, 1842–1851.

Hendrischk, A. K., Frühwirth, S. W., Moldt, J., Pokorny, R., Metz, S., Kaiser, G., et al. (2009a). A cryptochrome-like protein is involved in the regulation of photosynthesis genes in *Rhodobacter sphaeroides*. *Molecular Microbiology*, *74*, 990–1003.

Hendrischk, A. K., Moldt, J., Frühwirth, S. W., & Klug, G. (2009b). Characterization of an unusual LOV domain protein in the α-proteobacterium *Rhodobacter sphaeroides*. *Photochemistry and Photobiology*, *85*, 1254–1259.

Huang, J. W., & Kaplan, S. (1973). Membrane proteins of *Rhodopseudomonas spheroides*. IV. Characterization of Chromatophore Proteins. *Biochimica Biophysica Acta (BBA)-Biomembranes*, *307*, 317–331.

Hunter, C. N., Daldal, F., Thurnauer, M., & Beatty, J. T. (2008). *The purple phototrophic bacteria*. (Vol. 28). Dordrecht, The Netherlands: Springer.

Hynes, A. P., Mercer, R. G., Watton, D. E., Buckley, C. B., & Lang, A. S. (2012). DNA packaging bias and differential expression of gene transfer agent genes within a population during production and release of the *Rhodobacter capsulatus* gene transfer agent, RcGTA. *Molecular Microbiology*, *85*, 314–325.

Igarashi, N., Harada, J., Nagashima, S., Matsuura, K., Shimada, K., & Nagashima, K.V. (2001). Horizontal transfer of the photosynthesis gene cluster and operon rearrangement in purple bacteria. *Journal of Molecular Evolution*, *52*, 333–341.

Imam, S., Yilmaz, S., Sohmen, U., Gorzalski, A. S., Reed, J. L., Noguera, D. R., et al. (2011). iRsp1095: a genome-scale reconstruction of the *Rhodobacter sphaeroides* metabolic network. *BMC Systems Biology*, *5*, 116.

Joshi, H. M., & Tabita, F. R. (1996). A global two component signal transduction system that integrates the control of photosynthesis, carbon dioxide assimilation, and nitrogen fixation. *Proceedings of the National Academy of Sciences of the United States of America*, *93*, 14515–14520.

Kiley, P. J., & Beinert, H. (1998). Oxygen sensing by the global regulator, FNR; the role of the iron–sulfur cluster. *FEMS Microbiology Reviews*, *22*, 341–352.

Kim, Y. J., Ko, I. J., Lee, J. M., Kang, H.Y., Kim, Y. M., Kaplan, S., et al. (2007). Dominant role of the *cbb*$_3$ oxidase in regulation of photosynthesis gene expression through the PrrBA system in *Rhodobacter sphaeroides* 2.4.1. *Journal of Bacteriology*, *189*, 5617–5625.

Kontur, W. S., Ziegelhoffer, E. C., Spero, M. A., Imam, S., Noguera, D. R., & Donohue, T. J. (2011). Pathways involved in reductant distribution during photobiological H_2 production by *Rhodobacter sphaeroides*. *Applied and Environmental Microbiology*, *77*, 7425–7429.

Körner, H., Sofia, H. J., & Zumft, W. G. (2003). Phylogeny of the bacterial superfamily of Crp–Fnr transcription regulators: exploiting the metabolic spectrum by controlling alternative gene programs. *FEMS Microbiology Reviews*, *27*, 559–592.

Laguri, C., Stenzel, R. A., Donohue, T. J., Phillips-Jones, M. K., & Williamson, M. P. (2006). Activation of the global gene regulator PrrA (RegA) from *Rhodobacter sphaeroides*. *Biochemistry*, *45*, 7872–7881.

Lee, J. K., & Kaplan, S. (1992). Isolation and characterization of *trans*-acting mutations involved in oxygen regulation of *puc* operon transcription in *Rhodobacter sphaeroides*. *Journal of Bacteriology*, *174*, 1158–1171.

Losi, A., & Gartner, W. (2011). Old chromophores, new photoactivation paradigms, trendy applications: flavins in blue light-sensing photoreceptors. *Photochemistry and Photobiology*, *87*, 491–510.

Mackenzie, C., Eraso, J. M., Choudhary, M., Roh, J. H., Zeng, X., Bruscella, P., et al. (2007). Postgenomic adventures with *Rhodobacter sphaeroides*. *Annual Reviews of Microbiology*, *61*, 283–307.

Mao, L., Mackenzie, C., Roh, J. H., Eraso, J. M., Kaplan, S., & Resat, H. (2005). Combining microarray and genomic data to predict DNA binding motifs. *Microbiology*, *151*, 3197–3213.

Masuda, S., & Bauer, C. E. (2002). AppA is a blue light photoreceptor that antirepresses photosynthesis gene expression in *Rhodobacter sphaeroides*. *Cell, 110*, 613–623.

Metz, S., Haberzettl, K., Frühwirth, S., Teich, K., Hasewinkel, C., & Klug, G. (2012a). Interaction of two photoreceptors in the regulation of bacterial photosynthesis genes. *Nucleic Acids Research, 40*, 5901–5909.

Metz, S., Jäger, A., & Klug, G. (2012b). Role of a short light, oxygen, voltage (LOV) domain protein in blue light- and singlet oxygen-dependent gene regulation in *Rhodobacter sphaeroides*. *Microbiology, 158*, 368–379.

Moskvin, O. V., Bolotin, D., Wang, A., Ivanov, P. S., & Gomelsky, M. (2011). Rhodobase, a meta-analytical tool for reconstructing gene regulatory networks in a model photosynthetic bacterium. *Biosystems, 103*, 125–131.

Moskvin, O.V., Gilles-Gonzalez, M. A., & Gomelsky, M. (2010). The PpaA/AerR regulators of photosynthesis gene expression from anoxygenic phototrophic proteobacteria contain heme-binding SCHIC domains. *Journal of Bacteriology, 192*, 5253–5256.

Moskvin, O. V., Gomelsky, L., & Gomelsky, M. (2005). Transcriptome analysis of the *Rhodobacter sphaeroides* PpsR regulon: PpsR as a master regulator of photosystem development. *Journal of Bacteriology, 187*, 2148–2156.

Moskvin, O.V., Kaplan, S., Gilles-Gonzalez, M. A., & Gomelsky, M. (2007). Novel heme-based oxygen sensor with a revealing evolutionary history. *Journal of Biological Chemistry, 282*, 28740–28748.

Moskvin, O.V., Lar, O., Suwansaard, M., Gomelsky, M. & Zeilstra-Ryalls, J. H. New pathway of photosynthesis regulation in Rhodobacter sphaeroides 2.4.1 involving partially redundant Crp-Fnr transcription factors, in preparation.

Mouncey, N. J., Gak, E., Choudhary, M., Oh, J., & Kaplan, S. (2000). Respiratory pathways of *Rhodobacter sphaeroides* 2.4.1(T): identification and characterization of genes encoding quinol oxidases. *FEMS Microbiology Letters, 192*, 205–210.

Mouncey, N. J., & Kaplan, S. (1998a). Cascade regulation of dimethyl sulfoxide reductase (*dor*) gene expression in the facultative phototroph *Rhodobacter sphaeroides* $2.4.1^{T}$. *Journal of Bacteriology, 180*, 2924–2930.

Mouncey, N. J., & Kaplan, S. (1998b). Redox-dependent gene regulation in *Rhodobacter sphaeroides* $2.4.1^{T}$: effects on dimethyl sulfoxide reductase (*dor*) gene expression. *Journal of Bacteriology, 180*, 5612–5618.

Myllykallio, H., Zannoni, D., & Daldal, F. (1999). The membrane-attached electron carrier cytochrome c_y from *Rhodobacter sphaeroides* is functional in respiratory but not in photosynthetic electron transfer. *Proceedings of the National Academy of Sciences of the United States of America, 96*, 4348–4353.

Naylor, G. W., Addlesee, H. A., Gibson, L. C.D., & Hunter, C. N. (1999). The photosynthesis gene cluster of *Rhodobacter sphaeroides*. *Photosynthesis Research, 62*, 121–139.

Nuss, A. M., Glaeser, J., Berghoff, B. A., & Klug, G. (2010). Overlapping alternative sigma factor regulons in the response to singlet oxygen in *Rhodobacter sphaeroides*. *Journal of Bacteriology, 192*, 2613–2623.

Oh, J. I., Ko, I. J., & Kaplan, S. (2004). Reconstitution of the *Rhodobacter sphaeroides* *cbb*$_3$-PrrBA signal transduction pathway in vitro. *Biochemistry, 43*, 7915–7923.

Pappas, C. T., Sram, J., Moskvin, O.V., Ivanov, P. S., MacKenzie, R. C., Choudhary, M., et al. (2004). Construction and validation of the *Rhodobacter sphaeroides* 2.4.1 DNA microarray: transcriptome flexibility at diverse growth modes. *Journal of Bacteriology, 186*, 4748–4758.

Penfold, R. J., & Pemberton, J. M. (1994). Sequencing, chromosomal inactivation, and functional expression in *Escherichia coli* of *ppsR*, a gene which represses carotenoid and bacteriochlorophyll synthesis in *Rhodobacter sphaeroides*. *Journal of Bacteriology, 176*, 2869–2876.

Peuser, V., Remes, B., & Klug, G. (2012). Role of the Irr protein in the regulation of iron metabolism in *Rhodobacter sphaeroides*. *Public Library of Science One*, 7, e42231.

Phillips-Jones, M. K., & Hunter, C. N. (1994). Cloning and nucleotide sequence of *regA*, a putative response regulator gene of *Rhodobacter sphaeroides*. *FEMS Microbiology Letters, 116*, 269–275.

Ponnampalam, S. N., Buggy, J. J., & Bauer, C. E. (1995). Characterization of an aerobic repressor that coordinately regulates bacteriochlorophyll, carotenoid, and light harvesting-II expression in *Rhodobacter capsulatus*. *Journal of Bacteriology, 177*, 2990–2997.

Potter, C. L., Jeong, E. L., Williamson, M. P., Henderson, P. J., & Phillips-Jones, M. K. (2006). Redox-responsive in vitro modulation of the signalling state of the isolated PrrB sensor kinase of *Rhodobacter sphaeroides* NCIB 8253. *FEBS Letters, 580*, 3206–3210.

Ranson-Olson, B., & Zeilstra-Ryalls, J. H. (2008). Regulation of the *Rhodobacter sphaeroides* 2.4.1 *hemA* gene by PrrA and FnrL. *Journal of Bacteriology, 190*, 6769–6778.

Roh, J. H., & Kaplan, S. (2002). Interdependent expression of the *ccoNOQP-rdxBHIS* loci in *Rhodobacter sphaeroides* 2.4.1. *Journal of Bacteriology, 184*, 5330–5338.

Roh, J. H., Smith, W. E., & Kaplan, S. (2004). Effects of oxygen and light intensity on transcriptome expression in *Rhodobacter sphaeroides* 2.4.1. *Journal of Biological Chemistry, 279*, 9146–9155.

Rott, M. A., & Donohue, T. J. (1990). *Rhodobacter sphaeroides spd* mutations that allow cytochrome c_2-independent photosynthetic growth. *Journal of Bacteriology, 172*, 1954–1961.

Sganga, M. W., & Bauer, C. E. (1992). Regulatory factors controlling photosynthetic reaction center and light-harvesting gene expression in *Rhodobacter capsulatus*. *Cell, 68*, 945–954.

Shapleigh, J. P., Hill, J. J., Alben, J. O., & Gennis, R. B. (1992). Spectroscopic and genetic evidence for two heme-Cu-containing oxidases in *Rhodobacter sphaeroides*. *Journal of Bacteriology, 174*, 2338–2343.

Swem, L. R., Kraft, B. J., Swem, D. L., Setterdahl, A. T., Masuda, S., Knaff, D. B., et al. (2003). Signal transduction by the global regulator RegB is mediated by a redox-active cysteine. *EMBO Journal, 22*, 4699–4708.

Tao, Y., Liu, D., Yan, X., Zhou, Z., Lee, J. K., & Yang, C. (2012). Network identification and flux quantification of glucose metabolism in *Rhodobacter sphaeroides* under photoheterotrophic H(2)-producing conditions. *Journal of Bacteriology, 194*, 274–283.

Tavano, C. L., Podevels, A. M., & Donohue, T. J. (2005). Identification of genes required for recycling reducing power during photosynthetic growth. *Journal of Bacteriology, 187*, 5249–5258.

Tebaldi, T., Re, A., Viero, G., Pegoretti, I., Passerini, A., Blanzieri, E., et al. (2012). Widespread uncoupling between transcriptome and translatome variations after a stimulus in mammalian cells. *BMC Genomics, 13*, 220.

Tsuzuki, M., Moskvin, O.V., Kuribayashi, M., Sato, K., Retamal, S., Abo, M., et al. (2011). Salt stress-induced changes in the transcriptome, compatible solutes, and membrane lipids in the facultative phototrophic bacterium *Rhodobacter sphaeroides*. *Applied and Environmental Microbiology*, 77, 7551–7559.

Woronowicz, K., Olubanjo, O. B., Sung, H. C., Lamptey, J. L., & Niederman, R. A. (2012). Differential assembly of polypeptides of the light-harvesting 2 complex encoded by distinct operons during acclimation of *Rhodobacter sphaeroides* to low light intensity. *Photosynthesis Research, 111*, 125–138.

Wu, J., & Bauer, C. E. (2008). RegB/RegA, a global redox-responding two-component system. *Advances in Experimental Medicine and Biology, 631*, 131–148.

Wu, J., & Bauer, C. E. (2010). RegB kinase activity is controlled in part by monitoring the ratio of oxidized to reduced ubiquinones in the ubiquinone pool. *MBio, 1*. pii: e00272–10.

Yin, L., Dragnea, V., & Bauer, C. E. (2012). PpsR, a regulator of heme and bacteriochlorophyll biosynthesis, is a heme-sensing protein. *Journal of Biological Chemistry, 287*, 13850–13858.

Zeilstra-Ryalls, J. H. (2008). Chapter 39: regulation of the tetrapyrrole biosynthetic pathway, p. 777–798. In C. N. Hunter, F. Daldal, M. Thurnauer & J. T. Beatty (Eds.), *The purple phototrophic bacteria*. (Vol. 28). Dordrecht, The Netherlands: Springer.

Zeilstra-Ryalls, J. H., Gabbert, K., Mouncey, N. J., Kranz, R. G., & Kaplan, S. (1997). Analysis of the *fnrL* gene and its function in *Rhodobacter capsulatus*. *Journal of Bacteriology*, *179*, 7264–7273.

Zeilstra-Ryalls, J. H., & Kaplan, S. (1995). Aerobic and anaerobic regulation in *Rhodobacter sphaeroides* 2.4.1: the role of the *fnrL* gene. *Journal of Bacteriology*, *177*, 6422–6431.

Zeilstra-Ryalls, J. H., & Kaplan, S. (2004). Oxygen intervention in the regulation of gene expression: the photosynthetic bacterial paradigm. *Cellular and Molecular Life Sciences*, *61*, 417–436.

Zeilstra-Ryalls, J. H., & Schornberg, K. (2006). Analysis of *hemF* gene function and expression in *Rhodobacter sphaeroides* 2.4.1. *Journal of Bacteriology*, *188*, 801–804.

Zeller, T., Moskvin, O. V., Li, K., Klug, G., & Gomelsky, M. (2005). Transcriptome and physiological responses to hydrogen peroxide of the facultatively phototrophic bacterium *Rhodobacter sphaeroides*. *Journal of Bacteriology*, *187*, 7232–7242.

Zeller, T., Mraheil, M., Moskvin, O.V., Li, K., Gomelsky, M., & Klug, G. (2007). Regulation of hydrogen peroxide-dependent gene expression in *Rhodobacter sphaeroides*: regulatory functions of OxyR. *Journal of Bacteriology*, *189*, 3784–3792.

Zeng, X., Choundhary, M., & Kaplan, S. (2003). A second and unusual *pucBA* operon of *Rhodobacter sphaeroides* 2.4.1: genetics and function of the encoded polypeptides. *Journal of Bacteriology*, *185*, 6171–6178.

Zeng, X., Roh, J. H., Callister, S. J., Tavano, C. L., Donohue, T. J., Lipton, M. S., et al. (2007). Proteomic characterization of the *Rhodobacter sphaeroides* 2.4.1 photosynthetic membrane: identification of new proteins. *Journal of Bacteriology*, *189*, 7464–7474.

Zhu, S., & Kaplan, S. (1985). Effects of light, oxygen, and substrates on steady-state levels of mRNA coding for ribulose-1,5-bisphosphate carboxylase and light-harvesting reaction center polypeptides in *Rhodopseudomonas sphaeroides*. *Journal of Bacteriology*, *162*, 925–932.

Ziegelhoffer, E. C., & Donohue, T. J. (2009). Bacterial responses to photo-oxidative stress. *Nature Reviews Microbiology*, 7, 856–863.

CHAPTER SEVEN

The Evolution of the Purple Photosynthetic Bacterial Light-Harvesting System

Sarah L. Henry, Richard J. Cogdell[1]
Institute of Molecular, Cell and Systems Biology, College of Medical, Veterinary and Life Sciences, Glasgow Biomedical Research Centre, University of Glasgow, Glasgow, UK
[1]Corresponding authors: E-mail: Richard.Cogdell@Glasgow.ac.uk

Contents

Abstract

This chapter presents a discussion of what can be determined about how the purple photosynthetic bacterial light-harvesting complexes (LHs) evolved, based on a detailed phylogenetic analysis. The starting hypothesis was that the light-harvesting complex 2 (LH2) evolved from the highly structurally homologous light-harvesting complex 1. This hypothesis could not be substantiated from the phylogenetic analyses, comparing the data from the 16S ribosomal RNA genes with those of the light-harvesting apoprotein genes (*puc* and *puf*). Several possible reasons for this are suggested. Proliferation of the *puc* genes is associated with the ability to synthesis LH2 complexes with different absorption spectra. This appears to enable those bacteria with multiple *puc* genes to optimise their light harvesting to suit a wide variety of ecological niches.

1. INTRODUCTION

The light-harvesting system found in purple photosynthetic bacteria has been studied extensively and is now very well understood. The X-ray

Advances in Botanical Research, Volume 66
ISSN 0065-2296, http://dx.doi.org/10.1016/B978-0-12-397923-0.00007-2

crystal structures of several different light-harvesting complexes (LHs) have been determined (Koepke, Hu, Muenke, Schulten, & Michel, 1996; McDermott et al., 1995; Papiz, Prince, Howard, Cogdell, & Isaacs, 2003; Roszak et al., 2003), and their energy transfer reactions have been time resolved from a few femtoseconds out to longer times (Fleming & van Grondelle, 1997). In contrast, there have been relatively few attempts to investigate the evolution of these LHs except by Lang, Harwood, and Beatty (2011). In this chapter, we investigate whether there is enough information in the sequences of the genes encoding the light-harvesting apoproteins to allow a detailed picture to be developed that will accurately describe their evolution.

The photosynthetic unit in purple bacteria is made up of two types of pigment–protein complexes; the LHs and the reaction centre. The light reactions of purple bacterial photosynthesis begin when a photon is absorbed by a bacteriochlorophyll or carotenoid contained in the LHs. Once excited, the LHs transfer the excitation energy to the reaction centre, where it is trapped and used to drive a transmembrane charge separation reaction. This process is illustrated in Fig. 7.1.

When the reaction centre has been through two cycles of excitation, a molecule of ubiquinone is fully reduced to ubiquinol. This ubiquinol delivers its reducing equivalents to the cytochrome b/c_1 complex and a rather simple cyclic electron transport pathway begins. As a result of this cyclic

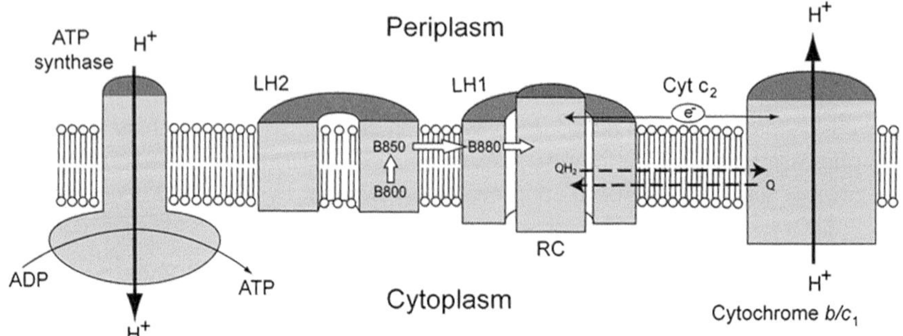

Figure 7.1 ***Schematic representation of light reactions in the purple bacterial membrane.*** The light reactions of purple photosynthetic bacteria represent a simple example of photosynthesis. The white arrows represent the transfer of excitation energy within and between the two types of LH complex, and to the single type of reaction centre. The reaction centre reduces ubiquinone, which delivers its reducing equivalents to the cytochrome b/c_1 complex that pumps protons across the photosynthetic membrane generating a proton motive force. This proton motive force is used by the ATP synthase to generate ATP.

electron pathway, protons are pumped across the photosynthetic membrane. The electrochemical gradient of protons is then used by the ATP synthase to make Adenosine triphosphate (ATP) (Wraight, Codgell, & Chance, 1978).

1.1. The Light-Harvesting Complexes

The reaction centre is surrounded by the light-harvesting complex 1 (LH1) and this complex is called the 'core' complex because of its central role in purple bacterial photosynthesis. Many species of purple bacteria also synthesise a second type of LH called LH2. LH2 is arranged peripherally around the core complex as illustrated in Fig. 7.2.

The position of the long wavelength (Q_y) absorption band of bacteriochlorophyll in the LHs is sensitive to its local environment (Cogdell, Howard, Isaacs, McLuskey, & Gardiner, 2002). The absorption

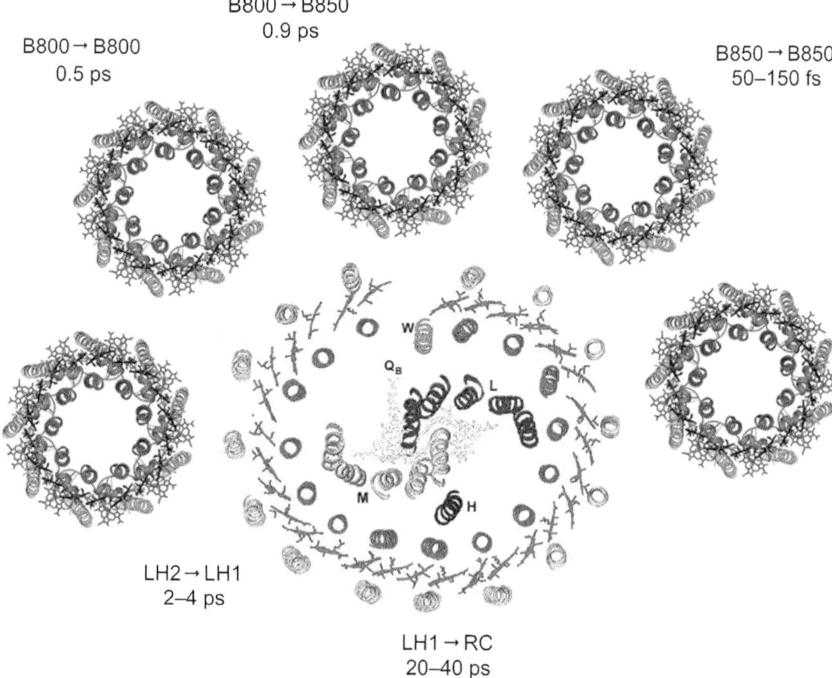

Figure 7.2 ***Illustration of a typical photosynthetic unit from a species of purple bacteria that contains both LH1 and LH2.*** The photosynthetic unit consists of the 'core' complex comprising a reaction centre surrounded by the LH1, and the peripheral LH2 complexes. This figure also shows the typical times for the major energy transfer reactions that take place between LH2 to LH1, and from LH1 to the reaction centre. *(Modified from Cogdell et al. (2006)).*

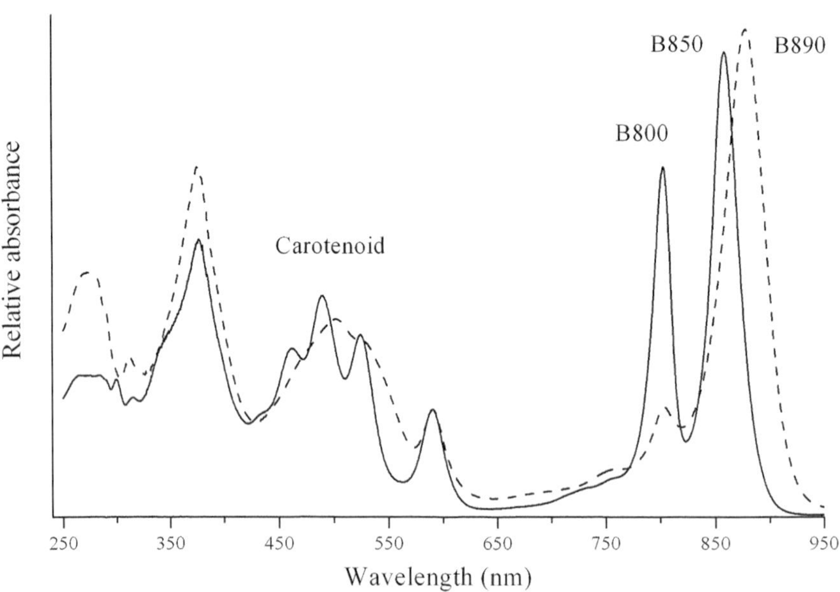

Figure 7.3 ***Absorption spectra of purified RC–LH1 and LH2 complexes from*** **Rhodopseudomonas palustris.** The solid line shows the absorption spectrum of LH2 and the dashed line represents the LH1 core complex spectrum. The spectra have been normalised to the bacteriochlorophyll Q_x band (~590 nm). The complexes were solubilised from a photosynthetic membrane using the detergent *N,N*-Dimethyldodecylamine *N*-oxide (LDAO). The absorption spectra were recorded with the complexes in 20 mM Tris–HCl buffer (pH 8) containing 0.1% LDAO. *(Gabrielsen et al., 2009)*

spectra of typical LH2 and LH1 core complexes are shown in Fig. 7.3. LH2 has two strong absorption bands in the near infrared (NIR) at 800 and 850 nm. These absorption bands reflect the two groups of bacteriochlorophyll molecules seen in the crystal structure (McDermott et al., 1995). The 800 nm-absorbing bacteriochlorophylls are monomeric and the 850 nm-absorbing ones form a tightly coupled aggregate. LH1 has a single strong NIR absorption band at 875 nm that arises from a similar, although larger, tightly coupled ring of bacteriochlorophylls. These spectroscopic differences between the different groups of bacteriochlorophylls are functionally important. The presence of LH2 broadens the wavelength range over which NIR light can be absorbed and thus, be available to drive photosynthesis. The further to the blue a molecule absorbs the higher the energy of its first excited singlet state, therefore, when a purple bacterial LH2 is excited with light of 800 nm, the 800 nm-absorbing bacteriochlorophyll molecules will naturally transfer the excitation energy downhill to the 850 nm-absorbing ones. Similarly, the energy gradient will also favour downhill energy transfer from LH2 to LH1. It is clear, therefore, that the presence of LH2 will both enhance the bacteria's capacity for light harvesting and allow the downhill transfer of absorbed light energy to the core complexes.

The LH1 and LH2 complexes are both constructed on the same modular principle (Law et al., 2004). Dimers of alpha and beta apoproteins non-covalently bind small numbers of bacteriochlorophyll and carotenoid molecules, and oligomerise to form circular/elliptical holocomplexes. These oligomers represent the intactly pigmented LHs. A comparison of the amino acid sequences of the LH1 and LH2 polypeptides reveals that they are structurally highly homologous. Each polypeptide is small, typically 50–60 amino acids in length, and shows a strong tripartite domain structure. The central hydrophobic domain forms a single transmembrane alpha helix. The more polar N and C termini are located on either side of the photosynthetic membrane. This domain structure is illustrated in Fig. 7.4.

More detailed comparisons of the amino acid sequences of a selection of alpha and beta apoproteins from both LH1 and LH2 complexes from purple bacterial are shown in Fig. 7.5 and 7.6. As well as the domain

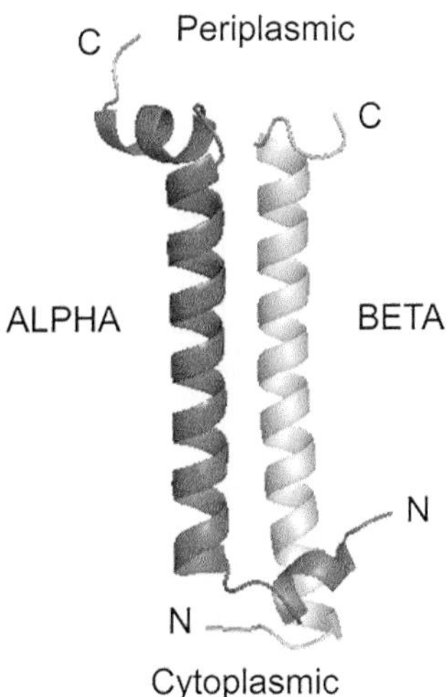

Figure 7.4 ***The structure of LH2 alpha and beta apoproteins from* Rps. acidophila *showing the typical three domain structure.*** The apoproteins have a central transmembrane helix domain containing the conserved histidine residues important for binding bacteriochlorophyll. The N and C termini consist of polar residues, which lay on either side of the photosynthetic membrane, the C on the periplasmic and the N on the cytoplasmic side *(McDermott et al., 1995).*

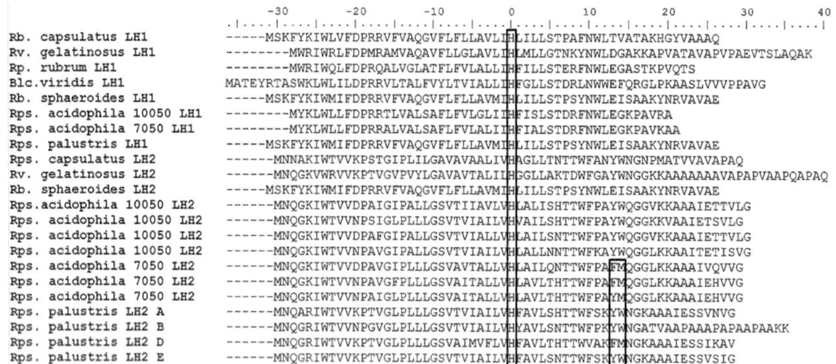

Figure 7.5 ***Protein sequences of alpha apoproteins from LH1 and LH2, from a number of different purple bacterial species.*** The conserved histidine is shown with an outline. The two boxed residues at position +13 and +14 indicate LH2 complexes where the tyrosine and tryptophan have been altered to phenylalanine and methionine, yielding an LH2 complex that has bacteriochlorophyll Q_y absorption peaks at 800 and 820 nm. Sequences were obtained from the NBCI website (Fowler et al., 1992; McLuskey et al., 2001). *(Adapted from Brunisholz and Zuber (1992)).*

```
                             -20       -10        0        10        20        30        40
Rb. capsulatus               -MADKNDLSFTGLTDEQAQELHAVYMSGLSAFIAVAVLAHLAVMIWRPWF
Rv. gelatinosus              --MAERKGSISGLTDDEAQEFHKFWVQGFVGFTAVAVVAHFLVWVWRPWL
Rb. sphaeroides LH1          -MADKSDLGYTGLTDEQAQELHSVYMSGLWLFSAVAIVAHLAVYIWRPWF
Rp. rubrum LH1               -MAEVKQESLSGITEGEAKEFHKIFTSSILVFFGVAAFAHLLVWIWRPWVPGPNGYS
Blc.viridis LH1              --MADLKPSLTGLTEEEAKEFHGIFVTSTVLYLATAVIVHYLVWTARPWIAPIPKGWVNLEGVQSALSYLV
Rps. acidophila 10050 LH1    --MAEDRSSLSGVSDAEAKEFHALFVSSFTAFIVIAVLAHVLAWAWRPWIPGPKGWA
Rps. acidophila 7050 LH1     --MAEDRSSLSGVSDAEAKEFHALFVSSFMGFMVVAVLAHVLAWAWRPWIPGPKGWA
Rps. palustris LH1           -MADKSDLGYTGLTDEQAQELHSVYMSGLWLFSAVAIVAHLAVYIWRPWF
Rps. capsulatus LH2          --MTDDKAGPSGLSLKEAEEIHSYLIDGTRVFGAMALVAHILSAIATPWLG
Rv. gelatinosus LH2          MADDANKVWPSGLTTAEAEELQKGLVDGTRIFGVIAVLAHILAYAYTPWLH
Rb. sphaeroides LH2          MTDDLNKVWPSGLTVAEAEEVHKQLILGTRVFGGMALIAHFLAAAATPWLG
Rps. acidophila 10050 LH2    ---------MATLTPEQSEELHKYVIDGTRVFLGLALVAHFLAFSATPWLH
Rps. acidophila 10050 LH2    ---------MATLTAEQSEELHKYVIDGTRVFLGLALVAHFLAFSATPWLH
Rps. acidophila 10050 LH2    -----MRWTMATLTAEQSEELHKYVIDGTRVFLGLALVAHFLAFSATPWLH
Rps. acidophila 10050 LH2    ---------MAVLNEAQAEELHKHVIDGARVFGVIALFAHVLAYSLTPWLH
Rps. acidophila 7050 LH2     -------MADKPLTADQAEELHKYVIDGARAFVAIAAFAHVLAYSLTPWLH
Rps. acidophila 7050 LH2     ---------MAVLSPEQSEELHKYVIDGARVFLGIALVAHFLAFSMTPWMH
Rps. acidophila 7050 LH2     ---------MAVLTPEQSEELHKYVIDGARVFLGVALVAHFLAFSATPWLH
Rps. palustris LH2           MADDPNKVWPTGLTIAESEELHKHVIDGTRIFGAIAIVAHFLAYVYSPWLH
Rps. palustris LH2           ----MADKTLTGLTVEESEELHKHVIDGTRIFGAIAIVAHFLAYVYSPWLH
Rps. palustris LH2           MVDDSKKVWPTGLTIAESEEIHKHVIDGARIFVAIAIVAHFLAYVYSPWLH
Rps. palustris LH2           MVDDPNKVWPTGLTIAESEELHKHVIDGSRIFVAIAIVAHFLAYVYSPWLH
Rps. palustris LH2           MADDPNKVWPTGLTIAESEELHKHVIDGTRIFGAIAIVAHFLAYVYSPWLH
```

Figure 7.6 ***Protein sequence for beta apoproteins from LH1 and LH2, from several of different purple bacterial species.*** The central conserved histidine residue responsible to binding the magnesium ion in bacteriochlorophyll ring is highlighted by a black box. Sequences were obtained from the NCBI website. *(Adapted from Brunisholz and Zuber (1992)).*

structure indicated by the distribution of hydrophilic and hydrophobic residues, several highly conserved amino acids can be seen from this comparison. The most important of these are the conserved histidine residues that are involved in binding the central magnesium atoms present in the bacteriochlorin rings of the bacteriochlorophyll molecules. Other key conserved amino acids have been shown to affect the spectroscopic properties of the tightly coupled ring of bacteriochlorophyll molecules. For example, in the case of *Rhodopseudomonas acidophila*, an alteration of the tyrosine and tryptophan on the alpha polypeptide (position +13 and 14) to phenylalanine and methionine is responsible for shifting an absorption peak of the LH2 complex from 850 nm to 820 nm (Fowler, Visschers, Grief, van Grondelle, & Hunter, 1992; McLuskey, Prince, Cogdell, & Isaacs, 2001).

1.2. The Photosynthesis Gene Cluster

Purple photosynthetic bacteria contain a photosynthesis gene cluster that encodes pigment biosynthesis enzymes and the proteins of the core complex. The LH1 alpha and beta apoproteins are encoded in the *puf* operon (*pufB* and *pufA*) (Donohue, Kiley, & Kaplan, 1988). The operon (illustrated in Fig. 7.7) contains *pufB* and *pufA* followed by the structural genes for the L and M subunits of the reaction centre, and the *pufX* protein, which is involved in the structure of the LH1 complex (Bullough, Qian, & Hunter, 2009). The LH2 alpha and beta apoproteins are encoded by the *pucB* and *pucA* genes (Ashby, Coomber, & Hunter, 1987; Youvan & Ismail, 1985). Unlike the *puf* genes, the *puc* genes are typically located outside the well known photosynthesis gene cluster. The *puc* operon also usually contains a gene called *puc*C that encodes a protein that is required for the assembly of LH2 complexes (Tichy, Albien, Gadon, & Drews, 1991).

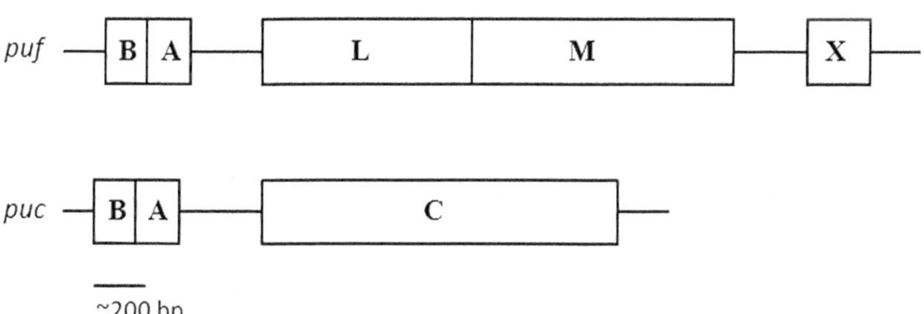

Figure 7.7 ***Diagram showing the typical organisation of the photosynthetic* puf *and* puc *operons within the genome.*** The *puf* operon contains the *pufB* and *pufA* genes that encode the two apoproteins in the LH core complex (Donohue et al., 1988), *pufL* and *pufM* that encode apoproteins in the reaction centre, and *pufX* is involved in the structure of the LH1 complex (Bullough et al., 2009). The *puc* operon encodes the proteins of the alpha and beta apoproteins of LH2 as well as the *pucC* gene which is involved in the assembly of the LH2 complex *(Tichy et al., 1991; Youvan & Ismail, 1985).*

1.3. Variation in Light-Harvesting Constituents Present in the Purple Bacterial Photosynthetic Unit

The purple bacteria can be conveniently divided on the basis of their complement of LHs into three groups (a partial list of species in these groups is shown in Table 7.1). The first group, which includes species such as *Rhodospirillum rubrum* and *Blastochloris viridis*, only contain LH1 as part of the core complex. The second group, which includes species such *Rhodobacter capsulatus* and *Rubrivivax gelatinosus*, contain both LH1 and a single type of LH2 complex. The third group, which includes *Rhodopseudomonas palustris*, *Rps. acidophila* and *Allochromatium vinosum*, contain both LH1 core complexes and multiple types of LH2 complexes (Gabrielson, Gardiner, & Codgell, 2009; Oda et al., 2008; Zuber & Codgell, 1995) (Table 7.1).

In species such as *R. rubrum*, the amount of the LH1 core complexes synthesised depends on the light intensity at which the cells are grown (Aagaard & Sistrom, 1972). In general, the lower the light intensity the more core complexes are present in the cell and the more highly those cells are pigmented. However, in LH2-containing species such as *Rb. capsulatus*, not only are the numbers of LH1 core complexes per cell regulated by light intensity but also the ratio of LH2 to LH1 (Katsuda et al., 2004; Reidl, Golecki, & Drews, 1983). The lower the light intensity the higher the ratio of LH2 to LH1. This means that in *Rb. capsulatus* not only are the numbers of photosynthetic units per cell regulated by light intensity but also the size of those units.

Species that are able to synthesise multiple spectroscopic forms of LH2 also appear to regulate the type and amount of the different LH2 complexes in response to environmental cues (Gardiner, Cogdell, & Takaichi, 1993;

Table 7.1 A partial list of groups of purple bacteria illustrating the variability in the presence of one or more types of LH2 complex

Species without LH2	Species with only one LH2	Species with multiple LH2
Blastochloris viridis	*Rubrivivax gelatinosus*	*Rhodopseudomonas palustris*
Rhodospirillum rubrum	*Rhodobacter capsulatus*	*Allochromatium vinosum*
Rubrivivax benzoatilyticus	*Dinoroseobacter shibae*	*Marichromatium purpuratum*
Erythrobacter litoralis	*Halorhodospira halophila*	*Rhodospirillum molischianum*
		Rhodopseudomonas acidophila

Tadros & Waterkamp, 1989; Tharia, Nightingale, Papiz, & Lawless, 1999). Figure 7.8 shows the B800–820 form of LH2 from *Rps. acidophila*. This absorption spectrum should be compared with the B800–850 form in Fig. 7.3. Usually these species that can produce such different spectroscopic forms of LH2 respond to a wide variety of cues, including light quality, light intensity, temperature, and in the case of *Al. vinosum*, the type of reduced sulphur compounds present in the growth medium (Mechler & Oelze, 1978a, 1978b, 1978c). It has been assumed that the ability to synthesise these different types of LH2 complexes is somehow correlated with the ability of those species to optimise their growth in a wide variety of different ecological niches. However, there have been very few direct experiments carried out that demonstrate that this hypothesis is indeed correct.

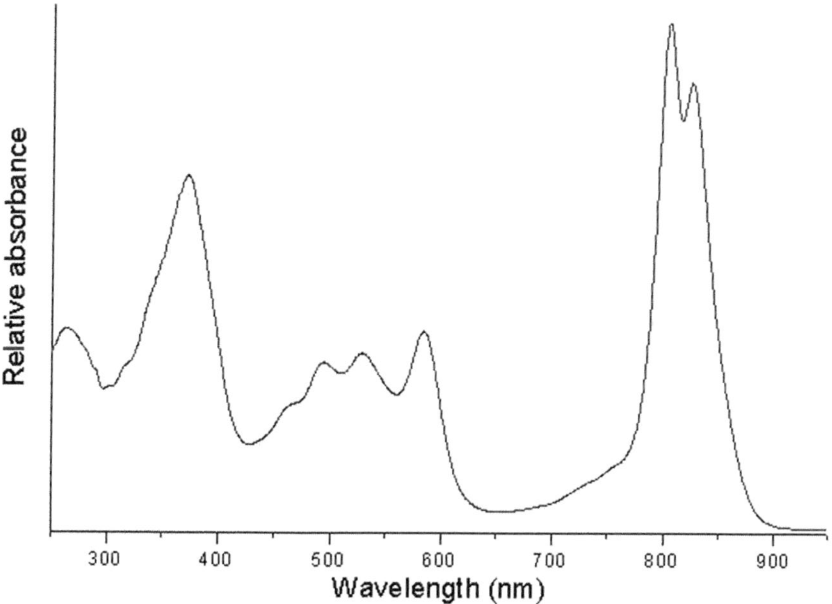

Figure 7.8 ***Absorption spectrum of the LH2 complex purified from* Rps. acidophila *cells grown in low light.*** The structure of the low light LH2 shows that the two amino acids at positions +13 and 14 are responsible for the shift in a bacteriochlorophyll Q_y peak from 850 to 820 nm, in response to growth under a low-light intensity. The blue shift in bacteriochlorophyll absorption allows for more efficient light capture and an increase in the efficiency of energy transfer towards the LH1 core complex. The membranes were solubilised in LDAO and LH2 B800–820 complexes purified in 20 mM Tris, 0.1% LDAO (McLuskey et al., 2001). (For interpretation of the references to colour in the figure legend, the reader is referred to the online version of this book.)

1.4. Hypothesis for the Evolution of the Genes Encoding LH2

It is interesting now to consider how this elegant purple bacterial light-harvesting system may have evolved. The simplest hypothesis is to suggest that initially there were only LH1 core complexes. Indeed, it may have been that the first LH1 complex was a homodimer due to the highly conserved domain structure of the LH1 apoproteins. If so, then following a gene duplication this evolved into the heterodimer system that is present today. It is tempting to suggest that LH2 is evolved from LH1 by further gene duplication. This would then allow the LH2 complexes to evolve independently of LH1 to produce smaller rings, where the hole in the middle was too small to accommodate the reaction centre, and there was space between the beta apoproteins to contain an additional bacteriochlorophyll molecule in each alpha/beta dimer (the B800 bacteriochlorophylls). Using this line of reasoning, subsequent gene duplication events allowed the evolution of multiple spectroscopic forms of LH2. Interestingly, the presence of multiple *puc* genes has allowed the LH2 complexes to move from completely homogeneous rings, in which all the alpha apoproteins are the same and beta apoproteins are the same, to the possibility of having heterogeneous rings, where in a single ring different alpha and beta apoproteins encoded by different *puc* genes may be accommodated (Brotosudarmo et al., 2009).

This hypothesis of how the LHs present in purple bacteria may have evolved is very appealing. However, is it correct? In this chapter, we investigate this hypothesis by comparing the evolutionary tree of the purple bacteria constructed from their 16S RNA sequences and looking to see if it is possible to identify when the *puf* and *puc* genes arose.

2. METHOD

A list of all the purple photosynthetic bacteria available on 15/10/12 from the Deutsche Sammlung von Mikroorganismen und Zellkulturen GmbH (DSMZ) bacterial database website was compiled (http://www.dsmz.de/bacterial-diversity/bacterial-nomenclature-up-to-date.html). This list of 225 species represents all the currently described taxons of the Rhodospirillaceae, Rhodobacteraceae, Acetobacteraceae, Sphingomonadaceae, Phyllobacteriaceae, Chromatiaceae and Ectothiorhodospiraceae. 16S ribosomal RNA (rRNA) sequences obtained from the National Center for Biotechnology Information (NCBI) website (http://www.ncbi.nlm.nih.gov/) were available for only 206 of the species present in this list. The DNA sequences corresponding to the 16S rRNA sequences were aligned using the ClustalW algorithm through the BioEdit program (Hall, 1999). Molecular Evolutionary Genetics Analysis (MEGA) (Kumar, Nei, Dudley, & Tamura, 2008) and BioEdit were then used to generate the phylogenetic trees that are discussed below. It is possible to generate phylogenetic trees in two quite different ways. The first approach calculates the distance the current sequence is from a putative common ancestor, whereas the second approach uses a sequence-based method to calculate a tree with the fewest evolutionary steps by generating a tree with the fewest number of sequence changes. Phylogenetic trees were generated using both methods to test the robustness of the conclusions. The distance-based analyses used the neighbour joining method, while for the sequence-based analyses the sequences were evaluated using the maximum likelihood model. The amino acid sequences of the beta and alpha subunits encoded by the *pucBA* genes were concatenated and the extraneous amino acids between the stop of the beta and start of the alpha was removed. This connected the beta and alpha sequences, resulting in a longer sequence more useful for phylogenetic analysis. The concatenated sequences were then analysed in MEGA, drawing trees for both distance and sequence divergence. The three species *Chloroflexus aggregans*, *Chloroflexus aurantiacus* and *Roseiflexus castenholzii* were included as an out-group to anchor our phylogenetic analysis using the 16S rRNA data, since they are thought to be more primitive photosynthetic bacterial species that do not contain *puc* genes (Hohmann-Marriott & Blankenship, 2011).

Unfortunately, there are only a few species where the sequence for the *puc* apoproteins has been determined. For most species in the list, absorption spectra were used to identify whether LH2 was present. However, there are also significant numbers of photosynthetic bacteria in the list without any published absorption spectra that would allow us to determine whether or not they contained LH2. This last group was excluded from the more detailed analyses presented below. We acknowledge and thank Bob Blankenship for his advice and suggestions for these analyses.

3. RESULTS OF THE PHYLOGENETIC ANALYSIS

The phylogenetic trees generated by both the neighbour joining and maximum likelihood methods for the 16S rRNA sequences of the 207 species showed consistent patterns, including the grouping of the species within each taxon. This was an important validation of the phylogenetic trees. Unfortunately, with so many species included in these trees they are far too big to be shown in a legible single figure. These complete trees, therefore, have been made available only on request from the authors.

Figure 7.9 shows a truncated version of the complete phylogenetic tree, which includes only the subgroup of purple bacterial species that the

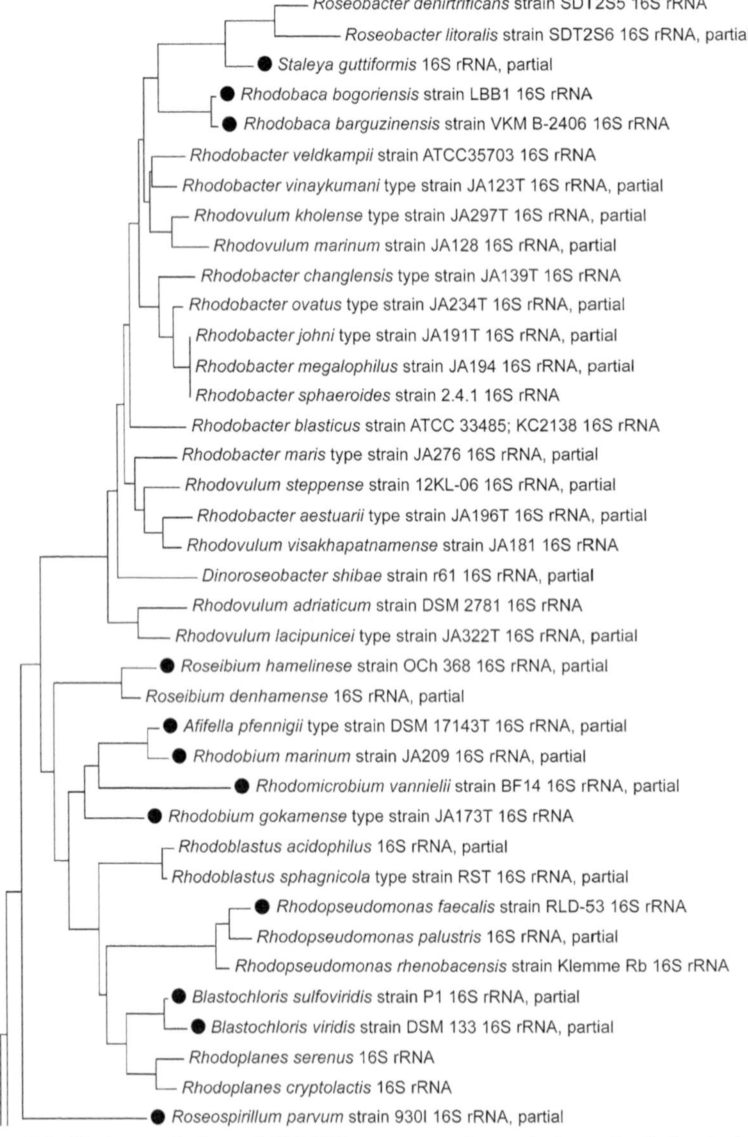

Figure 7.9 ***Phylogenetic tree of 16S DNA sequences from photosynthetic bacteria that can unequivocally be shown to contain either LH1, or LH1 and LH2, calculated using the neighbour joining method.*** The species known to contain only LH1 are indicated with a black circle.

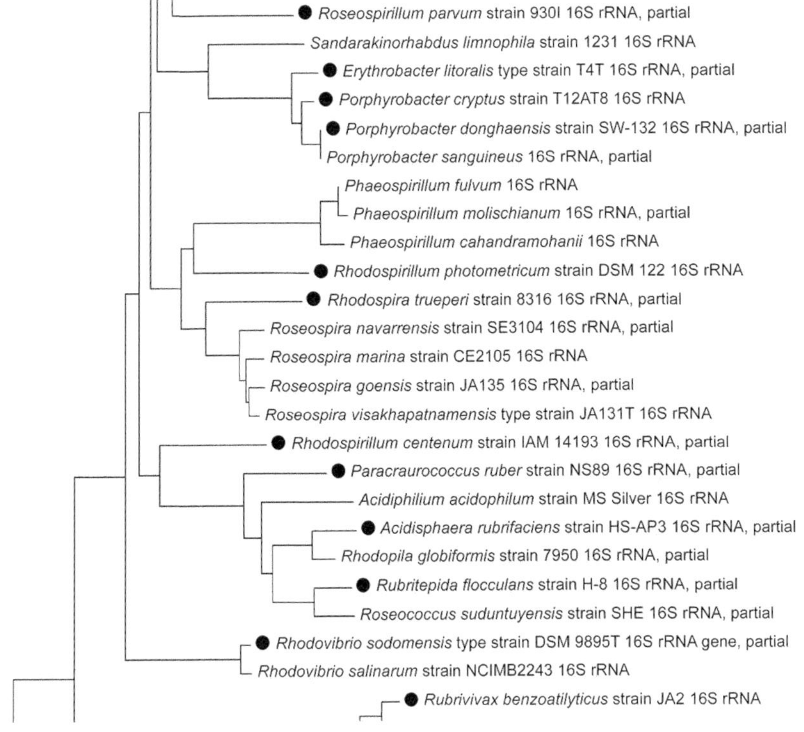

Figure 7.9 Cont'd

data show contain only LH1, or LH1 and LH2. All the references where these species were first described and their absorption spectra shown can be found from the DSMZ web resource (http://www.dsmz.de/bacterial-diversity/bacterial-nomenclature-up-to-date.html). There is no evidence after a careful inspection of this figure to suggest that those species that contain both LH1 and LH2 evolved after those that contain only LH1. Indeed, the occurrence in the phylogenetic tree of the species that contain only LH1, shown in Fig. 7.9 appears to be random.

Unfortunately, due to the relatively few species of purple bacteria that have had their complete genome sequenced, it has not been possible to investigate whether the species that contain multiple *puc* genes have evolved from progenitors of species that contain single *puc* genes. To try answering this question, another phylogenetic analysis was made, comparing the species for which concatenated amino acid sequences of both the alpha and beta LH2 apoproteins was available.

Figure 7.10 presents the results of this analysis. The species containing multiple *puc* genes encoding apoproteins are indicated with a black triangle.

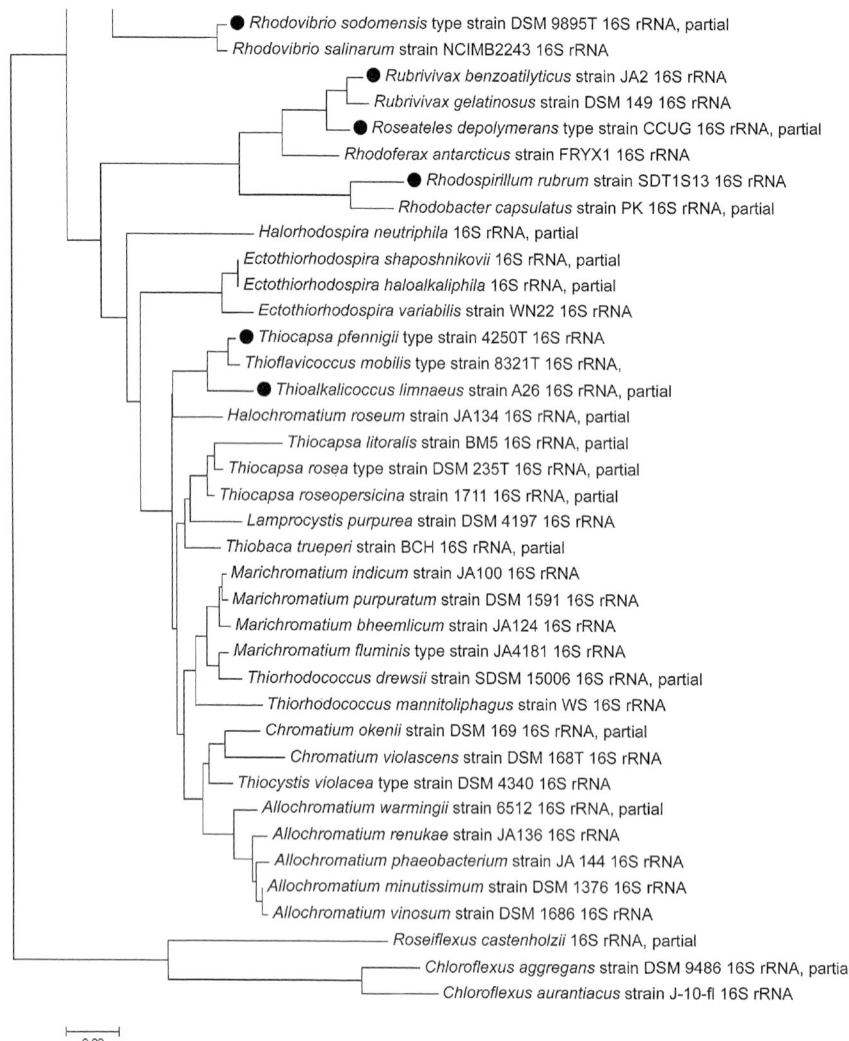

Figure 7.9 Cont'd

As in the analysis shown in Fig. 7.9, there is no consistent pattern. The species that have single LH2 *puc* genes are not clustered at earlier stages in the phylogenetic tree, as suggested in the hypothesis presented in the introduction section.

The phylogenetic tree shown in Fig. 7.10 indicates that there is no evidence for species with multiple *puc* genes to have evolved after those with a single *puc* gene. This interpretation is because those species with one or

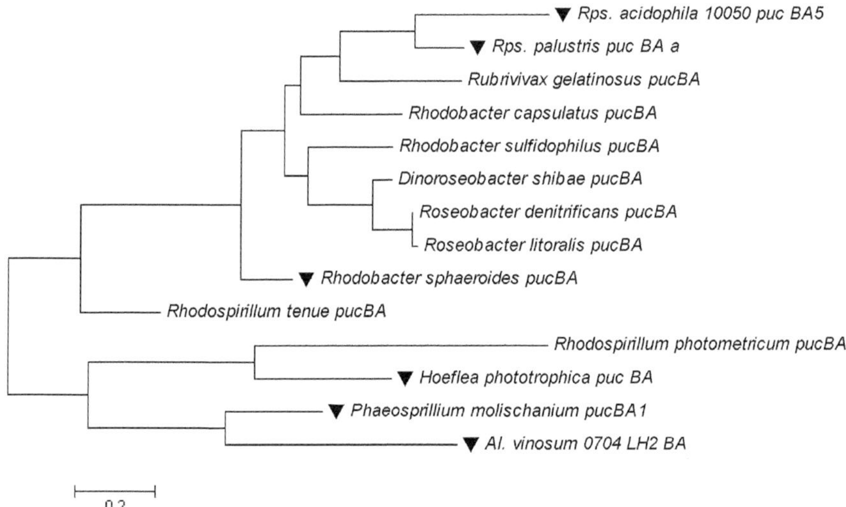

Figure 7.10 ***Phylogenetic tree based on concatenated beta and alpha amino acid sequences for the LH2 apoproteins.*** For those species where there are multiple *puc* genes a single LH2 beta–alpha pair was randomly selected for this analysis. When all of the available *puc* gene sequences from those species that contain multiple *puc* genes were used for this analysis, the phylogenetic tree obtained was not significantly different from the simplified version presented here. The black triangle indicates the species that have multiple genes for LH2. (All the *puc* gene sequences were obtained from the NCBI website.)

multiple *puc* genes do not occur at more derived positions in the phylogenetic trees than the species without *puc* genes.

There are not many species for which LH2 apoprotein sequences are available and these sequences are rather short. Therefore, the question of whether LH2 genes arose from LH1 genes, and whether genomes containing multiple LH2 operons arose from a genome containing a single LH2 operon was further investigated by comparing a phylogenetic tree constructed using the 16S DNA sequences from all the species of purple photosynthetic bacteria where genomic sequence data is available for the *puc* genes. It also includes the 16S DNA sequences from those species, which only contain LH1 core complexes. This phylogenetic tree (Fig. 7.11) shows an even distribution of the species that contain only the LH1 core complex amongst those species, which also contain both single and multiple LH2 *puc* genes.

Again there is no evidence to support any aspects of the hypothesis for the evolution of the LH genes presented in the introduction section.

There have been several other studies looking at the evolution of purple photosynthetic bacteria based on the analyses of the sequences of a

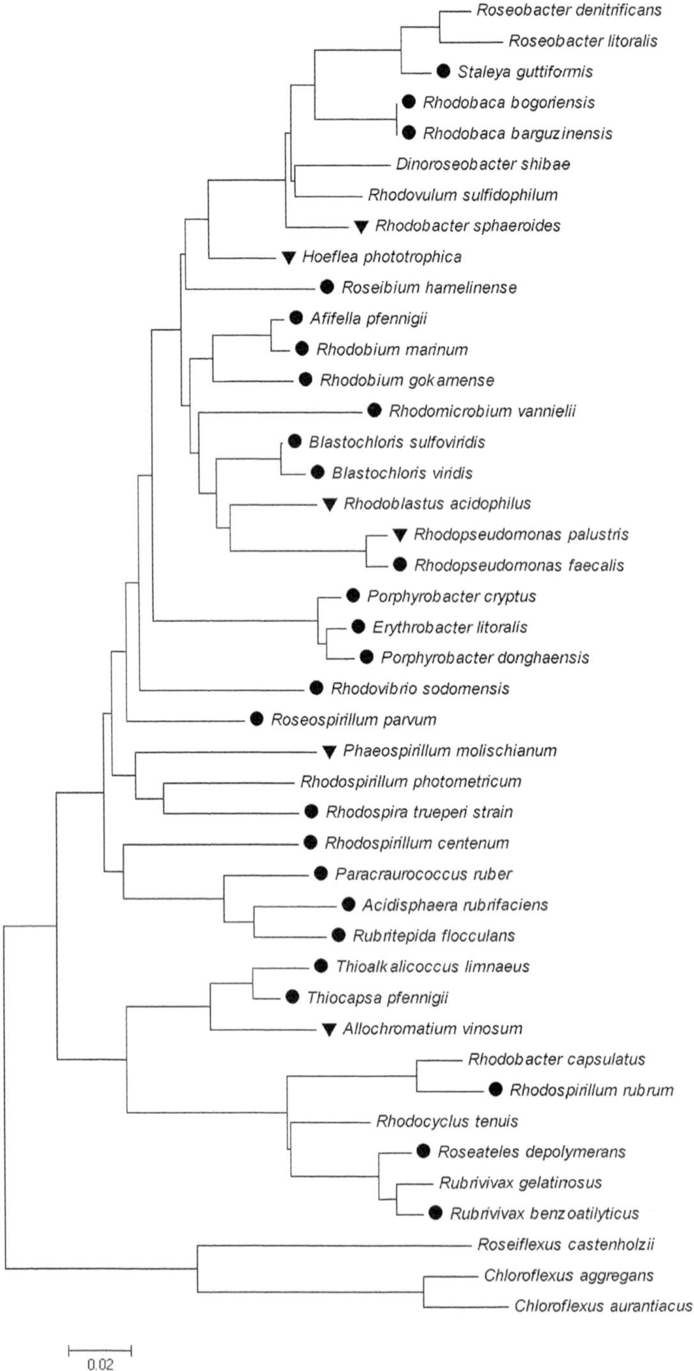

Figure 7.11 ***Phylogenetic tree calculated from 16S sequences of only the species where the*** puc ***genes have been sequenced and the species that are known to only contain the LH1 core complex (indicated with a black dot).*** The black triangle indicates the species that has multiple LH2 genes whereas those that are unmarked have only one *puc* operon, except the species of *Chloroflexus* and *Roseiflexus*, which are included as an out-group.

range of genes encoding proteins involved in the biosynthesis of carotenoids (Igarashi et al., 2001; Klassen, 2009), bacteriochlorophylls (Xiong, Fischer, Inoue, Nakahara, & Bauer, 2000), and reaction centre polypeptides (Kawasaki, Hoshino, & Yamasato, 1993; Tank, Thiel, & Imhoff, 2009). The typical phylogenetic trees constructed using these sequences are broadly consistent with the trees presented here (reviewed in Lang et al., 2011; Swingley, Blankenship, & Raymond, 2009).

If the hypothesis outlined above in the introduction section cannot be supported by the phylogenetic trees presented here, what may be suggested as to how the LH genes evolved? One possibility is that the hypothesis is correct, but that there is insufficient residual information remaining in the sequences of these genes in the extant purple bacterial species to corroborate it. However, it is more likely that the evolution of the LH genes was more complicated than the simple linear evolutionary pathway suggested by our hypothesis. For example, Nagashima has presented convincing evidence for horizontal gene transfer across the species barriers in purple bacteria (Igarashi et al., 2001; Nagashima, Hiraishi, Shimada, & Matsuura, 1997). Their work focused on the photosynthesis gene cluster including the genes coding for the synthesis of bacteriochlorophyll and carotenoids. They showed that although there was striking sequence homology amongst the corresponding genes within the cluster from species to species, the order of these genes within the cluster was highly variable. This was taken as evidence for the extensive lateral gene transfer. If such lateral gene transfer has occurred with the LH genes then it is not surprising that their evolutionary history appears to be so scrambled.

4. DISCUSSION

4.1. Possible Consequences of the Presence of Multiple *puc* Genes

It appears that multiple spectroscopic forms of LH2 have evolved to allow the different species of purple bacteria to be better able to absorb light in a wide range of different ecological niches (Oda et al., 2008). The presence of multiple *puc* genes allows for LH2 complexes with different NIR absorption spectra to be synthesised (Zuber & Codgell, 1995). In order to understand how this occurs it is necessary to understand what features control the position of the NIR Q_y transition of bacteriochlorophylls when they are packaged into the LH2 complexes (Cogdell et al., 2002). Imagine a single molecule of bacteriochlorophyll in its binding site. It will have a site energy that is just specified by the interactions within the protein in that binding site. For the B800 bacteriochlorophyll molecules this results in a small shift in the position of the Q_y band relative to its position in a solvent such as acetone. In this case, the shift is from about 770 nm to 800 nm. When the tightly coupled ring of bacteriochlorophyll molecules interacts then the red shift seen is more marked due to their excitonic coupling (Cogdell, Gall, & Koehler, 2006). This is illustrated in Fig. 7.12.

In the case of the LH2 complex from *Rps. acidophila* strain 10050, the ring of bacteriochlorophylls is symmetrical with respect to both the structure and distribution of site energies. As a consequence of this all of the oscillator strength resides in the $k^{+/-}$ states. This gives rise to the well known B850 band. Multiple *puc* genes allow two types of spectroscopic variation. The first is illustrated by the B800–820 LH2 complex from *Rps. acidophila* (Fig. 7.8). In this case, the ring is still symmetrical but contains apoproteins encoded by a different pair of *puc* genes (McLuskey et al., 2001).

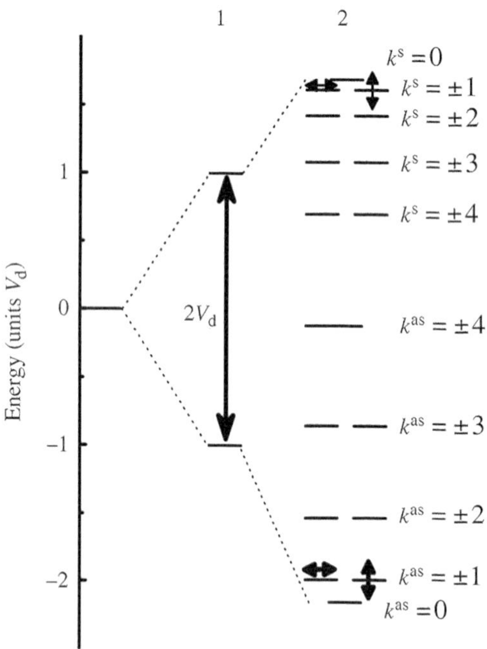

Figure 7.12 ***A schematic representation of the energy level manifold of a bacteriochlorophyll dimer (manifold 1) and the B850 ring of the LH2 (manifold 2).*** V_d is the energy splitting between the two exciton bands of a dimer. The short black arrows represent the polarisation of the two k states. A full description of how this representation was calculated is presented in Cogdell et al. (2006). *(Modified from Cogdell et al. (2006)).*

These alternative apoproteins have binding sites for the pair of tightly coupled bacteriochlorophyll molecules that shift the pigments' site energies to the blue. As a result, even though the extent of the exciton coupling is not changed, the site energies produce a blue shift relative to the B850 form of LH2 and the resultant position of the long wavelength band is now only at 820 nm, not 850 nm. The key residues in the protein that cause this change in site energy are phenylalanine and methionine (see Fig. 7.5 sequences for *Rps. palustris* and *Rps. acidophila*). When the residues at these positions change from tyrosine and tryptophan, it results in a change in the pattern of H-bonds to carbonyl groups present on the edge of the bacteriochlorin rings. As a result of this, the carbonyl groups rotate out of conjugation of the bacteriochlorin bacteriochlorophyll ring and hence there is a blue shift in their site energies.

The presence of multiple *puc* genes also allows another mechanism to operate to change the absorption spectrum of LH2 complexes. Some species such as *Rps. palustris* synthesise heterogeneous LH2 rings, where apoproteins encoded by different *puc* genes coexist in a single ring (Brotosudarmo et al., 2009). In that case, the distribution of site energies of the different bacteriochlorophyll molecules present in the tightly coupled ring can be unsymmetrical. These changes mean that the different excitonic states in the ladder of states illustrated in Fig. 7.12 can gain oscillator strength. As a result, the absorption spectra can become much more complicated, with multiple bands rather than the rather simple '850 or 820' bands. The final spectral outcome depends on the exact distribution of the site energies around the LH2 ring. The extent of the red shift due to the strong exciton coupling also depends on the ring size. In principle, the larger the ring the larger the red shift. This explains why the red shift seen in LH1 complexes is larger than that seen in a typical LH2 complex. The presence of multiple *puc* genes would appear to confer an advantage to the bacteria by allowing more flexible adaptation to the specific light conditions found in a wide range of different ecological niches.

Exactly how the genes encoding the purple bacterial LHs have evolved is not yet clear. There are no definite patterns on how that evolution may have occurred present in any of the phylogenetic trees described in this chapter. This may appear to be a disappointing outcome. However, the lack of a consistent evolutionary record does not reduce the importance of the evolution of the LH2 complexes on the ability of purple bacterial species to compete in a wide range of ecological niches where light intensity may be limiting for growth. It may be worthwhile to repeat this analysis when there are *puf* and *puc* gene sequences available for all the 225 species of purple photosynthetic bacteria described in the DSMZ website.

SUPPLEMENTARY DATA

Supplementary data related to this chapter can be found at http://dx.doi.org/10.1016/B978-0-12-397923-0.00007-2.

ACKNOWLEDGEMENTS

SLH and RJC would like to acknowledge Prof. Bob Blankenship for his help with the phylogenetic trees and Dr Alastair T. Gardiner for help with the figures. RJC and SLH also thank the Human Frontiers of Science Program and the BBSRC for financial support.

REFERENCES

Aagaard, J., & Sistrom, W. R. (1972). Control of synthesis of reaction center bacteriochlorophyll in photosynthetic bacteria. *Photochemistry and Photobiology, 15*, 209.

Ashby, M. K., Coomber, S. A., & Hunter, C. N. (1987). Cloning, nucleotide-sequence and transfer of genes for the B800–850 light harvesting complex of *Rhodobacter sphaeroides*. *FEBS Letters, 213*, 245–248.

Brotosudarmo, T. H., Kunz, R., Boehm, P., Gardiner, A. T., Moulisova, V., Cogdell, R. J., et al. (2009). Single-molecule spectroscopy reveals that individual low-light LH2 complexes from *Rhodopseudomonas palustris* 2.1.6. have a heterogeneous polypeptide composition. *Biophysical Journal, 97*, 1491–1500.

Brunisholz, R. A., Zuber, H. (1992). Structure, Function and Organization of Antenna Polypeptides and Antenna Complexes from the 3 Families of Rhodospirillaneae. *Journal of Photochemistry and Photobiology B-Biology, 15*, 113–140.

Bullough, P. A., Qian, P., & Hunter, C. N. (2009). Reaction centre-light-harvesting core complexes of purple bacteria. In C. N. Hunter, F. Daldal & J. T. Beatty (Eds.), *Advances in photosynthesis and respiration The purple photosynthetic bacteria* (Vol. 28, pp. 155–179). Dordrecht: Springer.

Cogdell, R. J., Gall, A., & Koehler, J. (2006). The architecture and function of the light-harvesting apparatus of purple bacteria: from single molecules to in vivo membranes. *Quarterly Reviews of Biophysics, 39*, 227–324.

Cogdell, R. J., Howard, T. D., Isaacs, N. W., McLuskey, K., & Gardiner, A. T. (2002). Structural factors which control the position of the Q(y) absorption band of bacteriochlorophyll a in purple bacterial antenna complexes. *Photosynthesis Research, 74*, 135–141.

Donohue, T. J., Kiley, P. J., & Kaplan, S. (1988). The *Puf* operon region of *Rhodobacter sphaeroides*. *Photosynthesis Research, 19*, 39–61.

Fleming, G. R., & van Grondelle, R. (1997). Femtosecond spectroscopy of photosynthetic light-harvesting systems. *Current Opinion in Structural Biology*, 7, 738–748.

Fowler, G. J.S., Visschers, R. W., Grief, G. G., Vangrondelle, R., & Hunter, C. N. (1992). Genetically modified photosynthetic antenna complexes with blueshifted absorbency bands. *Nature, 355*, 848–850.

Gabrielson, M., Gardiner, A. T., & Codgell, R. J. (2009). Peripheral complexes of purple bacteria. In C. N. Hunter, F. Daldal & J. T. Beatty (Eds.), *Advances in photosynthesis and respiration The purple photosynthetic bacteria* (Vol. 28, pp. 135–153). Dordrecht: Springer.

Gardiner, A. T., Cogdell, R. J., & Takaichi, S. (1993). The effect of growth-conditions on the light-harvesting apparatus in *Rhodopseudomonas acidophila*. *Photosynthesis Research, 38*, 159–167.

Hall, T. A. (1999). BioEdit: a user-friendly biological sequence alignment editor and analysis program for windows 95/98/NT. *Nucleic Acids Symposium Series, 41*, 95–98.

Hohmann-Marriott, M. F., & Blankenship, R. E. (2011). Evolution of photosynthesis. *Annual Review of Plant Biology, 62*, 515–548.

Igarashi, N., Harada, J., Nagashima, S., Matsuura, K., Shimada, K., & Nagashima, K. V.P. (2001). Horizontal transfer of the photosynthesis gene cluster and operon rearrangement in purple bacteria. *Journal of Molecular Evolution, 52*, 333–341.

Katsuda, T., Yegani, R., Fujii, N., Igarashi, K., Yoshimura, S., & Katoh, S. (2004). Effects of light intensity distribution on growth of *Rhodobacter capsulatus*. *Biotechnology Progress, 20*, 998–1000.

Kawasaki, H., Hoshino, Y., & Yamasato, K. (1993). Phylogenetic diversity of phototrophic purple non-sulfur bacteria in the Proteobacteria alpha group. *FEMS Microbiology Letters, 112*, 61–66.

Klassen, J. L. (2009). Pathway evolution by horizontal transfer and positive selection is accommodated by relaxed negative selection upon upstream pathway genes in purple bacterial carotenoid biosynthesis. *Journal of Bacteriology, 191*, 7500–7508.

Koepke, J., Hu, X. C., Muenke, C., Schulten, K., & Michel, H. (1996). The crystal structure of the light-harvesting complex II (B800–850) from *Rhodospirillum molischianum*. *Structure, 4*, 581–597.

Kumar, S., Nei, M., Dudley, J., & Tamura, K. (2008). MEGA: a biologist-centric software for evolutionary analysis of DNA and protein sequences. *Briefings in Bioinformatics, 9*, 299–306.

Lang, A. S., Harwood, C. S., & Beatty, J. (2011). Evolutionary relationships among antenna proteins of purple phototrophic bacteria. In R. L. Burnap & W. F.J. Vermaas (Eds.), *Functional genomics and evolution of photosynthetic systems Advances in photosynthesis and respiration 33* (pp. 253–264). : Springer Science.

Law, C. J., Roszak, A. W., Southall, J., Gardiner, A. T., Isaacs, N. W., & Cogdell, R. J. (2004). The structure and function of bacterial light-harvesting complexes (review). *Molecular Membrane Biology, 21*, 183–191.

McDermott, G., Prince, S. M., Freer, A. A., Hawthornthwaite-Lawless, A. M., Papiz, M. Z., Cogdell, R. J., et al. (1995). Crystal structure of an integral membrane light-harvesting complex from photosynthetic bacteria. *Nature, 374*, 517–521.

McLuskey, K., Prince, S. M., Cogdell, R. J., & Isaacs, N.W. (2001). The crystallographic structure of the B800–820 LH3 light-harvesting complex from the purple bacteria *Rhodopseudomonas acidophila* strain 7050. *Biochemistry, 40*, 8783–8789.

Mechler, B., & Oelze, J. (1978a). Differentiation of photosynthetic apparatus of *Chromatium vinosum*, strain-D.1. Influence of growth-conditions. *Archives of Microbiology, 118*, 91–97.

Mechler, B., & Oelze, J. (1978b). Differentiation of photosynthetic apparatus of *Chromatium vinosum*, strain-D.2. Structural and functional differences. *Archives of Microbiology, 118*, 99–108.

Mechler, B., & Oelze, J. (1978c). Differentiation of photosynthetic apparatus of *Chromatium vinosum*, strain-D.3. Analyses of spectral alterations. *Archives of Microbiology, 118*, 109–114.

Nagashima, K. V.P., Hiraishi, A., Shimada, K., & Matsuura, K. (1997). Horizontal transfer of genes coding for the photosynthetic reaction centers of purple bacteria. *Journal of Molecular Evolution, 45*, 131–136.

Oda, Y., Larimer, F. W., Chain, P. S.G., Malfatti, S., Shin, M.V., Vergez, L. M., et al. (2008). Multiple genome sequences reveal adaptations of a phototrophic bacterium to sediment microenvironments. *Proceedings of the National Academy of Sciences, 105*, 18543–18548.

Papiz, M. Z., Prince, S. M., Howard, T., Cogdell, R. J., & Isaacs, N. W. (2003). The structure and thermal motion of the B800–850 LH2 complex from *Rps. acidophila* at 2.0 (A)overcircle resolution and 100 K: new structural features and functionally relevant motions. *Journal of Molecular Biology*, *326*, 1523–1538.

Reidl, H., Golecki, J. R., & Drews, G. (1983). Energetic aspects of photophosphorylation capacity and reaction center content of *Rhodopseudomonas capsulata*, grown in a turbidostat under different irradiances. *Biochimica et Biophysica Acta*, *725*, 455–463.

Roszak, A. W., Howard, T. D., Southall, J., Gardiner, A. T., Law, C. J., Isaacs, N. W., et al. (2003). Crystal structure of the RC–LH1 core complex from *Rhodopseudomonas palustris*. *Science*, *302*, 1969–1972.

Swingley, W. D., Blankenship, R. E., & Raymond, J. (2009). Evolutionary relationships among purple photosynthetic bacteria and the origin of proteobacterial photosynthetic systems. In C. N. Hunter, F. Daldal & J. T. Beatty (Eds.), *Advances in photosynthesis and respiration The purple photosynthetic bacteria* (Vol. 28, pp. 17–29). Dordrecht: Springer.

Tadros, M. H., & Waterkamp, K. (1989). Multiple copies of the coding regions for the light-harvesting B800–850 alpha-polypeptide and beta-polypeptide are present in the *Rhodopseudomonas palustris* genome. *EMBO Journal*, *8*, 1303–1308.

Tank, M., Thiel, V., & Imhoff, J. F. (2009). Phylogenetic relationship of phototrophic purple sulfur bacteria according to pufL and pufM genes. *International Microbiology*, *12*, 175–185.

Tharia, H. A., Nightingale, T. D., Papiz, M. Z., & Lawless, A. M. (1999). Characterisation of hydrophobic peptides by RP-HPLC from different spectral forms of LH2 isolated from *Rps. palustris*. *Photosynthesis Research*, *61*, 157–167.

Tichy, H.V., Albien, K. U., Gadon, N., & Drews, G. (1991). Analysis of the *Rhodobacter capsulatus* puc operon – the pucC gene plays a central role in the regulation of LH2 (B800–850 complex) expression. *EMBO Journal*, *10*, 2949–2955.

Wraight, C. A., Codgell, R. J., & Chance, B. (1978). In R. K. Clayton & W. R. Sistrom (Eds.), *Ion transport and electrochemical gradients in photosynthetic bacteria* (pp. 471–512). New York: Plenum Press.

Xiong, J., Fischer, W. M., Inoue, K., Nakahara, M., & Bauer, C. E. (2000). Molecular evidence for the early evolution of photosynthesis. *Science*, *289*, 1724–1730.

Youvan, D. C., & Ismail, S. (1985). Light-harvesting-2 (B800–B850 complex) structural genes from *Rhodopseudomonas capsulata*. *Proceedings of the National Academy of Sciences of the United States of America*, *82*, 58–62.

Zuber, H., & Codgell, R. J. (1995). Structure and organization of purple bacterial antenna complexes. In R. E. Blankenship, M. T. Madigan & C. E. Bauer (Eds.), *Anoxygenic photosynthetic bacteria* (pp. 315–348). Dordrecht: Kluwer Academic Publishers.

CHAPTER EIGHT

Role and Evolution of Endogenous Plasmids in Photosynthetic Bacteria

John C. Willison*[,1], Jean-Pierre Magnin†[,1]

*Laboratoire de Chimie et Biologie des Métaux, CEA Grenoble, Grenoble-INP, France
†Laboratoire d'Electrochimie et de Physicochimie des Matériaux et des Interfaces, St Martin d'Hères, Grenoble, France
[1]Corresponding authors: E-mails: john.willison@cea.fr

Contents

Advances in Botanical Research, Volume 66
ISSN 0065-2296, http://dx.doi.org/10.1016/B978-0-12-397923-0.00008-4

Abstract

Endogenous plasmids are found in most strains of *Rhodobacter sphaeroides*, *Rhodobacter capsulatus* and *Rhodospirillum rubrum*, but appear to be less common in other species of purple non-sulphur bacteria. They have also been found in some species of green and purple sulphur bacteria. Until recently, there was little conclusive evidence for the functions of these plasmids, and most were considered to be cryptic. The advent of whole genome sequencing has enabled predictions of possible plasmid functions that can then be tested experimentally. In addition, transcriptomic and proteomic studies in *R. sphaeroides* 2.4.1 have shown that plasmids play an active role in genome function and may code for essential metabolic processes. This chapter summarises experimental and genomic evidence for plasmid function in the anaerobic anoxygenic phototrophs, with particular emphasis on photosynthetic metabolism, nitrogen oxide reduction, cell wall biosynthesis and heavy metal resistance, and new experimental evidence is presented for the role of the endogenous plasmid in the *R. capsulatus* strains B10 and SB1003. The evolution of these plasmids will also be considered, and insights from nucleotide sequence comparisons will be combined with those from phylogenetic analysis of plasmid replication modules in the *Roseobacter* clade of the α-*Proteobacteria*, which includes aerobic anoxygenic phototrophs.

1. INTRODUCTION

Endogenous plasmids are widespread in bacteria, including photosynthetic bacteria. Although many plasmids code for specific, adaptive functions, such as resistance to antibiotics or heavy metals, the ability to degrade xenobiotic compounds, or pathogenicity, the function of many others is not apparent, and these plasmids are often referred to as 'cryptic'. Most plasmids from photosynthetic bacteria fall into this category.

The discovery of 'satellite' DNA in photosynthetic bacteria, indicating the presence of extrachromosomal elements, led at first to suggestions that this extrachromosomal DNA might play a role similar to that of plastid DNA in the chloroplasts of eukaryotes (Gibson & Niederman, 1970). Although it was later considered unlikely that such an essential function as photosynthesis would be plasmid-encoded (Saunders, 1978), a persistent correlation was observed between either plasmid rearrangements in *Rhodobacter sphaeroides* (Nano & Kaplan, 1984; Saunders, Saunders, & Bennett, 1976) or plasmid curing in *Rhodospirillum rubrum* (Kuhl, Nix, & Yoch, 1983, Kuhl, Wimer, & Yoch, 1984) and the loss of the ability to grow photosynthetically. On the other hand, plasmid-cured strains of *Rhodobacter capsulatus* were photosynthetically competent (Willison, 1990; Willison, Magnin, & Vignais, 1987). Genetic analysis in *R. sphaeroides* and *R. capsulatus*, followed by the genome sequencing of *R. rubrum* and many other phototrophic species has shown that, in the anaerobic anoxygenic phototrophs, the genes for photosynthesis, which are usually clustered, are chromosomally located. The apparent association between plasmid loss or rearrangement and photosynthesis minus phenotype in *R. sphaeroides* and *R. rubrum* has so far not been satisfactorily explained.

Before the advent of whole genome sequencing, the number of specific functions that could be assigned to endogenous plasmids in photosynthetic bacteria was very small. The temperate bacteriophage Rϕ6P from *R. sphaeroides* RS601, which exists as a plasmid in the prophage state, encodes penicillin resistance (Pemberton, Cooke, & Bowen, 1983). A 14-kb endogenous plasmid in strains of *Chlorobium limicola* f. sp. *thiosulfatophilum* confers the ability to utilise thiosulphate as electron donor for photosynthesis (Méndez-Alvarez, Pavón, Esteve, Guerrero, & Gaju, 1994, 1995). Curing of the 115-kb endogenous plasmid in *R. capsulatus* AD2 resulted in the loss of the ability to both assimilate and dissimilate nitrate (Willison, 1990) and the *napA* gene coding for a periplasmic nitrate reductase has been localised to this plasmid (Koch & Klemme, 1994). The availability of the nucleotide sequence of a plasmid should enable predictions to be made about its function, particularly when compared with the gene content of the chromosome, and allow experimental approaches to be designed to test these predictions. Additional insights can come from global approaches, such as transcriptomics and proteomics. Finally, the availability of a large number of plasmid sequences allows inter- and intra-species comparisons that can shed light on the evolutionary origin of these replicons.

The purpose of the present chapter is to summarise data from the literature concerning the occurrence and distribution of plasmids in the anaerobic anoxygenic photosynthetic bacteria and their plausible functions. Particular attention will be given to the potential roles of endogenous plasmids in photosynthetic metabolism, nitrogen oxide reduction and cell wall biosynthesis. With the exception of *R. sphaeroides* 2.4.1, which has been extensively studied in Sam Kaplan's laboratory, genomic approaches to plasmid function have not been carried out in these organisms. A survey of genome sequences available in the databases will be given, and suggestions made for further work in this area. In addition, new experimental data will be presented concerning the role of the 134-kb endogenous plasmid from *R. capsulatus* B10 and SB1003. The *Roseobacter* clade, which includes species of aerobic anoxygenic phototrophic bacteria, has attracted much attention recently since it is highly abundant in marine environments, accounting for up to 25% of the microbial biomass (Petersen et al., 2012). Most species in this group contain multiple plasmids and a wealth of genomic data is available, with more than 40 genomes sequenced or in progress. Although the physiology of the aerobic phototrophs is very different from that of the anaerobes, the extensive bioinformatic analysis of plasmid function and replication in the former has shed light on the origin and evolution of plasmids in the latter and will therefore be considered in this chapter. It is also remarkable that the photosynthesis gene cluster (PGC), which is chromosomally located in all of the anaerobic phototrophs so far sequenced, has been shown to be plasmid-located in two species of aerobic anoxygenic phototrophic bacteria (Petersen et al., 2012; Pradella et al., 2004).

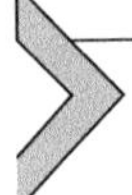

2. OCCURRENCE AND DISTRIBUTION OF ENDOGENOUS PLASMIDS IN THE ANAEROBIC ANOYXGENIC PHOTOTROPHIC BACTERIA

Extrachromosomal DNA in photosynthetic bacteria was first detected by Suyama and Gibson (1966) using CsCl density gradient analysis. The so-called satellite DNA was detected in *R. sphaeroides* ATCC 14690 and *Chromatium* D, but not in *R. rubrum* ATCC 11170 or *Rhodopseudomonas palustris*. However, this method will only detect plasmids with a buoyant density (G + C content) significantly different from that of the bulk genomic DNA. Using ethidium bromide in buoyant density centrifugation to separate supercoiled, circular DNA from linear DNA, Gibson and Niederman (1970) detected two plasmids of similar size (75 and 70 MDa) but differing buoyant densities in *R. sphaeroides* strain NCIB 8327. Saunders et al. (1976) then detected three species of extrachromosomal DNA by electron microscopy in *R. sphaeroides* strain 2.4.1, with sizes of 75, 66 and 28 MDa determined from their contour length. Molecular analysis subsequently showed that *R. sphaeroides* 2.4.1 contains five distinct plasmid species (Fornari, Watkins, & Kaplan, 1984), which was confirmed by whole genome sequencing (Mackenzie, Choudhary, Larimer, & et al., 2001). Other publications reported that numerous species and strains of purple non-sulphur bacteria (PNSB) contain endogenous plasmids, as do some strains of green and purple sulphur bacteria (Table 8.1). This is confirmed by a survey of the

Table 8.1 Literature reports of endogenous plasmids in the anaerobic anoxygenic phototrophic bacteria

Species/strain	Plasmid designation (if any)	Size (kb)*	Method of size determination†	Reference
Rhodobacter sphaeroides				
NCIB 8327	Two plasmids of similar size	105–112*	BDC	Gibson and Niederman (1970)
2.4.1 (NRS)‡		112* 99* 42*	EM (CL)	Saunders et al. (1976)
2.4.1	pRS2.4.1.a pRS2.4.1.b pRS2.4.1.c pRS2.4.1.d pRS2.4.1.e	113.6 104 100 99 42	VGE	Fornari et al. (1984)
2.4.1		113 104 100 94	VGE	Michalski and Nicholas (1988)
2.4.1	pRS2.4.1.a pRS2.4.1.b pRS2.4.1.c pRS2.4.1.d pRS2.4.1.e	110 105 97 94 42	PGFE	Suwanto and Kaplan (1989)
2.4.1	pRS2.4.1.a pRS2.4.1.b pRS2.4.1.c pRS2.4.1.d pRS2.4.1.e	114 113 104 101 37	WGS	Mackenzie et al. (2001)
DSM159		114 107§ 38	PGFE	Pradella et al. (2004)
RS2	pRS2a pRS2b pRS2c pRS2d	110 105 98 43	VGE	Fornari et al. (1984)
Y	pRSYa pRSYb	104 103	VGE	Fornari et al. (1984)

Continued

Table 8.1 Literature reports of endogenous plasmids in the anaerobic anoxygenic phototrophic bacteria—cont'd

Species/strain	Plasmid designation (if any)	Size (kb)*	Method of size determination†	Reference
L	pRSLa	110	VGE	Fornari et al. (1984)
	pRSLb	105		
	pRSLc	98		
	pRSLd	92		
	pRSLe	43		
2.4.7	pRS247a	106.5	VGE	Fornari et al. (1984)
8253	pRS8253a	106.5	VGE	Fornari et al. (1984)
RS601	Rf6P (prophage)	50‖	EM (CL)	Tucker and Pemberton (1978)
RS630	pRS630a	140	VGE	Fornari et al. (1984)
	pRS630b	89		
WS20	pRSWSa	130	VGE	Fornari et al. (1984)
	pRSWSb	100		
SWL	pRSSWLa	90	VGE	Fornari et al. (1984)
	pRSSWLb	89		
81-1		113	VGE	Michalski and Nicholas (1988)
		104		
		100		
		94		
KD131 (KCTC12085)	Plasmid A	157	WGS	Lim et al. (2009)
	Plasmid B	103		
WS8N (nalidixic acid-resistant derivative of WS8)		200	WGS	Munk et al. (2011)
		110		
Rhodobacter sphaeroides *f. sp.* denitrificans				
IL106		108	VGE	Michalski and Nicholas (1988)
81-3		113	VGE	Michalski and Nicholas (1988)
		104		
		100		
		94		
2.4.3		113	VGE	Michalski and Nicholas (1988)
		104		
		100		
		94		
IL106		115	TAFE	Schwintner et al. (1998)
		102		

Rhodobacter capsulatus				
SB1003 (rifampicin-resistant derivative of B10)	pRCB133	133	WGS	Strnad et al. (2010)
B10		129	VGE	Willison et al. (1987)
AD2		111	VGE	Willison (1990)
AD2		115	VGE	Witt and Klemme (1991)
A1		134	VGE	Witt and Klemme (1991)
BK5		134 115	VGE	Witt and Klemme (1991)
BK5 and BK5NIT		135 115	VGE	Richardson et al. (1994)
Kb1		115	VGE	Witt and Klemme (1991)
R10		106	VGE	Witt and Klemme (1991)
DSM 1710		99	VGE	Witt and Klemme (1991)
Fc101		115 32 3	VGE	Witt and Klemme (1991)
R1		93	VGE	Witt and Klemme (1991)
BH9 (rifampicin-resistant mutant of strain H9)	pBH91 pBH92	141* 111	SGC	Hu and Marrs (1979).
St. Louis		99*	VGE	Willison et al., 1987
JH1		144* 109*	VGE	Willison et al., 1987
ATCC 11166		150	PFGE	Jumas-Bilak et al. (1998)

Continued

Table 8.1 Literature reports of endogenous plasmids in the anaerobic anoxygenic phototrophic bacteria—cont'd

Species/strain	Plasmid designation (if any)	Size (kb)*	Method of size determination†	Reference
Other* Rhodobacter *sp.				
R. veldkampii DSM11550^{T}		59¶ 115¶	PFGE	Pradella et al. (2004)
R. blasticus TCRI-14 (marine strain)	pMG160	3.4	HGE	Inui et al., 2003).
Marine *Rhodobacter* sp. NKPB 002106	pRD06S pRD06L	5.8* 7*	HGE	Matsunaga et al. (1986)
Marine *Rhodobacter* sp. (nine strains)	2–4 plasmids/strain	4.7–17*	HGE	Matsunaga et al. (1986)
Marine *Rhodobacter* sp. NKPB 043402	pRD31 pRD32 pRD33	3.1 8.3 16	HGE	Matsunaga et al. (1990)
Rhodospirillum rubrum				
S1-G S1-A G-1980 ATCC 11170 FR1 S-4	pKY1 pKY2	55 (54)	HGE (RA)	Kuhl et al. (1983)
S1 G-9 IFO3986	pRR1	55	RA	Kawamukai et al. (1990)
S-1		54	WGS	Munk et al. (2011)
Marine *Rhodospirillum* sp. (two strains)	4–6 plasmids/strain	4.9–15*	HGE	Matsunaga et al. (1986)
Rhodopseudomonas palustris				
S55	pMG101	15	NS	Inui et al. (2000)
CGA009		8.4	WGS	Larimer et al. (2004)

Chlorobium limicola *f. sp.* thiosulfatophilum			
DSM 249 (Tio$^+$)	14	NS	Méndez-Alvarez et al. (1995)
BF8000 (Tio$^+$)	14		
UdG6038	650		

*An asterisk indicates that the size was given in Daltons × 10^6 (MDa) in the original publication.
†BDC, buoyant density centrifugation; EM (CL), electron microscopy (contour length); HGE, horizontal agarose gel electrophoresis; NS; not specified; PFGE, pulsed-field gel electrophoresis; RA; restriction analysis; SGC, sucrose gradient centrifugation; TAFE, transverse alternating field electrophoresis; VGE, vertical agarose gel electrophoresis; WGS, whole genome sequencing.
‡Mutant resistant to nalidixic acid, rifampicin and streptomycin.
§The 107-kb band was broad and may have included more than one plasmid.
‖The prophage DNA is a dimer of the circular DNA found in phage particles.
¶Linear plasmids (although see text).

genome database (Table 8.2), although the strains that have been sequenced are not always the same as those in which the endogenous plasmids have been physically characterised. Plasmids have so far not been shown to occur in the filamentous, green non-sulphur bacteria (*Chloroflexaciae*) and are not always present in PNSB and the green and purple sulphur bacteria.

2.1. Purple Non-sulphur Bacteria

2.1.1. *R. sphaeroides*

Using high-resolution gel electrophoresis, Fornari et al. (1984) showed that *R. sphaeroides* strain 2.4.1 contains five endogenous plasmids, which were designated pRS2.4.1.A (114 kb), pRS2.4.1.B (104 kb), pRS2.4.1.C (100 kb), pRS2.4.1.D (99 kb), and pRS2.4.1.E (42 kb). pRS2.4.1.A presumably corresponds to the 75 MDa (114 kb) plasmid identified by Saunders et al. (1976) and pRS2.4.1.E to the 28 MDa (42 kb) plasmid, whereas the 66 MDa (99 kb) species may have included the three intermediate-sized plasmids. It should be noted that Pradella et al. (2004) were only able to detect three plasmid species of 38, 107 and 114 kb in *R. sphaeroides* strain DSM 159, but the three intermediate-sized plasmids may have been poorly resolved. These authors also screened nine other strains of *R. sphaeroides* for plasmids and all were found to contain at least one endogenous plasmid (Table 8.1). Restriction analysis and hybridisation indicated that plasmids from different strains were related (Fornari et al., 1984). Using a strain with a Tn*5* insertion in plasmid E, Suwanto and Kaplan (1992) showed that this plasmid is self-transmissible and can replicate in strains L and RS2, but not in 630, 2.4.7, *R. capsulatus* B10 or *Paracoccus denitrificans*. In strains L and RS2, plasmid E presumably displaces the endogenous plasmid of similar size. Plasmid D was also shown to be transmissible.

Whole genome sequencing of strain 2.4.1 gave estimated sizes of the five plasmids of 114 kb, 113 kb, 104 kb, 101 kb and 37 kb, with G + C contents of 69.3%, 70.1%, 64.1%, 63.9% and 67.6%, respectively (Mackenzie et al., 2001). Ribeiro, Przyblski, Yin, and et al. (2012) applied a new automated sequencing and assembly approach to 16 bacterial samples, including *R. sphaeroides* and made a large number of corrections to the reference sequence. As can be seen in Table 8.2, there is still some uncertainty as to the exact sizes of plasmids A, B and D, and the definitive sequences may not yet be available.

In addition to strain 2.4.1, several other strains of *R. sphaeroides* have been sequenced, including strain KD131 and strain WS8N. Strain KD131 contains two plasmids, A and B, of size 157 and 103 kb, respectively (Lim et al., 2009). Strain WS8N, a spontaneous nalidixic acid-resistant derivative of WS8, which has been much used for studies on chemotaxis, also contains two large plasmids of size 200 kb and 110 kb (Munk, Copeland, Lucas, & et al., 2011). The gene organisation of both the strains is similar to that of the strain 2.4.1, so they differ from the type strain mostly in terms of their plasmid content. Also in the nucleotide sequence database are the whole genome sequences of the denitrifying strain 2.4.3, which contains four plasmids of size 289 kb, 122 kb, 36 kb and 14 kb, and strain ATCC 17029, which contains a single plasmid of 123 kb. In addition to its plasmid content, the genome organisation of strain ATCC 17025 differs significantly from that of the other strains at the chromosomal level, although two chromosomes are still present (Choudhary, Zanhua, Fu, & Kaplan, 2007). Chromosome 2 of this strain is designated both in the NCBI nucleotide database and the Integrated Microbial Genomes database as a plasmid (pRSPA01) and this should be corrected (Table 8.2).

2.1.2. *R. capsulatus*

Plasmids in *R. capsulatus* were first detected by Hu and Marrs (1979) who showed that strain BH9 (a rifampicin-resistant mutant of the wild type strain H9) contains two plasmids of size 94 MDa (140 kb) and 74 MDa (111 kb). The copy numbers of these plasmids were estimated to be one to two copies per cell by renaturation kinetics. Y262, a gene transfer agent (GTA)-overproducing mutant of strain B10, was shown to contain DNA that hybridised with the BH9 plasmids, suggesting that the plasmid DNA shares sequences with genomic DNA from B10. Using vertical agarose gel electrophoresis, Willison et al. (1987) showed that wild type strain B10 contains a single plasmid of 86 MDa (129 kb) and plasmids were also detected in several other strains (Table 8.1). Witt and

Table 8.2 Survey of plasmid sequences in the NCBI nucleotide database (August, 2012). For complete genome sequences of anaerobic anoxygenic photosynthetic bacteria, the chromosomes are also shown for comparison

Species/strain	Replicon	Length (bp)	Accession number(s)
Rhodobacter sphaeroides			
2.4.1^T (ATH 2.4.1^T, ATCC 17023^T, DSM 158^T)	Chromosome 1	3,188,609	CP000143; NC_007493
	Chromosome 2	943,016	CP000144; NC_007494
	Plasmid A (partial)	114,045	DQ232586; NC_009007
	Plasmid Ax	124,310	AKVW01000003 AKBU01000003
	Plasmid B	114,178	CP000145; NC_007488 AKVW01000004; AKBU01000004
	Plasmid C	105,284	CP000146; NC_007489 AKVW01000005 AKBU01000005
	Plasmid D	100,828	CP000147; NC_007490
	Plasmid D (shot-gun)	100,819	AKVW01000006
	Plasmid Dx	52,135	AKVW01000007
	Plasmid E (partial)	37,100	DQ232587
	Plasmid E	37,100	NC_009008
WS8N	Chromosome I	3,139,378	CM001161
	Chromosome 01	3,139,278	AFER01000001
	Chromosome II	968,208	CM001162
	Chromosome 02	968,108	AFER01000002
	Plasmid pWS8N_A	110,310	CM001161; AFER01000003
	Plasmid pWS8N_B	199,892	CM001164; AFER01000004
ATCC 17025 (2.4.3)	Chromosome (=chromosome 1)	3,217,726	CP000661; NC_009428
	Plasmid pRSPA01 (=chromosome 2)	877,879	CP000662; NC_009429
	Plasmid pRSPA02	289,489	CP000663; NC_009430

Continued

Table 8.2 Survey of plasmid sequences in the NCBI nucleotide database (August, 2012). For complete genome sequences of anaerobic anoxygenic photosynthetic bacteria, the chromosomes are also shown for comparison—cont'd

Species/strain	Replicon	Length (bp)	Accession number(s)
	Plasmid pRSPA03	121,962	CP000664; NC_009431
	Plasmid pRSPA04	36,198	CP000665; NC_009432
	Plasmid pRSPA05	13,873	CP000666; NC_009433
KD131 (KCTC 12085)	Chromosome 1	3,152,792	CP001150; NC_011963
	Chromosome 2	1,297,647	CP001151; NC_011958
	Plasmid pRSKD131A	157,345	CP001152; NC_011962
	Plasmid pRSKD131B	103,355	CP001153; NC_011960
ATCC 17029	Chromosome 1	3,147,721	CP000577; NC_009049
	Chromosome 2	1,219,053	CP000578; NC_009050
	Plasmid pRSPH01	122,606	CP000579; NC_009007
Rhodobacter capsulatus			
SB1003	Chromosome	3,738,958	CP001312
	Plasmid pRCB133	132,962	CP001313; NC_014035
Rhodobacter blasticus			
TCRI-14	Plasmid pMG160	3431	AB082959
Rhodospirillum rubrum			
S1^T (ATCC 11170^T)	Chromosome	4,325,825	CP000230; NC_007643
	Plasmid	53,732	CP000231; NC_007641
F11*	Chromosome	4,325,825	CP003046; NC_017584
Rhodopseudomonas palustris			
CGA009	Chromosome	5,459,213	BX571963; NC_005296

	Plasmid pRPA	8427	BX571964; NC_005297
AS1.2352†	Plasmid pRPSZY	2306	DQ318958; NC_013548
Chlorobium limicola			
DSM 249	Plasmid pCL1	14,636	U77780; NC_002095
Allochromatium vinosum			
DSM 180	Chromosome	3,526,903	CP001896; NC_013851
	Plasmid pALVIN01	102,242	CP001897; NC_013852
	Plasmid pALVIN02	39,929	CP001898; NC_013862

*Strain F11 is a rhodoquinone-deficient mutant derived from strain S1.
†The molecule sequenced is actually a circularised gene coding for an RNA O-methylase that shows 97% sequence identity with a gene from the *E. coli* plasmid pEC157 (AF432497). This entry should be removed from the database.

Klemme (1991) screened several wild type isolates for plasmids. Most of these strains contained one or two plasmids with sizes ranging between 93 kb and 134 kb, but strain Fc101 was unusual in containing, in addition to a 115-kb species, two smaller plasmids of 32 and 3 kb. Whole genome sequencing showed that *R. capsulatus* strain SB1003 (a rifampicin-resistant derivative of strain B10) contains a single chromosome of 3.7 Mb and a circular plasmid of 133 kb, designated pRCB133 with 154 open reading frames (ORFs) (Strnad et al., 2010). *Rhodobacter capsulatus* ATCC 11166 has been shown by pulsed field gel electrophoresis (PFGE) contain a single chromosome as well as a 150-kb plasmid (Jumas-Bilak, Michaux-Charachon, Bourg, Ramuz, & Allardet-Servent, 1998).

2.1.3. *Other* Rhodobacter *sp*

Matsunaga, Matsunaga, Tsubaki, and Tanaka (1986) screened a collection of marine photosynthetic bacterial isolates for plasmids. Ten strains provisionally identified as *Rhodopseudomonas* (probably *Rhodobacter*) sp. contained between two and four plasmids ranging in size from 3.1 to 11 MDa (4.5–16.5 kb) and two strains of *Rhodospirillum* sp. contained either four or six plasmids of size 3.6–9.9 MDa (4.9–15 kb), four of which were of identical size in both the strains. Marine photosynthetic bacteria, therefore, appear to contain numerous small plasmids

(<15 kb) in contrast to freshwater strains that generally harbour large plasmids (>40 kb). There are exceptions to this as small plasmids have been found in *Rhodobacter blasticus* TCRI-14, *R. capsulatus* C11 and Fc101, *R. palustris* S55 and CGA009, and *R. sphaeroides* 2.4.3, while *R. sphaeroides* KD131, which was isolated from a marine environment, contains two large plasmids (Table 8.2). Unfortunately, apart from the replication region of pRD31 (Matsunaga, Mihashita, Miyake, & Burgess, 1991), none of these marine photosynthetic bacterial plasmids have been sequenced.

A 3.4-kb cryptic plasmid, pMG160, from *R. blasticus* TCRI-14 was found to be mobilisable into *R. sphaeroides*, *R. capsulatus* and *R. palustris*, but did not replicate in *R. rubrum*, *Rhodocyclus gelatinosus* or phototrophic *Bradyrhizobium* species (Inui, Nakata, Roh, Vertès, & Yukawa, 2003). This plasmid was used to construct two-6.1-kb *Escherichia coli–R. sphaeroides* shuttle vectors, designated pMG170 and pMG171 (Inui et al., 2003), and pMG170 has been used to develop an isopropyl β-D-1-thiogalactopyranoside-inducible, controlled expression vector for use in *R. sphaeroides* WS8N and *P. denitrificans* PD 1222 (pIND4) (Ind et al., 2009).

In addition to *R. sphaeroides* DSM 159, Pradella et al. (2004) also examined the plasmid content of Rhodobacter *veldkampii* DSM11550^T, which appeared to contain two linear plasmids of 59 and 115 kb. If this is confirmed, it will be the first reported occurrence of linear plasmids in the anaerobic anoxygenic photosynthetic bacteria. However, the DNA preparations from this strain gave a large background smear on gels, suggesting the presence of a strong nuclease activity, so circular conformations of plasmids may not remain intact. In halophilic bacteria, this problem can be overcome by prior washing of cells with 0.1% SDS (Mouné et al., 1999; Vargas, Fernandez-Castillo, Canovas, Ventosa, & Nieto, 1995).

2.1.4. *R. rubrum*

A single plasmid of 55 kb, designated pKY1, was found in crude cell lysates of each of the nine strains of *R. rubrum*, including the wild type isolates S1^T, G-1980, ATCC 11170^T, FR1, and S-4 (Kuhl et al., 1983). Restriction endonuclease analysis showed identical fragment patterns with a given nuclease for all plasmids except one, from a laboratory variant of S1, named S1-A, for which an additional *Eco*RI site was observed. Kawamukai et al. (1990) also found 55-kb plasmids in strains S1, G-9 and IFO3986. The size of 54 kb estimated by restriction mapping (Kuhl et al., 1983) was identical to that determined by complete genome sequencing of *R. rubrum* type strain S1, which contains a 4.4 Mb chromosome and a single, 54-kb circular plasmid (Munk et al., 2011). The G + C content of the plasmid was 60%, compared to 65% for the chromosome. *Rhodospirillum rubrum* strain F11 has also been sequenced (Lonjers et al., 2012). The size and sequence of the chromosome is identical to that of the strain S1, but no plasmid sequence was reported.

2.1.5. Other Rhodospirillaceae

As mentioned above, two marine strains of *Rhodospirillum* sp. contained several small plasmids of <15 kb, which have not been characterized (Matsunaga et al., 1986). Whole genome sequences are available for *Rhodospirillum centenum* (NC_011420; Lu, Marden, Han, & et al., 2010) and *Rhodospirillum phototometricum* DSM 122 (HE663493), but these strains apparently do not contain plasmids.

2.1.6. R. palustris

Out of 400 strains of PNSB (environmental isolates) screened for plasmids smaller than 20 kb, Inui, Roh, Zahn, and Yukawa (2000) identified one isolate, *R. palustris* S55, which contained a 15-kb cryptic plasmid named pMG101. The shuttle vector pMG101Km derived from this plasmid was transferred into other strains by electroporation and shown to replicate in other *R. palustris* strains and phototrophic *Bradyrhizobium* species, but not in *R. sphaeroides* ATCC 17023 or in non-phototrophic *Rhizobium* species. A 3.0-kb restriction fragment containing the replication region of pMG101, including the *repA* and *parA* genes, has been sequenced (Inui et al., 2000; accession no. 031076) and shown to have a lower G + C content (57.9%) than the overall *R. palustris* genome.

The complete genome sequence of *R. palustris* CGA009 revealed a chromosome of 5.5 Mb, and a small circular plasmid of 8.4 kb, pRPA (Larimer, Chain, Hauser, & et al., 2004). However, other sequenced strains do not contain plasmids (Oda, Larimer, Chain, & et al., 2008; Simmons, Isokpehi, Brown, & et al., 2011).

2.1.7. Other PNSB

Brown, Kysela, Buechlein, Hemmerich, and Brun (2011) sequenced the genome of *Rhodomicrobium vannielii* ATCC 17100 (accession no. NC_014664), but no plasmid was found in this strain. No plasmids have been reported in members of the genera *Rhodovulum* or *Rhodocyclus*, either in the literature or in the database.

2.2. Purple Sulphur Bacteria

Suyama and Gibson (1966) first showed the presence of plasmid (satellite) DNA in *Chromatium* D, but Kobayashi and Akazawa (1983) were unable to detect plasmids in a different *Chromatium* sp., using several methods. Nevertheless, the complete genome sequence of *Allochromatium* (formerly *Chromatium*) *vinosum* DSM 180 showed the presence of two endogenous plasmids of size 102 and 40 kb (Table 8.2). However, *Thiocystis violascens* contains a single chromosome and no plasmids. More data are needed to determine whether plasmids are a rare or common occurrence among the purple sulphur bacteria.

2.3. Green Sulphur Bacteria

Chlorobium limicola f. sp. *thiosulphatophilum* strains DSM 249 and BF 8000 were both found to contain a 14-kb endogenous plasmid that was not present in strains that were unable to utilise thiosulphate, such as *C. limicola* DSM 245 (Méndez-Alvarez et al., 1994) (see below).

Similarly, *C. limicola* f. sp. *thiosulphatophilum* strain UdG6038 lacks the 14-kb plasmid, but contains a 650-kb extrachromosomal element that hybridises with the 14-kb plasmid, whereas *C. limicola* strain UdG6041 lacks extrachromosomal elements (Méndez-Alvarez et al., 1995) (Table 8.1).

Endogenous plasmids were not detected in *Chlorobium tepidum* during studies of conjugational transfer of IncQ-group plasmids (Wahlund & Madigan, 1995), or natural transformation with pUC19-derived plasmids (Frigaard & Bryant, 2001). The absence of endogenous plasmids was confirmed by the complete genome sequencing of *C. tepidum* (now called *Chlorobaculum tepidum*) TLS (Eisen, Nelson, Paulsen, & et al., 2002).

Other *Chlorobiaceae* whose genomes have been completely sequenced include *C. limicola* DSM 245, *Chlorobium phaeovibriodes* DSM 265, which can use both thiosulphate and tetrathionate as electron donors, and *Chlorobaculum parvum* (formerly *Chlorobium vibrioforme* subsp. *thiosulfatophilum*) NCIB 8327. None of these strains contain plasmids. The 14-kb endogenous plasmid from *C. limicola* DSM 249 has been sequenced (Table 8.2) but the whole genome sequence is not available.

2.4. Green Non-sulphur Bacteria

The filamentous green non sulphur bacteria (*Chloroflexaceae*) are thermophilic microorganisms that grow in microbial mats in hot springs. Unlike the green sulphur bacteria, which form a coherent phylogenetic group, the green non-sulphur bacteria include both phototrophs and non-phototrophs. Complete genome sequences have been determined for a number of species, including *Chloroflexus aurantiacus* J-10-fl, *Chloroflexus aggregans* DSM 9485, *Chloroflexus* sp.Y-400-fl, *Roseiflexus castenholzii* DSM 13941, and *Roseiflexus* sp. RS-1, but none of these strains contain plasmids.

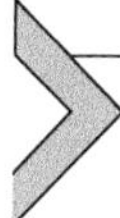

3. FUNCTIONAL ROLES OF ENDOGENOUS PLASMIDS IN PHOTOTROPHIC BACTERIA

Horizontal gene transfer (HGT) is considered to play a major role in bacterial genome evolution, and the ability of plasmids to mediate intra- and inter-species gene transfer, as well as to promote genome rearrangements, suggests that they are important agents of genome plasticity (Aminov, 2011; Treangen & Rocha, 2010). Indeed, almost any plasmid may play a role in this respect. Even very small plasmids, with only sufficient coding capacity to ensure their own replication and maintenance or transfer, are potential vectors for gene transfer or sources of replication modules. For this reason, we prefer to use the term "endogenous", indicating that a plasmid exists naturally in a wild type isolate rather than having been introduced artificially, instead of "cryptic". The substantial coding capacity of larger plasmids suggests they should play a role in cellular function. Indeed, in strains carrying multiple plasmids, the combined coding capacity of the extrachromosomal elements may exceed one third of the total genome (Pradella, Päuker, & Petersen, 2010).

Evidence is becoming available that some plasmids may encode essential functions and are not therefore exclusively 'accessory' elements. For instance, essential genes for pantothenate biosynthesis (*panBC*) are located on endogenous plasmids in *Rhizobium etli* and *Rhizobium leguminosarum* (Villaseñor et al., 2011). Phylogenetic analysis suggests a common origin for plasmid-located and chromosomal *panBC* genes, suggesting that they are orthologues rather than xenologues, i.e. that the plasmid location arose by gene capture from chromosome rather than HGT. Transcriptomic analysis in *R. sphaeroides* 2.4.1 showed that endogenous plasmids may carry genes that, while not essential, are nevertheless important for photosynthetic metabolism (Callister, Nicora, Zeng, & et al., 2006; Tavano, Podevels, & Donohue, 2005). In this section, the evidence for the functional roles of plasmids in different species will be summarised, with particular emphasis on four areas for which experimental evidence is available: photosynthetic metabolism, nitrogen oxide reduction, cell wall biosynthesis, and tolerance to heavy metals.

3.1. Photosynthetic Metabolism

3.1.1. *R. sphaeroides*

Saunders et al. (1976) found that, in the photosynthetically incompetent strain SLS obtained after treatment of strain 2.4.1(NRS) with SDS, the size of the smallest plasmid had increased from 28 MDa (42 kb) to 34 MDa (51 kb). The mutant did not revert to the photosynthesis plus phenotype and retained the multiple antibiotic resistance of the parental strain. Nano and Kaplan (1984) confirmed these observations and showed that plasmid rearrangement involves an interaction between the 42-kb plasmid and a 99-kb plasmid (presumably either C or D), giving rise to 50-kb and 96-kb forms and a high correlation with photosynthesis minus phenotype. Extensive homology between the 42-kb and 99-kb plasmids was observed, and no chromosomal sequences were detected on the rearranged 50-kb plasmid. The authors concluded that the loss of photosynthesis might have been due to a general genomic instability that affected both plasmid stability and the photosynthetic phenotype. This idea was supported by the results of Suwanto and Kaplan (1989, 1992) who found that Tn5 insertions in plasmids A, B and E were frequently accompanied by point mutations in one of the chromosomes. Furthermore, the curing of plasmid E had no effect on photosynthetic growth (Suwanto & Kaplan, 1992). The degree of genetic instability in strain 2.4.1 appears to be unusual, and indeed, Pellerin and Gest (1983) suggested that strain 81-1 would be more useful as a reference strain than the neotype strain 2.4.1, on the basis of the stability of its characteristic morphological features.

Comparison of the aerobic and photosynthetic proteomes of strain 2.4.1 grown on succinate medium led to the identification of 1675 gene products, of which only 50 were unique to the photosynthetic proteome and none to the aerobic proteome (Callister et al., 2006). Of the 50 polypeptides detected exclusively under photosynthetic growth conditions, two are encoded by plasmid B, one by plasmid C and two by plasmid D (Table 8.3). All of the plasmids carry genes that are expressed under aerobic or photosynthetic growth conditions, and the proportion of plasmid-encoded genes detected in the total transcriptome was equal to or higher than that of chromosome 2.

The two unique, photosynthetically expressed proteins from plasmid D were RSP4158 and RSP4159, which code for a methyltransferase and isopropyl-malate synthase. These genes appear to be co-transcribed with a

Table 8.3 Number and percentage of predicted ORFs shown to be expressed in the total proteome or transcriptome of *R. sphaeroides* 2.4.1 grown under photosynthetic or aerobic conditions in succinate medium*

Replicon	No. of predicted ORFs	Expressed in transcriptome (% total)	Expressed in proteome (% total)	Unique to photosynthetic proteome
Chromosome 1	3037	2027 (67%)	1197 (39%)	39
Chromosome 2	846	354 (42%)	185 (22%)	6
Plasmid A	103	50 (48%)	21 (20%)	0
Plasmid B	96	40 (42%)	15 (16%)	2
Plasmid C	79	39 (49%)	6 (8%)	1
Plasmid D	87	51 (59%)	18 (21%)	2
Plasmid E	21	13 (62%)	3 (14%)	0

*Data from Callister et al. (2006).

third gene, RSP4157, coding for a radical SAM enzyme, and all the three genes are upregulated during photosynthesis (Tavano et al., 2005). A mutant containing a polar insertion in RSP4157 was unable to grow photosynthetically in a succinate-based medium unless compounds that could be used to recycle reducing power (Dimethyl sulphoxide (DMSO) or CO_2) were provided. This suggests that the insertion in RSP4157 caused a defect in recycling reducing power during photosynthetic growth when an organic carbon source was present.

3.1.2. *R. rubrum*

Elimination of the 55-kb endogenous plasmid from *R. rubrum* strain S1-G required that the cells be sub-cultured several times in 25 mM Ca^{2+}-containing medium, then at least twice in low-Ca^{2+} medium under photosynthetic growth conditions before treatment with ethyl methanesulphonate (Kuhl et al., 1983). The curing rate was 83–90% and all cured strains were incapable of pigment formation and photosynthetic growth. Furthermore, no hybridisation was observed between plasmid DNA and chromosome in the cured strains, showing that the plasmid had been eliminated from the cells and had not become integrated into the chromosome (Kuhl, Wimer, & Yoch, 1984). Plasmid loss was also accompanied by a number of other phenotypic properties that suggested modification of the cell membrane.

The sequence of the endogenous plasmid from strain S1 (CP000231) does not contain genes whose loss would obviously lead to photosynthesis minus phenotype. On the other hand, it contains numerous genes that are potentially involved in cell wall biosynthesis, so their absence could explain the changes in permeability and membrane composition. The non-photosynthetic phenotype might be due to an additional mutational event, possibly induced by plasmid loss, as in *R. sphaeroides*; being non-reversible, this would have to involve a deletion or stable insertion. The most direct approach to solving this problem would be to sequence the genome of a cured strain. This is not unrealistic, since the chromosome of strain F11 was sequenced for the sole purpose of identifying a gene required for rhodoquinone synthesis (Lonjers et al., 2012).

3.1.3. Other Species

Curing of small endogenous plasmids from marine *Rhodobacter* sp. NKPB 002106 (Matsunaga et al., 1986), *R. palustris* S55 (Inui et al., 2000) and *R. palustris* CGA009 (Berne, Allainmat, & Garcia, 2005) gave no detectable phenotype. Furthermore, *R. capsulatus* strains B10 and AD2 cured of their single endogenous plasmids were found to be photosynthetically competent (Willison, 1990; Willison et al., 1987). *R. capsulatus*, therefore, seems not to share the genetic hypervariability found in *R. sphaeroides* 2.4.1 and *R. rubrum* S1.

Pradella et al. (2004) showed by hybridisation that the reaction centre *pufLM* genes are located on a linear plasmid of 91 kb in *Roseobacter litoralis* DSM 6996^T and on a linear plasmid of 120 kb in *Staleya* (now *Sulphitobacter*) *guttiformis*, but are chromosomally located in the other strains tested, including *Roseobacter denitrificans*. These plasmids contain the entire PGC of around 45 kb (Kalhoefer et al., 2011), and the replication region of each plasmid is located at precisely the same position within the PGC, adjacent to *bchO* (Petersen et al., 2012). As will be discussed below, phylogenetic analysis suggests that the plasmid-located PGCs were derived from the chromosomes of the host bacteria and not introduced by HGT. Nevertheless, their occurrence on plasmids indicates that plasmid-mediated lateral transfer of photosynthesis genes may have occurred in the past.

3.2. Nitrogen Oxide Reduction

Nitrate can be used either as a source of nitrogen (nitrate assimilation) or as a source of energy (nitrate dissimilation). In both the cases, nitrate is first reduced to nitrite, which, in the case of assimilation, is reduced to ammonia by an assimilatory nitrite reductase. In the case of dissimilation, nitrite either accumulates (nitrate respiration), or is reduced further (via NO), to N_2O or N_2 in the process known as denitrification. Nitrate utilisation in the anaerobic anoxygenic phototrophs is restricted to the PNSB, particularly those belonging to the α subgroup of *Proteobacteria*. In these organisms, the ability to utilise nitrate is unevenly distributed, which would be consistent with a plasmid-mediated dissemination and localisation of the corresponding genes. There is anecdotal evidence that plasmid-mediated transfer of nitrate reductase activity can occur in the environment, and experimental and genomic evidence for the localisation on endogenous plasmids of genetic determinants for nitrate assimilation, nitrate respiration and denitrification.

3.2.1. *R. capsulatus*

Czichos and Klemme (1982) reported that of 14 strains of *R. capsulatus* in their laboratory collection, only seven were able to use nitrate as N source. They also stated that 'occasionally, newly isolated strains of *Rps. capsulata* contained two types of colonies when grown aerobically on KNO_3–malate–agar plates: small white ones unable to assimilate nitrate, and normally sized brown-red ones effectively assimilating nitrate'. This observation is consistent with the presence of assimilatory nitrate reductase genes on an extrachromosomal element that is initially unstable in fresh isolates.

Rhodobacter capsulatus strain AD2 both assimilates and dissimilates nitrate (Alef, 1987). Like many strains containing a periplasmic dissimilatory nitrate reductase, strain AD2 is unable to grow anaerobically with nitrate as electron acceptor, and nitrate reduction in these strains serves primarily as an electron sink to support photoheterotrophic growth on reduced carbon sources (Richardson et al., 1988). A mutant strain of *R. capsulatus* AD2 unable to utilise nitrate was isolated that is non-reversible and whose phenotype was not restored by incubation with GTA (Czichos & Klemme, 1982), suggesting the presence of a large deletion or two widely spaced mutations. Willison (1990) showed that this mutant, strain C2, had lost the 74 MDa (111-kb) endogenous plasmid present in the wild type and that this plasmid could be eliminated by conjugation with a broad-host-range plasmid containing a cloned 3.4-kb *Hind*III restriction fragment from the endogenous plasmid. This curing method resembles that used to cure plasmid E in *R. sphaeroides* 2.4.1 (Suwanto & Kaplan, 1992), indicating that the cloned fragment contains incompatibility determinants. Cured strains were unable either to assimilate or dissimilate nitrate, suggesting that the genes required for both processes are located on the endogenous plasmid (Willison, 1990). It was subsequently shown that an 11-kb *Eco*RI fragment of the ~115 kb endogenous plasmid (Witt & Klemme, 1991) hybridised with the *napA* gene from *Alcaligenes eutrophus*, confirming that the structural gene for periplasmic nitrate reductase is located on the plasmid (Koch & Klemme, 1994).

Castillo, Dobao, Reyes, and et al. (1996) suggested that the gene for assimilatory nitrate reductase (*nasA*) is located on an endogenous plasmid in strain E1F1, but this plasmid has not been characterised.

The genome sequence of *R. capsulatus* SB1003 shows that this strain contains a complete nitrous oxide reductase operon (*nosRZDFYLX*), but the genes for the other denitrification enzymes, including nitrate reductase, appear to be absent (Haselkorn, Lapidus, Kogan, & et al., 2001). The *nos* genes are located on the endogenous plasmid pRCB133 (CP001313, genes rcp00070–rcp00076) and the closest homologues on a protein basis (80% sequence identity) are from other α-*Proteobacteria*. N_2O reductase activity and growth in the dark with N_2O as electron acceptor have been shown for several strains of *R. capsulatus* (McEwan, Greenfield, Wetzstein, Jackson, & Ferguson, 1985), but strains SB1003 and B10 have not so far been tested. Although wild type strains of *R. capsulatus* are generally non-denitrifying, a gain-of-function mutant that is able to denitrify due to the acquisition of nitrite reductase activity was isolated from strain BK5 (Richardson, Bell, Moir, & Ferguson, 1994). This strain contains two plasmids of 115 and 135 kb, but the genomic location of the nitrite reductase genes is not known.

3.2.2. *R. sphaeroides*

Some subspecies of *R. sphaeroides* are able to denitrify. The first to be isolated was *R. sphaeroides* f. sp. *denitrificans* IL106. Satoh, Hoshino, and Kitamura (1976) and Pellerin and Gest (1983) subsequently identified two strains (out of nine tested) that were able to grow anaerobically in the dark on nitrate and produce gas.

When the denitrifying strains *R. sphaeroides* f. sp. *denitrificans* IL106, 81-3 and 2.4.3 were compared with the non-denitrifying strains 81-1 and 2.4.1, all the strains were found to have nitrate reductase activity, but only denitrifying strains had nitrite reductase and N_2O reductase (Michalski & Nicholas, 1988). In that study, strain IL106 was found to contain one plasmid of 108 kb, while the others had similar plasmid profiles, with sizes of 113, 104, 100 and 94 kb (Table 8.1). Schwintner, Sabaty, Berna, Cahors, and Richaud (1998), using transverse alternating field electrophoresis, which has a higher resolution than vertical agarose gel electrophoresis, later showed that strain IL106 contains two large plasmids of 102 kb and 115 kb. It was shown that the *napA* gene, the *nirK* gene and the *norCB* genes, encoding the nitrate, nitrite and nitric oxide reductases, respectively, are located on different *Ase*I- and *Sna*BI-digested chromosomal DNA fragments, whereas the *nosZ* gene, encoding the nitrous oxide reductase, was located on the 115-kb plasmid.

In *R. sphaeroides* 2.4.1, the *napA* probe hybridised with a 97-kb plasmid (probably plasmid C), *nirK* with CII, *norC* with CI, and no hybridisation was observed with *nosZ*. The result with *napA* confirms the observations of Castillo et al. (1996), who suggested a plasmid location for this gene in strain DSM 158^T. The hybridisation with *nirK* may have been non-specific, since strain 2.4.1 does not have nitrite reductase activity and nitrite reductase genes were absent from the genome sequence (Mackenzie et al., 2001).

Strain DSM 158^T (2.4.1) is non-denitrifying, but can reduce nitrate for redox balance using a periplasmic nitrate reductase encoded by *napABC* (Reyes, Roldán, Klipp, Castillo, & Moreno-Vivian, 1996). The genome of strain 2.4.1 contains a single nitrate reductase gene, *napA*, which is located on plasmid C (Mackenzie et al., 2001), confirming the findings of Schwintner et al. (1998). There is no evidence for an assimilatory nitrate reductase and the genome lacks a gene for either assimilatory or dissimilatory nitrite reductase. Comparison with the 2.4.3 genome, which has a similar gene organisation to strain 2.4.1, suggests that the latter may have possessed dissimilatory nitrite reductase genes on chromosome I, downstream from a gene encoding pseudoazurin, but that these were lost during evolution. Genes for nitric oxide reductase (*nor*) are located within the PGC, and a regulatory *nnrR* gene is located within the *nor* gene cluster, confirming the results of Tosques, Shi, and Shapleigh (1996), who showed that a *norB–lacZ* gene fusion from strain 2.4.3 was expressed in strain 2.4.1. Genes for nitrous oxide reduction (*nos*), on the other hand, are absent (Mackenzie et al., 2001).

The *napEFDABC* operon located on plasmid C of 2.4.1 (RSP4113-4118) is also found on plasmid pRSPA02 of the denitrifying strain ATCC 17025 (2.4.3), unlike strain IL106 in which the *napA* gene is chromosomal. N_2O reductase (*nos*) genes are located on pWS8N_B in strain WS8N and on pKD131A in strain KD131, but are on CII (pRSPA01) in ATCC 17025 (2.4.3), unlike IL106, in which they are plasmid-located. Alignment of plasmid B from strain 2.4.1 with pWS8N_B and pKSD131A indicates a possible deletion of the region containing *nos* in plasmid B (Fig. 8.1A). The *nos* region from pRSKD131A (136511–141929) showed 97% nucleotide sequence identity with the partial N_2O reductase operon (*nosZDFYL*) from *R. sphaeroides* f. sp. *denitrificans* IL106 (AF125260).

3.2.3. Other Species

The complete genome sequence of *R. palustris* CGA009 contains two nitrite reductase genes, NO reductase genes and N_2O reductase genes on the chromosome, but apparently no nitrate reductase gene (Larimer et al.,

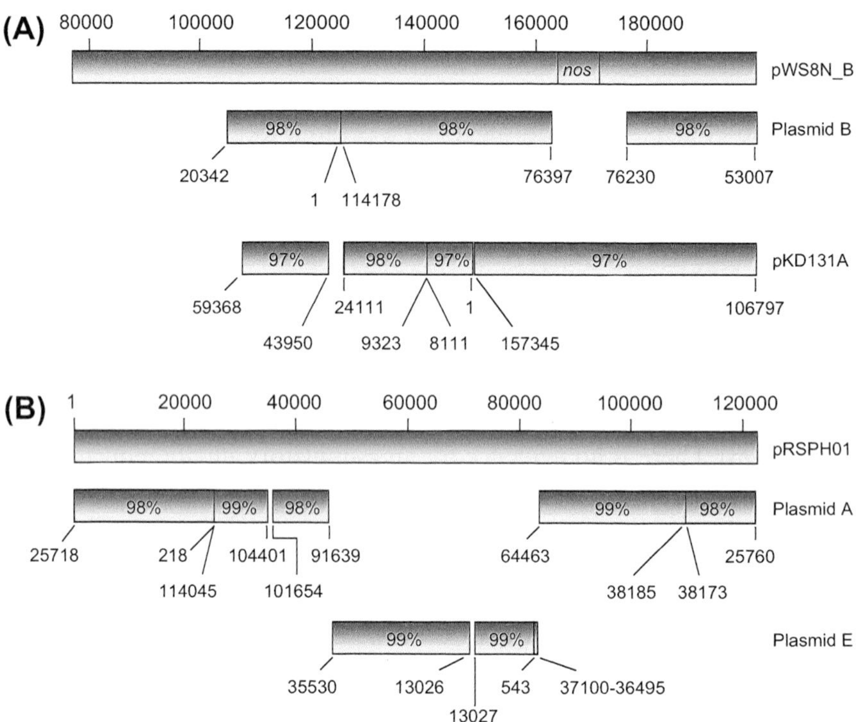

Figure 8.1 A. Alignment of the nucleotide sequences of *R. sphaeroides* plasmids pWS8N-B, plasmid B (strain 2.4.1) and pKD131A, showing an apparent deletion of the region containing the N_2O reductase (*nos*) genes in plasmid B. The percentage sequence identity with pWS8N_B is shown for each segment. For clarity, only part of the alignment is shown. B. Alignment of the nucleotide sequence of plasmid pRSPH01 with plasmids A and E from strain 2.4.1. The gap in the alignment between pRSPH01 and plasmid A is almost completely filled by the sequence of plasmid E. See the color plate.

2004). Some strains of *R. palustris* are known to denitrify or carry out nitrate respiration (Ferguson, Jackson, & McEwan, 1987) and a strongly denitrifying photosynthetic bacterium isolated from sludge was identified as *R. palustris* (Kim, Lee, Kim, & Moon, 1999).

The relationship between photosynthesis and denitrification is clearly complex, as witnessed by the fact that there are relatively few denitrifying strains of photosynthetic bacteria, and endogenous plasmids may play role in the evolution of this process. A plausible scenario is that denitrification genes are acquired by HGT and then become stabilised by integration into the chromosome, although even in the bona fide denitrifying strains IL106 and 2.4.3, some of the denitrification genes are plasmid-located. Even then, a fully denitrifying strain may lose this capacity, either by gene inactivation, as in *R. capsulatus* BK5, or by loss of genes, as may have occurred in *R. sphaeroides* 2.4.1. Although fewer data are available, the same logic may apply to nitrate assimilation, as suggested by the observations of Czichos and Klemme (1982).

3.3. Cell Wall Biosynthesis

3.3.1. R. capsulatus

Magnin, Willison, and Vignais (1987) discovered that B10 derivatives containing broad-host-range, antibiotic resistance, P1-group plasmids could be cured of these plasmids by sub-culturing in yeast extract-peptone (YP) medium. YP medium differs from the standard YP salts (YPS) medium in that it lacks Ca^{2+} and Mg^{2+} salts, and no plasmid curing was observed during photosynthetic growth in YPS medium (Magnin et al., 1987). Weaver, Wall, and Gest (1975) recommended using YPS rather than YP medium for routine cultivation of *R. capsulatus*, since some strains may form spheroplasts in the latter. This suggests that cell wall biosynthesis may be affected by a deficiency in divalent cations, and the curing of antibiotic resistance plasmids in YP medium may therefore be related to changes in the cell wall or membrane structure that affect plasmid partitioning.

The wild type strain B10 was not cured of its endogenous plasmid by sub-culturing in YP medium, and conventional curing techniques using SDS, ethidium bromide or acridine orange were also unsuccessful. However, by using strains containing the mutant R plasmid pTH10, which shows chromosome mobilising ability in *R. capsulatus*, or a temperature-sensitive derivative of pTH10, pMJP1, a very low rate of curing was observed among the kanamycin-resistant survivors after sub-culturing in YP medium (Willison et al., 1987). Some of the strains that had lost the endogenous plasmid contained chromosomal insertions of R plasmid DNA and showed phenotypes (loss of autotrophic growth or inability to utilise pyruvate) that could be attributed to these insertions. Subsequently, strain RC205, which has no chromosomal plasmid sequences, was cured of plasmid pTH10 by sub-culturing four times on YP medium, yielding two plasmid-free strains, RC219 and RC220, that have been used for functional studies on the role of the endogenous plasmid (J.C. Willison and J.P. Magnin, unpublished).

Inspection of the annotated nucleotide sequence of pRCB133 (CP001313) revealed two genes, with no counterparts in the chromosome, whose loss of functionality could lead to experimentally testable

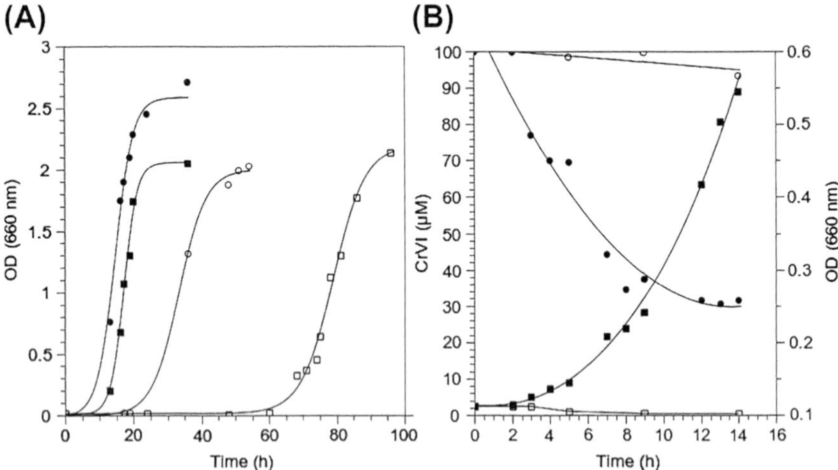

Figure 8.2 A. Effect of SDS on the growth of *R. capsulatus* strain B10 (circles) and the plasmid-cured strain RC219 (squares). Strains were grown photosynthetically in YPS medium (filled symbols) or in YPS medium containing 0.01% SDS (empty symbols). B. Effect of dichromate ions (CrVI) on the growth of *R. capsulatus* strain SB1003 (filled squares) and the plasmid-cured strain RC220 (empty squares). Strains were grown aerobically in the dark in YP medium containing 100 μM potassium dichromate. The concentration of hexavalent chromium remaining in solution was measured in cultures of strains SB1003 (filled circles) and RC220 (empty circles); the decrease observed was due to the reduction of Cr(VI) ions to Cr(III).

phenotypes. The first, *chrA* (rcp00138), predicted to code for a chromate transporter (efflux), will be discussed in the next section. The second gene, rcp00108, is annotated as coding for an ABC-transporter, ATP-binding permease, but it is also homologous to the *E. coli msbA* gene, which is involved in the export of lipid A into the outer membrane (Zhou, White, Polissi, Georgopoulos, & Raetz, 1998). This gene is essential in *E. coli*, but not in *Neisseria meningitidis*, although its inactivation in the latter results in severe growth defects (Tefsen, Bos, Beckers, Tommassen, & de Cock, 2005). If this gene were functional in *R. capsulatus*, then its loss by elimination of the endogenous plasmid would affect the permeability of the outer membrane, and possibly also the cell wall structure. This was tested by comparing the effects of detergents and solvents on the wild type and cured strains. Growth of all the three strains (B10, RC219 and RC220) was completely inhibited by 0.1% SDS, but with 0.01% SDS, growth was observed after a lag period that was much more pronounced (60 h compared to 20 h) for the plasmid-cured strains than for the wild type (Fig. 8.2A). This effect was observed in YPS medium, but not in minimal *Rhodopseudomonas capsulata* medium V (RCV). Cured strains were also more sensitive to ethanol than the wild type, but no difference in sensitivity to sodium deoxycholate was observed (data not shown).

Further evidence of a modification of the cell envelope in plasmid-cured strains was provided by Zn-binding studies, which showed that the maximum absorption capacity of live biomass from strain B10 was approximately twice as high as that of strains RC219 and RC220 (M. Cheikh, J.P. Magnin, N. Gondrexon, J. Willison, A. Hasdsan, and M. Velasquez, unpublished).

3.3.2. *R. rubrum*

Plasmid-cured strains of *R. rubrum* S1-G showed increased sensitivity to streptomycin and increased resistance to $HgCl_2$ and cycloserine (Kuhl et al., 1984). These pleiotropic effects suggest that major changes had occurred in the permeability of the cell membrane and/or cell wall, and changes in the protein composition of the outer and inner membranes were reported. In addition, the cured strain KY11 had 21% less total lipid than S1-G and a much higher ratio of neutral lipid to phospholipid. The major phospholipid in the mutant was phosphatidylcholine, in contrast to phosphatidylethanolamine and phosphatidylglycerol in the wild type.

Although it cannot be affirmed that all these changes resulted from loss of the endogenous plasmid, a role in cell wall biosynthesis was indicated by Ideguchi et al. (1993), who cloned and sequenced a 3188-bp *Bgl*II fragment from pKY1 and showed that it contained a gene homologous to *algA* from *Pseudomonas aeruginosa*, encoding phosphomannose isomerase–guanosine diphospho-D-mannose pyrophosphorylase (PMI–GMP), designated *pssM*. Inspection of the annotated nucleotide sequence of the endogenous plasmid from strain S1 (CP000231) shows that, of the 50 ORFs present on this plasmid, 13 are potentially involved in cell wall biosynthesis or carbohydrate metabolism. These include four genes for glycosyl transferase, two genes for NAD-dependent epimerase/dehydratase, and genes for dTDP-4-dehydrorhamnose 3,5-epimerase, phosphomannomutase, mannose-1-phosphate guanylyltransferase (GDP), glucose-1-phosphate thymidyltransferase, dTDP-4-dehydrorhamnose reductase, and dTDP-glucose-4,6-dehydratase. Of particular interest is the *sapC* gene (Rru_B0024), which codes for a crystalline protein forming part of the S layer in some pathogenic bacteria (Thompson et al., 1998). If this gene were expressed, a role could be suggested for the endogenous plasmid in conferring resistance to grazing by protozoa, to explain the unusual persistence of this plasmid in *R. rubrum* strains.

3.3.3. *Other Species*

An *rfb*-gene cluster essential for the development of O antigens, i.e. lipopolysaccharides of the outer membranes of Gram-negative bacteria, was identified on the 63-kb plasmid pRLO149_63 of *R. litoralis* OCh149 and on a plasmid of 69 kb (pTB2) in *R. denitrificans* (Kalhoefer et al., 2011), and these plasmids also carried genes associated with cell envelope biosynthesis. Genome analysis showed that 11/12 organisms with chromosomal *rfb*-genes were isolated from the water column, whereas 9/12 organisms with plasmid-located *rfb*-gene clusters were isolated from the surfaces of marine animals or algae. This suggests that a plasmid location of the *rfb*-genes in *Roseobacter* clade bacteria is correlated with the host association (Kalhoefer et al., 2011).

3.4. Heavy Metal Resistance

High-level resistance to heavy metals is frequently associated with the presence of self-transmissible plasmids (Silver & Phung, 1996). High-level resistance is rare in photosynthetic bacteria, but many strains are tolerant to certain levels (the case of cured strains in *R. rubrum* is unusual, since resistance to 100 μM Hg^{2+} can be considered to be high-level, but it results from the *loss* of a plasmid). In *R. capsulatus*, we have shown that the strains lacking the endogenous plasmid, unlike the wild type, are completely inhibited by 100 μM chromate during aerobic growth on YP medium (Fig. 8.2B), suggesting that the *chrA* transporter gene is active in conferring chromate-resistance (J.C. Willison and J.P. Magnin, unpublished). Protein BLAST of the *chrA* gene product (ADE87393) shows that the closest relative in the database is ChrA from *R. capsulatus* E1F1, the gene for which is located within the *nas* operon (Cabello et al., 2004; AY273169), which is thought to be plasmid-encoded (Castillo et al., 1996).

Genes potentially involved in zinc tolerance are located on the chromosome of *R. capsulatus* SB1003, including MMT1 (*zitB*; rcc00089), which encodes a heavy metal-transporter and is involved in zinc-resistance in *E. coli* (Grass et al., 2001) and two homologues of *zntA*, *zntA*1 (rcc02064) and *zntA2* (rcc02190), which encode a heavy metal-translocating P-type ATPase. Two copies of *zntA* are also found on chromosome 1 of *R. sphaeroides* 2.4.1, but homologues of *zitB* are located on plasmid E in strain 2.4.1, and on plasmids pRSPH01 and pWS8N_A in strains ATCC 17029 and WS8N, respectively. The predicted protein products of the plasmid-located genes are almost identical and show 95% sequence identity with a protein from *Sagitulla stellata* E-37 (annotated as CzcD), suggesting the possibility of lateral gene transfer.

The comparative genome study of *R. litoralis* OCh149 and *R. denitrificans*, carried out by Kalhoefer et al. (2011), identified several potential heavy metal-resistance genes, including a *czcD* homologue, on the 83-kb plasmid pRLO149_83 of *R. litoralis* that were absent from the genome of *R. denitrificans*, and these genes were homologous to plasmid-encoded resistance genes in other *Roseobacter* clade members. *R. litoralis* was found to be more Zn-tolerant than *R. denitrificans*, while the latter strain was more Cu-tolerant.

3.5. Other Functions

Additional roles supported by experimental evidence include membrane transport and sulphur metabolism. The growth yield of a plasmid-cured strain of *R. rubrum* S1 on a complex medium was about 50% lower than that of the wild type, whereas the two strains had similar growth yields on a minimal medium supplemented with 0.3% yeast extract (Kuhl et al., 1984). Similarly, the growth yields of cured *R. capsulatus* B10 strains are lower than the wild type on complex YPS medium (Fig. 8.2A), but not on minimal RCV medium. These results suggest that transport functions encoded by the plasmids may be necessary for the uptake of small molecules (sugars, amino acids) used as C and N sources, or, alternatively, that extracellular enzymes needed for the degradation of high molecular weight substrates are plasmid-encoded.

Regarding sulphur metabolism, the sequence of the endogenous plasmid from *C. limicola* DSM 249 (Tio^+) (accession no. U77780) does not contain any genes likely to be involved in thiosulphate utilisation. The ability to confer the thiosulphate utilisation (Tio^+) phenotype on transformed strains may be due to a mutator phenotype induced by the presence of the plasmid, by analogy to the hypothetical mechanism that has been proposed for the loss of photosynthetic capability in *R. sphaeroides* and *R. rubrum*.

Plasmid-encoded penicillin resistance was suggested for *R. capsulatus* strain SP108, which produces a diffusible penicillinase and loses resistance to penicillin at high frequency (Weaver et al., 1975), and also for *R. sphaeroides* isolates in which penicillin-resistance is cured by mitomycin C but is not phage-encoded (Pemberton et al., 1983). Unusual genes that are located on various plasmids in *R. sphaeroides* include the circadian clock genes *kaiC1* and *kaiB1* (Mackenzie et al., 2001), and genes coding potentially for an F_0F_1-type ATP synthase (Pappas, Sram, Moskvin, & et al., 2004).

4. EVOLUTION OF ENDOGENOUS PLASMIDS

4.1. R. sphaeroides

Fornari et al. (1984) used restriction profiles and DNA–DNA hybridisation to show that plasmids from different strains of *R. sphaeroides* are interrelated. The smallest plasmids of strains 2.4.1, RS2 and L showed almost identical *Pvu*II digestion patterns and also hybridised with each other, and total plasmid DNA from strains 2.4.7,Y, and WS20 hybridised to the largest plasmid of strains L, RS2 and 2.4.1. Plasmid E from strain 2.4.1 also hybridised to one of the largest plasmids in 2.4.1 and L (but not RS2). It was suggested that the large plasmid homologous to plasmid E in 2.4.1 was plasmid B (Fornari et al., 1984). However, BLASTn comparisons show that plasmid E has very little homology with plasmid B, and shares the most sequences with plasmid D (Table 8.4). There is some uncertainty here, because the plasmid Ax sequence obtained by shotgun sequencing contains approximately 30 kb of DNA with >99% identity to plasmid E, and the plasmid D (shotgun), and plasmid Dx sequences contain about 20 kb of DNA homologous to plasmid A.

Analysis of the nearly complete genome sequence of strain 2.4.1 showed that G + C content of plasmids C and D was slightly lower than that of the chromosomes and the other plasmids, and the dinucleotide and trinucleotide frequencies were also different (Mackenzie et al., 2001). This suggested that plasmids C and D might have been acquired more recently than the other three, possibly by HGT from a lower G + C organism. Plasmid genomic signatures (nucleotide frequencies) have been used to predict the host range of transmissible plasmids, since the genomic signature of plasmids tends to evolve to match that of the host chromosome (Suzuki, Yano, Brown, & Top, 2010). The percentage of predicted ORFs with counterparts in the GenBank database was also lower for plasmids C and D (Mackenzie et al., 2001).

The extensive bioinformatic and phylogenetic analysis of plasmid replicons in the *Roseobacter* clade has given some insights into the evolution of endogenous plasmids in *R. sphaeroides*. Phylogenetic analysis of *repABC* replicons, which include plasmids B and D from strain 2.4.1, suggested extensive evolution by HGT (Cevallos, Cervantes-Rivera, & Gutiérrez-Rios, 2008). The RepA and RepC proteins of plasmids B and D were distinct but in the same phylogenetic cluster, whereas the RepB fell into different clusters. RepB from plasmid B was most closely related to *Roseovarius* HTCC 2601 (copy c) and *Oceanicola* HTCC 2597 (copy a), while RepB from plasmid D clustered with *Roseovarius* sp. 217 (copy c). A more refined

Table 8.4 Interrelationships between plasmids from different strains of *R. sphaeroides* as revealed by nucleotide BLAST. The accession numbers are given in Table 8.2

Strain	Query sequence	#	Target sequence (% coverage of query sequence)*													
			1	2	3	4	5	6†	7†	8	9	10	11	12	13	14
2.4.1	Plasmid A	1		1%		8%		ND	ND	10%	1%	1%			**63%**	**74%**
	Plasmid B	2			2%			ND	ND					**87%**		
	Plasmid C	3		2%		5%	1%	ND	ND	**15%**	9%					
	Plasmid D	4	9%		5%		4%	ND	ND		6%				4%	3%
	Plasmid E	5	2%	2%	3%	10%		ND	ND		5%				**47%**	**97%**
WS8N	pWS8N_A	6	**76%**			3%	**18%**		ND	10%	3%				**74%**	**94%**
	pWS8N_B	7	1%	**55%**	5%	3%		ND		1%	2%			**66%**		
ATCC 17025	pRSPA02	8	4%		5%			ND	ND		1%				4%	4%
	pRSPA03	9	1%		8%	7%	1%	ND	ND	2%					3%	3%
	pRSPA04	10	3%			4%		ND	ND							
	pRSPA05	11						ND	ND	**14%**						
KD131	pKD131A	12		**63%**				ND	ND							
	pKD131B	13	**69%**				**16%**	ND	ND							**84%**
ATCC 17029	pRSPH01	14	**69%**				**29%**	ND	ND	9%					**71%**	

*>70% sequence identity (percent coverage greater than 10% is shown in bold).
†pWS8N_A and pWS8N_B did not appear as target sequences in the nucleotide BLAST search.

phylogenetic analysis of RepC indicated the presence of nine incompatibility groups (C1–C9), which would be expected to stably coexist within the same cell, and this prediction was supported by RepA and RepB phylogenies (Petersen, Brinkmann, & Pradella, 2009). RepC from plasmid D clustered with *Paracoccus versutus* and *Paracoccus pantotrophus* in group C1, and plasmid B with *S. stellata* in C3; RepA from plasmid D clustered with *Roseovarius* sp. 217 in group A2 and plasmid B with *Oceanicola granulosus* and *S. stellata* in A3 (Petersen et al., 2009).

Another plasmid replication type found in the Rhodobacterales contains a *parAB*-type partitioning operon and a novel, DnaA-like replication initiator (Petersen et al., 2011). This replication region was also found to comprise nine incompatibility groups, giving a total of 18 for *repABC* and 'DNA-like' replicons. Many members of the *Roseobacter* clade contain multiple extrachromosomal replicons, and strains of the (non-photosynthetic) species *Marinovum algicola* contain up to 12 plasmids (Pradella et al., 2010). Phylogenetic analysis of DnaA-like replicons showed that the replication regions of plasmid C from *R. sphaeroides* 2.4.1 and pRSPA03 from ATCC 17025 are closely related (Petersen et al., 2011).

The overall degree of relatedness between plasmids of *R. sphaeroides* was assessed by nucleotide BLAST of whole plasmid sequences against the NCBI nucleotide database (Table 8.4). The results show that the plasmids can be classified into three groups: (1) pRSPH01, pKD131B and pWS8N_A, which were homologous to plasmid A and also to plasmid E; (2) pKD131A, pWS8N_B and plasmid B, which shared more than 50% of closely related sequences; and (3) plasmids C and D and all the four plasmids from ATCC 17025, which shared less than 10% of homologous sequences. Interestingly, the gap in the alignment between pRSPH01 and plasmid A was almost entirely filled with sequences from plasmid E (Fig. 8.1B). Plasmids with closely related replication regions (see above) are therefore not necessarily homologous overall. However, pKD131B and pRSPH01 both have repA-I replication regions, and pKD131A and plasmid B both belong to repABC-type 3 (Petersen, 2011).

Among the plasmids of strain ATCC 17025, the 289-kb plasmid pRSPA02 is particularly interesting, since it contains a ribosomal RNA (*rrn*) operon, tRNA genes, and genes for CO_2 fixation, including genes for the large and small subunits of ribulose bis-phosphate carboxylase (Rubisco). Many species of *Rhizobiaceae* (α-*Proteobacteria*) have multiple (two or three) chromosomes, and endogenous plasmids are thought to play a role in the formation of secondary chromosomes, since these often have plasmid-like replication regions (Cooper et al., 2010). For example, chromosome 2 of *R. sphaeroides* strains all have repB-I replication regions (Petersen, 2011). pRSPA02 might therefore be an 'embryonic' third chromosome, and this atypical strain may be of interest in studying genome evolution.

4.2. Other Species

The lack of multiple plasmid sequences from species other than *R. sphaeroides* means that few conclusions can be drawn about plasmid evolution in the anaerobic anoxygenic phototrophs. Phylogenetic analysis is needed to determine whether the chromosomal-type genes on many of these plasmids are derived by gene capture from the chromosome or HGT. The *R. capsulatus* plasmid pRCB133 contains sequences that are highly similar to sequences from the chromosome (Willison et al., 1987), and these presumably arose by gene duplication and transposition. The genome of *R. capsulatus* SB1003 contains numerous transposase genes, insertion sequences and phage-related genes (Haselkorn et al., 2001; Strnad et al., 2010). Nucleotide BLAST of pALVIN01 shows that this plasmid also contains islands of strong similarity to the *A. vinosum* DSM 180 chromosome (9% coverage, compared to 6% for pRCB133). Two of the duplicated regions in pRCB133, one of 2.7 kb containing the gene for an ArsR family transcription regulator, and the other, also of 2.7 kb, containing genes for iron transport, show more than 99% identity with the corresponding sequences on the chromosome, suggesting that the transposition events are fairly recent.

An interesting feature of the *R. rubrum* plasmid pKY1 is the presence of a resolvase/invertase gene (*rin*) flanked by two copies of a mutator-type transposase. The transposase genes are homologous at the nucleotide sequence level with genes from other Rhodospirillaceae, including *Tistrella mobilis* pTM1 and *Magnetospirillum* sp., whereas the *rin* gene shows >95% identity with sequences from *Delftia acidovorans* and *Acidovorax delafieldii*, which are β-*Proteobacteria*. In addition, the *R. rubrum repA* gene product is phylogenetically close to that of *Burkholderia cenocepacia* (Petersen et al., 2011). This suggests that at least part of the plasmid pKY1 may have originated in a β-proteobacterium.

5. CONCLUSIONS AND PERSPECTIVES

The availability of complete genome sequences enables predictions to be made about plasmid function that can be tested experimentally. If enough sequences are available for comparative study, then bioinformatic analysis can shed light on plasmid structure, function and evolution, as has been done with bacteria belonging to the *Roseobacter* clade. Among the anaerobic anoxygenic phototrophs, *R. sphaeroides* is the only species for which sufficient sequence information is available to enable such a study. Unfortunately, with the exception of the neotype strain 2.4.1, the strains that have been sequenced are different from those that have been characterised molecularly in terms of plasmid interrelationships (Fornari et al., 1984). This area would therefore benefit from a defined, plasmid-based sequencing project. The same applies to *R. capsulatus*. The endogenous plasmids have been identified and characterised in many strains, but only one strain has been sequenced.

The rationale for studying endogenous plasmids in the photosynthetic bacteria is twofold. Firstly, some of the plasmid-related phenomena described in this chapter are still unexplained, and their elucidation would shed light on aspects of the biology of this interesting group of microorganisms. Secondly, biotechnological applications have been proposed for various anaerobic anoxygenic phototrophs, in areas such as bio-hydrogen production (McKinlay & Harwood, 2010; Vignais, Magnin, & Willison, 2006), heavy metal bioremediation (Moore & Kaplan, 1992; Panwichian, Kantachote, Wittayaweerasak, & Mallavarapu, 2010) and metal nanoparticle production (Bai, Zhang, & Gong, 2006; Simmons et al., 2011). Strain improvement by genetic and metabolic engineering will require a sound knowledge of the structure and dynamics of the corresponding genomes. Although *R. sphaeroides* and *R. palustris* are being actively studied in this respect, we suggest that *R. capsulatus* B10, which has a relatively small genome and a single, stable, well-characterised extrachromosomal element, may be advantageous from this point of view.

ACKNOWLEDGEMENTS

We would like to thank Paulette Vignais, in whose laboratory our work on endogenous plasmids in *R. capsulatus* was initiated, for support and encouragement. The financial assistance of the Centre National de la Recherche Scientifique (CNRS) and the Commisariat à l'Energie Atomique (CEA) is gratefully acknowledged.

REFERENCES

Alef, K. (1987). The interaction between dimethylsulfoxide and nitrate reducing pathways in *Rhodobacter capsulatus*. *FEMS Microbiology Letters*, *48*, 11–14.

Aminov, R. I. (2011). Horizontal gene exchange in environmental microbiota. *Frontiers in Microbiology*10.3389/fmicb.2011.00158.

Bai, H.-J., Zhang, Z.-M., & Gong, J. (2006). Biological synthesis of semiconductor zinc sulfide nanoparticles by immobilized *Rhodobacter sphaeroides*. *Biotechnology Letters*, *28*, 1135–1139.

Berne, C., Allainmat, B., & Garcia, D. (2005). Tributyl phosphate degradation by *Rhodopseudomonas palustris* and other photosynthetic bacteria. *Biotechnology Letters, 27*, 561–566.

Brown, P. J.B., Kysela, D. T., Buechlein, A., Hemmerich, C., & Brun, Y. V. (2011). Genome sequences of eight morphologically diverse *Alphaproteobacteria*. *Journal of Bacteriology, 193*, 4567–4568.

Cabello, P., Pino, C., Olmo-Mira, M. F., Castillo, F., Roldán, M. D., & Moreno-Vivián, C. (2004). Hydroxylamine assimilation by *Rhodobacter capsulatus* E1F1. *The Journal of Biological Chemistry, 44*, 45485–45494.

Callister, S. J., Nicora, C. D., Zeng, X., et al. (2006). Comparison of aerobic and photosynthetic *Rhodobacter sphaeroides* 2.4.1 proteomes. *Journal of Microbiological Methods, 67*, 424–436.

Castillo, F., Dobao, M. M., Reyes, F., et al. (1996). Molecular and regulatory properties of the nitrate reducing systems of *Rhodobacter*. *Current Microbiology, 33*, 341–346.

Cevallos, M. A., Cervantes-Rivera, R., & Gutiérrez-Rios, R. M. (2008). The *repABC* plasmid family. *Plasmid, 60*, 19–37.

Choudhary, M., Zanhua, X., Fu, Y. X., & Kaplan, S. (2007). Genome analyses of three strains of *Rhodobacter sphaeroides*: evidence of rapid evolution of chromosome II. *Nucleic Acids Research, 27*, 61–62.

Cooper, V. S., Vohr, S. H., Wrocklage, S. C., & Hatcher, P. J. (2010). Why genes evolve faster on secondary chromosomes in bacteria. *PLoS Computational Biology, 6*(4), e1000732.

Czichos, J., & Klemme, J. -H. (1982). Isolation of mutants of *Rhodopseudomonas capsulata* with a defective nitrate assimilation system (nit$^-$) and demonstration of genetic transfer of nit gene(s). *FEMS Microbiology Letters, 14*, 15–19.

Eisen, J. A., Nelson, K. E., Paulsen, I. T., et al. (2002). The complete genome sequence of *Chlorobium tepidum*, a photosynthetic, anaerobic, green-sulfur bacterium. *Proceedings of the National Academy of Sciences of the United States of America, 99*, 9509–9514.

Ferguson, S. J., Jackson, J. B., & McEwan, A. G. (1987). Anaerobic respiration in the *Rhodospirillaceae*: characterization of pathways and evaluation of roles in redox balancing during photosynthesis. *FEMS Microbiology Reviews, 46*, 117–143.

Fornari, C. S., Watkins, M., & Kaplan, S. (1984). Plasmid distribution and analyses in *Rhodopseudomonas sphaeroides*. *Plasmid, 11*, 39–47.

Frigaard, N. -U., & Bryant, D. A. (2001). Chromosomal gene inactivation in the green sulfur bacterium *Chlorobium tepidum* by natural transformation. *Applied and Environmental Microbiology, 67*, 2538–2544.

Gibson, K. D., & Niederman, R. A. (1970). Characterization of two circular satellite species of deoxyribonucleic acid in *Rhodopseudomas spheroides*. *Archives of Biochemistry and Biophysics, 141*, 694–704.

Grass, G., Fan, B., Rosen, B., Franke, S., Nies, D. H., & Rensing, C. (2001). ZitB (YbgR), a member of the cation diffusion facilitator family, is an additional zinc transporter in *Escherichia coli*. *Journal of Bacteriology, 183*, 4664–4667.

Haselkorn, R., Lapidus, A., Kogan, Y., et al. (2001). The *Rhodobacter capsulatus* genome. *Photosynthesis Research, 70*, 43–52.

Hu, N. T., & Marrs, B. L. (1979). Characterization of the plasmid DNAs of *Rhodopseudomonas capsulata*. *Archives of Microbiology, 121*, 61–69.

Ideguchi, T., Hu, C., Kim, B.-H., Nishise, H., Yamashita, J., & Kakuno, T. (1993). An open reading frame in the *Rhodospirillum rubrum* plasmid, pKY1, similar to *algA*, encoding the bifunctional enzyme phosphomannose isomerase-guanosine diphospho-D-mannose pyrophosphorylase (PMI-GMP). *Biochimica et Biophysica Acta, 1172*, 329–331.

Ind, A. C., Porter, S. L., Brown, M. T., Byles, E. D., de Beyer, J. A., Godfrey, S. A., et al. (2009). *Applied and Environmental Microbiology, 75*, 6613–6615.

Inui, M., Nakata, K., Roh, J. H., Vertès, A. A., & Yukawa, H. (2003). Isolation and molecular characterization of pMG160, a mobilizable cryptic plasmid from *Rhodobacter blasticus*. *Applied and Environmental Microbiology, 69*, 725–733.

Inui, M., Roh, J. H., Zahn, K., & Yukawa, H. (2000). Sequence analysis of the cryptic plasmid pMG101 from *Rhodopseudomonas palustris* and construction of stable cloning vectors. *Applied and Environmental Microbiology, 66*, 54–63.

Jumas-Bilak, E., Michaux-Charachon, S., Bourg, G., Ramuz, M., & Allardet-Servent, A. (1998). Unconventional genomic organization in the alpha subgroup of the *Proteobacteria. Journal of Bacteriology, 180*, 2749–2755.

Kalhoefer, D., Thole, S., Voget, S., Lehmann, R., Liesegang, H., Wolher, A., et al. (2011). Comparative genome analysis and genome-guided physiological analysis of *Roseobacter litoralis. BMC Genomics, 12*, 324–339.

Kawamukai, M., Matsuzaki, S., Omura, H., Takata, A., Nakagawa, T., & Matsuda, H. (1990). Physical map of the cryptic 55 kilobase plasmid from the photosynthetic bacterium *Rhodospirillum rubrum. Agricultural and Biological Chemistry, 54*, 1317–1318.

Kim, J. K., Lee, B.-K., Kim, S.-H., & Moon, J.-H. (1999). Characterization of denitrifying photosynthetic bacteria isolated from photosynthetic sludge. *Aquacultural Engineering, 19*, 179–193.

Kobayashi, H., & Akazawa, T. (1983). Biosynthetic mechanism of ribulose-1,5-bisphosphate carboxylase in the purple photosynthetic bacterium, *Chromatium vinosum. Archives of Biochemistry and Biophysics, 224*, 152–160.

Koch, H.-G., & Klemme, J.-H. (1994). Localization of nitrate reductase genes in a 115-kb plasmid of *Rhodobacter capsulatus* and restoration of NIT^+ character in nitrate reductase negative mutant or wild-type strains by conjugative transfer of the endogenous plasmid. *FEMS Microbiology Letters, 118*, 193–198.

Kuhl, S. A., Nix, D. W., & Yoch, D. C. (1983). Characterization of a *Rhodospirillum rubrum* plasmid: loss of photosynthetic growth in plasmidless strains. *Journal of Bacteriology, 156*, 737–742.

Kuhl, S. A., Wimer, L. T., & Yoch, D. C. (1984). Plasmidless, photosynthetically incompetent mutants of *Rhodospirillum rubrum. Journal of Bacteriology, 159*, 913–918.

Larimer, F. W., Chain, P., Hauser, L., et al. (2004). Complete genome sequence of the metabolically versatile photosynthetic bacterium *Rhodopseudomonas palustris. Nature Biotechnology, 22*, 55–61.

Lim, S.-K., Kim, S. J., Cha, S. H., Oh, Y.-K., Rhee, H.-J., Kim, M.-S., et al. (2009). Complete genome sequence of *Rhodobacter sphaeroides* KD131. *Journal of Bacteriology, 191*, 1118–1119.

Lonjers, Z. T., Dickson, E. L., Chu, T. T., Kreutz, J. E., Neacsu, F. A., Anders, K. R., et al. (2012). Identification of a new gene required for the biosynthesis of rhodoquinone in *Rhodosprillum rubrum. Journal of Bacteriology*10.1128/JB.06319-11.

Lu, Y.-K., Marden, J., Han, M., et al. (2010). Metabolic flexibility revealed in the genome of the cyst-forming α-1 proteobacterium *Rhodospirillum centenum. BMC Genomics, 11*, 325.

McEwan, A. G., Greenfield, A. J., Wetzstein, H. G., Jackson, J. B., & Ferguson, S. J. (1985). Nitrous oxide reduction by members of the family *Rhodospirillaceae* and the nitrous oxide reductase of *Rhodopseudomonas capsulata. Journal of Bacteriology, 164*, 823–830.

Mackenzie, C., Choudhary, M., Larimer, F. W., et al. (2001). The home stretch, a first analysis of the nearly completed genome of *Rhodobacter sphaeroides* 2.4.1. *Photosynthesis Research, 70*, 19–41.

McKinlay, J. B., & Harwood, C. S. (2010). Photobiological production of hydrogen gas as a biofuel. *Current Opinion in Biotechnology, 21*, 244–251.

Magnin, J.-P., Willison, J. C., & Vignais, P. M. (1987). Elimination of R plasmids from the photosynthetic bacterium *Rhodobacter capsulatus. FEMS Microbiology Letters, 41*, 157–161.

Matsunaga, T., Matsunaga, N., Tsubaki, K., & Tanaka, T. (1986). Development of a gene cloning system for the hydrogen-producing marine photosynthetic bacterium *Rhodopseudomonas* sp. *Journal of Bacteriology, 168*, 460–463.

Matsunaga, T., Tsubaki, K., Miyashita, H., & Burgess, J. G., (1990). Chloramphenicol acetyltransferase expression in marine *Rhodobacter* sp. NKPB 0021 by use of shuttle vectors containing the minimal replicon of an endogenous plasmid. *Plasmid, 24*, 90–99.

Matsunaga, T., Mihashita, H., Miyake, M., & Burgess, J. G. (1991). Nucleotide sequence of the replication region of the marine *Rhodobacter* plasmid pRD31. *FEBS Letters, 283*, 263–266.

Méndez-Alvarez, S., Pavón, V., Esteve, I., Guerrero, R., & Gaju, N. (1994). Transformation of *Chlorobium limicola* by a plasmid that confers the ability to utilize thiosulfate. *Journal of Bacteriology, 176*, 7395–7397.

Méndez-Alvarez, S., Pavón, V., Esteve, I., Guerrero, R., & Gaju, N. (1995). Genomic heterogeneity in *Chlorobium limicola*: chromosomic and plasmidic differences among strains. *FEMS Microbiology Letters, 134*, 279–285.

Michalski, W. P., & Nicholas, D. J.D. (1988). Identification of two new denitrifying strains of *Rhodobacter sphaeroides*. *FEMS Microbiology Letters, 52*, 239–244.

Moore, M. D., & Kaplan, S. (1992). Identification of intrinsic high-level resistance to rare-earth oxides and oxyanions in members of the class *Proteobacteria*: characterization of tellurite, selenite and rhodium sesquioxide reduction in *Rhodobacter sphaeroides*. *Journal of Bacteriology, 174*, 1505–1514.

Mouné, S., Manac'h, N., Hirschler, A., Caumette, P., Willison, J. C., & Matheron, R. (1999). *Haloanaerobacter salinarius* sp. nov., a novel halophilic fermentative bacterium that reduces glycine-betaine to trimethylamine with hydrogen or serine as electron donors; emendation of the genus *Haloanaerobacter*. *International Journal of Systematic Bacteriology, 49*, 103–112.

Munk, A. C., Copeland, A., Lucas, S., et al. (2011). Complete genome sequence of *Rhodospirillum rubrum* type strain (S1T). *Standards in Genomic Sciences, 4*, 293–302.

Nano, F. E., & Kaplan, S. (1984). Plasmid rearrangements in the photosynthetic bacterium *Rhodopseudomonas sphaeroides*. *Journal of Bacteriology, 158*, 1094–1103.

Oda, Y., Larimer, F. W., Chain, P. S.G., et al. (2008). Multiple genome sequences reveal adaptations of a phototrophic bacterium to sediment microenvironments. *Proceedings of the National Academy of Sciences of the United States of America, 105*, 18543–18548.

Panwichian, S., Kantachote, D., Wittayaweerasak, B., & Mallavarapu, M. (2010). Isolation of purple nonsulfur bacteria for the removal of heavy metals and sodium from contaminated shrimp ponds. *Electronic Journal of Biotechnology*, 13. doi:10.2225/vol13-issue4-fulltext-8.

Pappas, C. T., Sram, J., Moskvin, O. V., et al. (2004). Construction and validation of the *Rhodobacter sphaeroides* 2.4.1 DNA microarray: transcriptome flexibility at diverse growth modes. *Journal of Bacteriology, 186*, 4748–4758.

Pellerin, N. B., & Gest, H. (1983). Diagnostic features of the photosynthetic bacterium *Rhodopseudomonas sphaeroides*. *Current Microbiology, 9*, 339–344.

Pemberton, J. M., Cooke, S., & Bowen, A. R. St. G. (1983). Gene transfer mechanisms among members of the genus *Rhodopseudomonas*. *Annales de l'Institut Pasteur Microbiology, 134B*, 195–204.

Petersen, J. (2011). Phylogeny and compatibility: plasmid classification in the genomics era. *Archives of Microbiology, 193*, 313–321.

Petersen, J., Brinkmann, H., Berger, M., Brinkhoff, T., Päuker, O., & Pradella, S. (2011). Origin and evolution of a novel DnaA-like plasmid replication type in *Rhodobacterales*. *Molecular Biology and Evolution, 28*, 1229–1240.

Petersen, J., Brinkmann, H., Bunk, B., Michael, V., Päuker, O., & Pradella, S. (2012). Think pink: photosynthesis, plasmids and the *Roseobacter* clade. *Environmental Microbiology*. doi:10.1111/j.1462-2920.2012.02806.x.

Petersen, J., Brinkmann, H., & Pradella, S. (2009). Diversity and evolution of repABC type plasmids in *Rhodobacterales*. *Environmental Microbiology, 11*, 2627–2638.

Pradella, S., Allgaier, M., Hoch, C., Päuker, O., Stackebrandt, E., & Wagner-Döbler, I. (2004). Genome organization and localization of the *pufLM* genes of the photosynthesis reaction center in phylogenetically diverse marine *Alphaproteobacteria*. *Applied and Environmental Microbiology, 70*, 3360–3369.

Pradella, S., Päuker, O., & Petersen, J. (2010). Genome organization of the marine *Roseobacter clade* member *Marinovum algicola*. *Archives of Microbiology, 192*, 115–126.

Reyes, F., Roldán, M. D., Klipp, W., Castillo, F., & Moreno-Vivian, C. (1996). Isolation of periplasmic nitrate reductase genes from *Rhodobacter sphaeroides* DSM 158: structural and functional differences among prokaryotic nitrate reductases. *Molecular Microbiology, 19*, 1307–1318.

Ribeiro, F., Przyblski, D., Yin, S., et al. (2012). Finished bacterial genomes from shotgun sequence data. *Genome Research*. doi:10.1101/gr.141515.212.

Richardson, D. J., Bell, L. C., Moir, J. W.B., & Ferguson, S. J. (1994). A denitrifying strain of *Rhodobacter capsulatus*. *FEMS Microbiology Letters, 120*, 323–328.

Richardson, D. J., King, G. F., Kelly, D. J., McEwan, A. G., Ferguson, S. J., & Jackson, J. B. (1988). The role of auxiliary oxidants in maintaining redox balance during phototrophic growth of *Rhodobacter capsulatus* on propionate or butyrate. *Archives of Microbiology, 150*, 131–137.

Satoh, T., Hoshino, Y., & Kitamura, H. (1976). *Rhodopseudomonas sphaeroides* forma sp. denitrificans, a denitrifying strain as a subspecies of *Rhodopseudomonas sphaeroides*. *Archives of Microbiology, 108*, 265–269.

Saunders, V. A. (1978). Genetics of *Rhodospirillaceae*. *Microbiological Reviews, 42*, 357–384.

Saunders, V. A., Saunders, J. R., & Bennett, P. M. (1976). Extrachromosomal deoxyribonucleic acid in wild-type and photosynthetically incompetent strains of *Rhodopseudomonas sphaeroides*. *Journal of Bacteriology, 125*, 1180–1187.

Schwintner, C., Sabaty, M., Berna, B., Cahors, S., & Richaud, P. (1998). Plasmid content and localization of the genes encoding the denitrification enzymes in two strains of *Rhodobacter sphaeroides*. *FEMS Microbiology Letters, 165*, 313–321.

Silver, S., & Phung, L. T. (1996). Bacterial heavy metal resistance: new surprises. *Annual Review of Microbiology, 50*, 753–789.

Simmons, S. S., Isokpehi, R. D., Brown, S. D., et al. (2011). Functional annotation analytics of *Rhodopseudomonas palustris* genomes. *Bioinformatics and Biology Insights, 5*, 115–129.

Strnad, H., Lapidus, A., Paces, J., Ulbrich, P., Vlcek, C., Paces, V., et al. (2010). Complete genome sequence of the photosynthetic purple nonsulfur bacterium *Rhodobacter capsulatus* SB 1003. *Journal of Bacteriology, 192*, 3545–3546.

Suwanto, A., & Kaplan, S. (1989). Physical and genetic mapping of the *Rhodobacter sphaeroides* 2.4.1 genome: genome size, fragment identification, and gene localization. *Journal of Bacteriology, 171*, 5840–5849.

Suwanto, A., & Kaplan, S. (1992). A self-transmissible, narrow-host-range endogenous plasmid of *Rhodobacter sphaeroides* 2.4.1: physical structure, incompatibility determinants, origin of replication, and transfer functions. *Journal of Bacteriology, 174*, 1124–1134.

Suyama, Y., & Gibson, J. (1966). Satellite DNA in photosynthetic bacteria. *Biochemical and Biophysical Research Communications, 24*, 549–553.

Suzuki, H., Yano, H., Brown, C. J., & Top, E. M. (2010). Predicting plasmid promiscuity based on genomic signature. *Journal of Bacteriology, 192*, 6045–6055.

Tavano, C. L., Podevels, A. M., & Donohue, T. J. (2005). Identification of genes required for recycling reducing power during photosynthetic growth. *Journal of Bacteriology, 187*, 5249–5258.

Tefsen, B., Bos, M. P., Beckers, F., Tommassen, J., & de Cock, H. (2005). MsbA is not required for phospholipid transport in *Neisseria meningitidis*. *The Journal of Biological Chemistry, 280*, 35961–35966.

Thompson, S. A., Shedd, O. L., Ray, K. C., Beins, M. H., Jorgensen, J. P., & Blaser, M. J. (1998). *Campylobacter fetus* surface layer proteins are transported by a type I secretion system. *Journal of Bacteriology, 180*, 6450–6458.

Tosques, I. E., Shi, J., & Shapleigh, J. P. (1996). Cloning and characterization of *nnrR*, whose product is required for the expression of proteins involved in nitric oxide metabolism in *Rhodobacter sphaeroides* 2.4.3. *Journal of Bacteriology, 178*, 4958–4964.

Treangen, T. J., & Rocha, E. P.C. (2010). Horizontal transfer, not duplication, drives the expansion of protein families in prokaryotes. *PLoS Genetics*, 7, e1001284.

Tucker, W. T., & Pemberton, J. M. (1978). Viral R plasmid Rϕ6P: properties of the penicillinase plasmid prophage and the supercoiled, circular encapsidated genome. *Journal of Bacteriology*, *135*, 207–214.

Vargas, C., Fernandez-Castillo, R., Canovas, D., Ventosa, A., & Nieto, J. J. (1995). Isolation of cryptic plasmids from moderately halophilic eubacteria of the genus *Halomonas*. Characterization of a small plasmid from *H. elongata* and its use for shuttle vector construction. *Molecular and General Genetics*, *246*, 411–418.

Vignais, P. M., Magnin, J.-P., & Willison, J. C. (2006). Increasing biohydrogen production by metabolic engineering. *International Journal of Hydrogen Energy*, *31*, 1478–1483.

Villaseñor, T., Brom, S., Dávalos, A., Lozano, L., Romero, D., & García-de los Santos, A. (2011). Housekeeping genes essential for pantothenate biosynthesis are plasmid-encoded in *Rhizobium etli* and *Rhizobium leguminosarum*. *BMC Microbiology*, *11*, 66.

Wahlund, T. M., & Madigan, M. T. (1995). Genetic transfer by conjugation in the thermophilic green sulfur bacterium *Chlorobium tepidum*. *Journal of Bacteriology*, *177*, 2583–2588.

Weaver, P. F., Wall, J. D., & Gest, H. (1975). Characterization of *Rhodopseudomonas capsulata*. *Archives of Microbiology*, *105*, 207–216.

Willison, J. C. (1990). Derivatives of *Rhodobacter capsulatus* strain AD2 cured of their endogenous plasmid are unable to utilize nitrate. *FEMS Microbiology Letters*, *66*, 23–28.

Willison, J. C., Magnin, J. P., & Vignais, P. M. (1987). Isolation and characterization of *Rhodobacter capsulatus* strains lacking endogenous plasmids. *Archives of Microbiology*, *147*, 134–142.

Witt, A., & Klemme, J.-H. (1991). No correlation between plasmid content and ability to reduce nitrate in wild-type strains of *Rhodobacter capsulatus*. *Zeitschrift für Naturforschung*, *46c*, 703–705.

Zhou, Z., White, K. A., Polissi, A., Georgopoulos, C., & Raetz, C. R.H. (1998). Function of *Escherichia coli* MsbA, an essential ABC family transporter, in lipid A and phospholipid biosynthesis. *The Journal of Biological Chemistry*, *273*, 12466–12475.

CHAPTER NINE

Evolution of Bacteriophytochromes in Photosynthetic Bacteria

Miroslav Papiz[1], Dom Bellini
Institute of Integrative Biology, University of Liverpool, Liverpool, UK
[1]Corresponding author: E-mail: mpapiz@liverpool.ac.uk

Contents

Abstract

Phytochromes are photoreceptors that respond to environmental light conditions and control a variety of photomorphogenic responses. Phytochromes contain three key elements: an N-terminal chromophore-binding domain (CBD), a 'middle' signal-transducing phytochrome-associated (PHY) domain and a C-terminal output-transducing domain (OTD). The light sensing chromophore, a linear tetrapyrrole, reversibly photoconverts between the Pr (red) and Pfr (far-red) absorbing states by isomerisation of the chromophore D-ring causing a light signal to be transferred through the PHY domain and into the OTD. This alters interactions, between the phytochrome OTD and a transcriptional response regulator, which results in differential expression

Advances in Botanical Research, Volume 66
ISSN 0065-2296, http://dx.doi.org/10.1016/B978-0-12-397923-0.00009-6

of target genes. Bacteriophytochrome photoreceptors (BphPs) are bacterial homologues and over 50 have been found in purple bacteria. While CBD and PHY domains are well conserved, several OTDs have been identified indicating that BphPs have evolved a range of functions. Some purple bacteria, such as *Rhodopseudomonas palustris*, have several BphPs whereas others have none. It is likely that all phytochromes have evolved from BphPs and this raises some important questions such as, why do all cyanobacteria and plants possess phytochromes but only some purple bacteria, and which BphPs have been important for the evolution of phytochromes in higher organisms.

1. INTRODUCTION

Bacteriophytochromes (BphPs) are bacterial photoreceptors belonging to the phytochrome family which control a variety of light-stimulated responses (Bhoo, Davis, Walker, Karniol, & Vierstra, 2001; Davis, Vener, & Vierstra, 1999; Giraud et al., 2002; Hughes et al., 1997; Lamparter, Michael, Mittmann, & Esteban, 2002). Initially phytochrome photoreceptors (PhyPs) were found in plants (Quail, 2002) but now they have been found in fungi (FphPs), cyanobacteria (CphPs) and bacteria (BphPs). They are part of a two component signalling system interacting with a response regulator that mediates differential expression of target genes. They utilize a photoactive chromophore covalently linked to a cysteine residue (Lagarias & Lagarias, 1989). The chromophore is a linear tetrapyrrole, which is linearized by haem oxidase, and in BphPs and FphPs it is the molecule biliverdin IXα. In cyanobacteria and plants additional enzymes are required to convert biliverdin into phycocyanobilin and phytochromobilin, respectively (Rockwell, Su, & Lagarias, 2006). The lack of enzymes such as bilin reductases strongly suggests that BphPs evolved from an ancestral bilin photoreceptor in bacteria (Montgomery & Lagarias, 2002).

BphPs can exist in either a Pr or Pfr dark stable form, the latter is sometimes referred to as bathy-BphP (Giraud & Vermeglio, 2008; Rottwinkel et al., 2010). Although the common photoconversion pathway is between the red Pr and far-red Pfr states, other light-induced states are known such as the near red-absorbing Pnr (Evans et al., 2005; Giraud et al., 2005) and the orange-absorbing Po (Chen et al., 2012; Jaubert et al., 2007). The N-terminal photosensory core domain (PCD) is a conserved structure within BphPs and is composed of a Per/Arnt/Sim (PAS) followed by the cyclic di-GMP phosophodiesterase/adenyl cyclase/Fhla (GAF) and the phytochrome-associated (PHY) domains. Because the chromophore is covalently bound to a Cys residue in the PAS domain and buried in a pocket within the GAF domain the combined PAS–GAF domains are also referred to as the chromophore-binding domain (CBD). Various C-terminal output domains (OTD) have evolved to interact with different response regulators. The archetypal BphP OTD is related to the cytoplasmic histidine kinase (HK) domain but several other types have been observed in genomic studies and by biochemical characterisation. Although BphPs are found in both photosynthetic and non-photosynthetic bacteria they probably evolved originally to control photosynthesis. In the general context of photosynthetic processes, there is overwhelming evidence that plant photosystems have evolved from cyanobacteria, however the first photosynthetic organisms are likely to have been anoxygenic bacteria (Blankenship, 2010), and BphPs are therefore likely to be the progenitors of all phytochromes.

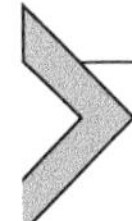

2. VARIATION IN THE BACTERIOPHYTOCHROME OUTPUT DOMAINS

Genomic analysis has revealed more than 50 BphPs in purple bacteria with at least eight different OTDs (Fig. 9.1). These are distributed amongst 14 species and *Rhodopseudomonas palustris* strains represent around 50% of BphPs. Each strain of *Rps. palustris* has several BphPs, for example CGA009 has six BphPs that rival the number of phytochromes in the plant model genome *Arabidopsis thaliana*. The phylogenetic diagram (Fig. 9.2) is over-represented with BphPs from *Rps. palustris* due to a concerted sequencing effort of seven different bacterial strains funded by the US DOE Joint Genome Institute. However there is no reason to believe that BphP types and frequency have been biased by this. BphPs with different OTDs have representatives in several species and so the large variety of different OTDs are not isolated to *Rps. palustris*.

2.1. Histidine Kinase Output Domain

All BphPs contain the CBD plus PHY domains, which make the PCD, and the archetypal BphPs are followed by an OTD that is a cytoplasmic histidine kinase (HK) domain (Bhoo et al., 2001; Davis et al., 1999) (Fig. 9.1A). The HK domain is formed from a dimerisation domain (Dhp) and the ATP-dependant kinase domain (KD). The Dhp has a histidine site for autophosphorylation by the KD. This BphP participates in phosphotransferase activity to an aspartate residue in a response regulator, in so doing converting a light signal into a classical phosphorelay signalling mechanism. Of all the BphP types it has been the most studied, and X-ray structures have been determined for the CBD and the CBD–PHY domains

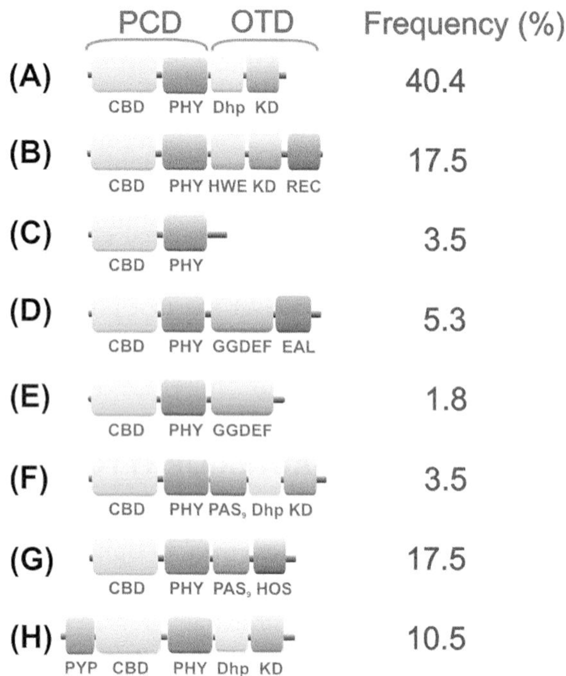

Figure 9.1 ***The domain organization of BphPs in purple bacteria can be classified into eight types (A–H).*** Their primary structure can be divided into the N-terminal photosensory core domain (PCD) and the output-transducing domain (OTD). The PCD is composed of the chromophore-binding domain (CBD) which contains the chromaphore biliverdin IXα and the phytochrome-associated (PHY) domains. The CBD is composed of the Per/Arnt/Sim (PAS) and the cyclic di-GMP phosphodiesterase/adenyl cyclase/Fhla (GAF) domains. The PCD is conserved in all BphPs but type H BphPs also have a blue sensing photoactive yellow protein domain (PYP) at the N-terminal that utilizes the 4-hydroxycinnamoyl chromophore. The histidine kinase OTD domain is formed from the dimerisation (Dhp) and ATP-dependent kinase (KD) domains. A variant of this is the B type HWE histidine kinase which, based on sequence homology, is also composed of Dhp and KD. A response receiver (REC) domain is found at the C-terminal end of the HWE domain. Type D OTDs are composed of a GGDEF and an EAL domain, which respectively synthesise and degrade the second messenger cyclic di-GMP. Type E do not possess the EAL domain, thus lack phosphodiesterase activity. Type F BphPs contain a PAS/PAC or PAS-9 domain between PHY and Dhp domains. Type G BphPs have an OTD made from PAS-9 and HOS domains, the latter is distantly related to Dhp domains in sequence and is functionally distinct. Type C BphPs have truncated OTD domains and it is not known if these are active. Frequency values represent the percentage of different BphP types occurring in purple bacteria strains listed in Fig. 9.2. (For interpretation of the references to colour in the figure legend, the reader is referred to the online version of this book.)

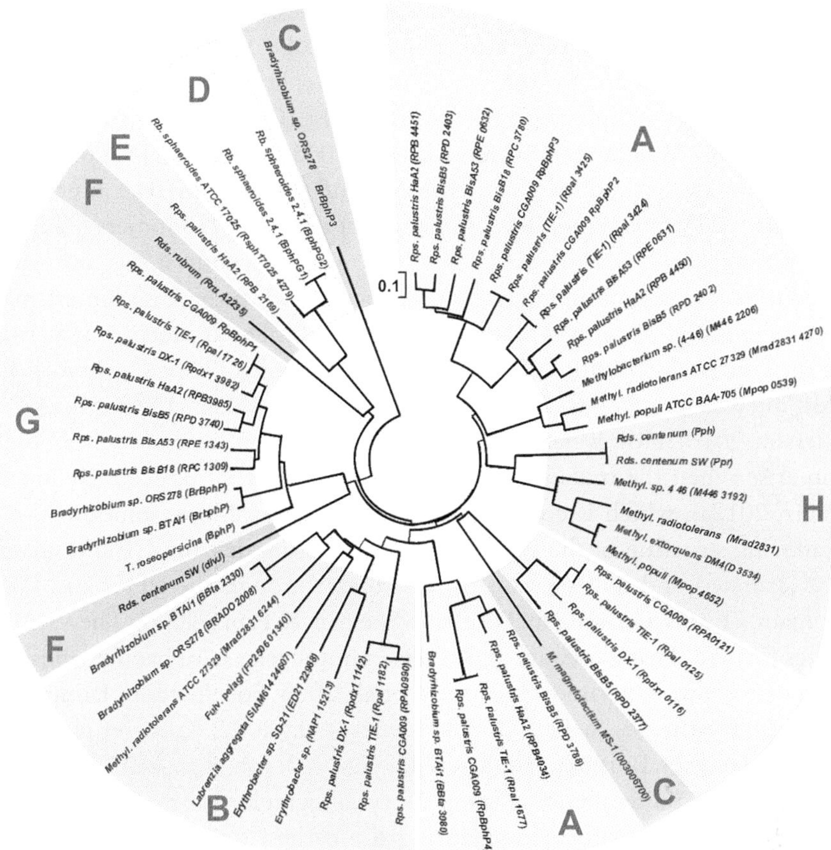

Figure 9.2 ***The phylogenic tree based on 55 PCD amino acid sequences of BphPs in purple bacteria.*** The evolutionary history was inferred with the neighbour-joining method using the bootstrap test (1000 replicates). The tree is drawn to scale, with branch lengths in the same units as those of the evolutionary distances used to infer the phylogenetic tree. The analyses were conducted in MEGA5 (Tamura et al., 2011). Bacterial species are *Rhodopseudomonas palustris (Rps. palustris), Rhodobacter sphaeroides (Rb. sphaerodies), Rhodospirillum centenum (Rsp. centenum), Bradyrhizobium* sp., *Thiocapsa roseopersicina (T. roseopersicina), Methylobacterium (Methyl.), Magnetospirillum magnetotacticum (M. Megnetotacticum), Rhodospirillum rubrum (Rsp. rubrum), Fulvimarina pelagi (Fulv. pelagi), Labrenzia aggregate* and *Erythrobacter* sp. The different types of BphPs based on domain architectures are labelled A–H as in Fig. 9.1. (For colour version of this figure, the reader is referred to the online version of this book.)

(Bellini & Papiz, 2012a; Essen, Mailliet, & Hughes, 2008; Wagner, Brunzelle, Forest, & Vierstra, 2005; Yang, Kuk, & Moffat, 2008, Yang, Stojkovic, Kuk, & Moffat, 2007) but not for the whole molecule; however, crystal structures of homologous cytoplasmic HK domains exist (Casino, Rubio, & Marina, 2009; Marina, Waldburger, & Hendrickson, 2005; Yamada et al., 2009). Phytochromes are homodimers and type A BphP dimers may have functional importance for *trans* autophosphorylation of opposing monomers (Aravind & Ponting, 1999; Marina et al., 2005); however recently it has also been shown that *cis* autophosphorylation can occur and in this case dimers may not be required (Casino, Rubio, & Marina, 2010). The quaternary structure of the complete *Dr*BphP from *Deinococcus radiodurans*, obtained by cryo electron microscopy, is a dimer interacting along its whole length (Li, Zhang, & Vierstra, 2010). However, the CBD of *Rp*BphP2 from *Rps. palustris* can only dimerise when the mutation Asn 136 to Arg 136 is introduced (Bellini & Papiz, 2012a), which is consistent with the low resolution solution X-ray scattering structure of the whole dimer, suggesting that the CBD domains are separated and dimerisation occurs solely through the PHY and Dhp domains (Evans, Grossmann, Fordham-Skelton, & Papiz, 2006). The various states formed during forward and reverse photocycles have been established in spectroscopy experiments (Borucki et al., 2005; Foerstendorf, Lamparter, Hughes, Gartner, & Siebert, 2000; Heyne et al., 2002; Kneip et al., 1999; Otto, Lamparter, Borucki, Hughes, & Heyn, 2003; Toh et al., 2011; Van Thor et al., 2001), and temperature trapped crystallography has revealed changes of chromophore conformation during the earliest stages of isomerisation (Yang, Ren, Kuk, & Moffat, 2011). Therefore, the photocycle is well characterised in this A type BphP but as the PCD is conserved in BphPs, the initial light mechanism maybe very similar in all types of BphP.

2.2. HWE Histidine Kinase Output Domain

A variant of the HK OTD domain is the HWE histidine kinase (Fig. 9.1B) that is characterised by an absence of a distinct F box and the presence of several conserved residues, including a histidine in the N box and a tryptophan-X-glutamic acid sequence in the G1 box (Karniol & Vierstra, 2004). Although a crystal structure of HWE has not been determined, sequence comparisons suggest that, like HK, they can be divided into Dhp and KD domains. In purple bacteria, all HWE type B BphPs have a CheY-like receiver domain (REC) following the KD domain. Sequence homology databases, such as SMART and Pfam, identify HWE sequence signatures in the Dhp domain, as shown in Fig. 9.1, but it should be noted that the kinase domain is only distantly similar to KD in HK and it would be more accurate to extend the HWE assignment to both Dhp and KD. It has been demonstrated that *At*BphP2, from *Agrobacterium tumefaciens*, a type B BphP, has autophosphorylation activity and so is likely to take part in phosphorelay transfer (Karniol & Vierstra, 2003). Presumably it phosphotransfers to its internal REC domain. *At*BphP2 is a bathy-BphP showing maximal kinase activity in its Pfr dark stable state, whereas *At*BphP1 has optimal kinase activity in its Pr state and phosphotransfers specifically to a neighbouring RR in the genome (Karniol & Vierstra, 2003). It is interesting to note that the domain organization with a C-terminal REC domain is the canonical form in fungal phytochromes (Rockwell et al., 2006).

2.3. The N-terminal PYP and C-terminal HK Output Domain

Another variant of a HK OTD domain is type H (Fig. 9.1H) which has a blue light receptor, the photoactive yellow protein (PYP) domain, at the N-terminal preceding the CBD. PYP is a PAS-like domain with the chromophore *p*-coumaric acid covalently linked to a Cys residue. A functional response to blue light photostimulation has been observed for Ppr in *Rhodospirillum centenum* that elicits phototactic motility in the bacterium (Jiang et al., 1999; Ragatz, Jiang, Bauer, & Gest, 1995). Type H BphPs have been observed in the genomes of a number of *Methylobacterium* sp. and in the sulphur purple bacterium *Thermochromatium tepidum* (Kyndt, Fitch, Meyer, & Cusanovich, 2005). The crystal X-ray structure of the PYP domain from Ppr has been determined (Rajagopal & Moffat, 2003), and time-resolved crystallography has shown global changes of the domain on illumination with blue light on the timescales of 10 ns–100 ms (Ren et al., 2001). Time-resolved spectroscopy has also been employed to determine the photocycle of PYP from *Ectothiorhodospira halophila* (Ng, Getzoff, & Moffat, 1995).

2.4. The PAS-HK Output Domain

A third HK OTD variant contains a PAS domain inserted between the PHY and Dhp domains (Fig. 9.1F). Although not common in purple bacteria, examples are the A2235 gene in *Rhodospirillum rubrum* and *divJ* in *Rsp. centenum* SW. On closer examination these PAS domains belong to one of the PAS subfamilies reported as PAS/PAC by the SMART database or PAS-9 by the Pfam database. The PAC domain was originally defined as a C-terminal sequence extension to the PAS domain, but it is now understood to be a PAS structural element rather than a sequence signature (Ponting & Aravind, 1997; Taylor & Zhulin, 1999). The structure of the PAS–HK domain from *Thermotoga maritime*, belonging to a HK, has been determined with a RR attached showing that the PAS domain interacts with the catalytic domain while the Dhp to RR and that this is suitable for intermolecular phosphotransfer (Yamada et al., 2009).

2.5. The GGDEF–EAL Output Domain

The BphPs of *Rhodobacter sphaeroides* are encoded within two plasmids. The OTDs of these BphPs are composed of two domains, the GGDEF (a diguanylate cyclase) that synthesises cyclic di-GMP, and the EAL (a phosphodiesterase) that degrades cyclic di-GMP. A construct lacking the EAL domain exhibited light-dependant diguanylate cyclase activity, optimal in the Pfr state, while the full length construct was locked in an inactive state. It seems that the EAL domain regulates activity between an inactive and active state (Tarutina, Ryjenkov, & Gomelsky, 2006). An example of a similar BphP is found in *Rps. palustris* strain HaA2 which has only the GGDEF domain, and it would be predicted that it is always active and turns over cyclic di-GMP in a light dependant manner. The second messenger cyclic di-GMP has recently emerged as a key regulator in a number of bacterial processes, including swarming behaviours (Williamson, Fineran, Ogawa, Woodley, & Salmond, 2008), thus suggesting the possibility that in purple bacteria this small molecule could play a role in shade avoidance in a way that is analogous to plants. Crystal structures of GGDEF and EAL domains also exist (Navarro, De, Bae, Wang, & Sondermann, 2009; Yang, Chin, & et al., 2011).

2.6. The PAS/PAC–HOS Output Domain

An unusual OTD has been identified amongst the BphPs of *Bradyrhizobium* ORS278 and *Rps. palustris*. These control the expression of a large number of photosynthesis genes on far-red light stimulation (Evans et al., 2005; Giraud et al., 2002; Giraud & Vermeglio, 2008). This G type BphP has an OTD composed of a PAS/PAC domain followed by an undefined 90 amino acid C-terminal segment. Recently it was suggested that the 90 amino acid segment, in *Rp*BphP1 of *Rps. palustris*, is distantly related to a Dhp domain with a sequence identity of ~10%. The domain has lost the His residue associated with autophosphorylation and does not possess a KD. Instead it has been shown to directly associate with a cognate repressor *Rp*PpsR2 (Bellini & Papiz, 2012b). Because the domain has a low sequence identity and is functionally distinct from Dhps, it has been named the 2-helix output sensor or HOS domain. Canonical BphPs, as observed by low-resolution cryo-EM, form dimer interfaces at Dhp domains which associate to make 4 helix bundles (Li et al., 2010). The crystal structure of the CBD–PHY–PAS/PAC fragment of *Rp*BphP1 forms antiparallel dimers. This arrangement does not allow HOS domains to interact with one another. Instead the HOS domains interact with a CBD–PHY domain on neighbouring monomers (Bellini & Papiz, 2012b). This BphP does not function by phosphorelay or second messenger cyclic di-GMP signalling, but instead directly inhibits the gene repressor *Rp*PpsR2 on illumination with far-red light.

2.7. Phytochrome-like Sequences

Sequence analysis of two BphPs in *Bradyrhizobium* ORS278 (*Br*BphP3) and *Magnetospirillum magnetotacticum* indicates that they are formed from only CBD–PHY domains (Fig. 9.1C). It has not been shown if these BphPs are active or whether they are inactive phytochrome-like fragments. Although their C-terminal segments, following the PHY domain, are short and composed of only 47 and 4 residues respectively they may yet be found to have functional roles. Recently it has been shown that the HOS domain, in *Rp*BphP1, is necessary for binding to the repressor *Rp*PpsR2 and so it maybe possible that these other BphPs also function in an analogous way. *Br*BphP3 is also unusual because it binds phycocyanobilin rather than biliverdin IXα and it has been proposed that it has been acquired by lateral gene transfer from a cyanobacterial species (Jaubert et al., 2007).

3. DISTRIBUTION OF THE DIFFERENT TYPES OF BACTERIOPHYTOCHROMES

BphPs are found in a number of photosynthetic bacteria such as *Rb. sphaeroides*, *Rps. palustris*, *Bradyrhizobium* sp. and *Methylobacterium* sp., but a notable absentee from the list is the model purple photosynthetic bacterium *Rhodobacter capsulatus*. This bacterium uses a blue light sensing photoreceptor protein AppA to control photosynthesis rather than red/far-red sensing BphPs (Masuda & Bauer, 2002). A phylogenetic tree, calculated using PCD sequences, reveals a split in the BphP population between the HK/HWE and those such as the PAS/PAC–HOS and GGDEF–EAL

(Fig. 9.2). Exceptions to this are the PAS–HK containing type F BphPs DivJ, from *Rsp. centenum*, and A2235 from *Rsp. rubrum*, which cluster around type G BphPs in the tree. Both type F and G BphPs have PAS domains, following PHY, which belong to the PAS/PAC or PAS-9 variant. This combined with tree proximity points to a close relationship between the two. It is reasonable to propose that G type BphPs evolved from type F by the loss of KD, followed by evolution of Dhp into the HOS domain. The PCD domains of G BphPs have a closer relationship with plant PhyPs than any other BphPs, including the most common A or B type BphPs (Montgomery & Lagarias, 2002), and this may explain why canonical plant PhyPs have two PAS domains following their PHY domain. Similarities between photosynthetic apparatus in cyanobacteria and plants suggest that plants evolved through a cyanobacterial line (Blankenship, 2010). However phylogenetic analysis indicates that there is closer similarity between some BphPs and PhyPs than between CphPs and PhyPs. For example, based on PCD sequence trees, these are *Rs*BphG1 from *Rb. sphaeroides* and *Rp*BphP1 from *Rps. palustris* (Montgomery & Lagarias, 2002), and on GAF sequences alone the Ppr from *Rsp. centenum* and *Rr*BphP from *Rsp. rubrum* (Karniol, Wagner, Walker, & Vierstra, 2005). Again it is interesting that these BphPs have unusual domain organizations that do not include the most common types A and B. CphPs are almost entirely of type A apart from one example from *Synechoccocus* sp. PCC7335 that has a PAS-9 domain following PHY, which may provide the missing link between cyanobacteria and plants although it appears that the F type CphPs are now rare.

Although the most common BphPs are the HK/HWE type A/B, the more unusual ones (C to H) combine to make a significant contribution in purple bacteria of approximately 30%. The species *Rps. palustris* stands out in the number and types it contains, and it is believed that the variability in BphPs between *Rps. palustris* strains may have its origins in gene deletion rather than by lateral gene acquisition (Giraud & Vermeglio, 2008). Some duplication has occurred amongst BphP of purple bacteria; for example the genes for BphPG1 and BphPG2 are found in separate plasmids of *Rb. sphaeroides*, and the adjoining *Rp*BphP2 and *Rp*BphP3 in *Rps. palustris* show close sequence similarity. The unusual *Br*BphP3 in *Bradyrhizobium* sp. ORS278 is found within a genomic island believed to have been acquired by lateral transfer from another species (Jaubert et al., 2007). This is supported by the phylogenic distance of this BphP from other BphPs and is one of the few likely acquisitions by lateral gene transfer.

4. BACTERIOPHYTOCHROMES IN *RHODOPSEUDOMONAS PALUSTRIS*

4.1. General Comments on BphPs in *Rps. palustris*

Rhodopseudomonas palustris is rich in the variety of biochemical systems and this biochemical complexity is further magnified by the relatively large genome difference between variants. These variants can be better described as ecotypes which have evolved to adapt to specific environmental conditions (Oda et al., 2008). Even within a small ecological space (~10 m) a relatively large number of genotypes (~30) have been observed, which suggests that this bacterium is capable of exploiting differences on the microenvironmental scale (Bent, Gucker, Oda, & Forney, 2003). The diversity also extends to the number and types of BphPs that this bacterium contains: the first strain to be sequenced (CGA009) was found to contain six BphPs, including one frame-shifted gene (Larimer et al., 2004). The number of BphPs varies between 2 and 6 in strains, suggesting that different environmental niches are being exploited (Oda et al., 2008). This must be viewed in the context of the plant model organism *A. thaliana* which has five phytochrome genes in its genome (2000), and demonstrates the large resources invested by *Rps. palustris* in monitoring changes in light conditions. Photosynthesis appears to be the primary system controlled by BphPs in *Rps. palustris*, and at least four BphPs have been found that perform this role: *Rp*BphP1 controls a large photosynthesis gene cluster and *Rp*BphP4 controls the high-light adapted light harvesting complexes LH2. An adjacent pair of BphPs, *Rp*BphP2 and *Rp*BphP3, control the production of low-light adapted light harvesting complexes LH3/LH4. The other BphPs in *Rps. palustris* are found in non-photosynthetic genomic contexts suggesting that non-photosynthetic biochemical processes are also under the control of light in *Rps. palustris* (Table 9.1).

4.2. Major Photosynthesis Gene Cluster and *Rp*BphP1

Most of the *pucBA* gene pairs, encoding for the peripheral light harvesting complex LH2, are scattered throughout the *Rps. palustris* genome, whereas the majority of photosynthesis genes responsible for the biosynthesis of pigments, reaction centre and light harvesting complex LH1 peptides are in one super-operon (Giraud et al., 2002; Larimer et al., 2004). This super-operon is under the control of the redox sensitive repressor *Rp*PpsR2 and *Rp*BphP1 which is a light sensitive anti-repressor (Giraud et al., 2002). This species is closely related to *Bradyrhizobium* ORS278 which has a similar

Table 9.1 Distribution of BphPs in strains of the bacterium *Rps. palustris*. BphPs found in the same genomic context are on the same row and their domain types are also shown as in Fig. 9.1. Type G BphPs control the major photosynthesis cluster of genes (>30) that code for reaction centre and LH1 peptides, bacteriochlorophyll *a*, and carotenoid biosynthesis enzymes, as well as the bilverdin IXα synthesising haem oxygenase. Genes are named according to the genomic numbering schemes in strains CGA009, HaA2, BisB5, BisB18, BisA53, TIE-1 and DX-1 as well as their common names found in the literature. In the laboratory strain CGA009, *Rp*BphP1* is frame shifted but has been found intact and active in other strains. The *Rp*BphP4 clade is found in two forms: † chromo-BphP form which is a normal light photoreceptor and the ‡ achromo-BphP form which has evolved to be a redox sensor and has lost the ability to bind BV

BphP	Strain							
Type	CGA009	HaA2	BisB18	BisB5	BisA53	TIE-1	DX-1	Function
G	RPA1537* (*Rp*BphP1)	RPB3985	RPC1309	RPD3740	RPE1343	RPA1_1726	RPDX1_3982	Photosynthetic super-operon
A	RPA3015 (*Rp*BphP2)	RPB4450	Disrupted	RPD2402	RPE0631	RPA1_3424		Low light LH4/LH3
A	RPA3016 (*Rp*BphP3)	RPB4451	RPC3780	RPD2403	RPE0632	RPA1_3425		Low light LH4/LH3
A	RPA1490 (*Rp*BphP4)‡	RPB4034†		RPD3788†		RPA1_1677†	RPDX1_4031†	LH2 and PucC
A	RPA0122							?
B	RPA0990						RPDX1_1142	?
E		RPB2169						?
A				RPD2377			RPDX1_0116	?

*Gene encoding *Rp*BphP1 is DNA frame shifted.
†Chromo-BphP.
‡Achromo-BphP.

cluster of photosynthesis genes (Giraud et al., 2002). *Rp*BphP1, a type G BphP, is present in all wild type strains and seems to be an important BphP. Gene *bphP1*, in the laboratory strain CGA009, is frame shifted and therefore inactive. It was proposed that another mutation in the transcriptional repressor *Rp*PpsR2 prevented its binding to DNA negating the effects of the frame-shifted *bphP1* (Giraud et al., 2004). However it was subsequently shown that a repaired *bphP1*, which makes an active *Rp*BphP1, can increase photosynthetic apparatus production over and above low oxygen conditions. It seems that the frame-shifted *bphP1* is the only hindrance to far-red light stimulated biosynthesis of photosynthetic complexes and only control of *Rp*PpsR2 by O_2 tension remains in this strain (Braatsch, Johnson, Noll, & Beatty, 2007). It is interesting to note that *Rp*PpsR2 in some strains do not possess Cys residues, which suggests that these are only controlled by light and not also by redox conditions. The *bphP1* super-operon also contains a haem oxygenase (*hemO*) gene which synthesises BV from haem and is often found alongside a *bphP* gene in other species. The gene *ppsR1*, like *ppsR2*, is a repressor of photosynthesis but does not respond to light through interaction with *RpBphP1* and instead is only a redox sensor (Braatsch et al., 2006, 2007). It is likely that redox control is shared by *Rp*PpsR1 and *Rp*PpsR1 in some strains while in others this is done by *Rp*PpsR1 alone. The X-ray crystal structure of a major 70 kDa fragment of *Rp*BphP1 (PCD–PAS/PAC) and model building of the HOS domain suggests that the HOS domain interacts with CBD and PHY domains in opposing monomers. It has also been shown that *Rp*BphP1 binds to *Rp*PpsR2 on illumination with 750 nm light but not in the dark, and that the HOS domain is essential for complex formation (Bellini & Papiz, 2012b). It was proposed that the HOS domain is in intimate contact with the CBD domain, and the mechanism for HOS domain activation involves 760 nm light stimulated detachment of the HOS domain, followed by complex formation with *Rp*PpsR2 involving protomer swapping between *Rp*BphP1 and *Rp*PpsR2 dimers. The formation of the complex *Rp*BphP1–*Rp*PpsR2 prevents these *Rp*PpsR2 molecules from binding to promoter regions, and in doing so *Rp*BphP1 acts as an antagonist of repressor activity. Recently it has been shown that *Rp*BphP1 and *Rp*PpsR2 also down regulate respiration, as observed in the reduced expression of the Krebs cycle enzyme α-ketoglutarate dehydrogenase. The regulation of respiration may fine tune bioenergy production between photosynthesis and respiration by the lowering of the redox coenzyme NADH while still allowing the production of key precursors important for the photosynthetic apparatus (Kojadinovic et al., 2008).

4.3. Operon for Low-light Adapted Light Harvesting Complexes

In *Rps. palustris* strain CGA009 and the closely related strain 2.1.6 it has been shown that there are two BphPs (*Rp*BphP2, *Rp*BphP3) which control both the induction of the low-light adapted LH4 complex peptides and the repression of high-light adapted LH2 complexes (Evans et al., 2005; Giraud et al., 2005; Hartigan, Tharia, Sweeney, Lawless, & Papiz, 2002; Tharia, Nightingale, Papiz, & Lawless, 1999). The absorption spectra of these BphPs differ, suggesting that they perform different roles in monitoring environmental light conditions: whereas *Rp*BphP2 exhibits a classical Pr to Pfr transition, the Pr state of *Rp*BphP3 is only quenched and a weak near-red band (Pnr) appears at ~650 nm (Evans et al., 2005; Giraud et al., 2005). All *Rps. palustris* genomes contain this operon, apart from strain DX-1, indicating that it is an important operon although not essential. The operon gene composition varies slightly between strains, but the consensus architecture includes the two *bph*P genes *luxR*, *pucC* as well as low-light adapted *pucBA* genes. The *luxR* gene has been found elsewhere in the genome belonging to a quorum sensing system which is sensitive to the concentration of *p*-coumaroyl-homoserine and measures bacterial cell density (Hirakawa et al., 2011). This could be important for eliciting cell motility into regions with better light properties as high cell concentrations result in attenuation of light at important photosynthetic wavelengths. The PucC protein is a member of the major facilitator superfamily of membrane transporters and is believed to transport bacteriochlorophyll *a* pigments to facilitate light harvesting complex assembly (Jaschke, Leblanc, Lang, & Beatty, 2008; Leblanc & Beatty, 1996; Tichy, Albien, Gad'on, & Drews, 1991). There are two pairs of *pucBA* genes within the operon; in strains CGA009 and 2.1.6 one pair codes for the low-light adapted LH4 complex while the other is disrupted at *pucA* and is not expressed (Tharia et al., 1999). The LH4 complex is characterized by a near-infrared spectrum with a single absorption band at 800 nm rather than the typical LH2-like absorption, which has bands at 800 and 850 nm. In several strains both *pucBA* pairs are intact and code for LH4 peptides but in strain BisA53 the *pucBA* pair of genes encode for LH3 low-light complexes (Kotecha, Georgiou & Papiz, 2012) similar to those found in *Rhodopseudomonas acidophila* strain 7050 (Mcluskey, Prince, Cogdell, & Isaacs, 2001): *pucA* has the sequence α-$F_{44}M_{45}$ rather than α-$Y_{44}W_{45}$, and is present in both LH3 and LH4, but lacks M_{26} which is only found in LH4 complexes

(Hartigan et al., 2002). Strain BisB18 is unusual because it has a disrupted gene *rpbph*P2 but an intact *rpbphP3* gene. The operon segment which contains *pucBA*, *luxR* and *pucC* is found upstream of *rpbphP3* rather than downstream, as is the case in all other strain. The remnants of an *rpbphP2* sequence can be found in a large section of non-coding DNA preceding *rpbphP3*. This strain can only produces LH2 complexes under high and low-light (Kotecha, Georgiou & Papiz, 2012) confirming other results which indicate that both BphPs must be active to make LH3 or LH4 (Giraud et al., 2005).

4.4. The *Rp*BphP4 Operon and LH2 Biosynthesis

The gene for *Rp*BphP4 is adjacent to *pucBA* encoding for a LH2 complexe. It has evolved to be an achromo form (biliverdin nonbinding) in the strains 2.1.6 and CGA009 by mutation of a Cys residue responsible for Biliverdin IXalpha (BV) ligase reaction (Evans et al., 2005; Vuillet et al., 2007). Despite this, *Rp*BphP4 is able to autophosphorylate and phosphotransfer to a cognate response regulator (*rpa1489* in CGA009). The gene encoding *Rp*BphP4 in strains CEA001, HaA2 and BisB5 is a chromo-BphP capable of binding BV and also reacts to light (Vuillet et al., 2007). This work indicates two clades of BphPs, one a redox sensing clade and the other a light-sensing clade. Both use the same phosphotransfer pathway to a response regulator encoded in the same operon belonging to the *luxR* family. This response regulator binds to the nearby promoter region for $pucBA_e$ which codes for LH2, and it can also bind to a distant promoter region preceding the LH2 genes $pucBA_b$ (Vuillet et al., 2007). Interestingly these promoter regions are similar to the promoter sequences recognized by PpsR2, and it was shown that purified PpsR2 can also bind to $pucBA_e$ and $pucBA_b$ promoter regions. These experiments indicate that these *pucBA* LH2 genes are under the control of LuxR, PpsR2 and probably PpsR1. Two rare genes (O1 and O2) are present in both the LH2 operon containing *Rp*BphP4 and the LH3/LH4 operon that contains *Rp*BphP2/*Rp*BphP3. These are marker genes that provide strong evidence for gene transfer between these operons (Fig. 9.3). The gene transfer, from LH2 to LH3/LH4 operons, was probably followed by the duplication of the *bph*P gene that evolved into *Rp*BphP2 and *Rp*BphP3 while the LH2 genes evolved into the low-light genes encoding LH3 or LH4. It is likely that the evolution of the LH2 to LH3 or LH4 came about only when the novel *Rp*BphP3 BphP evolved as this is sensitive to absolute light intensity rather than the ratio of intensities at the Pr and Pfr absorption wavelengths. The unique low-light sensing capability of this

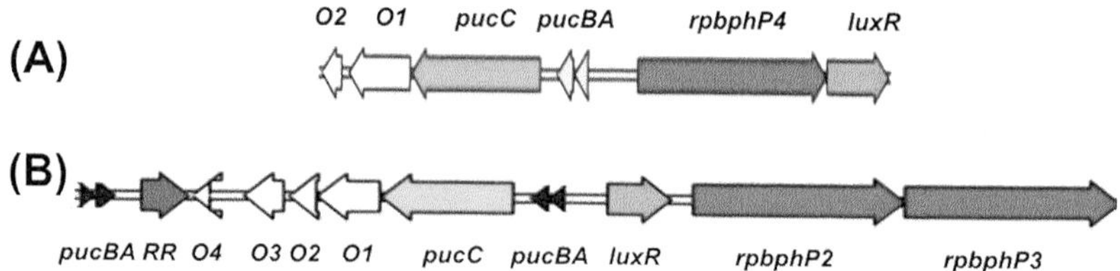

Figure 9.3 ***Operons in*** **Rps. palustris** ***strain BisB5 that control (A) LH2 and (B) low-light LH4 complex synthesis.*** (A) contains the *Rp*BphP4 which is a chromo-BphP in this strain and contains a unique pair of open reading frame genes (O1 and O2) which are also found in (B) along with two adjacent BphPs, *Rp*BphP2 and *Rp*BphP3 which control the expression of low-light pucBA genes that encode for LH4 in this strain. The similarity in gene cluster indicates that operons A and B are related by a transfer of this operon followed by separate evolution to a low-light LH4 gene cluster. RR is the response regulator belonging to the luxR family which is a gene activator, pucC is a transporter belonging to the major facilitator membrane transporter superfamily and may facilitate light-harvesting complex assembly by transportation of the pigment bacteriochlorophyll *a*. (For colour version of this figure, the reader is referred to the online version of this book.)

operon created a close link between low-light environmental conditions and expression of the *puc*BA genes in this operon that allowed low-light selection and evolution of these genes. This operon transfer may explain how the regulation of LH2 and LH4 are coupled as these operons share promoter sequences and response regulators arising from a common origin (Vuillet et al., 2007): LH4 is up-regulated and LH2 down-regulated (Hartigan et al., 2002; Tharia et al., 1999). An important point to note is that there are two *pucBA* gene pairs, in the low-light operon which encode LH3 or LH4, and these belong to different clades. This implies that although one *pucBA* originated from the operon transfer the other was recruited into the operon from elsewhere and then coevolved into LH3 or LH4 under the tight control of *RpBph2* and *RpBphP3* (Kotecha, Georgiou & Papiz, 2012).

4.5. Miscellaneous BphPs in *Rps. palustris*

Strains CGA009, HaA2, BisB5 and DX-1 possess six other BphPs of types A, B and E. Only the type B BphP is found in a similar genomic context, in strains CGA009 and DX-1, suggesting that there are in total 5 separate biochemical systems under the control of these BphPs. It has been suggested that these singular gene arrangements arose by deletion rather than gene acquisition (Giraud & Vermeglio, 2008), which implies that there maybe unknown strains that have more than six BphPs in a single genome. These BphPs have not been assigned a function but are probably not connected with photosynthesis for the following reasons. These BphPs are not in the neighbourhood of photosynthesis genes and other BphPs have been found in non-photosynthetic bacteria (Davis et al., 1999). Other roles have been found for BphPs and CphPs such as phototaxis (Fiedler, Borner, & Wilde, 2005; Ng, Grossman, & Bhaya, 2003; Ragatz et al., 1995). Moreover, polyketide synthase is regulated by the type H Ppr in *Rsp. centenum* (Jiang et al., 1999). *Br*BphP3 in the *Bradyrhizobium* strain ORS278 is found in an island of genes acquired by lateral transfer and it is involved in the synthesis of the pigment phycocyanobilin and gas vesicles. The latter gene transfer was probably from a cyanobacterium, hence the phycocyanobilin synthesis, and floatation sacks within gas vesicles enable movement towards the light in an aquatic environment and therefore forms part of a phototaxis system (Jaubert et al., 2007). However, there are probably many more biochemical systems controlled by BphPs that are unknown at the present time.

5. FUTURE PERSPECTIVES

BphP molecules are modular in structure with a conserved input light-sensing domain followed by a variable output domain. These molecules are flexible and can exploit different environmental conditions by channelling light signals to very different transcription regulatory systems. These can be triggered by phosphorelay, second messenger, or by direct inhibition of response regulators by protein–protein interactions. An important evolutionary question is why there is a proliferation of BphPs in some purple bacteria, such as *Rps. palustris*, whilst others such as *Rb. capsulatus* have none at all? This implies that changes in available light and other light correlated resources maybe important for some organisms but less so for others. Cyanobacteria and plants are phototropic organisms in that they grow towards the light source, whereas proteobacteria are mostly non-phototropic (Blankenship, 2010). This may explain why all cyanobacteria and plants contain phytochromes whereas many proteobacteria (purple bacteria) do not, and raises the question of whether phototropic responses and their control occurred before photosynthetic control. Most BphPs in purple bacteria have unknown functions and the genes they control are yet to be determined. It has been observed that genes controlled by BphPs in *Rps. palustris* are often in the same operon as the relevant BphP, although in principle BphPs can control genes scattered throughout the genome. To determine the role of many BphPs will require an 'omic' approach to the problem. Interactions between light conditions and cell biochemical systems are complex and can arise through correlated factors such as light, redox, and nutrients. For example attenuation of light by water is greater at longer wavelengths (>750 nm) and is therefore more pronounced at wavelengths important for photosynthesis between 800 and 890 nm (Evans et al., 2005). At increasing water depths anoxic conditions become more prevalent and nutrient concentrations can increase. However light flux is strongly attenuated and is less available for photosynthesis, so BphPs may aid in determining optimal water depths for parameters such as, light, O_2 concentration, and nutrients.

It is thought that phytochromes in cyanobacteria, fungi, and plants have their origins in BphPs, and as noted earlier the similarity between some BphPs and PhyPs suggests the possibility for an early divergence of PhyPs through a common ancestor rather than a linear evolution through cyanobacteria. Cph1 has the domain organization of type A while plant phytochromes contain a pair of PAS domains between PHY and HK domains.They may have evolved from type F BphPs that are currently found only in the purple bacteria *Rsp. centenum* and *Rsp. rubrum*. Type G BphPs are also plausible candidates as a precursor to plant PhyPs because *Rp*BphP1 lies close to plant PhyPs in the phylogenetic tree (Montgomery & Lagarias, 2002). We note that even though PhyPs have a canonical domain organization PCD–PAS–PAS–HK, there are over 200 plant genes that have the type G organization PCD–PAS-(X), which have generally been annotated as PhyP fragments. Because *Rp*BphP1, in *Rps. palustris,* has been shown to bind a cognate repressor, there is the possibility that G type plant PhyPs may also be functioning molecules rather than aberrant fragments.

ACKNOWLEDGEMENTS

We would like to thank the Biotechnology and Biological Sciences Research Council and the Science and Technology Facilities Council for supporting the work.

REFERENCES

Arabidopsis Genome Initiative. (2000). Analysis of the genome sequence of the flowering plant *Arabidopsis thaliana*. *Nature*, *408*, 796–815.

Aravind, L., & Ponting, C. P. (1999). The cytoplasmic helical linker domain of receptor histidine kinase and methyl-accepting proteins is common to many prokaryotic signalling proteins. *FEMS Microbiology Letters*, *176*, 111–116.

Bellini, D., & Papiz, M. Z. (2012a). Dimerization properties of the RpBphP2 chromophore-binding domain crystallized by homologue-directed mutagenesis. *Acta Crystallographica. Section D, Biological Crystallography*, *68*, 1058–1066.

Bellini, D., & Papiz, M. Z. (2012b). Structure of a bacteriophytochrome and light-stimulated protomer swapping with a gene repressor. *Structure*, *20*, 1436–1446.

Bent, S. J., Gucker, C. L., Oda, Y., & Forney, L. J. (2003). Spatial distribution of *Rhodopseudomonas palustris* ecotypes on a local scale. *Applied and Environmental Microbiology*, *69*, 5192–5197.

Bhoo, S. H., Davis, S. J., Walker, J., Karniol, B., & Vierstra, R. D. (2001). Bacteriophytochromes are photochromic histidine kinases using a biliverdin chromophore. *Nature*, *414*, 776–779.
Blankenship, R. E. (2010). Early evolution of photosynthesis. *Plant Physiology*, *154*, 434–438.
Borucki, B., Von Stetten, D., Seibeck, S., Lamparter, T., Michael, N., Mroginski, M. A., et al. (2005). Light-induced proton release of phytochrome is coupled to the transient deprotonation of the tetrapyrrole chromophore. *Journal of Biological Chemistry*, *280*, 34358–34364.
Braatsch, S., Bernstein, J. R., Lessner, F., Morgan, J., Liao, J. C., Harwood, C. S., et al. (2006). *Rhodopseudomonas palustris* CGA009 has two functional ppsR genes, each of which encodes a repressor of photosynthesis gene expression. *Biochemistry*, *45*, 14441–14451.
Braatsch, S., Johnson, J. A., Noll, K., & Beatty, J. T. (2007). The O2-responsive repressor PpsR2 but not PpsR1 transduces a light signal sensed by the BphP1 phytochrome in *Rhodopseudomonas palustris* CGA009. *FEMS Microbiology Letters*, *272*, 60–64.
Casino, P., Rubio, V., & Marina, A. (2009). Structural insight into partner specificity and phosphoryl transfer in two-component signal transduction. *Cell*, *139*, 325–336.
Casino, P., Rubio, V., & Marina, A. (2010). The mechanism of signal transduction by two-component systems. *Current Opinion in Structural Biology*, *20*, 763–771.
Chen, Y., Zhang, J., Luo, J., Tu, J. M., Zeng, X. L., Xie, J., et al. (2012). Photophysical diversity of two novel cyanobacteriochromes with phycocyanobilin chromophores: photochemistry and dark reversion kinetics. *FEBS Journal*, *279*, 40–54.
Davis, S. J., Vener, A. V., & Vierstra, R. D. (1999). Bacteriophytochromes: phytochrome-like photoreceptors from nonphotosynthetic eubacteria. *Science*, *286*, 2517–2520.
Essen, L. O., Mailliet, J., & Hughes, J. (2008). The structure of a complete phytochrome sensory module in the Pr ground state. *Proceedings of the National Academy of Sciences of the United States of America*, *105*, 14709–14714.
Evans, K., Fordham-Skelton, A. P., Mistry, H., Reynolds, C. D., Lawless, A. M., & Papiz, M. Z. (2005). A bacteriophytochrome regulates the synthesis of LH4 complexesin *Rhodopseudomonas palustris*. *Photosynthesis Research*, *85*, 169–180.
Evans, K., Grossmann, J. G., Fordham-Skelton, A. P., & Papiz, M. Z. (2006). Small-angle X-ray scattering reveals the solution structure of a bacteriophytochrome in the catalytically active Pr state. *Journal of Molecular Biology*, *364*, 655–666.
Fiedler, B., Borner, T., & Wilde, A. (2005). Phototaxis in the cyanobacterium *Synechocystis* sp. PCC 6803: role of different photoreceptors. *Photochemistry and Photobiology*, *81*, 1481–1488.
Foerstendorf, H., Lamparter, T., Hughes, J., Gartner, W., & Siebert, F. (2000). The photoreactions of recombinant phytochrome from the cyanobacterium Synechocystis: a low-temperature UV–Vis and FT-IR spectroscopic study. *Photochemistry and Photobiology*, *71*, 655–661.
Giraud, E., Fardoux, J., Fourrier, N., Hannibal, L., Genty, B., Bouyer, P., et al. (2002). Bacteriophytochrome controls photosystem synthesis in anoxygenic bacteria. *Nature*, *417*, 202–205.
Giraud, E., & Vermeglio, A. (2008). Bacteriophytochromes in anoxygenic photosynthetic bacteria. *Photosynthesis Research*, *97*, 141–153.
Giraud, E., Zappa, S., Jaubert, M., Hannibal, L., Fardoux, J., Adriano, J. M., et al. (2004). Bacteriophytochrome and regulation of the synthesis of the photosynthetic apparatus in *Rhodopseudomonas palustris*: pitfalls of using laboratory strains. *Photochemical and Photobiological Sciences*, *3*, 587–591.
Giraud, E., Zappa, S., Vuillet, L., Adriano, J. M., Hannibal, L., Fardoux, J., et al. (2005). A new type of bacteriophytochrome acts in tandem with a classical bacteriophytochrome to control the antennae synthesis in *Rhodopseudomonas palustris*. *Journal of Biological Chemistry*, *280*, 32389–32397.
Hartigan, N., Tharia, H. A., Sweeney, F., Lawless, A. M., & Papiz, M. Z. (2002). The 7.5-A electron density and spectroscopic properties of a novel low-light B800 LH2 from *Rhodopseudomonas palustris*. *Biophysical Journal*, *82*, 963–977.

Heyne, K., Herbst, J., Stehlik, D., Esteban, B., Lamparter, T., Hughes, J., et al. (2002). Ultrafast dynamics of phytochrome from the cyanobacterium synechocystis, reconstituted with phycocyanobilin and phycoerythrobilin. *Biophysical Journal, 82*, 1004–1016.

Hirakawa, H., Oda, Y., Phattarasukol, S., Armour, C. D., Castle, J. C., Raymond, C. K., et al. (2011). Activity of the *Rhodopseudomonas palustris p*-coumaroyl-homoserine lactone-responsive transcription factor RpaR. *Journal of Bacteriology, 193*, 2598–2607.

Hughes, J., Lamparter, T., Mittmann, F., Hartmann, E., Gartner, W., Wilde, A., et al. (1997). A prokaryotic phytochrome. *Nature, 386*, 663.

Jaschke, P. R., Leblanc, H. N., Lang, A. S., & Beatty, J. T. (2008). The PucC protein of *Rhodobacter capsulatus* mitigates an inhibitory effect of light-harvesting 2 alpha and beta proteins on light-harvesting complex 1. *Photosynthesis Research, 95*, 279–284.

Jaubert, M., Lavergne, J., Fardoux, J., Hannibal, L., Vuillet, L., Adriano, J. M., et al. (2007). A singular bacteriophytochrome acquired by lateral gene transfer. *Journal of Biological Chemistry, 282*, 7320–7328.

Jiang, Z., Swem, L. R., Rushing, B. G., Devanathan, S., Tollin, G., & Bauer, C. E. (1999). Bacterial photoreceptor with similarity to photoactive yellow protein and plant phytochromes. *Science, 285*, 406–409.

Karniol, B., & Vierstra, R. D. (2003). The pair of bacteriophytochromes from *Agrobacterium tumefaciens* are histidine kinases with opposing photobiological properties. *Proceedings of the National Academy of Sciences of the United States of America, 100*, 2807–2812.

Karniol, B., & Vierstra, R. D. (2004). The HWE histidine kinases, a new family of bacterial two-component sensor kinases with potentially diverse roles in environmental signaling. *Journal of Bacteriology, 186*, 445–453.

Karniol, B., Wagner, J. R., Walker, J. M., & Vierstra, R. D. (2005). Phylogenetic analysis of the phytochrome superfamily reveals distinct microbial subfamilies of photoreceptors. *Biochemical Journal, 392*, 103–116.

Kneip, C., Hildebrandt, P., Schlamann, W., Braslavsky, S. E., Mark, F., & Schaffner, K. (1999). Protonation state and structural changes of the tetrapyrrole chromophore during the Pr → Pfr phototransformation of phytochrome: a resonance Raman spectroscopic study. *Biochemistry, 38*, 15185–15192.

Kojadinovic, M., Laugraud, A., Vuillet, L., Fardoux, J., Hannibal, L., Adriano, J. M., et al. (2008). Dual role for a bacteriophytochrome in the bioenergetic control of *Rhodopseudomonas palustris*: enhancement of photosystem synthesis and limitation of respiration. *Biochimica et Biophysica Acta, 1777*, 163–172.

Kotecha, A., Georgiou, T., & Papiz, M. Z., 2012. Evolution of low-light adapted peripheral light-harvesting complexes in strains of *Rhodopseudomonas palustris. Photosynthesis Research*. Dec 19. [Epub ahead of print].

Kyndt, J. A., Fitch, J. C., Meyer, T. E., & Cusanovich, M. A. (2005). *Thermochromatium tepidum* photoactive yellow protein/bacteriophytochrome/diguanylate cyclase: characterization of the PYP domain. *Biochemistry, 44*, 4755–4764.

Lagarias, J. C., & Lagarias, D. M. (1989). Self-assembly of synthetic phytochrome holoprotein in vitro. *Proceedings of the National Academy of Sciences of the United States of America, 86*, 5778–5780.

Lamparter, T., Michael, N., Mittmann, F., & Esteban, B. (2002). Phytochrome from *Agrobacterium tumefaciens* has unusual spectral properties and reveals an N-terminal chromophore attachment site. *Proceedings of the National Academy of Sciences of the United States of America, 99*, 11628–11633.

Larimer, F. W., Chain, P., Hauser, L., Lamerdin, J., Malfatti, S., Do, L., et al. (2004). Complete genome sequence of the metabolically versatile photosynthetic bacterium *Rhodopseudomonas palustris. Nature Biotechnology, 22*, 55–61.

Leblanc, H. N., & Beatty, J. T. (1996). Topological analysis of the *Rhodobacter capsulatus* PucC protein and effects of C-terminal deletions on light-harvesting complex II. *Journal of Bacteriology, 178*, 4801–4806.

Li, H., Zhang, J., & Vierstra, R. D. (2010). Quaternary organization of a phytochrome dimer as revealed by cryoelectron microscopy. *Proceedings of the National Academy of Sciences of the United States of America, 107*, 10872–10877.

Mcluskey, K., Prince, S. M., Cogdell, R. J., & Isaacs, N. W. (2001). The crystallographic structure of the B800-820 LH3 light-harvesting complex from the purple bacteria *Rhodopseudomonas acidophila* strain 7050. *Biochemistry, 40*, 8783–8789.

Marina, A., Waldburger, C. D., & Hendrickson, W. A. (2005). Structure of the entire cytoplasmic portion of a sensor histidine-kinase protein. *EMBO Journal, 24*, 4247–4259.

Masuda, S., & Bauer, C. E. (2002). AppA is a blue light photoreceptor that antirepresses photosynthesis gene expression in *Rhodobacter sphaeroides. Cell, 110*, 613–623.

Montgomery, B. L., & Lagarias, J. C. (2002). Phytochrome ancestry: sensors of bilins and light. *Trends in Plant Science, 7*, 357–366.

Navarro, M. V., De, N., Bae, N., Wang, Q., & Sondermann, H. (2009). Structural analysis of the GGDEF-EAL domain-containing c-di-GMP receptor FimX. *Structure, 17*, 1104–1116.

Ng, K., Getzoff, E. D., & Moffat, K. (1995). Optical studies of a bacterial photoreceptor protein, photoactive yellow protein, in single crystals. *Biochemistry, 34*, 879–890.

Ng, W. O., Grossman, A. R., & Bhaya, D. (2003). Multiple light inputs control phototaxis in *Synechocystis* sp. strain PCC6803. *Journal of Bacteriology, 185*, 1599–1607.

Oda, Y., Larimer, F. W., Chain, P. S.G., Malfatti, S., Shin, M. V., Vergez, L. M., et al. (2008). Multiple genome sequences reveal adaptations of a phototrophic bacterium to sediment microenvironments. *Proceedings of the National Academy of Sciences of the United States of America, 105*, 6.

Otto, H., Lamparter, T., Borucki, B., Hughes, J., & Heyn, M. P. (2003). Dimerization and inter-chromophore distance of Cph1 phytochrome from *Synechocystis*, as monitored by fluorescence homo and hetero energy transfer. *Biochemistry, 42*, 5885–5895.

Ponting, C. P., & Aravind, L. (1997). PAS: a multifunctional domain family comes to light. *Current Biology, 7*, R674–R677.

Quail, P. H. (2002). Phytochrome photosensory signalling networks. *Nature Reviews Molecular Cell Biology, 3*, 85–93.

Ragatz, L., Jiang, Z. Y., Bauer, C. E., & Gest, H. (1995). Macroscopic phototactic behavior of the purple photosynthetic bacterium *Rhodospirillum centenum. Archives of Microbiology, 163*, 1–6.

Rajagopal, S., & Moffat, K. (2003). Crystal structure of a photoactive yellow protein from a sensor histidine kinase: conformational variability and signal transduction. *Proceedings of the National Academy of Sciences of the United States of America, 100*, 1649–1654.

Ren, Z., Perman, B., Srajer, V., Teng, T. Y., Pradervand, C., Bourgeois, D., et al. (2001). A molecular movie at 1.8 A resolution displays the photocycle of photoactive yellow protein, a eubacterial blue-light receptor, from nanoseconds to seconds. *Biochemistry, 40*, 13788–13801.

Rottwinkel, G., Oberpichler, I., & Lamparter, T., 2010. Bathy phytochromes in rhizobial soil bacteria. *Journal Bacteriology, 192*, 5124–5133.

Rockwell, N. C., Su, Y. S., & Lagarias, J. C. (2006). Phytochrome structure and signaling mechanisms. *Annual Review of Plant Biology, 57*, 837–858.

Tamura, K., Peterson, D., Peterson, N., Stecher, G., Nei, M., & Kumar, S. (2011). MEGA5: molecular evolutionary genetics analysis using maximum likelihood, evolutionary distance, and maximum parsimony methods. *Molecular Biology and Evolution, 28*, 2731–2739.

Tarutina, M., Ryjenkov, D.A., & Gomelsky, M. (2006). An unorthodox bacteriophytochrome from Rhodobacter sphaeroides involved in turnover of the second messenger c-di-GMP. *Journal of Biological Chemistry, 281*, 34751–34758.

Taylor, B. L., & Zhulin, I. B. (1999). PAS domains: internal sensors of oxygen, redox potential, and light. *Microbiology and Molecular Biology Reviews, 63*, 479–506.

Tharia, H. A., Nightingale, T. D., Papiz, M. Z., & Lawless, A. M. (1999). Characterisation of hydrophobic peptides by RP-HPLC from different spectral forms of LH2 isolated from *Rps. palustris*. *Photosynthesis Research, 61*, 157–167.

Tichy, H.V., Albien, K. U., Gad'on, N., & Drews, G. (1991). Analysis of the *Rhodobacter capsulatus* puc operon: the pucC gene plays a central role in the regulation of LHII (B800-850 complex) expression. *EMBO Journal, 10*, 2949–2955.

Toh, K. C., Stojkovic, E. A., Rupenyan, A. B., Van Stokkum, I. H., Salumbides, M., Groot, M. L., et al. (2011). Primary reactions of bacteriophytochrome observed with ultrafast mid-infrared spectroscopy. *Journal of Physical Chemistry. A, 115*, 3778–3786.

Van Thor, J. J., Borucki, B., Crielaard, W., Otto, H., Lamparter, T., Hughes, J., et al. (2001). Light-induced proton release and proton uptake reactions in the cyanobacterial phytochrome Cph1. *Biochemistry, 40*, 11460–11471.

Vuillet, L., Kojadinovic, M., Zappa, S., Jaubert, M., Adriano, J. M., Fardoux, J., et al. (2007). Evolution of a bacteriophytochrome from light to redox sensor. *EMBO Journal, 26*, 3322–3331.

Wagner, J. R., Brunzelle, J. S., Forest, K. T., & Vierstra, R. D. (2005). A light-sensing knot revealed by the structure of the chromophore-binding domain of phytochrome. *Nature, 438*, 325–331.

Williamson, N. R., Fineran, P. C., Ogawa, W., Woodley, L. R., & Salmond, G. P. (2008). Integrated regulation involving quorum sensing, a two-component system, a GGDEF/EAL domain protein and a post-transcriptional regulator controls swarming and RhlA-dependent surfactant biosynthesis in Serratia. *Environmental Microbiology, 10*, 1202–1217.

Yamada, S., Sugimoto, H., Kobayashi, M., Ohno, A., Nakamura, H., & Shiro, Y. (2009). Structure of PAS-linked histidine kinase and the response regulator complex. *Structure, 17*, 1333–1344.

Yang, C.Y., Chin, K. H., Chuah, M. L., Liang, Z. X., Wang, A. H., & Chou, S. H. (2011). The structure and inhibition of a GGDEF diguanylate cyclase complexed with (c-di-GMP) (2) at the active site. *Acta Crystallographica. Section D, Biological Crystallography, 67*, 997–1008.

Yang, X., Kuk, J., & Moffat, K. (2008). Crystal structure of *Pseudomonas aeruginosa* bacteriophytochrome: photoconversion and signal transduction. *Proceedings of the National Academy of Sciences of the United States of America, 105*, 14715–14720.

Yang, X., Ren, Z., Kuk, J., & Moffat, K. (2011). Temperature-scan cryocrystallography reveals reaction intermediates in bacteriophytochrome. *Nature, 479*, 428–432.

Yang, X., Stojkovic, E. A., Kuk, J., & Moffat, K. (2007). Crystal structure of the chromophore binding domain of an unusual bacteriophytochrome, RpBphP3, reveals residues that modulate photoconversion. *Proceedings of the National Academy of Sciences of the United States of America, 104*, 12571–12576.

CHAPTER TEN

Iron Homeostasis in the *Rhodobacter* Genus

Sébastien Zappa, Carl E. Bauer[1]

Department of Molecular and Cellular Biochemistry, Indiana University, Bloomington, IN, USA

[1]Corresponding author: E-mail: bauer@indiana.edu

Contents

Abstract

Metals are utilized for a variety of critical cellular functions and are essential for survival. However cells are faced with the conundrum of needing metals coupled with the fact that some metals, iron in particular are toxic if present in excess. Maintaining metal homeostasis is therefore of critical importance to cells. In this review we have systematically analyzed sequenced genomes of three members of the *Rhodobacter* genus, *R. capsulatus* SB1003, *R. sphaeroides* 2.4.1 and *R. ferroxidans* SW2 to determine how these species undertake iron homeostasis. We focused our analysis on elemental ferrous and ferric iron uptake genes as well as genes involved in the utilization of iron from heme. We also discuss how *Rhodobacter* species manage iron toxicity through export and sequestration of iron. Finally we discuss the various putative strategies set up by these *Rhodobacter* species to regulate iron homeostasis and potential novel means of regulation. Overall, this genomic analysis highlights surprisingly diverse features involved in iron homeostasis in the *Rhodobacter* genus.

Advances in Botanical Research, Volume 66

ISSN 0065-2296, http://dx.doi.org/10.1016/B978-0-12-397923-0.00010-2

1. INTRODUCTION

The origin of oxygenic photosynthesis can be traced to ~2.9 10^9 years ago when cyanobacteria-driven photosynthesis created a Great Oxidizing Event that enriched atmospheric oxygen. Prior to photosynthetic oxidation of Earth, most iron was in a reduced ferrous state (Fe^{2+}) that is biologically available as it has a solubility of 0.1 M at pH 7. This form of iron is thought to been present in deep biotopes until ~1.8-1 billion years ago (Planavsky et al., 2011; Van Der Giezen & Lenton, 2012). Following that time the presence of atmospheric oxygen effectively oxidized most surface and oceanic iron to a ferric state (Fe^{3+}) that has an extremely low solubility (10^{-18}M at pH 7) (Andrews, Robinson, & Rodríguez-Quiñones, 2003). Consequently, the oxidation of Earth must have caused a crisis of iron availability necessitating that cells evolved a diverse array of Fe^{2+} and Fe^{3+} uptake systems.

Iron is an important cofactor in many enzymes where it can form mono- or di-iron centers, more complex iron-sulfur clusters, and bound to protoporphyrin IX to form heme, an important gas or electron carrier. Enzymes that utilize iron are involved in major biochemical processes such as photosynthesis, N_2 fixation, methanogenesis, H_2 production and consumption, respiration, the tricarboxylic acid cycle, oxygen transport, gene regulation and DNA biosynthesis. Iron is also an important actor in cellular events such as virulence, biofilm formation and quorum-sensing (Steele, O'Connor, Burpo, Kohler, & Johnston, 2012; Vasil, 2007; Wen, Kim, Son, Lee, & Kim, 2012). The role of iron in so many systems indicates that Life evolved enzymes that utilized iron when it was readily available, and as a result had to invent biochemical pathways to maintain iron homeostasis when Earth's oxidation caused iron to become scarce. Extremely rare are examples of organisms that solved the iron availability issue by circumventing the need for iron, such as the lactobacilli and the Lyme disease agent (Archibald, 1983; Posey, 2000; Weinberg, 1997).

Bacteria have developed systems to uptake both the Fe^{2+} and Fe^{3+} forms of elemental iron. They have also evolved iron scavenging pathways by excreting and transporting siderophores that function as iron chelators. Bacteria have also developed means of transporting heme synthesized by other organisms as a salvage pathway (Andrews et al., 2003; Braun & Hantke, 2011; Wandersman & Delepelaire, 2004). These diverse iron transport systems must be tightly regulated, as excess cellular iron is toxic.

Free iron can generate hydroxyl free radicals through Fenton's chemistry (Eqns (10.1)–(10.3)) that have deleterious effects on fatty acids and other biomacromolecules (Chiancone, Ceci, Ilari, Ribacchi, & Stefanini, 2004; Touati, 2000).

$$Fe^{2+} + O_2 \rightarrow Fe^{3+} + O_2^- \tag{10.1}$$

$$2O_2^- + 2H^+ \rightarrow H_2O_2 + O_2 \tag{10.2}$$

$$Fe^{2+} + H_2O_2 \rightarrow Fe^{3+} + OH^- + {}^{\bullet}OH \tag{10.3}$$

Challenged with balancing a need for iron with iron's toxicity, cells must maintain a tightly regulated iron homeostasis that controls the dynamic equilibrium between import, export and the storage of iron in proteins.

Purple nonsulfur bacteria are facultative phototrophs distributed among the α- and β-subclasses of proteobacteria. They have an extremely versatile metabolism that utilizes iron in ways that allows growth under multiple environmental conditions. The use of iron by purple nonsulfur bacteria can be exemplified by such processes as: i) Aerobic respiration where terminal cytochrome oxidase cbb_3 and b_{260} use heme as a cofactor; ii) Respiratory and photosynthetic electron transport where heme-containing cytochromes c_y, c_2 and bc_1 shuttle electrons to photosystem reaction centers as well as to respiratory terminal oxidases; iii) Enzymes such as coproporphyrinogen III oxidase, which contains a Fe-S cluster involved in synthesis of heme; iv) Enzymes involved in bacteriochlorophyll synthesis that utilize iron-sulfur clusters (Sarma et al., 2008; Sirijovski, Mamedov, Olsson, Styring, & Hansson, 2007); v) and in some species, the anaerobic oxidation of Fe^{2+} to facilitate phototrophic growth (Caiazza, Lies, & Newman, 2007; Croal, Jiao, & Newman, 2007; Ehrenreich & Widdel, 1994; Poulain & Newman, 2009; Widdel et al., 1993). These are just a few representative examples of the many processes used by this group of bacteria that rely on the use of iron as a cofactor, and that illustrate their desperate need for this metal. In this chapter, wewill discuss what is known about iron homeostasis in the *Rhodobacter* genus, focusing on iron homeostasis genes present in the genomes of *R. capsulatus* SB1003, *R. sphaeroides* 2.4.1 and *R. ferroxidans* SW2 (formerly known as *R.* sp. SW2 (Saraiva, Newman, & Louro, 2012)). Despite numerous studies on photosynthesis, respiration and general physiology, there have been surprisingly few studies on iron needs and regulation of iron homeostasis in *Rhododbacter.*

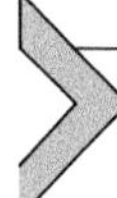

2. FERROUS IRON UPTAKE

2.1. Feo System

The Feo iron transport system is widespread among bacteria and thus appears to be a major route of Fe^{2+} acquisition (Cartron, Maddocks, Gillingham, Craven, & Andrews, 2006; Perry, Mier, & Fetherston, 2007). Since the first description of the Feo system in *Escherichia coli* in 1987 (Hantke, 1987), it has been shown to be involved in many iron-related phenotypes such as magnetosome formation (Rong et al., 2012) and virulence (Fetherston, Mier, Truszczynska, & Perry, 2012). The Feo system is encoded in many bacterial and some archaeal genomes. Interestingly, the Feo uptake system has significant sequence similarity to eukaryal G-proteins and thus has been referred to as a "living fossil" of this family of eukaryotic GTPases (Hantke, 2003).

Genes coding for the Feo system are present in each of the three genomes considered in this study, namely *R. capsulatus* SB1003, *R. sphaeroides* 2.4.1 and *R. ferroxidans* SW2. *R. sphaeroides* and *R. ferroxidans* contain only one *feo* gene cluster while *R. capsulatus* has two clusters. It is not unusual to find two (or more) *feo* loci in one organism. In such cases, it is hypothesized that one of the Feo systems is specialized in manganese uptake, and when proven to be involved in this process is called Meo (Cartron et al., 2006; Dashper et al., 2005; He et al., 2006). However, duplicate Feo systems can also be specialized in two different iron-related pathways, such as magnetosome formation and oxidative stress management (Rong et al., 2012).

In the *Rhodobacter* genomes, a putative four-gene operon is present in each strain, consisting of two *feoA* genes followed by *feoB* and *feoC* loci ($feoA_1A_2BC$) (Table 10.1). Other cases of multiple *feoA* genes in a *feo* operon have been reported (Cartron et al., 2006). Another putative operon, unique to *R. capsulatus* SB1003, named *feo2AB*, displays only one *feoA* gene followed by a *feoB* (Table 10.1). In ~80% of the genomes where a *feo* locus is present, it consists of a small *feoA* gene followed by a larger *feoB* gene in a *feoAB* operonal organization. Occasionally an additional *feoC* Open Reading Frame (ORF) is also present, particularly in the γ-proteobacteria, in a *feoABC* operon.

Alignment of translated $feoA_1$, $feoA_2$ and *feo2A* genes from these three *Rhodobacter* species highlights a higher sequence similarity/identity for $FeoA_1$ representatives with homologues from the other *Rhodobacter* species than to gene paralogues present in their own genome (Table 10.2). For example, *R. capsulatus* $FeoA_2$ shows 69.51% and 68.29% similarity with

Table 10.1 Ferrous iron uptake genes identified in the genomes of *R. capsulatus* SB1003, *R. sphaeroides* 2.4.1 and *R. ferroxidans* SW2

Putative operon name	Gene name	COG	Label *Rcap* SB1003*	Label *Rsph* 2.4.1*	Label *Rfer* SW2*	Product
efeUOB	*efeUO*	2822, 0672	03065	N.D.	N.D.	Fe^{2+} permease EfeU
	efeB	2837	03066	N.D.	N.D.	Dyp-type peroxidase family protein
	efeM	2822	03067	N.D.	N.D.	Protein of unknown function DUF451
feoA_1A_2*BC*	*feo*A_1	1918	00090	6020	3172	Fe^{2+} transport protein A
	feoA_1	1918	00091	1819	3171	Fe^{2+} transport protein A
	feoB	0370	00092	1818	3170	Fe^{2+} transport protein B
	feoC		00093	1817	3169	Fe–S dependent transcriptional regulator
feo2AB	*feo2A*	1918	02028	N.D.	N.D.	Fe^{2+} transport protein A
	feo2B	0370	02029	N.D.	N.D.	Fe^{2+} transport protein B

N.D., not detected. COG: Cluster of Orthologous Groups

**R. capsulatus* SB1003, *R. sphaeroides* 2.4.1 and *R. ferroxidans* SW2 ORFs are accessible in the Integrated Microbial Genomes of the DOE Joint Genome Institute (http://img.jgi.doe.gov/cgi-bin/w/main.cgi) using the following locus tag RCAP_rcc#####, RSP_#### and Rsw2DRAFT_####, respectively, where #### and ##### are the or digits indicated in the table. Regarding *R. capsulatus* numbering, genes on the plasmid are references as RCAP_rcp##### on the JGI database and p##### in this study.

$FeoA_2$ from *R. sphaeroides* and *R. ferroxidans*, respectively. This is in contrast to $FeoA_2$ from *R. capsulatus* exhibiting only 40.22% and 36.56% similarity with $FeoA_1$ and Feo2A from its own genome (Table 10.2). The same pattern occurs when comparing FeoB from these three species: FeoB sequences, translated from the *feo*A_1A_2*BC* operons, exhibit more similarity to other species than to FeoB present in other operons (Table 10.2). Taken together, gene organization and sequence homology indicate that the *feo*A_1A_2*BC* operon may be a general feature of the *Rhodobacter* genus, while the *feo2B* operon may be specific to *R. capsulatus*.

Table 10.2 Identity and similarity in FeoA, FeoB and FeoC from *R. capsulatus* SB1003, *R. shaeroides* 2.4.1 and *R. ferroxidans* SW2. Similarity is indicated using parenthesis

FeoA	Rcap $FeoA_1$	Rcap $FeoA_2$	Rcap Feo2A	Rsph $FeoA_1$	Rsph $FeoA_2$	Rfer $FeoA_1$	Rfer $FeoA_2$
Rcap $FeoA_1$	100 (100)						
Rcap $FeoA_2$	19.6 (40.2)	100 (100)					
Rcap Feo2A	15.1 (36.6)	19.6 (37.0)	100 (100)				
Rsph $FeoA_1$	30.0 (42.2)	27.6 (36.8)	18.9 (37.8)	100 (100)			
Rsph $FeoA_2$	18.5 (38.0)	57.3 (69.5)	19.1 (37.1)	23.0 (35.6)	100 (100)		
Rfer $FeoA_1$	31.1 (48.9)	22.1 (38.4)	23.9 (40.9)	52.4 (68.3)	22.1 (38.4)	100 (100)	
Rfer $FeoA_2$	20.7 (42.4)	50.0 (68.3)	15.7 (36.0)	18.4 (34.5)	65.8 (78.5)	19.8 (37.2)	100 (100)

FeoB	Rcap FeoB	Rcap Feo2B	Rsph FeoB	Rfer FeoB
Rcap FeoB	100 (100)			
Rcap Feo2B	30.1 (45.6)	100 (100)		
Rsph FeoB	55.6 (71.8)	29.1 (44.6)	100 (100)	
Rfer FeoB	56.9 (72.7)	30.0 (46.8)	71.1 (82.6)	100 (100)

FeoC	Rcap FeoC	Rsph FeoC	Rfer FeoC
Rcap FeoC	100 (100)		
Rsph FeoC	31.0 (41.7)	100 (100)	
Rfer FeoC	33.3 (44.1)	35.7 (52.4)	100 (100)

Sequences were aligned using the CLC Sequence Viewer software (CLC Bio, Denmark). Similarity and identity scores were calculated based on alignments using the "Ident and Sim" utility (http://www.bioinformatics.org/sms2/ident_sim.html) (Stothard, 2000). Similarity groups are: GAVLI, FYW, CM, ST, KRH, DENQ, P.

Although no experimental evidence exists for the role of Feo gene products in *Rhodobacter* species, it is likely that the general model for the Feo system will prevail. If this is the case then FeoB codes for a membrane bound Fe^{2+} permease with weak GTPase activity that is enhanced by FeoA. FeoA has a SH3 domain that is thought to be involved in protein-protein interactions. In addition, *feoA* mutants in *E. coli* have been shown to have hampered Fe^{2+} uptake (Kammler, Schön, & Hantke, 1993; Kim, Lee, & Shin, 2012). In operons that contain FeoC, this protein is hypothesized to be a transcription factor involved in controlling the *feoABC* operon.

However experimental confirmation of the role of FeoC is still lacking (Fetherston et al., 2012; Guo, Nair, Galván, Liu, & Schifferli, 2011). Multiple control of the FeoB permease by two different FeoA GTPase enhancers, a local transcription factor FeoC, and a global regulator such as Fur might be related to potential toxicity of iron and the need to maintain finely tuned regulation (Cartron et al., 2006; Kammler et al., 1993). A closer look at the FeoB protein sequences from *Rhodobacter* species confirms the presence of the GTPase domain, as out of the 5 G-motifs are conserved (Dashper et al., 2005). The only major divergence may be Feo2B from *R. capsulatus* where the G1 motif exhibits a GPPNCG sequence instead of GNPNCG. The hydrophobic C-terminal domain of FeoB, which consists of two GATE motifs in opposite orientation in the membrane, is similar to the Ftrp1 yeast iron permease. Good conservation of the *Rhodobacter* GATE motifs occurs especially with the presence of two key Cys residues located in the segment IV of GATEs 1 and 2, which are potentially involved in iron binding. Once again, the only striking substitution is observed for Feo2B from *R. capsulatus* where a consensus PC sequence is changed into QC in the GATE 2 segment IV. Preliminary transcription data with *R. capsulatus* show that both Feo systems are controlled by iron availability, but in a different manner. Indeed, while $FeoA_1A_2BC$ is induced ~70- and 200-fold under mild and harsh iron stress, respectively, Feo2AB is induced 3-fold and repressed ~4-fold under the same respective conditions (Zappa and Bauer, 2013). A dendrogram based on FeoB sequence alignment highlights the occurrence of two major clades: one containing the *Rhodobacter* FeoB sequences and the other containing *R. capsulatus* Feo2B (Fig. 10.1). Interestingly, the two FeoB from *Porphyromonas gingivalis* are distributed in each clade. In *P. gingivalis* FeoB1 was shown to be involved in iron uptake, while FeoB2 was shown to be the major manganese transporter (Dashper et al., 2005; He et al., 2006). *R. capsulatus* Feo2B may therefore be involved in manganese homeostasis.

2.2. EfeUOB(M) System

The *E*lemental *Fe*rrous iron (EfeUOB) system was recently identified as a transporter highly specific to Fe^{2+} (Cao, Woodhall, Alvarez, Cartron, & Andrews, 2007; Grosse et al., 2006). Orthologues of *E. coli* EfeOUB, formerly called YcdNOB, are found in many bacterial genomes {(Rajasekaran, Nilapwar, et al., 2010b). Fe^{2+} uptake systems similar to but distinct from EfeUOB (called EfeUOB-like) have also been identified in various bacteria, such as the P19-Ftr1P system in *Campylobacter jejuni* (Chan et al., 2010;

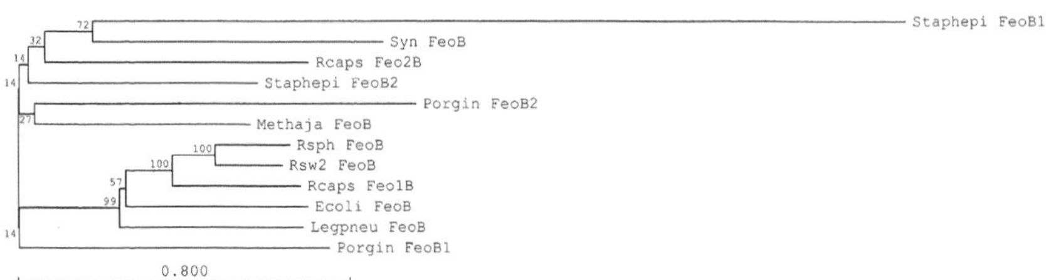

Figure 10.1 Dendrogram of FeoB from various organisms, based on protein sequence alignment. Alignment and dendrogram were built using CLC Sequence Viewer software (CLC Bio, Denmark). The bootstrap analysis algorithm was used, with 100 replicates. Bootstrap values are indicated at each knots and substitution rate at the bottom. Sequences were retrieved from the Integrated Microbial Genomes of the DOE Joint Genome Institute (http://img.jgi.doe.gov/cgi-bin/w/main.cgi). Ecoli: *Escherichia coli* DH1, Legpneu: *Legionella pneumophila* Paris, Methaja: *Methanocaldococcus jannaschii* DSM 2661, Porgin: *Porphyromonas gingivalis* ATCC 33227, Rcap: *Rhodobacter capsulatus* SB1003, Rsph: *Rhodobacter sphaeroides* 2.4.1, Rfer: *Rhodobacter ferroxidans* SW2, Staphepi: *Staphylococcus epidermidis* ATCC 12228, Syn: *Synechocystis sp.* PCC 6803. Accession numbers: Ecoli FeoB, BAJ45144; Legpneu FeoB, YP_125016; Methaja FeoB, NP_247545; Porgin FeoB1, YP_001929201; Porgin FeoB2, YP_001929425; Rcaps FeoB, YP_003576264; Rcaps Feo2B, YP_003578180; Rsph FeoB, ; YP_351866 Rsw2 FeoB, ZP_05845183; Staphepi FeoB1, NP_763744; Staphepi FeoB2, NP_765669; Syn FeoB, NP_440528.

van Vliet, Wooldridge, & Ketley, 1998), FetMP in *E. coli* (Koch et al., 2011) and FtrABCD in *Bordetella* species (Brickman & Armstrong, 2012), making the EfeUOB-type transporter a widely utilized iron uptake strategy. Among the *Rhodobacter* genomes, *R. capsulatus* displays an EfeUOB operon but no EfeUOB-like system (Table 10.1, Fig. 10.2a). In *R. sphaeroides* and in *R. ferroxidans* SW2 there is a surprising absence of EfeUOB and EfeUOB-like operons. Moreover, *R. capsulatus* also encodes a putative EfeU-EfeO fusion protein (Fig. 10.2b). Although unusual, a similar EfeU-EfeO fusion has been reported previously (Rajasekaran, Nilapwar, et al., 2010b).

In *E. coli*, transcription of the *efeUOB* operon is known to be induced under iron starvation, low pH or in the presence of exogenous copper. The transcription factor Fur and the phosphorelay CpxAR are involved in iron- and pH-dependent expression, respectively (Cao et al., 2007). Interestingly, expression occurs under aerobic conditions (Cao et al., 2007). In addition to *E. coli*, a few other studies indicate that *efeUOB* homologues from other species are also controlled by iron availability. For example, *efeUOB* homologues in *Bacillus subtilis*, (Baichoo, Wang, Ye, & Helmann, 2002; Ollinger, Song, Antelmann, Hecker, & Helmann, 2006), *Neisseria menigitidis* (Grifantini et al., 2003), the magnetotactic bacterium strain MV-1 (Dubbels et al., 2004) and *Magnetospirillum magneticum* AMB-1

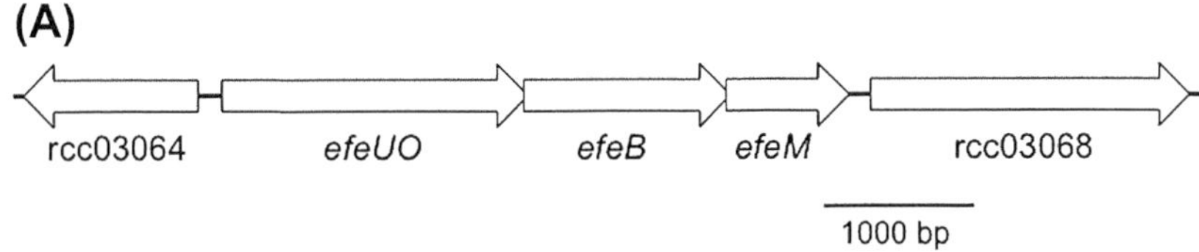

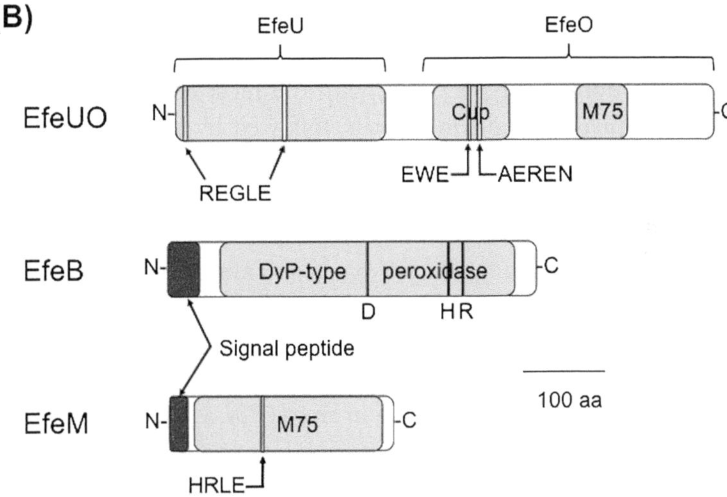

Figure 10.2 Genetic organization (a) and protein architecture (b) of *R. capsulatus* EfeUOBM system. Flanking genes orf 03064 and orf 03068 are predicted to encode a membrane protein (HPP family/CBS domain) and a cache sensor protein, respectively. Arrows indicate conserved sequences putatively involved in metal binding. Cup: cupredoxin domain; M75: M75 metallopeptidase domain.

(Suzuki, Okamura, Calugay, Takeyama, & Matsunaga, 2006) are regulated in response to iron.

Only a few experimental studies on the function of *efeUOB* proteins have been reported, so a large part of our understanding comes from sequence analysis. The EfeU protein is homologous to the yeast iron permease Ftr1p, with seven transmembrane helices (TMH). Two of these helices, TMH-I and TMH-IV, contain a iron transport REXXE motif (Grosse et al., 2006; Rajasekaran, Nilapwar, et al., 2010b). These motifs are conserved in the *R. capsulatus* EfeU domain of the EfeU-EfeO fusion protein (Fig. 10.2b).

EfeB is a periplasmic homodimeric heme-containing DyP-type peroxidase (Liu et al., 2011; Sturm, Schierhorn, Lindenstrauss, Lilie, & Brüser, 2006). EfeB has also been proposed to act as a deferrochelatase, providing iron to *E. coli* by extracting iron from heme (Létoffé, Heuck, Delepelaire, Lange, & Wandersman, 2009). However, a study with the related protein YfeX could not confirm such a role (Dailey et al., 2011). The *R. capsulatus* EfeB sequence shows the presence of a conserved TAT signal (Bendtsen, Nielsen, Widdick, Palmer, & Brunak, 2005) indicating that it is likely exported to the periplasm (Fig. 10.2b). Moreover, alignments with other DyP-type peroxidases (not shown) show conservation of residues involved in heme binding: D235, H330, R347 (Sugano, 2009).

The role of EfeO (COG 2822) is even less clear. EfeO proteins can present different domain organization, with the most common one comprising a N-terminal cupredoxin domain followed by a C-terminal peptidase M75 domain. However EfeOs consisting of only a cupredoxin domain or a peptidase M75 domain are also frequently encoded in genomes (Rajasekaran, Nilapwar, et al., 2010). Although nomenclature is still being defined, the trend is to name cupredoxin-containing members as EfeO, and cupredoxin-less members that consist of a solo peptidase M75 domain as EfeM (Rajasekaran, Mitchell, et al., 2010; Rajasekaran, Nilapwar, et al., 2010). The *R. capsulatus* putative operon shows two representatives of the EfeO/M family (COG 2822), resulting in an *efeUOBM* operon where *efeU* and *efeO* are fused (Table 10.1, Fig. 10.2a). The first one is fused to the EfeU domain mentioned above and has the signature of an EfeO protein, *i.e.* showing both the cupredoxin and M75 domains (Fig. 10.2b). On the other hand, the second copy (orf 03067) is a typical M75, containing only EfeM. Sequence motifs potentially involved in iron binding (EWE, EEREN) are conserved in the EfeUO cupredoxin domain (Fig. 10.2b). The EfeM peptide also contains the putative HXXE iron binding sequence while EfeUO M75 domain does not, highlighting probable functional differences between these two EfeO/M like proteins (Rajasekaran, Mitchell, et al., 2010; Rajasekaran, Nilapwar, et al., 2010). Finally, EfeO/M are thought to be periplasmic proteins (Rajasekaran, Nilapwar, et al., 2010; Sturm et al., 2006) as a signal peptide signature sequence is predicted to be present on EfeM (Petersen, Brunak, Heijne von, & Nielsen, 2011). The putative cellular location of EfeO is less clear. Presumably, export of the EfeO domain in the periplasm could be achieved during folding of EfeUO, but this has yet to be confirmed experimentally.

A proposed mechanism of Fe^{2+} uptake by the EfeUOB system is based on homology with the yeast permeation/ferroxidation Ftr1p/Fet3p system (Kosman, 2003; Rajasekaran, Nilapwar, et al., 2010; Stearman, Yuan, Yamaguchi-Iwai, Klausner, & Dancis, 1996). Briefly, Fe^{2+} in the periplasm binds to the M75 domain of EfeO and is subsequently oxidized to Fe^{3+} by the copper center of the cupredoxin domain. Fe^{3+} is then transferred first to the EfeO cupredoxin domain and then to the permease EfeU. Finally, the copper center is regenerated by the EfeB peroxidase (Rajasekaran, Nilapwar, et al., 2010). A similar mechanism could be possible in *R. capsulatus*, involving both the EfeO domain of EfeUO and EfeM instead of a unique EfeO. Moreover, according to the hypothetical mechanism, EfeU and EfeO are interacting. The fusion of these proteins in *R capsulatus* is compatible with such a hypothesis. In summary, *R. capsulatus* is unique among *Rhodobacter* species in that it seems to have an intact Fe^{2+} uptake system, with analysis of the sequence indicating that it is functional (Fig. 10.2a and b).

3. FERRIC IRON UPTAKE

The major Fe^{3+} uptake system involves mediation by siderophores. Such iron uptake pathways have been extensively studied and reviewed (Chu et al., 2010; Hider & Kong, 2010; Krewulak & Vogel, 2008, 2011; Köster, 2001; Sandy & Butler, 2009). Briefly, siderophores are small molecules secreted by bacteria, fungi and graminaceous plants that solubilize Fe^{3+} in aerobic environments, due to their high binding affinity for Fe^{+3} (K_d<10^{-20} M). In Gram-negative micro-organisms, siderophore transporters consist of a tonB-dependent outer membrane ferrisiderophore receptor and an ABC transporter cassette (a periplasmic siderophore binding protein, a permease and an ATPase). Such organization is well conserved among bacteria and archaea with similar systems used to import heme and vitamin B_{12}. The specificity of the outer membrane receptor is usually very high compared to the specificity of the ABC cassette. This sometimes enables the cross-use of the same ABC cassette to uptake different siderophores that have been imported into the periplasm by different outer membrane receptors.

According to *Rhodobacter* genome annotations, *R. capsulatus* and *R. sphaeroides* display an *exbB-exbD-tonB* operon while *R. ferroxidans* does not (Table 10.3). The inner membrane bound ExbB–ExbDTonB complex enables transfer of proton-motive force from the cytoplasmic membrane to the outer membrane receptor. As such, it provides the energy needed to import the ferrisiderophore through the outer membrane receptor to a periplasmic binding protein for delivery to the cytoplasmic membrane bound ABC transporter, which then imports the ferrisiderophore into the cytoplasm. Since the ExbB–ExbDTonB complex is absent from

Table 10.3 Ferrisiderophore uptake genes identified in the genomes of *R. capsulatus* SB1003, *R. sphaeroides* 2.4.1 and *R. ferroxidans* SW2

Gene product	COG	Label *Rcap* SB1003*	Label *Rsph 2.4.1**	Label *Rfer* SW2*
ExbB	0811	02375	0920	N.D.
ExbD	0848	02376	0921	N.D.
TonB	0810	02377	0922	N.D.

Uptake element	COG	Label in *Rcap* SB1003*	Comments from annotations
TOMR	4773	00111 (*fhuE*)	Ferrichrome uptake system
PBP	4607	00108 (*fhuD1*)	Presence of a putative esterase (orf00110)
IMP	4605	00106 (*fhuB1*)	
	4606	00107 (*fhuB2*)	
ATPase	4604	00105 (*fhu-1*)	
TOMR	4774	01049	Outer membrane receptor for monomeric catechols
PBP	0614	01047	Presence of a nearby enterochelin esterase (orf 01050) and an AraC-like regulator (orf 01048)
IMP	0609	01046	
ATPase	1120	01045	
TOMR	1629	01429	Presence of a protein from the major facilitator superfamily: potential role of PBP?
PBP	N.D.	N.D.	
IMP/ATPase	1132	01426/01427	
TOMR	4774	01433	Enterobactin uptake system
PBP	0614	01434 (*fepB1*)	Presence of a nearby siderophore interacting protein (orf 01438) and two AraC-like regulators (orf 01431 and 01432)
IMP	0609	01435 (*fepD1*)	
	4779	01436 (*fepG1*)	
ATPase	1120	01437 (*fepC1*)	
TOMR	1629	01445	Enterobactin uptake system
PBP	4592	01444 (*fepB2*)	Presence of a nearby siderophore interacting protein (orf 01438)
IMP	0609	01443 (*fepD2*)	
	4779	01442 (*fepG2*)	
ATPase	1120	01441 (*fepC2*)	

Table 10.3 Ferrisiderophore uptake genes identified in the genomes of *R. capsulatus* SB1003, *R. sphaeroides* 2.4.1 and *R. ferroxidans* SW2—cont'd

Uptake element	COG	Label in *Rcap* SB1003*	Comments from annotations
TOMR	1629	p00051	Presence of one TOMR with two ABC transport systems
PBP	0614	p00046	
		p00050	
IMP	0609	p00045	
		p00049	
ATPase	1120	p00044	
		p00048	
TOMR	4774	p00112 (*fhuA*)	Ferrichrome uptake system
PBP	0614	p00111 (*fhuD2*)	Presence of a nearby AraC-like regulator (orf p00113)
IMP	0609	p00110 (*fhuB3*)	
	1132	p00108	
ATPase	1132	p00108	
	1120	p00109 (*fhuC2*)	
Uptake element		**Label in *Rsph* 2.4.1***	**Comments from annotations**
TOMR	4774	1440	Hydroxamate –type siderophore uptake system
PBP	0614	1439	orf 1438 is a fused subunit permease
IMP	0609	1438	
	4779		
ATPase	1120	1437	
TOMR	4771	3223	Enterochelin/colicin uptake system
PBP		N.D.	A periplamic protein, annotated as histidine kinase, is in the same operon (orf 03225): a potential PBP?
IMP	4606	3220	
	4605	3221	
ATPase	4604	3222	
TOMR	4774	3417	Ferrichrome uptake system
PBP	0614	3416	
IMP	0609	3415	
	4779	3414	
ATPase	1120	3413	

Continued

Table 10.3 Ferrisiderophore uptake genes identified in the genomes of *R. capsulatus* SB1003, *R. sphaeroides* 2.4.1 and *R. ferroxidans* SW2—cont'd

Uptake element		Label in *Rsph* 2.4.1*	Comments from annotations
TOMR	4773	4273 (*fhuA*)	Fe^{3+} coprogen or Fe^{3+} rhotorulic acid uptake system. In the same putative operon of a FecRI system (orf 4274 and 4275)
PBP	06014	4272 (*fhuD*)	
IMP	0609	4271 (*fhuB*)	
ATPase	1120	7397	

N.D.: Not detected; TOMBR: TonB-dependent outer membrane receptor; PBP: Periplasmic binding protein; IMP: Inner membrane protein.
*See Table 10.1.

R. ferroxidans it is not surprising that no siderophore uptake system is annotated in the *R. ferroxidans* SW2 genome. On the other hand, and complete siderophore uptake systems are present in the *R. capsulatus* SB1003 and *R. sphaeroides* 2.4.1 genomes, respectively. Inspection of the sequences and the genomic environment indicates that the respective TonB-dependent uptake systems orf 3358 to orf 3362 (in *R. capsulatus*) and orf 2102 to orf 2105 (*R. sphaeroides*) are more likely to be involved in vitamin B_{12} uptake than siderophore transport. These strains still potentially encode to siderophore uptake systems as detailed in Table 10.3. Interestingly, no siderophore synthesis gene cluster are identified in either genome, meaning that *R. capsulatus* and *R. sphaeroides* likely scavenge siderophores from other species (termed xenosiderophores) to fulfill their iron needs. Such a situation is not unusual as some species even rely on xenosiderophores to provide enough iron for growth (D'Onofrio et al., 2010). This situation exists between bacteria as well as between bacteria and fungi (Kosman, 2003). Moreover, it has been shown that bacteria can use products of primary metabolism (α-keto acids) as siderophores (Reissbrodt et al., 1997).

Once a ferrisiderophore complex enters the cytoplasm, its fate is not well known. One of the most studied siderophore processing events is the involvement of the enterobactin esterase in *E. coli*. This enzyme hydrolyses enterobactin, producing trimers of dihydroxybenzoylserine thereby weakening the bond with Fe^{3+}. Enterobactin esterase also seems to act as an enterobactin-specific reductase that reduces Fe^{3+} into soluble Fe^{2+} which triggers iron release from the siderophore (Andrews et al., 2003;

Rudolph, Hennecke, & Fischer, 2006). Importantly, two putative esterases (both located in siderophore uptake gene clusters) are annotated in the *R. capsulatus* genome, orf 00110 and orf 01050. The latter is annotated as an enterobactin (enterochelin) esterase. The presence of this gene is very unusual among α-proteobacteria (Rudolph et al., 2006).

4. IRON ABC TRANSPORTERS

Some bacteria also contain metal-ABC transporters that exhibit specificity for iron. Unlike classical siderophore-based iron uptake systems, the metal-ABC transporters are not dependent on an outer membrane receptor for iron or siderophore transport into the periplasmic space. The best characterized version of this class of transporters is the FbpABC system (called sometimes AfuABC) studied in *Neisseria gonorrhoeae*, *N. meningitidis*, *C. jejuni*, *Bordetella pertussis*, *Marinobacter* species, *Vibrio cholerae*, *Pasteurella multocida* and *Actinobacillus pleuropneumoniae* (Amin, Green, Waheeb, Gärdes, & Carrano, 2012; Brickman, Cummings, Liew, Relman, & Armstrong, 2011; Chin, Frey, Chang, & Chang, 1996; Khun, Kirby, & Lee, 1998; Paustian, May, & Kapur, 2001; Strange, Zola, & Cornelissen, 2011; Tom-Yew, Cui, Bekker, & Murphy, 2005; Wyckoff, Mey, Leimbach, Fisher, & Payne, 2006). This transporter is specific to Fe^{3+} and consists of a Fe^{3+}-binding periplasmic protein, a membrane permease and an ATPase. In *Neisseria* and *Bordetella*, this transporter also undergoes tonB-independent uptake of iron from endogenous siderophores or xenosiderophores (Brickman et al., 2011; Strange et al., 2011). Similar transporters have been characterized in *Serratia marcescens* and *Haemophilus influenzae* under the respective names SfuABC and HitABC (Angerer, Klupp, & Braun, 1992; Sanders, Cope, & Hansen, 1994). The cyanobacterium *Synechocystis* sp. PCC6803 also contains this transport system, with the particularity of having two Fe^{3+}-binding proteins instead of one (Badarau et al., 2008; Katoh, Hagino, Grossman, et al., 2001; Katoh, Hagino, & Ogawa, 2001).

Present in the annotated *R. sphaeroides* genome is a putative homologue of FbpA, the periplasmic substrate-binding protein of the FbpABC system (Tom-Yew et al., 2005). Looking closer at the genome reveals that this ORF (orf 2913) is not part of an FbpABC operon. On the other hand, a second ORF matching FbpA's COG 1840 appears to share an operonal organization with three ORFs that are subunits of an ABC transporter: two inner membrane proteins and an ATPase. These four ORFs (orf 0346 to orf 0349) may thus constitute an FbpABC transport system in *R. sphaeroides*. Such a transport system is also found in *R. ferroxidans* and *R. capsulatus*, where similar gene organization is observed (Table 10.4).

Another metal-ABC transporter system, YfeABCD/SitABCD, is thought to be involved in iron uptake, although its specificity between iron and manganese is not clear. Indeed, the YfeABCD system has been shown to transport both Fe^{3+} and Mn^{2+} in *Yersinia pestis* and *Photorhabdus luminescens*, being involved in virulence mechanisms in both cases (Bearden & Perry, 1999; Watson, Millichap, Joyce, Reynolds, & Clarke, 2010). The SitABCD system has also been shown to transport both Fe^{2+}, instead of Fe^{3+}, and Mn^{2+} in *Sinorhizobium meliloti* and *Shigella flexneri* (Chao, Becker, Buhrmester, Pühler, & Weidner, 2004; Fisher et al., 2009; Platero, 2004). In an *E. coli* avian pathogenic strains, the transport of Fe^{2+}, Fe^{3+} or Mn^{2+} varied as a function of the strain genetic background (Sabri, Léveillé, & Dozois, 2006). Finally, SitABCD in *Salmonella enterica* Serovar *Typhimurium* was shown to uptake both Fe^{2+} and Mn^{2+} but with an affinity for Mn^{2+} that is stronger and physiologically more relevant (Janakiraman & Slauch, 2000; Kehres, Janakiraman, Slauch, & Maguire, 2002). An orthologous SitABCD transporter is annotated in the genome of *R. sphaeroides* 2.4.1, but not in *R. capsulatus* SB1003 and *R. ferroxidans* SW2 genomes (Table 10.4). However, there is no sequence signature in the SitABCD system in general, and in the *R. sphaeroides* SitABCD cluster in particular, that enables prediction of its involvement in either iron or/and manganese transport. Nevertheless, *R. sphaeroides* does not seem to have an NRAMP-type manganese dedicated transporter MntH, while *R. capsulatus* and *R. ferroxidans* SW2 contain the MntHR Mn transport system (Table 10.5). Assuming that *R. sphaeroides* has typical manganese needs, and that no other manganese transporters are present, it is reasonable to hypothesize that the SitABCD system in *R. sphaeroides* is dedicated to manganese uptake rather than to iron.

5. HEME IRON USAGE

Heme is often a crucial iron source for pathogens that scavenge it from their host. However, heme uptake is also frequently an iron acquisition mechanism that occurs in beneficial symbiotic bacteria (Anzaldi & Skaar, 2010; Nienaber, Hennecke, & Fischer, 2001; Runyen-Janecky, Brown, Ott, Tujuba, & Rio, 2010; Septer, Wang, Ruby, Stabb, & Dunn, 2011). A major hemin uptake system is the PhuRSTUVW-type heme transporter as described in *Pseudomonas aeruginosa*, where PhuR is a TonB-dependent outer membrane heme receptor. The PhuT subunit of the PhuRSTUVW heme transporter

Table 10.4 Iron-ABC transporter genes identified in the genomes of *R. capsulatus* SB1003, *R. sphaeroides* 2.4.1 and *R. ferroxidans* SW2

ABC transporter type	Gene name	COG	Label *Rcap* SB1003*	Label *Rsph* 2.4.1*	Label *Rfer* SW2*	Product
FbpABC		1840	01369	0346	1135	ABC transporter, substrate-binding protein
		4132	01370	0347	1134	ABC transporter, inner membrane subunit
		1177	01371	0348	1135	ABC transporter, inner membrane subunit
		3842	01371	0348	1135	ABC transporter, ATPase subunit
SitABCD	*sitA*	0803	N.D.	0904	N.D.	ABC Mn^{+2}/Fe^{+2} transporter, periplasmic substrate-binding protein SitA
	sitB	1121	N.D.	0905	N.D.	ABC Mn^{+2}/Fe^{+2} transporter, ATPase subunit SitB
	sitC	1108	N.D.	0906	N.D.	ABC Mn^{+2}/Fe^{+2} transporter, inner membrane subunit SitC
	sitD	1108	N.D.	0907	N.D.	ABC Mn^{+2}/Fe^{+2} transporter, inner membrane subunit SitD

N.D.: Not detected
*See Table 10.1.

Table 10.5 Putative Fe^{2+} efflux pumps (COG 0053) identified in the genomes of *R. capsulatus* SB1003, *R. sphaeroides* 2.4.1 and *R. ferroxidans* SW2, with the sequence corresponding to Fe/Zn binding sites dimerization in the *E. coli* FieF (Lu & Fu, 2007; Nies, 2011)

Organism/ gene locus	Fe/Zn-binding site**		
	Z1	**Z2 dimerisation**	**Z3/Z4**
E. coli K12 FieF*	$D_{45}X^3D_{49}$–$H_{153}X^3D_{157}$	$D_{68}DNHX^3H_{75}$	H_{232}–H_{261}–$H_{283}XD_{285}$
Rcaps/orf 00522	$D_{47}X^3D_{51}$–$H_{156}X^3D_{160}$	$D_{70}DDHX^3H_{77}$	H_{236}–H_{265}–$H_{287}XD_{299}$
Rsph/orf 0463	$E_{40}X^3N_{44}$–$H_{148}X^3D_{152}$	$D_{63}ANHX^3H_{70}$	H_{229}–H_{258}–$H_{280}XE_{282}$
Rfe/orf 1374	$D_{50}X^3D_{54}$–$H_{160}X^3D_{164}$	$D_{73}EDHX^3H_{80}$	H_{240}–H_{269}–$H_{291}XD_{293}$
Rfe/orf 2021	$H_{88}X^3D_{92}$–$H_{196}X^3D_{200}$	$S_{111}RTFX^3L_{118}$	H_{247}–L_{303}–$H_{419}XV_{421}$

**E. coli* K12 Fief accession number NP_418350.
**numbering relative to each sequence is given. X: random amino acid. X^n: succession of *n* random amino acids.

is the periplamic heme-binding protein, PhuUV is an ABC transporter and PhuS is a cytoplasmic protein (Anzaldi & Skaar, 2010). Extensive characterization of similar systems has been undertaken in the pathogenic bacteria *B. pertussis*, *Yersinia pestis*, *Yersinia enterolitica*, *Shigella dysenteriae*, *V. cholerae*, *C. jejuni*, *Bartonella quintana* and *E. coli* O157:H7 (Anzaldi & Skaar, 2010) with the transporters described using varying terminology. Homologous transporters have been described in the symbiotic *Bradyrhizobium japonicum*, *Rhizobium leguminosarum* and *S. meliloti* where they are called HmuQR-HmuTUV, HmuPSTUV and ShmR-HmuPSTUV, respectively (Amarelle et al., 2010; Amarelle, O'Brian, & Fabiano, 2008; Anzaldi & Skaar, 2010; Nienaber et al., 2001; Wexler et al., 2001).

R. capsulatus has an orthologous gene cluster annotated as *hmuRSTUV* (loci from orf 00094 to orf 00098). Interestingly, *R. sphaeroides* and *R. ferroxidans* do not seem to code for this heme transporter. Thus, unlike other *Rhodobacter* representatives, *R. capsulatus* is presumably able to use heme from exogenous sources. Unlike *B. japonicum* and *R. leguminosarum*, the *R. capsulatus* heme uptake gene cluster does not have a nearby TonB-ExBD subunit. It may therefore rely on a homologous gene located in a different region of the genome (Table 10.3) or contain an unidentified protein that provides this function. Also, the *R. capsulatus* Hmu system lacks the HmuP regulator that has been identified in *B. japonicum* and *S. meliloti*, which functions as a co-activator of this heme uptake system along with Irr (Amarelle et al., 2010; Escamilla-Hernandez & O'Brian, 2012). A BLAST analysis of the *Rhodobacter* clade indicates possible candidates for an HmuP transcription factor in *R. capsulatus* (orf 01112) and *R. sphaeroides* (orf 6006) but not in *R. ferroxidans*. However, it should be cautioned that these putative HmuP homologues have poor sequence conservation, and no genome context with heme transporters, so their actual function needs to be experimentally confirmed.

The heme uptake gene cluster in *R. capsulatus* also contains *hmuS* that in other organisms has been shown to be involved in heme degradation. The mechanism of action of HmuS orthologues is not clear but seems eclectic: i) Some HmuS were shown to enzymatically degrade heme; ii) Some seem to operate via a non-enzymatic process with H_2O_2; iii) Some may store and/or traffic heme to a heme oxygenase (Anzaldi & Skaar, 2010; Barker, Barkovits, & Wilks, 2012; Liu, Boulouis, & Biville, 2012; O'Neill, Bhakta, Fleming, & Wilks, 2012).

Based on COG analysis, it is notable that *R. capsulatus* contains a second putative HmuS locus (orf 03488) that is located next to a putative coproporphyrinogen oxidase III gene. *R. sphaeroides* only has one HmuS homolog (orf 0228), while *R. ferroxidans* has no HmuS annotated. The presence of a putative heme degrading system is thought to be crucial for the use of heme as an iron source as well as to prevent heme toxicity by ensuring that there is no pool of unbound free heme in the cell (Anzaldi & Skaar, 2010; Frankenberg-Dinkel, 2004). Moreover, a typical heme oxygenase (BphO-like) is not present in either *R. capsulatus* or *R. ferroxidans*, while there are two annotated heme oxygenase genes annotated in *R. sphaeroides*. These are associated with bacteriophytochrome encoding genes (orf 4191, orf 7212). As shown recently in *P. aeruginosa*, bacteriophytochrome-associated BphO is not involved in heme degradation from exogenous heme uptake, but only in holo-bacteriophytochrome synthesis from *de novo* synthesized heme (Barker et al., 2012). Atypical IsdG-like heme oxygenases HmuQ and HmuD were recently characterized in *B. japonicum* (Puri & O'Brian, 2006; Skaar, Gaspar, & Schneewind, 2006). A BLAST analysis against the *Rhodobacter* clades reveals the presence of a potential HmuQ/D-like protein in *R. sphaeroides* (orf 0826) however further work is needed to confirm this identity. Finally, as detailed in chapter II.B, the EfeUOB system has been shown to be able to extract iron from heme by a deferrochelatase activity, although such activity is debated (Dailey et al., 2011; Létoffé et al., 2009). Such an activity could provide iron to

R. capsulatus, the only representative of the *Rhodobacter* genus displaying the *efeUOB* gene cluster (Table 10.1).

A final point regarding heme usage is the presence of "heme exporter" gene clusters in the annotation of the three *Rhodobacter* strains of interest. A closer look reveals that they correspond to a CcmABCDG system used for *c*-type cytochrome synthesis. It has been shown in *R. capsulatus* that a mutation that disrupts cytochrome *c* maturation results in massive secretion of porphyrins (Biel and Biel, 1990). The authors did not confirm if a *ccmABCDG* operon mutation was the cause of this phenotype but did speculate that there is a connection between iron homeostasis, heme and cytochrome synthesis (Biel and Biel, 1990; Lascelles, 1956).

6. MANAGING IRON TOXICITY

As stated in the introduction, free iron can be extremely toxic. Thus, once cellular iron needs are fulfilled, being able to counterbalance iron intake to avoid overload is definitely of interest for a cell. There are two ways available in bacteria to counterbalance iron intake. One mechanism consists in reversing the uptake process by excreting excess iron using iron efflux pumps. The second process involves the binding or sequestration of iron in a dedicated peptide where it is stored in a harmless state. These peptides are called ferritins and/or bacterioferritins.

6.1. Iron Efflux Pump

The efflux of iron in bacteria is not well characterized although the last decade has highlighted a prominent role of Cation Diffusion Facilitators (CDF). The CDF family is ubiquitous in the three domains of life and grouped together as heavy metal transporters. They are classified into three subfamilies according to their metal substrate specificity: Zn-CDF, Fe/Zn-CDF and Mn-CDF (Montanini, Blaudez, Jeandroz, Sanders, & Chalot, 2007; Nies, 2011). Despite having a "favorite" metal substrate, they usually are capable of transporting a wide array of metals (Montanini et al., 2007; Munkelt, Grass, & Nies, 2004). One of the most studied representatives is the Fe^{2+} efflux protein (FieF, also known as YiiP). Its presence was shown to increase *E. coli*'s tolerance to iron and to lower the total iron cellular content (Grass et al., 2005). Although FieF was biochemically well characterized *in vitro* using zinc as a substrate, its main substrate *in vivo* is Fe^{2+} and, as such, is likely to play an important role in iron homeostasis (Nies, 2007). An ORF was annotated as FieF in *R. capsulatus* SB1003 (orf 00522). Using the same COG number (COG 0053), candidates were found in the two other *Rhodobacter* genomes: orf 0463 in *R. sphaeroides* 2.4.1, and orf 1374 and orf 2021 in *R. ferroxidans* SW2. *E. coli* FieF has four metal binding sites named Z1 through Z4 (Lu & Fu, 2007; Nies, 2011). Alignment of the putative *Rhodobacter* FieF sequences shows good conservation at most of these sites, with the exception of orf 2021 in *R. ferroxidans*. Other CDF genes were also found in these *Rhodobacter* genomes, under COG 1230, but their sequences align better with the *E. coli* zinc transporter ZitB then with the iron transporter FieF (data not shown). However, given the relaxed substrate specificity of CDFs, it is worth noting that the *R. capsulatus* ZitB homologue (orf 00089) is located next to the $feoA_1A_2BC$ putative Fe^{2+} uptake system.

6.2. Storage and Detoxification

The Ftn family consists of three types of protein that form distinct phylogenetic clades: the ferritins (Ftn), the bacterioferritins (Bfr) and the *DNA-binding proteins from starved cells* (DPS). Ftn and Bfr have strong structural homology with each consisting of homooligomers of 24 subunits. DPS are comprised of 12-mers of the same subunit (Andrews, 2010; Bou-Abdallah, 2010). Ftn and Bfr form ball-shaped complexes with an outer diameter of ~120 Å that can theoretically store up to 4500 iron atoms in a ~-Å diameter cavity. The iron in this cavity consists of either amorphous iron with inorganic phosphate or crystalline ferrihydrite. DPS also form a ball-like structure, but they have an outer diameter of ~95 Å that could encapsulate up to ~500 molecules of iron (Andrews, 2010; Bou-Abdallah, 2010; Carrondo, 2003). Cellular iron concentration is estimated at ~10^{-4}M, which is far above the solubility of this metal. This concentration is reached due to the presence of these iron sequestration storage proteins, which concentrate and store the metal in non-reactive form (Theil & Goss, 2009). By isolating the toxic iron from cellular machinery, storing it and releasing it as needed, ferritins function almost as cellular organelles.

The ferritin family is ubiquitous with Ftn found in all three domains of life while Bfr and DPS are specific to Bacteria and Archaea (Andrews, 2010). Their activity relies on ferroxidase centers consisting of *intra*subunit di-iron centers in Ftn and Bfr and *inter*subunit di-iron centers in DPS (Andrews, 2010). These centers channel Fe^{2+} into the cavity by oxidizing it into Fe^{3+}, which generates the insoluble oxidized form of iron stored inside the complex core. It seems that Ftn and Bfr use mostly O_2 as the iron oxidant while DPS use H_2O_2 (Andrews, 2010; Bou-Abdallah, 2010). The main role Ftn and Bfr is for iron storage, while the function of DPS is thought to be as a detoxifier that protects DNA from redox stress generated by the iron mediated Fenton reaction (Eqns (10.1)–(10.3)) (Andrews, 2010). It is quite common for bacteria to have each type of ferritin as well as several copies of one type of ferritin. For example, *E. coli* is typical with the presence of two Ftn, one Bfr and one DPS (Andrews et al., 2003; Chiancone et al., 2004). The major difference between Ftn and Bfr is the presence of 12 b-type heme moieties in the Bfr complex (Andrews, 2010; Bou-Abdallah, 2010; Carrondo, 2003; Cobessi et al., 2001). After decades of mystery about the role of the heme groups, it has recently been shown that they are involved in the mobilization of iron back to the cytosol. Thus while ferroxidase centers oxidize the iron during the mineralization process that internalizes the iron into the Bfr, the heme moieties reduce the core iron for its export into the cytoplasm. Ferroxidase centers and heme moieties function independently (Yasmin, Andrews, Moore, & Brun, 2011).

Analysis of the *Rhodobacter* genomes reveals the presence of Ftn and Bfr proteins (Table 10.6). *R. sphaeroides* has a membrane-bound ferritin (orf 0850). A BLAST analysis on the genome of *R. capsulatus* also reveals the presence of a homologue in this species (orf 03466). Interestingly, no homologues were found in the *R. ferroxidans* genome in the NCBI database, but an ORF with the same COG number was found in the JGI database (orf 0178). It is annotated as rubrerythrin, which is likely to be the ancestral form of the ferritin-like protein family (Andrews, 2010). The annotation is probably incorrect as the three putative Ftn proteins share strong homology (more than 70% identity and more than 80% similarity, as shown in Table 10.7). Indeed, they are all 325 amino acids long while Bfr proteins fall in the 160 amino acid range. Conserved domain analysis showed that they all consist of an Ftn-like N-terminal domain of ~140–150 amino acids

Table 10.6 Ferritin-family proteins identified in the genomes of *R. capsulatus* SB1003, *R. sphaeroides* 2.4.1 and *R. ferroxidans* SW2

Ftn type	Gene name	COG	Label *Rcap* SB1003*	Label *Rsph* 2.4.1*	Label *Rfer* SW2*	Comments
Ftn	*mbfA*	1633	03466	0850	0178	Membrane-bound ferritin
Bfr	*bfr*	2193	00913	1546	N.D.	
	bfr2	2193		3342		
DPS		0783	N.D.	N.D.	N.D.	

N.D.: Not detected
*See Table 10.1.

followed by a C-terminal CCC1 domain of ~120-140 amino acids (not shown). CCC1 domains are involved in iron and manganese transport. In yeast, CCC1 is a vacuole transmembrane protein responsible for iron and manganese accumulation in vacuoles (Li, Chen, McVey Ward, & Kaplan, 2001). Finally, these genes share similarity in terms of genomic organization, as seen in Fig. 10.3, indicating an ancient gene cluster that was conserved during evolution.

Beside a membrane-bound ferritin, Bfr homologues are present in the genomes of *R. capsulatus* and *R. sphaeroides*, but not in *R. ferroxidans* (Table 10.6). In fact, *R. sphaeroides* displays two Bfr, one on each chromosome, while *R. capsulatus* has only one. While *R. capsulatus* Bfr share almost 80% similarity with each of the *R. sphaeroides* Bfr, the latter proteins are close to 90% similar to each other (Table 10.7). This suggests that a single *bfr* gene was duplicated in *R. sphaeroides*. Unlike membrane-bound ferritin, Bfr show weaker genomic conservation. One *R. sphaeroides* Bfr (orf 1446) is organized in a putative operon with a Bfr-associated-ferredoxin (orf 1447), an iron-regulated protein (orf 1448) and a hypothetical protein (orf 6006). Interestingly, the latter shows homology with HmuP, a heme uptake regulator described in *B. japonicum* and *S. meliloti* as reported in Section 5. No iron homeostasis genes are found in the vicinity of the other *R. sphaeroides bfr* gene (not shown). Regarding the *R. capsulatus* Bfr, it seems to be part of an operon that contains a hypothetical protein that aligns well with *R. sphaeroides* Bfr-associated-ferredoxin (orf 1447). Association of a *bfr* gene with a ferrodoxin is very common in bacteria (Rodionov, Gelfand, Todd, Curson, & Johnston, 2006). This genomic colocalization may suggest that orf 1446 was the "original" *bfr* in *R. sphaeroides* before being duplicated without the ferredoxin gene.

Table 10.7 Identity and similarity in MbfA and Bfr from *R. capsulatus* SB1003, *R. shaeroides* 2.4.1 and *R. ferroxidans* SW2. Similarity is indicated using parenthesis*

MbfA	Rcap	Rsph	Rfer
Rcap	100 (100)		
Rsph	73.5 (81.9)	100 (100)	
Rfer	79.1 (86.1)	73.5 (82.4)	100 (100)
Bfr	**Rcap**	**Rsph Bfr**	**Rsph Bfr2**
Rcap	100 (100)		
Rsph Bfr	67.7 (79.5)	100 (100)	
Rsph Bfr2	66.7 (78.8)	82.4 (87.3)	100 (100)

*See Table 10.2.

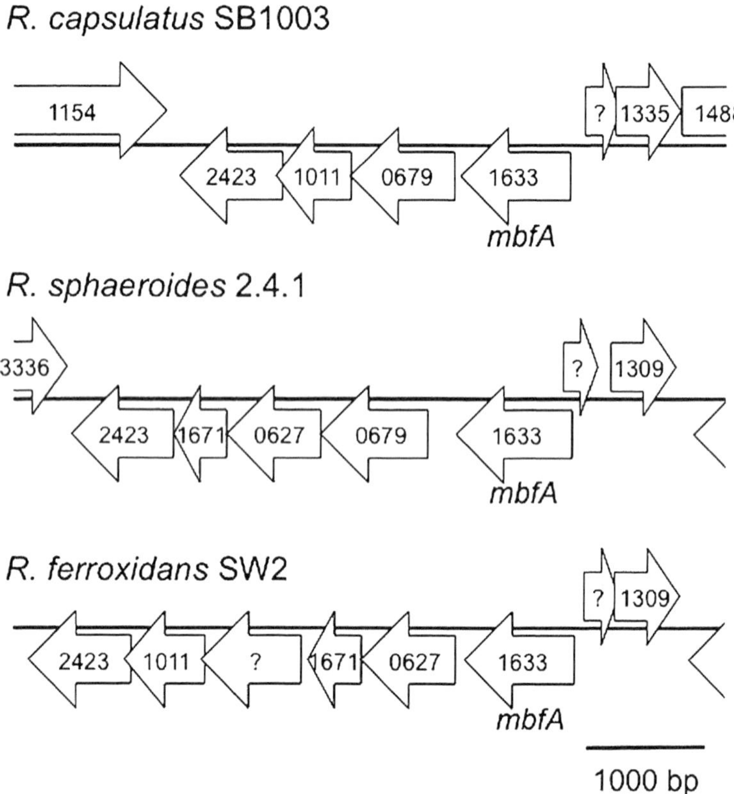

Figure 10.3 Genomic environment of membrane-bound ferritin encoding genes, *mbfA*, in *R. capsulatus* SB1003, *R. sphaeroides* 2.4.1, *R. ferroxidans* SW2. COG numbers of each reading frame are indicated.

Some experimental work has been undertaken with the *Rhododacter* Bfr. The first Bfr to be studied was from *R. sphaeroides*, although most work has been done on the *R. capsulatus* homologue (Meyer & Cusanovich, 1985). Between 900 and 1000 amorphous iron atoms along with 600 phosphate molecules are contained in each Bfr complex. The complex is located in the cytoplasm and its expression in "normal" growth medium is stable. Nevertheless, moderate control by iron is observed, as iron starvation induces a decrease in Bfr cellular content, while iron replete conditions promote accumulation of Bfr in the cell (Penfold et al., 1996; Ringeling et al., 1994). The *Rhodobacter* Bfr structure was the second Bfr structure to be elucidated after that of *E. coli* Bfr (Cobessi et al., 2001).

Regarding iron storage in *R. sphaeroides*, some transcriptional mechanisms are reported. Transcription of *bfr* encoded by orf 1546, along with its associated ferredoxin appears to be controlled by iron availability as their mRNA levels increase when iron is depleted. The transcription factor Irr is implicated in this control (Peuser, Remes, & Klug, 2012). The other Bfr (orf 3342), which lacks a linked ferredoxin, is almost insensitive to iron levels and shows a weak Irr regulation profile (Peuser et al., 2012). Finally, the Ftn-like MbfA displays weak iron control but a strong Irr-dependence. In fact, as predicted by Rodionov et al., MbfA was shown to be under the direct regulation of Irr (Peuser et al., 2012; Rodionov et al., 2006).

7. IRON HOMEOSTASIS REGULATORS

The regulation of iron homeostasis involves a complex overlapping set of global regulators as well as more specialized regulators dedicated to the control of specific iron homeostasis genes, such as the control of siderophore synthesis and Fe^{2+} uptake. Such local regulators are diverse and belong to the following families: i) AraC-type transcription factor (Ducey, Carson, Orvis, Stintzi, & Dyer, 2005; Fantappiè, Scarlato, & Delany, 2011; Hollander, Mercante, Shafer, & Cornelissen, 2011; Pradel, Guiso, & Locht, 1998); ii) two component systems (Steele et al., 2012); iii) extracytoplasmic function (ECF) σ factors (Braun, 1997; Braun, Mahren, & Ogierman, 2003; Koster, van Klompenburg, Bitter, Leong, & Weisbeek, 1994); iv) LysR-type transcription factor (Litwin & Quackenbush, 2001; Vanderpool & Armstrong, 2003); v) small RNA regulators (Ducey, Jackson, Orvis, & Dyer, 2009; Huang et al., 2012; Massé, Salvail, Desnoyers, & Arguin, 2007; Metruccio et al., 2009; Smaldone et al., 2012). *P. aeruginosa* is an example of organism that combines all the aforementioned regulators as well as several global regulators (Cornelis, Matthijs, & Van Oeffelen, 2009; Vasil, 2007). Computational analysis showed that *R. sphaeroides* possesses FecRI (orf 4274 and 4275, Table 10.3) where FecI is a homologue of the ECF-type σ factor and FecR is a periplasmic regulator of FecI (Rodionov et al., 2006). Based on its genomic environment, the *fecRI* operon may be co-transcribed with a siderophore uptake gene cluster that could be under its control. FecRI is involved in Fe^{3+}-citrate uptake in *E. coli* but homologues such as PupBI and HurRI have been characterized in other species. PupBI and HurRI are involved in pseudobactin siderophore and heme uptake, respectively (Koster et al., 1994; Vanderpool & Armstrong, 2003). Genome analysis shows also that *R. capsulatus* has four AraC-like transcription factors (Table 10.3) located next to siderophore uptake gene clusters (Rodionov et al., 2006).

In many cases, transcription factors dedicated to the control of specific iron homeostasis genes are themselves under the control of a global transcription factor (Braun, 1997; Fantappiè et al., 2011; Pradel et al., 1998), or act as co-regulators with a global regulator (Escamilla-Hernandez & O'Brian, 2012). For a long time, the paradigm of iron homeostasis in bacteria was associated with the Ferric Uptake Regulator (Fur) that was first characterized in *E. coli.* Additional studies quickly showed that Fur homologues are widespread in both gram-negative (proteobacteria) and gram-positive (Firmicutes) bacteria, as well as in some cyanobacteria (Carpenter, Whitmire, & Merrell, 2009; Lee & Helmann, 2007). The basic mechanism of Fur regulation consists of transcriptional repression of iron uptake genes under iron replete conditions by Fur that contains a bound Fe^{2+} (Fe^{2+}-Fur). However, under conditions of iron limitation, Fur without bound iron (apo-Fur) is incapable of binding these promoters, derepressing the expression of these iron uptake genes (Carpenter et al., 2009; Lee & Helmann, 2007; Rudolph et al., 2006). The regulation by Fur is now known to be more complex since Fe^{2+}-Fur can also activate genes in an indirect manner, via the derepression of a small regulatory RNA (Ducey et al., 2009; Huang et al., 2012; Massé et al., 2007; Metruccio et al., 2009; Smaldone et al., 2012). Finally, direct transcriptional activation by apo-Fur on a target gene promoter has also been observed (Carpenter et al., 2009; Lee & Helmann, 2007).

Members of the Rhizobiales and Rhodobacterales regulate iron homeostasis genes by the global regulator Irr (Iron response regulat (Johnston et al., 2007; Rodionov et al., 2006; Rudolph et al., 2006). It was proposed that upon the appearance of Irr, Fur evolved into a manganese uptake regulator. Consequently the Fur homologues in Rhizobiales and Rhodobacterales have been renamed Mur. [Another evolutionary event is the inclusion of a third iron regulator, RirA, that is present in the Rhizobiales (Rhodobacterales that have only Fur/Irr) (Johnston et al., 2007; Rodionov et al., 2006; Rudolph et al., 2006)]. Fur and Irr belong to the same Fur-superfamily of metal regulators and thus share strong sequence homology. Nevertheless, their mechanisms of action are very different. Fur directly binds elemental iron which subsequently affects the ability of Fur to activate or repress gene expression. Like Fur, Irr can both activate and repress iron-dependent gene expression. However, Irr does not bind free iron but instead monitors the iron level indirectly by sensing the level of heme biosynthesis. Specifically, Irr interacts with and monitors the activity of ferrochelatase, the last enzyme of the heme synthesis pathway. Irr obtains a heme from ferrochelatase which results in the targeting of Irr for degradation (Rodionov et al., 2006; Rudolph et al., 2006; Small, Puri, & O'Brian, 2009).

Members of the Fur family are represented in each of the sequenced *Rhodobacter* genomes. However, a closer look reveals that: i) Fur is missing in *R. capsulatus*, while present in *R. sphaeroides* and *R. ferroxidans*; ii) Each genome contains a copy of Irr; iii) Each genome also has a Zur-encoding gene (Table 10.8). The latter is also a Fur-family member but specialized in zinc uptake (Lee & Helmann, 2007). Zur is readily identifiable because it is consistently located within a *znuA'RCB* gene cluster that encodes a zinc transporter. Although sharing a high degree of sequence similarity, Fur and Irr can be discriminated on the basis of sequence analysis. A main feature distinguishing Fur and Irr is the HHDH Fe^{2+} binding motif that is a signature of Fur, which becomes a HHH (or HQH) in Irr. This aspartate deletion transforms the site from a Fe^{2+} to a heme binding site (Rudolph et al., 2006). Thus, while *R. sphaeroides* and *R. ferroxidans* display the classical Rhodobacterales iron homeostasis transcription factor features, *R. capsulatus* has a very unusual set of regulators. Moreover, *R. capsulatus*, along with *Mesorhizobium loti*, is among the very rare α-proteobacteria that also have a manganese MntHR uptake system, most likely acquired through horizontal gene transfer (Rodionov et al., 2006). Based on *in silico* analysis and studies on Rhizobiales, it has been proposed that the acquisition of Irr as global iron regulator may have pushed Fur into a more marginal role where it evolved into manganese homeostasis regulation, as Mur (Johnston et al., 2007; Rodionov et al., 2006). Following on this hypothesis, *R. capsulatus* appears to be an extreme example of this evolutionary trend.

Table 10.8 Iron regulators identified in the genomes of *R. capsulatus* SB1003, *R. sphaeroides* 2.4.1 and *R. ferroxidans* SW2

Regulator type	Gene name	COG	Label *Rcap* SB1003*	Label *Rsph* 2.4.1*	Label *Rfer* SW2*	Product
Fur family	*fur/mur*	0735	N.D.	2494	2373	Fe^{3+}/Mn^{2+} uptake regulator
	irr	0735	02670	3179	0819	Iron response regulator
	zur	0735	01134	3569	1799	Zinc uptake regulator

N.D.: Not detected
*See Table 10.1.

In addition to genome analysis, some experimental work has been undertaken on the role of Fur/Mur and Irr in *R. sphaeroides* 2.4.1. In this organism, Fur/Mur is involved in iron homeostasis and oxidative stress response upon iron scarcity, where it acts as a repressor. Interestingly, the Δ*fur/mur* mutant shows a more hampered growth profile during manganese rather than iron limitation. Furthermore the Δ*fur/mur* mutant also has no activation of the putative Mn^{2+}/Fe^{2+} uptake system SitABCD, highlighting this operon as a potential target for Fur (Peuser, Metz, & Klug, 2011). The role of Irr in *R. sphaeroides* is also poorly defined. A Δ*irr* strain displays little growth deficiency in an iron-depleted medium, suggesting that Irr does not have a major function in iron homeostasis. Moreover, Irr in *R. sphaeroides* was shown to bind heme and to activate many genes beyond those of iron homeostasis, such as stress response, oxidative phosphorylation, transport and photosynthesis genes. Consequently the effect of Irr on iron homeostasis may be indirect. Oxidative stress also appears to be an important part of the Irr regulon as Δ*irr* is more resistant to this stress (Peuser et al., 2012). Such resistance could be achieved by Irr through indirect control of a catalase and possibly a small RNA involved in singlet oxygen/superoxide response. Evidence for direct Irr control of *mbfA* and *ccpA*, encoding the membrane-bound ferritin and a cytochrome *c* peroxidase, respectively, was found. Although transcription data showed a weak effect, Irr was proven to directly interact with the respective promoter of these genes, confirming computational prediction of Irr regulon in *R. sphaeroides* (Peuser et al., 2012; Rodionov et al., 2006). This computational study of iron and manganese regulons predicted that *mbfA* and *ccpA* are Irr-regulated genes in both *R. capsulatus* and *R. sphaeroides*. Beside Irr-specific regulation, the iron regulon in *R. capsulatus* and *R. sphaeroides* was predicted to consist of 18 and 8 genes, respectively (Rodionov et al., 2006).

Based on data from *R. sphaeroides* experiments, neither Fur/Mur nor Irr appears as a master regulator of iron homeostasis in this species. Rodionov et al. hypothesized a potentially major role for IscR, a regulator of Fe-S cluster synthesis in *E. coli* (Rodionov et al., 2006). While *R. capsulatus* and *R. sphaeroides* genomes have an IscR-encoding gene (orf 01853 and 0443, respectively), none could be found in the *R. ferroxidans* genome. Nevertheless, the absence of Fur and presence of MntR in *R. capsulatus* indicate that *R. capsulatus* regulates iron homeostasis differently from *R. sphaeroides* and *R. ferroxidans*. Sequence alignments (not shown) of Fur, Irr, Zur sequences show very high similarity and identity, on the order of

Table 10.9 Identity and similarity in Fur/Mur, Irr and Zur from *R. capsulatus* SB1003, *R. shaeroides* 2.4.1 and *R. ferroxidans* SW2. Similarity is indicated using parenthesis*

	Rcap	Rsph	Rfer
Fur/Mur			
Rcap	No Rcap Fur	–	–
Rsph	No Rcap Fur	100 (100)	–
Rfer	No Rcap Fur	74.6 (84.8)	100 (100)
Irr			
Rcap	100 (100)	–	–
Rsph	62.6 (76.8)	100 (100)	–
Rfer	60.0 (73.5)	66.7 (78.2)	100 (100)
Zur			
Rcap	100 (100)	–	–
Rsph	63.4 (71.5)	100 (100)	–
Rfer	68.6 (75.2)	73.2 (80.4)	100 (100)

*See Table 10.2.

60 to 80%, respectively (Table 10.9). But, consistently, *R. sphaeroides* and *R. ferroxidans* yield higher values with each other than with *R. capsulatus*.

8. CONCLUSION

Our analyses of the genomes of three model *Rhodobacter* species highlight the diversity of strategies used by these organisms to maintain iron homeostasis (Table 10.10). The two most extreme strategies appeared to be employed by *R. capsulatus* and *R. ferroxidans*. On the one hand, *R. capsulatus* presents a large battery of putative Fe^{2+}, Fe^{3+} and heme iron uptake proteins (between 10 and 12!) in addition to an iron efflux pump and several storage genes. On the other hand, *R. ferroxidans* displays a minimalist arsenal with one Fe^{2+} uptake system, an efflux pump and one storage system. The case of *R. sphaeroides* seems intermediate with a balanced distribution of Fe^{2+} and Fe^{3+} uptake systems (between and 7). Also, it is notable that *R. capsulatus* is the only *Rhodobacter* with a clearly identified heme uptake system that has been mostly studied in pathogens and more recently with symbionts. Moreover, *R. capsulatus* is the only sequenced *Rhodobacter* genome that is missing the canonical iron global regulator Fur. Such a paradox suggests novel regulation of iron homeostasis in this organism. Finally, among *Rhodobacter* species, *R. capsulatus* is the only organism displaying a large portion of its genome putatively dedicated to iron

Table 10.10 Summary of iron/manganese homeostasis features identified in the genomes of *R. capsulatus* SB1003, *R. sphaeroides* 2.4.1 and *R. ferroxidans* SW2

Feature type	Feature name	*Rcap* SB1003	*Rsph* 2.4.1	*Rfer* SW2
Fe/Mn uptake	Feo	2	1	1
	Efe	1	0	0
	MntH	1	0	2
	ABC-type	1	2	1
	Siderophore	7	4	0
	Haem	1	0	0
	Total	13	7	4
Efflux pump	FieF	1	1	1
Iron storage	Bfr	1	2	0
	MbfA	1	1	0
Fur-type regulators	Fur	0	1	1
	Irr	1	1	1
	Zur	1	1	1
Manganese regulator	MntR	1	0	2

uptake. This "iron island" is a ~22 kb gene cluster consisting of a Fe^{2+} transport cassette, a heme uptake system, an ABC transporter and a siderophore import system (Fig. 10.4).

Among the questions that remain to be solved is the mechanism of iron transport through the periplasmic membrane. Indeed, unlike siderophores and heme, which have specific outer membrane receptors, the access of elemental iron to the inner membrane acquisition systems (Feo, Efe or ABC-type) has not been characterized. The paradigm for such a small compound consists of passive diffusion, but recent research indicates that porin-like channels in the outer membrane transfer specific divalent metal cations to the periplasm (Hohle, Franck, Stacey, & O'Brian, 2011).

The field of metal homeostasis, and the regulation of iron uptake in particular, is complex. Indeed, after decades of research on iron regulation in a well-studied organism such as *E. coli*, iron uptake systems are still being discovered (Koch et al., 2011). Decades of studies on Fur do not resolve debates about its mechanism of action, nor differentiate between related representatives such as Mur, Zur, Nur (Lee & Helmann, 2007). In addition, a new layer of complexity is emerging in the field, concerning the overlap between the homeostasis of iron and manganese, two metals that have crucial and yet opposite effects on oxidative stress (Horsburgh, Wharton, Karavolos, & Foster, 2002; Jakubovics & Jenkinson, 2001; Puri, Hohle, & O'Brian, 2010). Depending on whether cells are challenged with

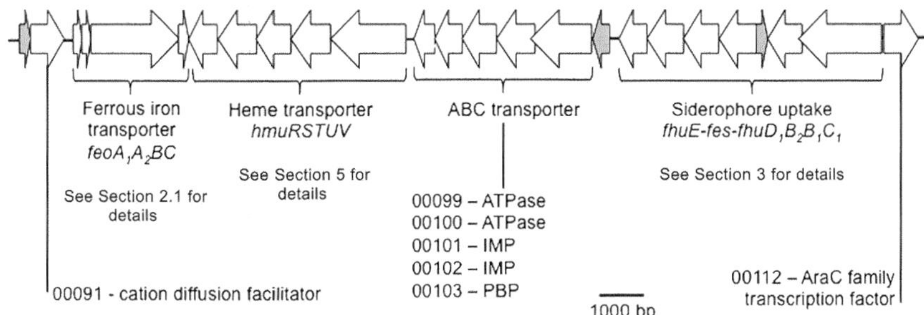

Figure 10.4 Putative "Iron island" in *R. capsulatus* SB1003. Genes encoding "hypothetical proteins" are shown in grey. IMP: Inner membrane protein; PBP: Periplasmic binding protein. ORF number is given with the digits only of the nomenclature of *R. capsulatus* SB1003 (for example, 12345 correspond to locus RCAP_rcc12345 for a chromosomal gene and p12345 to RCAP_rcp12345 for a plasmid gene).

an oxidative stress response or DNA synthesis under iron scarcity, organisms have been shown to be able to switch from an Fe-based enzyme activity to an Mn-based equivalent (Andrews, 2011; Andrews et al., 2003; Lee & Helmann, 2007). The various strategies to maintain metal homeostasis are present in the genomes of these *Rhodobacter* strains, and clearly highlight the diversity of this genus.

REFERENCES

Amarelle, V., Koziol, U., Rosconi, F., Noya, F., O'Brian, M. R., & Fabiano, E. (2010). A new small regulatory protein, HmuP, modulates haemin acquisition in *Sinorhizobium meliloti*. *Microbiology*, *156*, 1873–1882.

Amarelle, V., O'Brian, M. R., & Fabiano, E. (2008). ShmR is essential for utilization of heme as a nutritional iron source in *Sinorhizobium meliloti*. *Applied and Environmental Microbiology*, *74*, 6473–6475.

Amin, S. A., Green, D. H., Al Waheeb, D., Gärdes, A., & Carrano, C. J. (2012). Iron transport in the genus *Marinobacter*. *Biometals: An International Journal on the Role of Metal Ions in Biology, Biochemistry, and Medicine*, *25*, 135–147.

Andrews, S. C. (2010). The Ferritin-like superfamily: evolution of the biological iron storeman from a rubrerythrin-like ancestor. *Biochimica et Biophysica Acta*, *1800*, 691–705.

Andrews, S. C. (2011). Making DNA without iron – induction of a manganese-dependent ribonucleotide reductase in response to iron starvation. *FEMS Microbiology Letters*, *80*, 286–289.

Andrews, S. C., Robinson, A. K., & Rodríguez-Quiñones, F. (2003). Bacterial iron homeostasis. *FEMS Microbiology Reviews*, *27*, 215–237.

Angerer, A., Klupp, B., & Braun, V. (1992). Iron transport systems of *Serratia marcescens*. *The Journal of Bacteriology*, *174*, 1378–1387.

Anzaldi, L. L., & Skaar, E. P. (2010). Overcoming the heme paradox: heme toxicity and tolerance in bacterial pathogens. *Infection and Immunity*, *78*, 4977–4989.

Archibald, F. (1983). *Lactobacillus plantarum*, an organism not requiring iron. *FEMS Microbiology Letters*, *19*, 29–32.

Badarau, A., Firbank, S. J., Waldron, K. J., Yanagisawa, S., Robinson, N. J., Banfield, M. J., et al. (2008). FutA2 is a ferric binding protein from *Synechocystis* PCC6803. *The Journal of Biological Chemistry, 283*, 12520–12527.

Baichoo, N., Wang, T., Ye, R., & Helmann, J. D. (2002). Global analysis of the *Bacillus subtilis* Fur regulon and the iron starvation stimulon. *Molecular Microbiology, 45*, 1613–1629.

Barker, K. D., Barkovits, K., & Wilks, A. (2012). Metabolic flux of extracellular heme uptake in *Pseudomonas aeruginosa* is driven by the iron-regulated heme oxygenase (HemO). *The Journal of Biological Chemistry, 287*, 18342–18350.

Bearden, S. W., & Perry, R. D. (1999). The Yfe system of *Yersinia pestis* transports iron and manganese and is required for full virulence of plague. *Molecular Microbiology, 32*, 403–414.

Bendtsen, J. D., Nielsen, H., Widdick, D., Palmer, T., & Brunak, S. (2005). Prediction of twin-arginine signal peptides. *BMC Bioinformatics, 6*, 167.

Biel, S. W., & Biel, A. J. (1990). Isolation of a *Rhodobacter capsulatus* mutant that lacks c-type cytochromes and excretes porphyrins. *Journal of Bacteriology, 172*, 1321–1326.

Bou-Abdallah, F. (2010). The iron redox and hydrolysis chemistry of the ferritins. *Biochimica et Biophysica Acta, 1800*, 719–731.

Braun, V. (1997). Surface signaling: novel transcription initiation mechanism starting from the cell surface. *Archives of Microbiology, 167*, 325–331.

Braun, V., & Hantke, K. (2011). Recent insights into iron import by bacteria. *Current Opinion in Chemical Biology, 15*, 328–334.

Braun, V., Mahren, S., & Ogierman, M. (2003). Regulation of the FecI-type ECF sigma factor by transmembrane signalling. *Current Opinion in Microbiology, 6*, 173–180.

Brickman, T. J., & Armstrong, S. K. (2012). Iron and pH-responsive FtrABCD ferrous iron utilization system of *Bordetella* species. *Molecular Microbiology, 86*, 580–593.

Brickman, T. J., Cummings, C. A., Liew, S. -Y., Relman, D. A., & Armstrong, S. K. (2011). Transcriptional profiling of the iron starvation response in *Bordetella pertussis* provides new insights into siderophore utilization and virulence gene expression. *Journal of Bacteriology, 193*, 4798–4812.

Caiazza, N. C., Lies, D. P., & Newman, D. K. (2007). Phototrophic Fe(II) oxidation promotes organic carbon acquisition by *Rhodobacter capsulatus* SB1003. *Applied and Environmental Microbiology, 73*, 6150–6158.

Cao, J., Woodhall, M. R., Alvarez, J., Cartron, M. L., & Andrews, S. C. (2007). EfeUOB (YcdNOB) is a tripartite, acid-induced and CpxAR-regulated, low-pH Fe^{2+} transporter that is cryptic in *Escherichia coli* K-12 but functional in *E. coli* O157:H7. *Molecular Microbiology, 65*, 857–875.

Carpenter, B. M., Whitmire, J. M., & Merrell, D. S. (2009). This is not your mother's repressor: the complex role of fur in pathogenesis. *Infection and Immunity, 77*, 2590–2601.

Carrondo, M. A. (2003). Ferritins, iron uptake and storage from the bacterioferritin viewpoint. *The EMBO Journal, 22*, 1959–1968.

Cartron, M. L., Maddocks, S., Gillingham, P., Craven, C. J., & Andrews, S. C. (2006). Feo–transport of ferrous iron into bacteria. *Biometals: An International Journal on the Role of Metal Ions in Biology, Biochemistry, and Medicine, 19*, 143–157.

Chan, A. C. K., Doukov, T. I., Scofield, M., Tom-Yew, S. A.L., Ramin, A. B., Mackichan, J. K., et al. (2010). Structure and function of P19, a high-affinity iron transporter of the human pathogen *Campylobacter jejuni*. *Journal of Molecular Biology, 401*, 590–604.

Chao, T.-C., Becker, A., Buhrmester, J., Pühler, A., & Weidner, S. (2004). The *Sinorhizobium meliloti* fur gene regulates, with dependence on Mn(II), transcription of the *sitABCD* operon, encoding a metal-type transporter. *The Journal of Bacteriology, 186*, 3609–3620.

Chiancone, E., Ceci, P., Ilari, A., Ribacchi, F., & Stefanini, S. (2004). Iron and proteins for iron storage and detoxification. *Biometals: An International Journal on the Role of Metal Ions in Biology, Biochemistry, and Medicine, 17*, 197–202.

Chin, N., Frey, J., Chang, C. F., & Chang, Y. F. (1996). Identification of a locus involved in the utilization of iron by *Actinobacillus pleuropneumoniae*. *FEMS Microbiology Letters*, *143*, 1–6.

Chu, B. C., Garcia-Herrero, A., Johanson, T. H., Krewulak, K. D., Lau, C. K., Peacock, R. S., et al. (2010). Siderophore uptake in bacteria and the battle for iron with the host; a bird's eye view. *Biometals: An International Journal on the Role of Metal Ions in Biology, Biochemistry, and Medicine*, *23*, 601–611.

Cobessi, D., Huang, L. S., Ban, M., Pon, N. G., Daldal, F., & Berry, E. A. (2001). The 2.6 Å resolution structure of *Rhodobacter capsulatus* bacterioferritin with metal-free dinuclear site and heme iron in a crystallographic 'special position'. *Acta Crystallographica Section D Biological Crystallography*, *58*, 29–38.

Cornelis, P., Matthijs, S., & Van Oeffelen, L. (2009). Iron uptake regulation in *Pseudomonas aeruginosa*. *Biometals: An International Journal on the Role of Metal Ions in Biology, Biochemistry, and Medicine*, *22*, 15–22.

Croal, L. R., Jiao, Y., & Newman, D. K. (2007). The fox operon from *Rhodobacter* strain SW2 promotes phototrophic Fe(II) oxidation in *Rhodobacter capsulatus* SB1003. *The Journal of Bacteriology*, *189*, 1774–1782.

D'Onofrio, A., Crawford, J. M., Stewart, E. J., Witt, K., Gavrish, E., Epstein, S., et al. (2010). Siderophores from neighboring organisms promote the growth of uncultured bacteria. *Chemistry and Biology*, *17*, 254–264.

Dailey, H. A., Septer, A. N., Daugherty, L., Thames, D., Gerdes, S., Stabb, E. V., et al. (2011). The *Escherichia coli* protein YfeX functions as a porphyrinogen oxidase, not a heme dechelatase. *MBio*, *2*. e00248–11.

Dashper, S. G., Butler, C. A., Lissel, J. P., Paolini, R. A., Hoffmann, B., Veith, P. D., et al. (2005). A novel *Porphyromonas gingivalis* FeoB plays a role in manganese accumulation. *The Journal of Biological Chemistry*, *280*, 28095–28102.

Dubbels, B. L., DiSpirito, A. A., Morton, J. D., Semrau, J. D., Neto, J. N.E., & Bazylinski, D. A. (2004). Evidence for a copper-dependent iron transport system in the marine, magnetotactic bacterium strain MV-1. *Microbiology (Reading, England)*, *150*, 2931–2945.

Ducey, T. F., Carson, M. B., Orvis, J., Stintzi, A. P., & Dyer, D. W. (2005). Identification of the iron-responsive genes of *Neisseria gonorrhoeae* by microarray analysis in defined medium. *The Journal of Bacteriology*, *187*, 4865–4874.

Ducey, T. F., Jackson, L., Orvis, J., & Dyer, D. W. (2009). Transcript analysis of *nrrF*, a Fur repressed sRNA of Neisseria gonorrhoeae. *Microbial Pathogenesis*, *46*, 166–170.

Ehrenreich, A., & Widdel, F. (1994). Anaerobic oxidation of ferrous iron by purple bacteria, a new type of phototrophic metabolism. *Applied and Environmental Microbiology*, *60*, 4517–4526.

Escamilla-Hernandez, R., & O'Brian, M. R. (2012). HmuP is a coactivator of Irr-dependent expression of heme utilization genes in *Bradyrhizobium japonicum*. *Journal of Bacteriology*, *194*, 3137–3143.

Fantappiè, L., Scarlato, V., & Delany, I. (2011). Identification of the in vitro target of an iron-responsive AraC-like protein from *Neisseria meningitidis* that is in a regulatory cascade with Fur. *Microbiology*, *157*, 2235–2247.

Fetherston, J. D., Mier, I., Truszczynska, H., & Perry, R. D. (2012). The Yfe and feo transporters are involved in microaerobic growth and the virulence of *Yersinia pestis* in bubonic plague. *Infection and Immunity*, *80*, 3880–3891.

Fisher, C. R., Davies, N. M.L.L., Wyckoff, E. E., Feng, Z., Oaks, E. V., & Payne, S. M. (2009). Genetics and virulence association of the *Shigella flexneri sit* iron transport system. *Infection and Immunity*, *77*, 1992–1999.

Frankenberg-Dinkel, N. (2004). Bacterial heme oxygenases. *Antioxidants and Redox Signaling*, *6*, 825–834.

Grass, G., Otto, M., Fricke, B., Haney, C. J., Rensing, C., Nies, D. H., et al. (2005). FieF (YiiP) from *Escherichia coli* mediates decreased cellular accumulation of iron and relieves iron stress. *Archives of Microbiology, 183,* 9–18.

Grifantini, R., Sebastian, S., Frigimelica, E., Draghi, M., Bartolini, E., Muzzi, A., et al. (2003). Identification of iron-activated and -repressed Fur-dependent genes by transcriptome analysis of *Neisseria meningitidis* group B. *Proceedings of the National Academy of Sciences of the United States of America, 100,* 9542–9547.

Grosse, C., Scherer, J., Koch, D., Otto, M., Taudte, N., & Grass, G. (2006). A new ferrous iron-uptake transporter, EfeU (YcdN), from *Escherichia coli. Molecular Microbiology, 62,* 120–131.

Guo, J., Nair, M. K.M., Galván, E. M., Liu, S.-L., & Schifferli, D. M. (2011). Tn5AraOut mutagenesis for the identification of *Yersinia pestis* genes involved in resistance towards cationic antimicrobial peptides. *Microbial Pathogenesis, 51,* 121–132.

Hantke, K. (1987). Ferrous iron transport mutants in *Escherichia coli* K12. *FEMS Microbiology Letters, 44,* 53–57.

Hantke, K. (2003). Is the bacterial ferrous iron transporter FeoB a living fossil? *Trends in Microbiology, 11,* 192–195.

He, J., Miyazaki, H., Anaya, C., Yu, F., Yeudall, W. A., & Lewis, J. P. (2006). Role of *Porphyromonas gingivalis* FeoB2 in metal uptake and oxidative stress protection. *Infection and Immunity, 74,* 4214–4223.

Hider, R. C., & Kong, X. (2010). Chemistry and biology of siderophores. *Natural Product Reports, 27,* 637–657.

Hohle, T. H., Franck, W. L., Stacey, G., & O'Brian, M. R. (2011). Bacterial outer membrane channel for divalent metal ion acquisition. *Proceedings of the National Academy of Sciences of the United States of America, 108,* 15390–15395.

Hollander, A., Mercante, A. D., Shafer, W. M., & Cornelissen, C. N. (2011). The iron-repressed, AraC-like regulator MpeR activates expression of *fetA* in. *Neisseria Gonorrhoeae. Infection and Immunity, 79,* 4764–4776.

Horsburgh, M. J., Wharton, S. J., Karavolos, M., & Foster, S. J. (2002). Manganese: elemental defence for a life with oxygen. *Trends in Microbiology, 10,* 496–501.

Huang, S.-H., Wang, C.-K., Peng, H.-L., Wu, C.-C., Chen, Y.-T., Hong, Y.-M., et al. (2012). Role of the small RNA RyhB in the Fur regulon in mediating the capsular polysaccharide biosynthesis and iron acquisition systems in *Klebsiella pneumoniae. BMC Microbiology, 12,* 148.

Jakubovics, N. S., & Jenkinson, H. F. (2001). Out of the iron age: new insights into the critical role of manganese homeostasis in bacteria. *Microbiology (Reading, England), 147,* 1709–1718.

Janakiraman, A., & Slauch, J. M. (2000). The putative iron transport system SitABCD encoded on SPI1 is required for full virulence of *Salmonella typhimurium. Molecular Microbiology, 35,* 1146–1155.

Johnston, A. W.B., Todd, J. D., Curson, A. R., Lei, S., Nikolaidou-Katsaridou, N., Gelfand, M. S., et al. (2007). Living without Fur: the subtlety and complexity of iron-responsive gene regulation in the symbiotic bacterium *Rhizobium* and other alpha-proteobacteria. *Biometals: An International Journal on the Role of Metal Ions in Biology, Biochemistry, and Medicine, 20,* 501–511.

Kammler, M., Schön, C., & Hantke, K. (1993). Characterization of the ferrous iron uptake system of *Escherichia coli, 175,* 6212–6219.

Katoh, H., Hagino, N., Grossman, A. R., & Ogawa, T. (2001b). Genes essential to iron transport in the cyanobacterium *Synechocystis* sp. strain PCC 6803. *The Journal of Bacteriology, 183,* 2779–2784.

Katoh, H., Hagino, N., & Ogawa, T. (2001a). Iron-binding activity of FutA1 subunit of an ABC-type iron transporter in the cyanobacterium *Synechocystis* sp. strain PCC 6803. *Plant and Cell Physiology, 42,* 823–827.

Kehres, D. G., Janakiraman, A., Slauch, J. M., & Maguire, M. E. (2002). SitABCD is the alkaline Mn($^{2+}$) transporter of *Salmonella enterica* Serovar Typhimurium. *The Journal of Bacteriology, 184,* 3159–3166.

Khun, H. H., Kirby, S. D., & Lee, B. C. (1998). A *Neisseria meningitidis fbpABC* mutant is incapable of using nonheme iron for growth. *Infection and Immunity*, *66*, 2330–2336.

Kim, H., Lee, H., & Shin, D. (2012). The FeoA protein is necessary for the FeoB transporter to import ferrous iron. *Biochemical and Biophysical Research Communications*, *423*, 733–738.

Koch, D., Chan, A. C.K., Murphy, M. E. P., Lilie, H., Grass, G., & Nies, D. H. (2011). Characterization of a dipartite iron uptake system from uropathogenic *Escherichia coli* strain F11. *The Journal of Biological Chemistry*, *286*, 25317–25330.

Kosman, D. J. (2003). Molecular mechanisms of iron uptake in fungi. *Molecular Microbiology*, *47*, 1185–1197.

Köster, W. (2001). ABC transporter-mediated uptake of iron, siderophores, heme and vitamin B12. *Research in Microbiology*, *152*, 291–301.

Koster, M., van Klompenburg, W., Bitter, W., Leong, J., & Weisbeek, P. (1994). Role for the outer membrane ferric siderophore receptor PupB in signal transduction across the bacterial cell envelope. *The EMBO Journal*, *13*, 2805–2813.

Krewulak, K. D., & Vogel, H. J. (2008). Structural biology of bacterial iron uptake. *Biochimica et Biophysica Acta*, *1778*, 1781–1804.

Krewulak, K. D., & Vogel, H. J. (2011). TonB or not TonB: is that the question? *Biochemistry and Cell Biology/Biochimie et Biologie Cellulaire*, *89*, 87–97.

Lascelles, J. (1956). The synthesis of porphyrins and bacteriochlorophyll by cell suspensions of *Rhodopseudomonas spheroides*. *The Biochemical Journal*, *62*, 78–93.

Lee, J. -W., & Helmann, J. D. (2007). Functional specialization within the Fur family of metalloregulators. *Biometals: An International Journal on the Role of Metal Ions in Biology, Biochemistry, and Medicine*, *20*, 485–499.

Létoffé, S., Heuck, G., Delepelaire, P., Lange, N., & Wandersman, C. (2009). Bacteria capture iron from heme by keeping tetrapyrrol skeleton intact. *Proceedings of the National Academy of Sciences of the United States of America*, *106*, 11719–11724.

Li, L., Chen, O. S., McVey Ward, D., & Kaplan, J. (2001). CCC1 is a transporter that mediates vacuolar iron storage in yeast. *The Journal of Biological Chemistry*, *276*, 29515–29519.

Litwin, C. M., & Quackenbush, J. (2001). Characterization of a *Vibrio vulnificus* LysR homologue, HupR, which regulates expression of the haem uptake outer membrane protein, HupA. *Microbial Pathogenesis*, *31*, 295–307.

Liu, M., Boulouis, H.-J., & Biville, F. (2012). Heme degrading protein HemS is involved in oxidative stress response of *Bartonella henselae*. *PloS One*, 7, e37630.

Liu, X., Du, Q., Wang, Z., Zhu, D., Huang, Y., Li, N., et al. (2011). Crystal structure and biochemical features of EfeB/YcdB from *Escherichia coli* O157: ASP235 plays divergent roles in different enzyme-catalyzed processes. *The Journal of Biological Chemistry*, *286*, 14922–14931.

Lu, M., & Fu, D. (2007). Structure of the zinc transporter YiiP. *Science (New York, NY)*, *317*, 1746–1748.

Massé, E., Salvail, H., Desnoyers, G., & Arguin, M. (2007). Small RNAs controlling iron metabolism. *Current Opinion in Microbiology*, *10*, 140–145.

Metruccio, M. M. E., Fantappiè, L., Serruto, D., Muzzi, A., Roncarati, D., Donati, C., et al. (2009). The Hfq-dependent small noncoding RNA NrrF directly mediates Fur-dependent positive regulation of succinate dehydrogenase in *Neisseria meningitidis*. *Journal of Bacteriology*, *191*, 1330–1342.

Meyer, T. E., & Cusanovich, M. A. (1985). Soluble cytochrome composition of the purple phototrophic bacterium, *Rhodopseudomonas sphaeroides* ATCC 17023. *Biochimica et Biophysica Acta*, *807*, 308–319.

Montanini, B., Blaudez, D., Jeandroz, S., Sanders, D., & Chalot, M. (2007). Phylogenetic and functional analysis of the cation diffusion facilitator (CDF) family: improved signature and prediction of substrate specificity. *BMC Genomics*, *8*, 107.

Munkelt, D., Grass, G., & Nies, D. H. (2004). The chromosomally encoded cation diffusion facilitator proteins DmeF and FieF from *Wautersia metallidurans* CH34 are transporters of broad metal specificity. *The Journal of Bacteriology, 186*, 8036–8043.

Nienaber, A., Hennecke, H., & Fischer, H. M. (2001). Discovery of a haem uptake system in the soil bacterium *Bradyrhizobium japonicum*. *Molecular Microbiology, 41*, 787–800.

Nies, D. H. (2007). Biochemistry. How cells control zinc homeostasis. *Science (New York, NY), 317*, 1695–1696.

Nies, D. H. (2011). How iron is transported into magnetosomes. *Molecular Microbiology, 82*, 792–796.

O'Neill, M. J., Bhakta, M. N., Fleming, K. G., & Wilks, A. (2012). Induced fit on heme binding to the *Pseudomonas aeruginosa* cytoplasmic protein (PhuS) drives interaction with heme oxygenase (HemO). *Proceedings of the National Academy of Sciences of the United States of America, 109*, 5639–5644.

Ollinger, J., Song, K. -B., Antelmann, H., Hecker, M., & Helmann, J. D. (2006). Role of the Fur regulon in iron transport in *Bacillus subtilis*. *The Journal of Bacteriology, 188*, 3664–3673.

Paustian, M. L., May, B. J., & Kapur, V. (2001). *Pasteurella multocida* gene expression in response to iron limitation. *Infection and Immunity, 69*, 4109–4115.

Penfold, C. N., Ringeling, P. L., Davy, S. L., Moore, G. R., McEwan, A. G., & Spiro, S. (1996). Isolation, characterisation and expression of the bacterioferritin gene of *Rhodobacter capsulatus*. *FEMS Microbiology Letters, 139*, 143–148.

Perry, R. D., Mier, I., & Fetherston, J. D. (2007). Roles of the Yfe and Feo transporters of *Yersinia pestis* in iron uptake and intracellular growth. *Biometals: An International Journal on the Role of Metal Ions in Biology, Biochemistry, and Medicine, 20*, 699–703.

Petersen, T. N., Brunak, S., von Heijne, G., & Nielsen, H. (2011). SignalP 4.0: discriminating signal peptides from transmembrane regions. *Nature Methods, 8*, 785–786.

Peuser, V., Metz, S., & Klug, G. (2011). Response of the photosynthetic bacterium *Rhodobacter sphaeroides* to iron limitation and the role of a Fur orthologue in this response. *Environmental Microbiology Reports, 3*, 397–404.

Peuser, V., Remes, B., & Klug, G. (2012). Role of the Irr protein in the regulation of iron metabolism in *Rhodobacter sphaeroides*. *PloS One*, 7, e42231.

Planavsky, N. J., McGoldrick, P., Scott, C. T., Li, C., Reinhard, C. T., Kelly, A. E., et al. (2011). Widespread iron-rich conditions in the mid-Proterozoic ocean. *Nature, 477*, 448–451.

Platero, R., Peixoto, L., O'Brian, M., & Fabiano, E. (2004). Fur is involved in manganese-dependent regulation of mntA (sitA) expression in *Sinorhizobium meliloti*. *Applied and Environmental Microbiology, 70*, 4349–4355.

Posey, J. E. (2000). Lack of a role for iron in the Lyme disease pathogen. *Science (New York, NY), 288*, 1651–1653.

Poulain, A. J., & Newman, D. K. (2009). *Rhodobacter capsulatus* catalyzes light-dependent Fe(II) oxidation under anaerobic conditions as a potential detoxification mechanism. *Applied and Environmental Microbiology, 75*, 6639–6646.

Pradel, E., Guiso, N., & Locht, C. (1998). Identification of AlcR, an AraC-type regulator of alcaligin siderophore synthesis in *Bordetella bronchiseptica* and *Bordetella pertussis*. *The Journal of Bacteriology, 180*, 871–880.

Puri, S., Hohle, T. H., & O'Brian, M. R. (2010). Control of bacterial iron homeostasis by manganese. *Proceedings of the National Academy of Sciences of the United States of America, 107*, 10691–10695.

Puri, S., & O'Brian, M. R. (2006). The *hmuQ* and *hmuD* genes from *Bradyrhizobium japonicum* encode heme-degrading enzymes. *The Journal of Bacteriology, 188*, 6476–6482.

Rajasekaran, M. B., Mitchell, S. A., Gibson, T. M., Hussain, R., Siligardi, G., Andrews, S. C., et al. (2010a). Isolation and characterisation of EfeM, a periplasmic component of the putative EfeUOBM iron transporter of *Pseudomonas syringae* pv. *syringae*. *Biochemical and Biophysical Research Communications, 398*, 366–371.

Rajasekaran, M. B., Nilapwar, S., Andrews, S. C., & Watson, K. A. (2010b). EfeO-cupredoxins: major new members of the cupredoxin superfamily with roles in bacterial iron transport. *Biometals: An International Journal on the Role of Metal Ions in Biology, Biochemistry, and Medicine*, *23*, 1–17.

Reissbrodt, R., Kingsley, R., Rabsch, W., Beer, W., Roberts, M., & Williams, P. H. (1997). Iron-regulated excretion of alpha-keto acids by *Salmonella typhimurium*. *The Journal of Bacteriology*, *179*, 4538–4544.

Ringeling, P. L., Davy, S. L., Monkara, F. A., Hunt, C., Dickson, D. P., McEwan, A. G., et al. (1994). Iron metabolism in *Rhodobacter capsulatus*. Characterisation of bacterioferritin and formation of non-haem iron particles in intact cells. *European Journal of Biochemistry/ FEBS*, *223*, 847–855.

Rodionov, D. A., Gelfand, M. S., Todd, J. D., Curson, A. R.J., & Johnston, A. W.B. (2006). Computational reconstruction of iron- and manganese-responsive transcriptional networks in alpha-proteobacteria. *PLoS Computational Biology*, *2*, e163.

Rong, C., Zhang, C., Zhang, Y., Qi, L., Yang, J., Guan, G., et al. (2012). FeoB2 functions in magnetosome formation and oxidative stress protection in *Magnetospirillum gryphiswaldense* strain MSR-1. *Journal of Bacteriology*, *194*, 3972–3976.

Rudolph, G., Hennecke, H., & Fischer, H.-M. (2006). Beyond the Fur paradigm: iron-controlled gene expression in rhizobia. *FEMS Microbiology Reviews*, *30*, 631–648.

Runyen-Janecky, L. J., Brown, A. N., Ott, B., Tujuba, H. G., & Rio, R.V. M. (2010). Regulation of high-affinity iron acquisition homologues in the tsetse fly symbiont *Sodalis glossinidius*. *Journal of Bacteriology*, *192*, 3780–3787.

Sabri, M., Léveillé, S., & Dozois, C. M. (2006). A SitABCD homologue from an avian pathogenic *Escherichia coli* strain mediates transport of iron and manganese and resistance to hydrogen peroxide. *Microbiology (Reading, England)*, *152*, 745–758.

Sanders, J. D., Cope, L. D., & Hansen, E. J. (1994). Identification of a locus involved in the utilization of iron by *Haemophilus influenzae*. *Infection and Immunity*, *62*, 4515–4525.

Sandy, M., & Butler, A. (2009). Microbial iron acquisition: marine and terrestrial siderophores. *Chemical Reviews*, *109*, 4580–4595.

Saraiva, I. H., Newman, D. K., & Louro, R. O. (2012). Functional characterization of the FoxE iron oxidoreductase from the photoferrotroph *Rhodobacter ferrooxidans* SW2. *The Journal of Biological Chemistry*, *287*, 25541–25548.

Sarma, R., Barney, B. M., Hamilton, T. L., Jones, A., Seefeldt, L. C., & Peters, J. W. (2008). Crystal structure of the L protein of *Rhodobacter sphaeroides* light-independent protochlorophyllide reductase with MgADP bound: a homologue of the nitrogenase Fe protein. *Biochemistry*, *47*, 13004–13015.

Septer, A. N., Wang, Y., Ruby, E. G., Stabb, E.V., & Dunn, A. K. (2011). The haem-uptake gene cluster in *Vibrio fischeri* is regulated by Fur and contributes to symbiotic colonization. *Environmental Microbiology*, *13*, 2855–2864.

Sirijovski, N., Mamedov, F., Olsson, U., Styring, S., & Hansson, M. (2007). *Rhodobacter capsulatus* magnesium chelatase subunit BchH contains an oxygen sensitive iron-sulfur cluster. *Archives of Microbiology*, *188*, 599–608.

Skaar, E. P., Gaspar, A. H., & Schneewind, O. (2006). *Bacillus anthracis* IsdG, a heme-degrading monooxygenase. *The Journal of Bacteriology*, *188*, 1071–1080.

Smaldone, G. T., Revelles, O., Gaballa, A., Sauer, U., Antelmann, H., & Helmann, J. D. (2012). A global investigation of the *Bacillus subtilis* iron-sparing response identifies major changes in metabolism. *Journal of Bacteriology*, *194*, 2594–2605.

Small, S. K., Puri, S., & O'Brian, M. R. (2009). Heme-dependent metalloregulation by the iron response regulator (Irr) protein in *Rhizobium* and other alpha-proteobacteria. *Biometals: An International Journal on the Role of Metal Ions in Biology, Biochemistry, and Medicine*, *22*, 89–97.

Stearman, R., Yuan, D. S., Yamaguchi-Iwai, Y., Klausner, R. D., & Dancis, A. (1996). A permease-oxidase complex involved in high-affinity iron uptake in yeast. *Science (New York, NY)*, *271*, 1552–1557.

Steele, K. H., O'Connor, L. H., Burpo, N., Kohler, K., & Johnston, J. W. (2012). Characterization of a ferrous iron-responsive two-component system in nontypeable *Haemophilus influenzae*. *Journal of Bacteriology, 194*, 6162–6173.

Stothard, P. (2000). The sequence manipulation suite: JavaScript programs for analyzing and formatting protein and DNA sequences. *BioTechniques, 28*, 1102–1104.

Strange, H. R., Zola, T. A., & Cornelissen, C. N. (2011). The *fbpABC* operon is required for Ton-independent utilization of xenosiderophores by *Neisseria gonorrhoeae* strain FA19. *Infection and Immunity, 79*, 267–278.

Sturm, A., Schierhorn, A., Lindenstrauss, U., Lilie, H., & Brüser, T. (2006). YcdB from *Escherichia coli* reveals a novel class of Tat-dependently translocated hemoproteins. *The Journal of Biological Chemistry, 281*, 13972–13978.

Sugano, Y. (2009). DyP-type peroxidases comprise a novel heme peroxidase family. *Cellular and Molecular Life Sciences: CMLS, 66*, 1387–1403.

Suzuki, T., Okamura, Y., Calugay, R. J., Takeyama, H., & Matsunaga, T. (2006). Global gene expression analysis of iron-inducible genes in *Magnetospirillum magneticum* AMB-1. *The Journal of Bacteriology, 188*, 2275–2279.

Theil, E. C., & Goss, D. J. (2009). Living with iron (and oxygen): questions and answers about iron homeostasis. *Chemical Reviews, 109*, 4568–4579.

Tom-Yew, S. A.L., Cui, D. T., Bekker, E. G., & Murphy, M. E.P. (2005). Anion-independent iron coordination by the *Campylobacter jejuni* ferric binding protein. *The Journal of Biological Chemistry, 280*, 9283–9290.

Touati, D. (2000). Iron and oxidative stress in bacteria. *Archives of Biochemistry and Biophysics, 373*, 1–6.

Van Der Giezen, M., & Lenton, T. M. (2012). The rise of oxygen and complex life. *The Journal of Eukaryotic Microbiology, 59*, 111–113.

van Vliet, A. H., Wooldridge, K. G., & Ketley, J. M. (1998). Iron-responsive gene regulation in a *Campylobacter jejuni fur* mutant. *The Journal of Bacteriology, 180*, 5291–5298.

Vanderpool, C. K., & Armstrong, S. K. (2003). Heme-responsive transcriptional activation of *Bordetella bhu* genes. *The Journal of Bacteriology, 185*, 909–917.

Vasil, M. L. (2007). How we learnt about iron acquisition in *Pseudomonas aeruginosa*: a series of very fortunate events. *Biometals: An International Journal on the Role of Metal Ions in Biology, Biochemistry, and Medicine, 20*(3–4), 587–601.

Wandersman, C., & Delepelaire, P. (2004). Bacterial iron sources: from siderophores to hemophores. *Annual Review of Microbiology, 58*, 611–647.

Watson, R. J., Millichap, P., Joyce, S. A., Reynolds, S., & Clarke, D. J. (2010). The role of iron uptake in pathogenicity and symbiosis in *Photorhabdus luminescens* TT01. *BMC Microbiology, 10*, 177.

Weinberg, E. D. (1997). The *Lactobacillus* anomaly: total iron abstinence. *Perspectives in Biology and Medicine, 40*, 578–583.

Wen, Y., Kim, I. H., Son, J.-S., Lee, B.-H., & Kim, K.-S. (2012). Iron and quorum sensing coordinately regulate the expression of vulnibactin biosynthesis in *Vibrio vulnificus*. *The Journal of Biological Chemistry, 287*, 26727–26739.

Wexler, M., Yeoman, K. H., Stevens, J. B., de Luca, N. G., Sawers, G., & Johnston, A. W. (2001). The *Rhizobium leguminosarum tonB* gene is required for the uptake of siderophore and haem as sources of iron. *Molecular Microbiology, 41*, 801–816.

Widdel, F., Schnell, S., Heising, S., Ehrenreich, A., Assmus, B., & Schink, B. (1993). Ferrous iron oxidation by anoxygenic phototrophic bacteria. *Nature, 362*, 834–836.

Wyckoff, E. E., Mey, A. R., Leimbach, A., Fisher, C. F., & Payne, S. M. (2006). Characterization of ferric and ferrous iron transport systems in *Vibrio cholerae*. *The Journal of Bacteriology, 188*, 6515–6523.

Yasmin, S., Andrews, S. C., Moore, G. R., & Le Brun, N. E. (2011). A new role for heme, facilitating release of iron from the bacterioferritin iron biomineral. *The Journal of Biological Chemistry, 286*, 3473–3483.

Zappa, S., & Bauer, C. E. (2013). Hbrl controls the stoichiometry of all organic and inorganic components of cytochromes. In preparation.

CHAPTER ELEVEN

Genes Associated with the Peculiar Phenotypes of the Aerobic Anoxygenic Phototrophs

Vladimir Yurkov[1], Elizabeth Hughes[1]
Department of Microbiology, University of Manitoba, Winnipeg, MB, Canada
[1]Corresponding authors: E-mail: vyurkov@cc.umanitoba.ca; hughes.elizabethe@gmail.com

Contents

Abstract

The aerobic anoxygenic phototrophs (AAP) are an important group of bacteria constituting up to 80% of the microbial population in some habitats. They likely evolved from the purple non-sulfur bacteria (PNSB), to fill an environmental niche, carrying out anoxygenic photosynthesis aerobically. Genomic studies on these colourful and

Advances in Botanical Research, Volume 66
ISSN 0065-2296, http://dx.doi.org/10.1016/B978-0-12-397923-0.00011-4

intriguing microorganisms have only just begun in an attempt to discover what changes occurred to allow for the paradox, that is bacteriochlorophyll *a* synthesis in the presence of O_2. Genes of the AAP and the PNSB are so similar that there is no absolute means of distinguishing between them by culture-independent techniques alone. The presence of AAP in every environment tested for their presence, in addition to their unique photosynthetic nature, are mysteries that have garnered much attention in the scientific community since their discovery.

1. INTRODUCTION

Having been discovered only 35 years ago, it is not surprising that there is relatively little known about the aerobic anoxygenic phototrophs (AAP). It was long believed that the anaerobic purple and green bacteria, as well as the heliobacteria, were the only organisms capable of anoxygenic photosynthesis (Yurkov & Beatty, 1998). However, in 1978, everything changed as marine strains OCh 101 and OCh 114 were isolated (Shiba, Simidu, & Taga, 1979). These were the first aerobic bacteria found to contain bacteriochlorophyll *a* (BChl *a*) and were later characterised as *Erythrobacter longus* OCh 101 and *Roseobacter denitrificans* OCh 114 (Shiba, 1991; Shiba & Simidu, 1982). Several freshwater AAP such as *Erythrobacter sibiricus* (later re-classified as *Erythromicrobium sibiricum* (Yurkov, Lysenko, & Gorlenko, 1991) and again as *Sandaracinobacter sibiricus* (Yurkov et al., 1997)), *Erythromicrobium ursincola* (re-classified as *Erythromonas ursincola* (Yurkov et al., 1997)), *Erythromicrobium ezovicum*, *Erythromicrobium ramosum*, *Erythromicrobium hydrolyticum* and *Roseococcus thiosulfatophilus* were then discovered in the early 1990s (Yurkov & Gorlenko, 1990, 1992a, 1992b; Yurkov, Gorlenko, & Kompantseva, 1993). This was followed by much activity to search for new species and begin studying the conundrum, that is aerobic anoxygenic photosynthesis. Since that time, 58 species in 37 genera have been taxonomically characterised and the number is increasing every year. The AAP seem to have evolved to fill an environmental niche after global oxygenation occurred 2.5 GYa ago due to the activity of oxygenic phototrophic cyanobacteria (Beatty, 2005). Competition for habitats that were not in contact with this toxic O_2-rich atmosphere was fierce and exerted selection pressure for an aerobic counterpart to the anaerobic anoxygenic phototrophs. AAP very likely arose from and remain closely related to various purple bacterial lineages (Yurkov & Csotonyi, 2009). At the moment, this evolution remains poorly understood, although a likely explanation is that AAP progenitors lost their photosynthesis genes on moving to oxygenated habitats only to regain them later through lateral gene transfer from purple non-sulfur bacteria (PNSB), thus forming the AAP (Yurkov & Csotonyi, 2009).

AAP belong to the Proteobacteria, with 56 of the 59 taxonomically characterised strains being from the α-Proteobacteria, one from the β-Proteobacteria, and two from the γ-Proteobacteria (Csotonyi, Stackebrandt, Swiderski, Schumann, & Yurkov, 2011a; Rathgeber, Beatty, & Yurkov, 2004). All contain BChl *a* and display the characteristic bright colours red, purple, orange and yellow due to the rich presence of carotenoids (crt). Most of these pigments are also synthesised by anaerobic phototrophs, although AAP produce crt in much higher quantities and keep the number of BChl-*a*-containing photosynthetic units per cell to a minimum (Rathgeber et al., 2004). In addition, they retained a photosynthetic apparatus very similar to that of PNSB consisting of a reaction centre (RC) bound to a light harvesting (LH) 1 complex with the option for an additional peripheral LH2. Unlike the PNSB, they require O_2 for both growth and photosynthetic activity. This is quite a paradox as O_2 inhibits BChl synthesis in all tested anaerobic anoxygenic phototrophs. They are also unable to grow autotrophically by way of the Calvin–Benson–Bassham (CBB) cycle and are therefore obligate heterotrophs (Yurkov & Csotonyi, 2009).

AAP flourish in many environments and shotgun sequencing of conserved genes allows for the use of culture-independent studies to identify their presence in these locations. The group is also highly tolerant to toxins and environmental conditions that would kill the vast majority of life on our planet. These can be, for example, heavy metal(loid) oxides that pollute habitats, acid mine drainage sites, high temperatures such as those in marine hydrothermal vents and hot temperature springs as well as high salinity found in hypersaline lakes (Rathgeber et al., 2004). The genes that evolved to allow such tolerance and flexibility remain a mystery but could help to explain what was necessary for the evolution of life in the extreme conditions that our planet had in its early days.

This chapter explores the riddles that have plagued the scientific community for three decades regarding the evolution of the AAP genome. AAP are typical proteobacterial cells, constructed from the same building blocks as their relatives, expressing no known novel metabolic processes, possessing typical cell walls, synthesising mostly trivial pigments and electron carriers, etc. However, evolutionary forces have combined some of these common components in chimeric ways, permitting anoxygenic phototrophy to function under conditions contrary to those for which it originally evolved, thus opening for exploitation huge niches not available to their ancestors.

This principal evolutionary development necessarily caused cascades of corollary adaptations, such as major changes to the regulation of photosynthetic pigment synthesis, and proliferation of photoprotective crt (Yurkov & Csotonyi, 2009). This leaves many questions unanswered regarding the way in which selective pressure forced the genome to evolve from that of an anaerobic phototroph to that of an aerobe capable of employing an anaerobic photosynthesis gene cluster (PGC).

2. ABUNDANCE AND DIVERSITY THROUGH CULTURING AND SEQUENCING

2.1. Morphological Pleomorphism

AAP are a highly diverse group of bacteria. Morphologically, they come in many forms including rods, coccoid, and even nearly cyclic such as *Roseicyclus mahoneyensis* (Fig. 11.1a) (Rathgeber et al., 2005). Many species are highly pleomorphic, for example, *E. ramosum* produces branches (Fig. 11.1c), *Citromicrobium bathyomarinum* forms Y-shaped cells (Fig. 11.1f) and *Porphyrobacter meromictius* has many appendages (Fig. 11.1d) (Rathgeber et al., 2007; Yurkov & Csotonyi, 2009; Yurkov, Krieger, Stackebrandt, & Beatty, 1999). The only morphology yet to be cultured is true spirilloid, although uncultured examples attributed to AAP have been observed in the Sargasso Sea (Sieracki, Gilg, Their, Poulton, & Goericke, 2006). Certain isolates found in meromictic lakes and oceans aggregate to form complex rosettes, possibly to aid in buoyancy or for quorum sensing in cell populations. These include organisms such as *Stappia marina*, *Roseovarius mucosus*, the taxonomically uncharacterised ocean strain C8 and meromictic lake strains BL7 (Fig. 11.1g) and BL14 (Fig. 11.1e) (Yurkov & Csotonyi, 2009), which produce the most complex configurations resembling dandelion heads or corals. Some, including *Cmi. bathyomarinum* and *P. meromictius* have also been shown to form chains with bubble-like formations connecting the cells (Fig. 11.1b), which could aid in the exchange of cellular materials and nutrients in oligotrophic environments, although it has yet to be proven (Rathgeber et al., 2007; Yurkov et al., 1999).

2.2. Comments on Taxonomy

Although AAP are physiologically associated with PNSB in many ways and it is difficult to differentiate between the two through culture-independent methods, their closest relatives are frequently non-phototrophs (Yurkov & Csotonyi, 2009). Sometimes this causes problems for scientists trying to

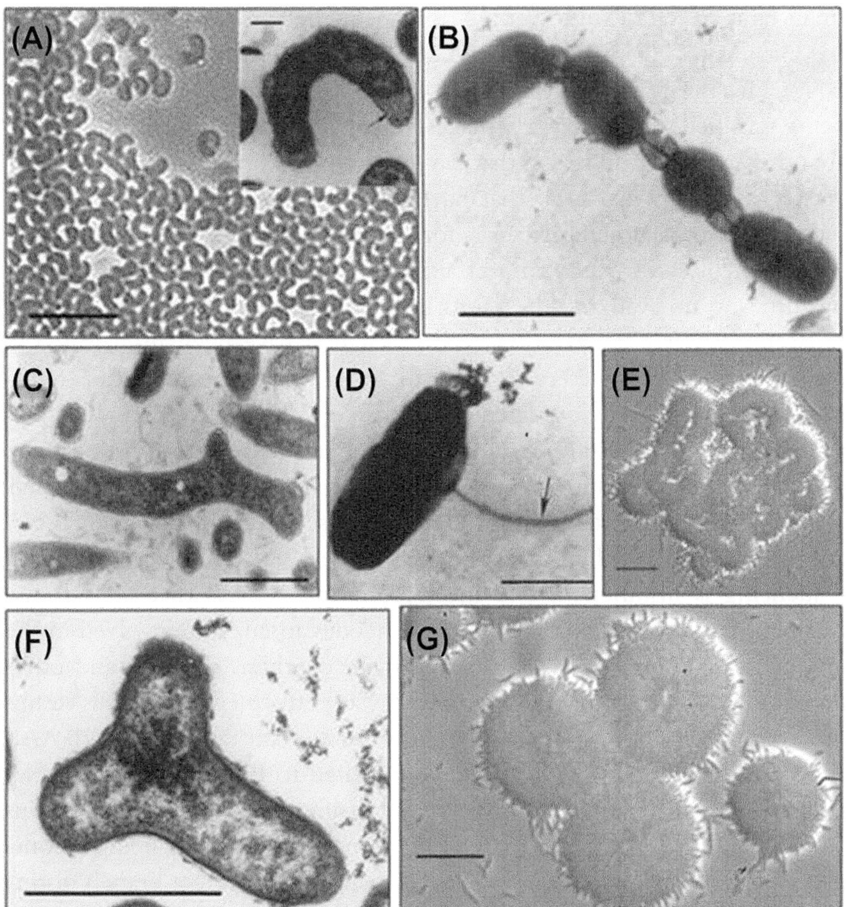

Figure 11.1 ***Morphological diversity of AAP.*** (A) *Roseicyclus mahoneyensis* (bar 5 μm, inset bar 0.25 μm); (B) *Porphyrobacter meromictius* (bar 1 μm); (C) *Erythromicrobium ramosum* (bar 1 μm); (D) *P. meromictius* (bar 0.5 μm); (E) Strain BL14 (bar 2.5 μm); (F) *Citromicrobium bathyomarinum* (bar 1 μm); and (G) Strain BL7 (bar 2.5 μm).

describe new species. Occasionally, 16S rRNA sequencing is insufficient to place new phototrophic strains in novel genera or species despite the fact that the proposed taxonomic group does not contain any other phototrophic representative. In these cases, physiological properties must be carefully analysed in addition to sequencing. The ability for phototrophy is a characteristic important enough to set them apart. However, there are a few strains that are still considered to be incorrectly classified. A recent example is that of *Erythrobacter litoralis*, strain HTCC2594. Although the genus *Erythrobacter* was originally devised to contain only AAP and *Erb. litoralis* type strain T4 is an AAP (Yurkov et al., 1994), sequencing revealed that all PGCs are missing in HTCC2594 (Oh, Giovannoni, Ferriera, Johnson, & Cho, 2009). This is only one of the several examples of non-phototrophic strains being classified as *Erythrobacter* species (Table 11.1) and, unfortunately, there are other examples of genera that have the same taxonomic problem. Conversely, there are also AAP that have been classified in taxa that do not represent photosynthetic bacteria, for example, *Hoeflea phototrophica* and *S. marina* (Biebl et al., 2006; Kim et al., 2006). It has been criticised numerous times to no avail and the knots in the system remain tangled (Yurkov & Beatty, 1998). Over the years, some bacteria have been re-classified as new information comes to light; however, many studies continue referring to these species by outdated rather than updated more-accurate names (Yurkov & Csotonyi, 2009). In these cases, it is important for reviewers to be more diligent so that mistakes can be corrected before articles are published.

There is research currently being done on sequences other than rRNA that can be used for the characterisation of new species when 16S rRNA sequencing is insufficient. The most promising gene is that for farnesyl pyrophosphate synthase (FPPS), which encodes an enzyme involved in the production of isoprenoids that feed into the synthesis of BChl and other important metabolites. FPPS sequencing was performed on various strains of *Erythrobacter* and *Roseobacter*. These groups contain tangles of both AAP and non-phototrophs and are so closely related to PNSB that 16S rRNA sequencing is inadequate for taxonomic classification. Separation of strains based on FPPS relatedness classified the non-phototrophs as being a completely different cluster from the AAP, which in turn were a branch distinguished from the PNSB. The data show that this gene evolved differently for the AAP than for its non-phototrophic relatives (Feng, Jiao, Du, & Zeng, 2010). Such research is very promising and will hopefully solve AAP taxonomic tangles.

2.3. Natural Abundance

AAP have been found in every environment tested. They thrive in oceans, lakes, rivers, and soil and can even survive in extreme habitats such as deep-sea hydrothermal vents, hot springs, mine tailing sites, meromictic lakes and hypersaline water bodies (Rathgeber et al., 2004; Yurkov & Csotonyi, 2009). The most recently discovered habitat for AAP is Canadian biological soil crusts, where they comprise up to 6% of cultivable bacteria (Csotonyi, Swiderski, Stackebrandt, & Yurkov, 2010). Through culturing techniques,

Table 11.1 Taxonomic confusion in *Erythrobacter*

Species described as	Metabolic category	Presence of BChl	Presence of PGC
Erythrobacter longus⋆ (Shiba & Simidu, 1982)	Ph	+	+
Erythrobacter litoralis T4 (Yurkov et al., 1994)	Ph	+	ND
Erythrobacter sp. NAP1 (Koblížek et al., 2011)	Ph	+	+
Erythrobacter sp. JL-475 (Feng et al., 2011)	Ph	+	ND
Erythrobacter litoralis HTCC2594 (Oh et al., 2009)	NPh	−	−
Erythrobacter sp. SD-21 (Denner, Vybiral, Koblížek, Busse, & Velimirov, 2002)	NPh	−	ND
Erythrobacter citreus (Denner et al., 2002)	NPh	−	ND
Erythrobacter flavus (Yoon, Kim, Kim, Kang, & Park, 2003)	NPh	−	ND
Erythrobacter vulgaris (Ivanova et al., 2005)	NPh	−	ND
Erythrobacter aquimaris (Yoon, Kang, Oh, & Park, 2004)	NPh	−	ND
Erythrobacter seohaensis (Yoon, Oh, et al., 2005, Yoon, Kang, et al., 2005)	NPh	−	ND
Erythrobacter gaetbuli (Yoon, Oh, et al., 2005)	NPh	−	ND
Erythrobacter luteolus (Yoon, Kang, et al., 2005)	NPh	−	ND
Erythrobacter gangjinensis (Lee, Lee, Kahng, Kim, & Jung, 2010)	NPh	−	ND
Erythrobacter nanhaisediminis (Xu et al., 2010)	NPh	−	ND
Erythrobacter pelangi (Wu et al., 2012)	NPh	−	ND
Erythrobacter marinus (Jung, Park, Oh, & Yoon, 2011)	NPh	−	ND
Erythrobacter jejuensis (Yoon, Lee, & Oh, 2012)	NPh	−	ND

ND, not determined; NPh, nonphototrophic; Ph, phototrophic.
⋆Type species of the genus.

it has been shown that AAP make up as much as 10% of the microbial population in the world's largest ecosystem, the ocean. High abundance generally means that an organism is important to that ecosystem. As the open ocean is oligotrophic, the role might be in the cycling of nutrients, and more specifically, cycling of the low amounts of carbon (Rathgeber et al., 2004). Culture-independent techniques, such as shotgun sequencing, infrared epifluorescence microscopy, fluorescence emission spectroscopy and high-performance liquid chromatography of the Mediterranean Sea and some oligotrophic lake waters, indicate that the abundance of AAP could be as high as 80% of the microbial population, although the vast majority of microbes have not been isolated or cultured (Eiler, Beier, Säwström, Karlsson, & Bertilsson, 2009). In addition, it is difficult to confirm these numbers as techniques have not yet been developed to allow for scientists to distinguish between AAP and PNSB through sequencing alone (Yurkov & Csotonyi, 2009). Although most studies on AAP abundance have been done in the open ocean, it has been suggested that this group is important in eutrophic habitats as well. They can make up 10.7–34% of the bacterial communities in nutrient-rich microbial mats, temperate estuaries and algae-rich waters (Yurkov & Csotonyi, 2009).

2.4. Culturing versus Sequencing

There is currently no differential enrichment growth medium that can be used for the specific isolation of AAP, as they grow best on a rich organic medium in the presence of O_2. Many types of heterotrophic bacteria can thrive under these conditions; therefore, AAP must be chosen for isolation based on their characteristic bright pigmentation that permits them to stand out on plates covered by other colonies with mundane colours. This is followed by spectrophotometric analysis to confirm the presence of BChl *a*, a crucial defining characteristic of the group. The disadvantage of culturing bacteria in this way is that only a fraction of naturally occurring microorganisms is able to grow in a controlled laboratory setting. It is a constant hurdle that leads scientists to underestimate the abundance and diversity of microorganisms.

Shotgun sequencing and other culture-independent methods indicate a much greater abundance of AAP in many habitats than was originally hypothesised. The most common sequences analysed are those of the *pufLM* genes, which encode subunits of the RC (Jiang et al., 2009; Salka, Cuperova, Masin, Koblizek, & Grossart, 2011; Zeng, Shen, & Jiao, 2009). However, these genes are highly similar in AAP and purple bacteria, making it impossible to differentiate between them. Recently, other genes have been proposed as appropriate for the enumeration of AAP, including three gene clusters, *chlN* or *bchN*, *chlB* or *bchB*, and *chlL* or *bchL*, which encode protochlorophyllide oxidoreductase and its homologue chlorophyllide oxidoreductase. These enzymes convert chlorophyllide to protochlorophyllide during the synthesis of chlorophyll (Chl) and BChl, allowing differentiation between anoxygenic and oxygenic phototrophs (Eiler et al., 2009). Other studies have estimated the amount of AAP by assessing the photosynthetic productivity, which is done by measuring levels of bacteriopheophytin present in the sample. Finally, infrared epifluorescence microscopy or high-performance liquid chromatography of the water itself have been used to quantify the amount of BChl (Yurkov & Csotonyi, 2009). The problem with all these techniques is that none are capable of discriminating between AAP and PNSB, so the dilemma continues. Therefore, a combination of culture-dependent and culture-independent techniques in conjunction is the best approach to improve our understanding of the ecology and distribution of the AAP.

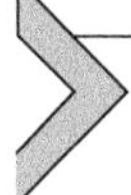

3. PHYLOGENY AND SPECULATIONS ON EVOLUTION

3.1. Phylogeny

Most characterised AAP belong to the α-Proteobacteria and are divided between the α-1 through α-4 subclasses (Yurkov & Csotonyi, 2009). The α-1 Proteobacteria include *Acidiphilium*, *Acidisphaera*, *Craurococcus*, *Geminicoccus*, *Paracraurococcus*, *Roseococcus* and *Rubritepida*, which were isolated from soil, hot springs and acidophilic environments. *Labrenzia*, *Roseibium*, *S. marina* and *H. phototrophica*, all marine isolates, belong to the α-2 Proteobacteria. The α-3 subclass consists of all members of the marine and halophilic *Roseobacter* clade. Finally, *Blastomonas natatoria*, *Citromicrobium*, *Erythrobacter*, *Erythromicrobium*, *Erythromonas*, *Sandaracinobacter* and *Sandarakinorhabdus* hail from the α-4 subclass, originating from various marine and freshwater environments (Boldareva et al., 2009; Labrenz, Lawson, Tindall, & Hirsch, 2009; Venkata Ramana, Sasikala, Takaichi, & Ramana, 2010; Yu, Yan, Li, & Zhang, 2011; Yurkov & Csotonyi, 2009). One species, *Roseatales depolymerans*, is classified as β-Proteobacteria and two, *Congregibacter litoralis* and *Chromocurvus halotolerans*, belong to the γ-Proteobacteria (Csotonyi et al., 2011a; Rathgeber et al., 2004) (Table 11.2). Although there are only three characterised species of AAP that do not belong to the α subclass, recent shotgun sequencing results have suggested that there are many more β-and

Table 11.2 Determinative characteristics of novel AAP described since 2009*

Species	Habitat, site of isolation	Phylogenetic affiliation**	Colour	Cell shape	*In vivo* BChl peaks (nm)
Roseibaca ekhonensis*** (Labrenz et al., 2009)	Ekho Lake, Antarctica	α-Proteobacteria, *Rhodobaca* (94%)	Pink	Rod	865–866
Roseococcus suduntuyensis (Boldareva et al., 2009)	Soda Lake, Russia	α-Proteobacteria, *Roseococcus thiosulfatophilus* (98.6%)	Pink	Cocci, short rods	865
Roseomonas aestuarii (Venkata Ramana et al., 2010)	Marine water, India	α-Proteobacteria, *Roseomonas cervicalis* (96.4%)	Orange	Coccobacilli	ND
Roseicitreum antarcticum (Yu et al., 2011)	Sandy sediment, Antarctic Ocean, China	α-Proteobacteria, *Rhodobacter veldkampii* (96.5%)	Pink to red	Rod	872–874
Chromocurvus halotolerans (Csotonyi et al., 2011a, 2011b)	East German Creek System, Canada	γ-Proteobacteria, *Haliea* (95.1–96.1%)	Pale pink	Pleomorphic	802, 877
Charonomicrobium ambiphototrophicum (Csotonyi et al., 2011a, 2011b)	East German Creek System, Canada	α-Proteobacteria, *Roseibacterium elongatum* (98.3%)	Deep brown	Coccoid, rods	802, 850, 879

ND, not determined.
*New species described before 2009 are summarised in Yurkov and Csotonyi, (2009); Rathgeber et al. (2004) and Yurkov and Beatty, (1998).
**Proteobacterial subclass, nearest 16S rRNA relative and phylogenetic distance to nearest relative.
***Names in bold indicate novel genera.

γ-proteobacterial AAP species (Cottrell, Ras, & Kirchman, 2010; Feng, Lin, Mao, & Zhu, 2011; Jiang et al., 2009; Lehours, Cottrell, Dahan, Kirchman, & Jeanthon, 2010; Zeng et al., 2009). What is most exciting is the emergence of a pattern. It seems that γ-Proteobacteria have evolved to dominate saltwater habitats, while β-Proteobacteria are found primarily in freshwater ecosystems and prefer alkaline pH. α-Proteobacteria appear to be present in high abundance in a variety of environments (Table 11.2) (Jiao et al., 2007; Salka et al., 2011). Whether these subclasses have an evolutionary advantage in their respective habitats is still unclear, but certainly warrants more research.

3.2. Evolution

There are many questions as to the nature of possible evolution. Did AAP branch from different PNSB lineages or did they all evolve from a single common ancestor? What genomic differences exist that allow them to aerobically produce BChl *a* and how did they come by these differences? Which genes have been changed through evolution to allow for a definitive physiological distinction between AAP and PNSB? None of these questions have been completely answered yet, although some investigations have begun to steer us in the right direction.

It is speculated that photosynthesis began solely as an anaerobic anoxygenic process performed by BChl-containing bacteria. This eventually evolved into an aerobic oxygenic process like that employed by cyanobacteria and vascular plants containing Chl. AAP could possibly be an evolutionary link between the two (Jiang et al., 2009; Yurkov & Csotonyi, 2009). As competition for habitats suitable for anaerobic anoxygenic photosynthesis became too fierce, the facultatively anaerobic PNSB were forced to evolve. It is thought that these organisms began by developing heterotrophic metabolic pathways. This allowed them to become independent of light for growth and therefore move into new environments. From there, it was unnecessary for the emerging species to retain their energetically costly photosynthetic properties, so the genes responsible were lost. The newly evolved non-phototrophs then became AAP by regaining a PNSB photosynthetic apparatus through lateral gene transfer (Jiang et al., 2009; Marrs, 1981; Yurkov & Csotonyi, 2009). This theory is supported by sequencing of AAP genomes, which show the presence of phage DNA directly associated with the photosynthesis operons. In *Cmi. bathyomarinum*, strain JL354, whole genome sequencing actually revealed the presence of two different photosynthesis operons, further supporting the theory of lateral gene transfer.

Isolated from the South China Sea, this organism was classified in the α-4 Proteobacteria, with 99.6% 16S rRNA sequence similarity to the type strain JF1. Sequencing showed the presence of the typical α-proteobacterial *puf* operon and, surprisingly, an incomplete γ-proteobacterial *puf* operon. Both gene clusters were closely associated with prophage DNA. The partial operon contained only *pufLMC–puhCBA*, which is missing several important genes for photosynthetic ability, suggesting incomplete gene transfer (Jiao, Zhang, & Zheng, 2010). There is still much speculation as to how lateral gene transfer of the entire PGC occurs.

Studies suggesting the evolution of PNSB to non-phototrophs before the progression to AAP may explain why strains are often most closely related to non-phototrophs (Cho et al., 2007; Igarashi et al., 2001). These heterotrophic species could be those from which AAP arose. The theory of lateral gene transfer of photosynthesis genes also supports the idea that AAP emerged separately from many lineages (Yurkov & Csotonyi, 2009). As this event seems to have occurred on many occasions, it can be assumed that the acquisition of the PGC is advantageous to heterotrophic bacteria as a means of supplementing energy production.

Contrarily, there is also a theory that AAP arose directly from a single PNSB ancestor (Woese, 1987). The AAP are dispersed throughout the Proteobacteria in a mix of both phototrophic and non-phototrophic organisms, suggesting that through evolution to aerobic metabolism, some organisms lost their PGC, while others retained them. It was proposed that all Proteobacteria, irrespective of their pathways for energy acquisition, evolved from a single ancestral lineage (Beatty, 2005; Woese, 1987). However, this argument is not very well supported yet.

More research must be done to confirm the evolutionary theories that may explain the change from PNSB to AAP. We must also find genes that can phylogenetically and taxonomically separate AAP from their relatives when 16S rRNA sequencing is insufficient.

4. NUTRITIONAL REQUIREMENTS, CARBON METABOLISM AND THE ASSOCIATED METABOLIC PATHWAYS

4.1. Organic Carbon Metabolism

As AAP are obligate heterotrophs, metabolism of organic carbon is very important. *Roseobacter denitrificans*, the most thoroughly studied AAP, *Dinoroseobacter shibae* and *Jannaschia* sp. CCS1 primarily employ the

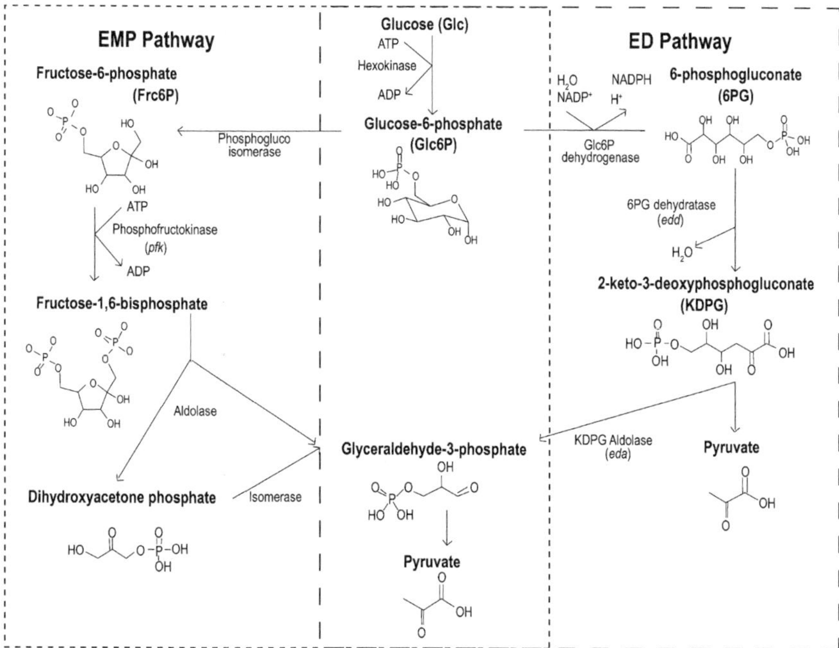

Figure 11.2 EMP and ED pathways.

Entner–Doudoroff (ED) pathway for the breakdown of carbohydrates from glucose to pyruvate through 6-phosphogluconate and 2-keto-3-deoxy-6-phosphogluconate (KDPG) (Fig. 11.2) (Fürch et al., 2009; Tang, Feng, Tang, & Blankenship, 2009). *Roseobacter denitrificans* also contains all genes of the non-oxidative pentose phosphate pathway (NOPP), which produces molecules that can then be used for biosynthesis of important co-factors, amino acids and nucleotides (Tang et al., 2009). Whole genome studies on *Rsb. denitrificans* showed that it contains both the *eda* gene encoding KDPG aldolase and *edd* gene encoding phosphogluconate dehydratase, the two key enzymes in the ED pathway (Fig. 11.2). The enzymes have also been observed in *Srb. sibiricus* (formerly *E. sibiricum*) and *Erb. longus*, supporting the idea that this is the predominant glucose metabolic pathway in AAP (Yurkov & Beatty, 1998). Presence of this path is quite typical of Gram-negative bacteria and it is often also present in anaerobic phototrophs (Yurkov & Beatty, 1998), although the latter primarily utilise the Embden–Meyerhof–Parnas (EMP) pathway also known as glycolysis. The EMP pathway converts glucose into two molecules of pyruvate through fructose-6-phosphate but is inactive in studied AAP because although most genes are present, the *pfk* gene that codes for the key enzyme phosphofructokinase is lacking (Tang et al., 2009; Tang, Tang, & Blankenship, 2011) (Fig. 11.2). Loss of *pfk* and gain of *eda* and *edd* suggests evolution from the purple bacteria's anaerobic lifestyle to that of AAP living in oxygenated environments. Although the *pfk* gene is missing in *Rsb. denitrificans*, *D. shibae* strain DFL12 contains a full complement of genes necessary for glycolysis, including *pfk*, despite the fact that it does not utilise this pathway (Fürch et al., 2009; Tang et al., 2009, 2011). This could be a bacterium that is not as far along on the evolutionary road as some AAP but is still far enough to not make use of the pathway, which allows us to theorise that it could be an intermediary in PNSB to AAP evolution.

Acetate-grown cultures of *Rsc. thiosulfatophilus*, *Erm. ursincola*, *E. ezovicum*, *E. hydrolyticum*, and *E. ramosum* revealed the presence of the tricarboxylic acid (TCA) cycle (Yurkov & Beatty, 1998; Yurkov & Gorlenko, 1992a, 1992b; Yurkov et al., 1994). In addition, whole genome sequencing of *Rsb. denitrificans* and *D. shibae* show that they also contain all genes of the TCA cycle (Fürch et al., 2009; Tang et al., 2009, 2011). The TCA cycle binds a molecule of acetyl-CoA to oxaloacetate to form citrate. From there, many important intermediates can be generated and used in other pathways for nucleotide, amino acid and co-factor synthesis. Eventually malate is formed, which is then converted back into oxaloacetate. The anapleurotic reactions fix carbon to molecules of pyruvate or phosphoenol pyruvate to replenish malate and oxaloacetate (Fig. 11.3).

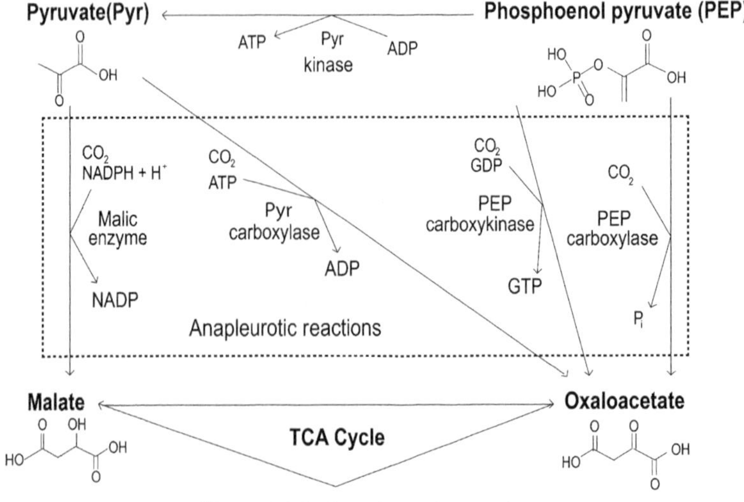

Figure 11.3 Anapleurotic reactions.

Genes for the NOPP pathway are present in *Rsb. denitrificans* and *D. shibae*, although the gene *pgd* encoding 6-phosphogluconate dehydrogenase, the key enzyme in the oxidative pentose phosphate (OPP) pathway, is missing (Tang et al., 2009). This is in keeping with many bacteria including anaerobic anoxygenic phototrophs (Tang et al., 2011), suggesting that this gene loss could have been occurred before relocation of the group to aerobic habitats. Conversely, sequencing revealed that the AAP *Jannaschia* sp. CCS1 contains the *pgd* gene, although studies have not yet been done to determine if the OPP pathway is in fact active (Tang et al., 2009).

Mechanisms and reasons for the evolution away from the EMP and OPP pathways are unclear, although it is consistent with the theory that AAP arose when anaerobic phototrophs migrated into aerobic habitats. It has also been observed that bacteria containing identical types of RC can have completely different metabolic pathways. This once again supports the theory of evolution from photosynthetic bacteria to non-phototrophic bacteria through lateral gene transfer of photosynthesis genes (Tang et al., 2011) as opposed to evolution from a single common ancestor.

4.2. Inorganic Carbon Fixation

Inorganic carbon in our atmosphere in the form of CO_2 is one of the largest sources of carbon in the world. The two most important sinks of CO_2 are through absorption by the ocean and photosynthesis by autotrophic organisms (Tang et al., 2009). The most common pathway for carbon fixation is the CBB cycle, which is used by both higher plants and bacteria including PNSB. However, unlike their photosynthetic ancestors, AAP are not autotrophic; they are obligate heterotrophs requiring the presence of organic carbon for growth. The absence of autotrophic growth and the CBB cycle key enzyme ribulose bisphosphate carboxylase/oxygenase (RUBISCO) was first reported in *Erb. longus* and *Srb. sibiricus* (originally *Erb. sibiricus*) (Yurkov, 1990). Later, whole genome sequence analysis of *Rsb. denitrificans* revealed the absence of genes *rbcS* and *rbcL* (encoding RUBISCO) and *cfxP* (phosphoribulokinase), which are key enzymes in the cycle (Sato-Takabe, Hamasaki, & Suzuki, 2011; Swingley et al., 2007; Tang et al., 2009).

Despite the lack of autotrophic genes, AAP from the *Roseobacter* clade, including *Rsb. denitrificans*, have been shown to fix up to 15% of protein carbon from CO_2 (Tang et al., 2009). Also, *Srb. sibiricus* tested with radiolabeled CO_2 indicated a low level (0.4%) of anapleurotic CO_2 fixation (Yurkov, 1990). The full gene complement of anapleurotic enzymes is present in the *Rsb. denitrificans* and *D. shibae* genomes (Fürch et al., 2009; Swingley et al.,

2007; Tang et al., 2009, 2011): phosphoenol pyruvate carboxykinase, phosphoenol pyruvate carboxylase, pyruvate carboxylase and the malic enzyme (Sato-Takabe et al., 2011; Tang et al., 2011). These enzymes form oxaloacetate and malate for use in the TCA cycle by binding CO_2 to pyruvate or phosphoenol pyruvate (Fig. 11.3). Although inorganic carbon is fixed, it does not allow for autotrophic growth because the required organic substrates cannot be formed from CO_2 without the use of the CBB or other cycles (Swingley et al., 2007; Tang et al., 2009).

Recent studies suggest that other pathways capable of CO_2 fixation may be present in AAP. One is the 3-hydroxypropionate (3-HP) cycle for which all genes were found in strain HTCC2080 of the NOR5/OM60 group. This strain may regenerate acetyl-CoA through 4-hydroxybutyrate (4-HB) as opposed to the more typical malyl-CoA suggesting that it may grow autotrophically by way of a hybrid 3-HP/4-HB cycle. Such pathways could provide this organism with a competitive advantage as it allows for mixotrophic assimilation of organic compounds (Hügler & Sievert, 2011). This is the only AAP found to date with this option and could be the exception to the rule. Every other AAP full genome sequence has only some of the required genes (Hügler & Sievert, 2011; Swingley et al., 2007). Also, the 12 species *Srb. sibiricus*, *Erm. ursincola*, *E. ezovicum*, *E. ramosum*, *E. hydrolyticum*, *Rsc. thiosulfatophilus*, *Erb. litoralis*, *Cmi. bathyomarinum*, *R. mahoneyensis*, *P. meromictius*, *Chc. halotolerans*, and *Charonomicrobium ambiphototrophicum*, were tested physiologically for autotrophic growth, and all produced negative results (Csotonyi et al., 2011a; Csotonyi, Stackebrandt, Swiderski, Schumann, & Yurkov, 2011b; Rathgeber et al., 2005, 2007; Yurkov & Gorlenko 1990; Yurkov, Gorlenko & Kompantseva 1993; Yurkov et al., 1991, 1994, 1997, 1999).

Finally, some AAP have been found to contain genes for the ethylmalonyl-CoA pathway, which can regenerate malate for the TCA cycle (Tang et al., 2011). Three molecules of acetyl-CoA and a molecule of CO_2 are used to form malate and propionyl-CoA. Once again, however, this pathway does not allow for true autotrophy.

4.3. Additional Nutrient Requirements

There are many nutrients required for biosynthesis of molecules such as amino acids, nucleic acids, adenosine triphosphate (ATP) and coenzymes. These nutrients include phosphorous, nitrogen, sulfur and many others. AAP have been found to have a wide range of utilisation of these nutrients, although the genomic evolution of the pathways has not been studied in detail. Some AAP such as *Rsb. denitrificans* can reduce nitrates to ammonium to remove excess reducing equivalents that many phototrophic microbes dispose of through carbon fixation (Tang et al., 2011). Some AAP require vitamins, whereas others can make their own. In addition, some are capable of utilising inorganic sulfur compounds as sources of sulfur, whereas others require organic compounds (Rathgeber et al., 2004; Yurkov & Beatty, 1998). As there have been no genomic studies into the nature of the pathways employed by AAP for these nutrients, they will not be considered further in this chapter.

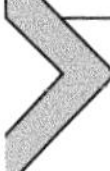

5. PHOTOSYNTHETIC PIGMENTS AND THEIR ASSOCIATION WITH THE PHOTOSYNTHETIC OPERON

5.1. Bacteriochlorophyll

BChl is used as the primary LH pigment in all types of anoxygenic phototrophic bacteria. The type used by AAP and most PNSB is BChl *a*. It is synthesised by cells and is integrated into the LH and RC complexes of the photosynthetic apparatus. Generally, exposure of BChl to light in the presence of O_2 generates highly reactive triplet BChl and singlet oxygen, which can both cause considerable oxidative damage to cells (Yurkov & Csotonyi, 2009). This problem is easily solved by anaerobic phototrophs as they will only carry out photosynthesis and therefore BChl *a* synthesis anoxically. For AAP, this is a more complicated task as they require O_2 for both growth and photosynthetic function. It seems that much of their pigment is produced under dark conditions (night) for use during periods of illumination (day) (Yurkov & Beatty, 1998; Yurkov & Van Germeden, 1993). In addition, it is believed that crt play a major role in photoprotection. High levels of non-photosynthetic crt may quench triplet BChl and singlet oxygen, consistent with the high amounts and variety of crt in AAP compared to PNSB (Yurkov & Csotonyi, 2009).

The pathway for BChl *a* synthesis is similar to that utilised for the production of Chl in cyanobacteria and plants (Yutin et al., 2009). The beginning of the pathway is identical to heme biosynthesis but branches off after the formation of protoporphyrin IX, which is followed by formation of chlorophyllide *a*. From this point, Chl and BChl synthesis is different. There are two directions the BChl pathway can take. The first uses an unknown enzyme, for which the gene is missing in AAP, to eventually produce BChl *c*, *d* and *e*. The second pathway is what AAP employ, which proceeds via chlorophyllide oxidoreductase (encoded by *chlL* or *bchL*) to eventually produce BChl *a*, *b* and *g* (Yutin et al., 2009) (Fig. 11.4). The *chlL* or *bchL* gene is the

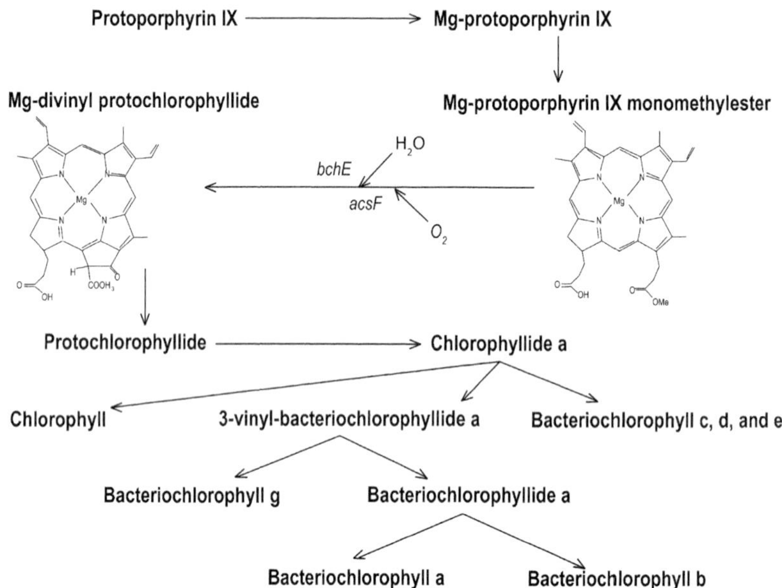

Figure 11.4 ***BChl*** **a** ***biosynthetic pathway.*** Conversion of Mg-protoporphyrin IX monomethylester is carried out by PNSB utilising the gene *bchE* with H_2O and by AAP utilising the *acsF* gene with O_2.

only one both specific and common to all anoxygenic phototrophs that is not present in any oxygenic phototrophs. This indicates that *chlL* or *bchL* was lost in the evolution from anoxygenic to oxygenic organisms, yet retained through evolution from anaerobic to aerobic habitats (Yutin et al., 2009).

The genes responsible for the BChl *a* pathway and its regulation are highly conserved among all anoxygenic phototrophs and are found in the PGC. The PGC consists of four conserved regions (*bchFNBHLM*, *bchCXYZ*, *bchIDO* and *bchOP*), which are identical to those found in PNSB species. In addition, the genes *bchEJ* are present in most members of *Rhodobacter* but have not been observed in any AAP tested other than *C. litoralis* (Zheng et al., 2011). It is possible that these genes are typical of γ-proteobacterial AAP, although *C. litoralis* is the only one that has been fully sequenced to date. The *bchEJ* genes may have been lost in most AAP because they function in processes that are not required or beneficial in an oxygenated environment (Raymond & Blankenship, 2004). The BchE enzyme catalyses the synthesis of Mg-divinyl protochlorophyllide by using an oxygen atom taken from H_2O (Fig. 11.4). Instead, AAP and other aerobic phototrophs have the gene *acsF*, which catalyses an analogous reaction using O_2 (Raymond & Blankenship, 2004; Zheng et al., 2011). This could be one of the reasons why AAP require O_2 for phototrophy. The *Roseobacter* clade and *Sphingomonadales* have retained two regulatory genes, *ppsR* and *ppsA*, which are sensitive to the concentration of O_2 and the intensity of light and are found in the PGC. The NOR5/OM60 clade has maintained the additional gene *crtJ*, which controls aerobic expression and repression of BChl *a*, crt and LH2 in PNSB (Ponnampalam, Elsen, & Bauer, 1998; Zheng et al., 2011).

A molecule of BChl *a* typically contains an Mg^{2+} ion; however, *Acidiphilum rubrum* has evolved to construct BChl *a* around a Zn^{2+} ion (Wakao et al., 1996). This difference could provide cells with an advantage as Zn–BChl *a* is more stable in acidic environments (Yurkov & Csotonyi, 2009). A preliminary study has shown that the absence of the gene *bchD* (encoding Mg-chetalase in BChl biosynthesis) results in the production of Zn–BChl *a* in PNSB, although it has not been confirmed in AAP (Jaschke, Hardjasa, Digby, Hunter, & Beatty, 2011).

5.2. Carotenoids

A feature of AAP is their diverse range of intense colours, conferred by abundant crt. Colours span the spectrum from watermelon red to brown, orange, yellow, pink, purple, and shades thereof (Yurkov & Csotonyi, 2009). Crt are found in remarkably high proportions and varieties in the AAP cells in contrast to the low levels of BChl (Rathgeber et al., 2004; Yurkov & Csotonyi, 2009). Crt may be associated with photosynthesis as an accessory LH molecule or be free of photosynthetic function. In the case of AAP, the majority of crt are non-photosynthetic and are thought to act as photoprotective molecules, as this is the most likely theory as to how AAP protect cells from the damaging effect of light in the presence of BChl *a* and O_2 (Klassen, 2009; Yurkov & Beatty, 1998; Zheng et al., 2011). In addition, crt aid in the assembly of the photosynthetic apparatus (Klassen, 2009).

Generally, the genes involved in crt biosynthesis are present in the PGC alongside those for BChl and photosynthetic apparatus protein synthesis, assembly and regulation (Zheng et al., 2011). However, the genes present vary greatly depending on the organism and which crt are produced. Additionally, although a PGC may lack a gene for synthesis of a specific crt, the gene is sometimes present elsewhere in the genome, suggesting independent regulation (Zheng et al., 2011).

Klassen (2009) divided PNSB into two groups based on photosynthetic crt content: (1) one that contains spheroidenone and (2) another that contains spirilloxanthin. The full gene complement, as described in *Rhodobacter capsulatus*, is *crtAIBKCDEFJ*. Similarly, the crt content of AAP cells has been broken into groups based on bacterial lineage, although in this case there are three. The first is the *Roseobacter* clade, which contains spheroidenone as the main crt. It is associated closely with the photosynthetic apparatus and acts as an LH pigment (Sato-Takabe et al., 2011; Zheng et al., 2011). This is consistent with their close relatives from the PNSB, *Rhodobacter sphaeroides* and *Rhodovulum marinum*, which have been shown to primarily produce spheroidenone when grown aerobically (Zheng et al., 2011). Both types of bacteria produce this pigment because of the presence of the gene *crtA*, which encodes the key enzyme of the biosynthetic pathway (Klassen, 2009). This supports the theory that the genes (*crtAIBCDF*) for the path are conserved among all Rhodobacterales (Zheng et al., 2011). The AAP genera *Roseobacter* and *Dinoroseobacter* are closest to having this full gene complement, *crtAIBK–hyp–crtCDEF*, and so lack only *crtJ* (Zheng et al., 2011). The order Sphingomonadales makes up the second group and has erythroxanthin sulfate as the predominant crt, although it is not associated with photosynthesis. Instead, this group uses the less abundant bacteriorubixanthinal, zeaxanthin and β-carotene as photosynthetic crt (Sato-Takabe et al., 2011; Yurkov & Beatty, 1998; Zheng et al., 2011). In general, this clade lacks the genes *crtACDF*. The loss of *crtA* is the reason why spheroidenone is not synthesised. Finally, the third group consists of the NOR5/OM60 clade, which produces spirilloxanthin for photosynthesis (Sato-Takabe et al., 2011). This crt also predominates in the PNSB *Rhodospirillum rubrum* and *Rhodopseudomonas palustris*, although AAP and PNSB may synthesise it by different pathways. *Rhodospirillum rubrum* uses a typical spirilloxanthin pathway for which the key enzyme is encoded by the gene *crtD*, missing in the PGC of *C. litoralis* KT71, implying the presence of an unusual biosynthetic pathway. Although the final crt manufactured is the same, the genomes of these two organisms may have evolved along very different lines (Zheng et al., 2011) (Fig. 11.5). Additionally, this group is typically missing the *crtCEF* genes.

6. PHOTOSYNTHETIC APPARATUS ORGANISATION AND GENES

The AAP photosynthetic apparatus consists of an RC bound to the LH1. Some species also contain a peripheral LH2. Although almost identical to their anaerobic counterparts, AAP are incapable of photosynthesis without the presence of O_2. Of course, as with any rule there are exceptions. *Charonomicrobium ambiphototrophicum*, a newly described AAP, is the first to be photosynthetically active anaerobically as well as aerobically using

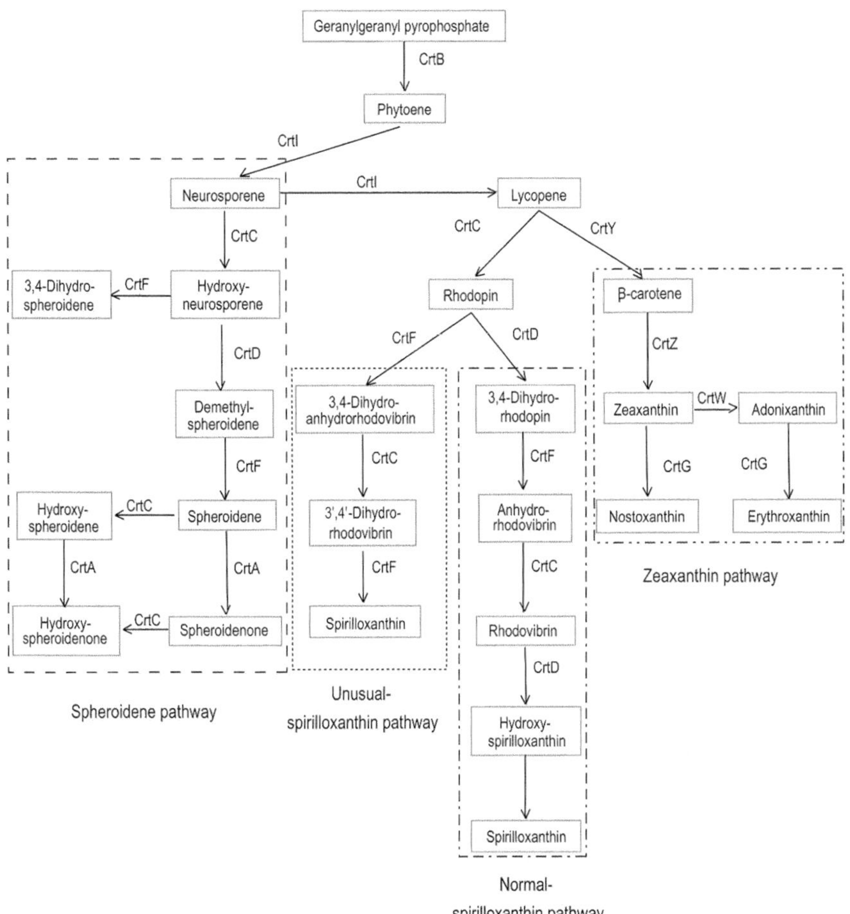

Figure 11.5 Tentative representation of crt biosynthesis in AAP proposed based on identification of genes in the analysed genomes (Zheng et al., 2011).

the same photosynthetic apparatus (Csotonyi et al., 2011b). In addition, *R. mahoneyensis* is photosynthetically most active under microaerophilic conditions (Rathgeber, Alric, Hughes, Vermeglio, & Yurkov, 2012). These could both represent intermediary stages in evolution from PNSB to AAP (Csotonyi et al., 2011b; Rathgeber et al., 2012). The photosynthetic genes in the AAP genome are almost identical to those of the PNSB, which once again supports the theory that purple bacteria are the AAP ancestors. As discussed above (Section 2.4, p. 10), the *pufLM* genes encoding for two of the main subunits of the RC are so similar that they cannot be used to distinguish between AAP and PNSB.

6.1. Reaction Centre

There are two types of bacterial RC; the one in green sulfur bacteria and heliobacteria is similar to photosystem 1 in cyanobacteria and higher plants. The other, referred to as a quinone-type RC, is present in green non-sulfur bacteria and Proteobacteria (Yutin et al., 2009), including the purple bacteria and AAP. This RC is an integral membrane pigment–protein complex bound to four molecules of BChl *a*, two molecules of bacteriopheophytin, two ubiquinones, a non-heme iron and a photosynthetic crt (Yurkov & Beatty, 1998; Yurkov & Csotonyi, 2009). The typical RC can be observed spectrophotometrically in vivo due to absorption peaks at around 760 nm (bacteriopheophytin), and 800 and 865 nm (BChl *a*), although the 865 nm peak is normally masked by a peak at 870 nm representing the LH1.

The *pufLM* and *puhA* genes encode the L (light), M (medium) and H (heavy) subunits of the RC (Liebetanz, Hornberger, & Drews, 1991). These genes are very similar to those of the PNSB and cause problems in studies attempting to identify AAP through shotgun sequencing methods in water columns.

6.2. LH Complexes

BChl *a* bound to the LH complexes allows for in vivo spectrophotometric identification. The LH1 typically shows a peak at 870 nm and the LH2 usually has two peaks, one at 805 nm and the other at 850 nm. However, several novel LH1 and LH2 complexes were discovered among AAP. For example, *R. mahoneyensis* has a monomodal LH2 absorbing only at 805 nm (Fig. 11.6c), which has also been observed in strains of *Roseobacter* and *Rubrimonas* (Rathgeber et al., 2005). In addition, some LH peaks can be shifted in the spectrum. In *Chc. halotolerans* and *Roseovarius tolerans* the LH1 is red-shifted to 879 nm (Fig. 11.7b) and in *Porphyrobacter dokdonensis* and *E. ramosum*, the LH2 major peak is blue-shifted to 835 nm and 832 nm, respectively. Because of the protein environment, *Rsc. thiosulfatophilus* has an LH1 absorption significantly blue-shifted to 855 nm (Fig. 11.6a) (Yurkov & Beatty, 1998; Yurkov et al., 1994). Blue-shifting was attributed to Zn–BChl *a* in *A. rubrum*, which has an in vivo LH1 absorption peak at 864 nm (Wakao et al., 1996).

Structurally the AAP LH1 is similar to that of PNSB, consisting of a ring of approximately 16 heterodimeric α/β proteins surrounding the RC (Tang, Zong, Zhang, Xiao, & Jiao, 2010) encoded by the *pufBA* genes (Kortluke, Breese, Gad'on, Labahn, & Drews, 1997; Zheng et al., 2011). The protein encoded by the *lhaA* gene is thought to help in assembly of the complex

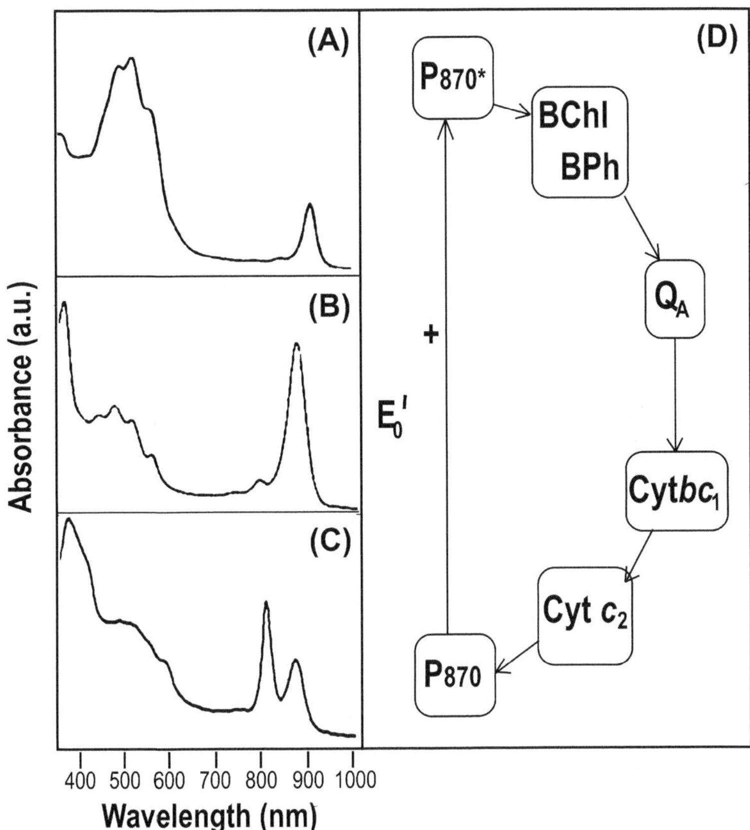

Figure 11.6 ***Variation in LH and RC complexes based on spectrophotometric analysis.*** (A) *Roseococcus thiosulfatophilus* displaying blue-shifted LH1 (855 nm); (B) *Chromocurvus halotolerans* displaying red-shifted LH1 (877 nm); (C) *Roseicyclus mahoneyensis* displaying monomodal LH2 (805 nm) and typical LH1 (870 nm); and (D) Suggested AAP electron transport chain.

(Zheng et al., 2011). The ring contains about 30 BChl *a* molecules (Yurkov & Beatty, 1998) and also contains crt, which may function in LH.

Although LH2 is similar to LH1 in that it also consists of a ring of α/β subunits, there are only eight to nine heterodimers (Tang et al., 2010). LH2 proteins are encoded in the *pucAB* operon, distant from the *puf* operon. Potentially this allows an independent regulation of the genes as is seen in *C. litoralis*, strain KT71, which does not always produce an LH2, but does consistently produce an LH1 (Spring, Lunsdorf, Fuchs, & Tindall, 2009). In addition, spectrophotometrically, the AAP LH2 can be surprisingly different from that of the PNSB as reviewed by Yurkov and Csotonyi (2009).

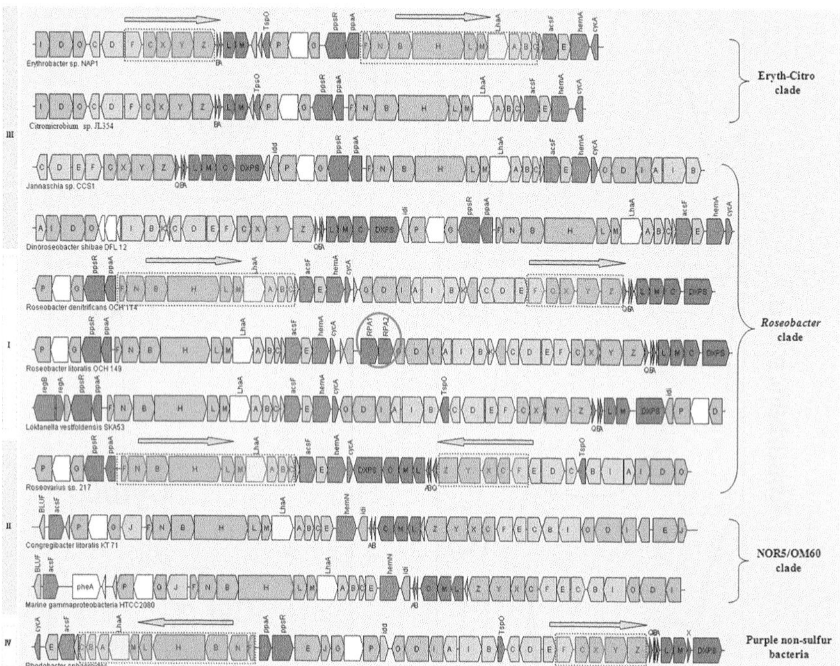

Figure 11.7 ***Photosynthetic gene cluster structure and arrangement in AAP.*** Green, *bch* genes; red, *puf* and regulators genes; pink, *puh* genes; orange, *crt* genes; blue, *hem* and *cyc* gene; yellow, *lhaA* gene; blank, uncertain or unrelated genes; grey, hypothetical protein. Horizontal arrows represent putative transcripts (Zheng et al., 2011). (For interpretation of the references to colour in this figure legend, the reader is referred to the online version of this book.)

6.3. Quinones, Cytochromes and Electron Transport

Once light energy is collected by the LH complexes and is converted into chemical energy by the RC, it is funnelled into a cyclic electron transport chain, which ultimately results in ATP production. This chain of reactions is very typical of purple bacteria. Photons trigger electron transfer within the RC from a BChl *a* to a Q_A and then to a Q_B quinone. Reduced Q_B quinones are oxidised in the membrane-bound cytochrome (cyt) bc_1 complex and cyts *c* in the periplasmic space transfer electrons back to the RC (Fig. 11.6d) (Yurkov & Beatty, 1998; Yurkov & Csotonyi, 2009). This is the same path taken by electrons in PNSB; nevertheless there are a few key differences. First, in PNSB, the Q_A has a negative midpoint redox potential. In the AAP RC, however, Q_A has a redox poise of +5 to +150 mV. This causes the AAP Q_A to be over-reduced under anaerobic conditions and incapable of accepting electrons (Rathgeber et al., 2004;

Yurkov & Csotonyi, 2009), although the quinone-binding pockets in the RC of the PNSB *Rba. sphaeroides* and the AAP *Rsb. denitrificans* seem to be very similar (Yurkov & Csotonyi, 2009). Second, AAP lack an alternative quinol oxidase pathway that in PNSB may be used to keep quinones in the proper redox state for efficient photosynthetic function (Yurkov & Beatty, 1998; Yurkov & Csotonyi, 2009). Third, O_2 is required for soluble cyts to transfer electrons from the cyt bc_1 complex to the RC-bound cyt *c* in *Rsb. denitrificans*. This is, therefore, another spot that could halt photosynthetic electron transfer under anaerobic conditions (Yurkov & Csotonyi, 2009). Apparently there are several reasons why AAP are unable to grow photosynthetically in the absence of O_2 and this phenomenon remains one of the major mysteries related to AAP.

Cyts used for photosynthesis in AAP can be broken into two groups. The first is seen in *E. ramosum*, *E. hydrolyticum*, *E. ezovicum*, *Erb. litoralis* and *Erb. longus* and contains a soluble cyt *c* that passes electrons from the cyt bc_1 complex directly to the RC. The second, which is the case for *Erm. ursincola*, *Srb. sibiricus*, *Rsc. thiosulfatophilus*, *R. mahoneyensis* and *Rsb. denitrificans*, has a tetraheme cyt *c* bound tightly to the RC, which accepts electrons from another cyt *c* before passing them to the RC.

Little work has been done on the genes encoding the quinones and cyt in AAP. Key genes are *pufQ* and *pufC*, respectively, and are sometimes found in the *puf* operon alongside those for the RC and LH1 subunits, such as in *Jannaschia* sp. CCS1, *D. shibae*, *Rsb. denitrificans*, *Loktanella vestifoldensis*, *Roseobacter litoralis* and *Roseovarius* sp., 217, which all belong to the *Roseobacter* clade. In other cases, such as in *Erythrobacter* sp. NAP1, *Citromicrobium* sp. JL354, *C. litoralis* and γ-proteobacterial strain HTCC2080, either can be found elsewhere in the genome not associated with the PGC (Zheng et al., 2011).

6.4. The PGC

The majority of photosynthesis genes in AAP are found in the PGC, which accounts for approximately 1% of the total genome. This includes *bch* genes for BChl *a* biosynthesis, the *puf* operon for protein subunits of the RC and LH1 complexes (and sometimes also a tetraheme cyt and quinones), the *puc* operon for LH2 subunits, the *puh* operon for an RC subunit and assembly of the photosynthetic apparatus and *crt* genes for crt production (Zheng et al., 2011). Most of the PGC is highly conserved among AAP and PNSB making it difficult, if not impossible, to distinguish between the two by sequence analysis alone. Interestingly, the GC content of the PGC is similar to the total GC content of the corresponding genomes in all sequenced strains. This suggests that despite the possibility of lateral gene transfer, the PGC has not been recently acquired from another species with a different GC content (Zheng et al., 2011).

There are only 27 genes present in all 10 of the AAP PGCs studied to date (Zheng et al., 2011). Most code for BChl *a* synthesis (*bchBCDF-GHILMNOPXYZ* and *ascF*) but only two code for crt. The others encode proteins such as the RC and LH1 subunits (*pufBALM*), transcriptional regulators (*ppsR*) and assembly factors (*puhABCE* and *lhaA*) (Fig. 11.7) (Zheng et al., 2011).

Although PGC gene sequences are highly conserved, the organisation varies depending on bacterial lineage, suggesting independent evolution (Zheng et al., 2011). For instance, there are four different organisations of the *puf* operon. The first, *pufQBALMC*, is found in the *Roseobacter* clade. The exception is that of *L. vestifoldensis*, which belongs to the clade but has the second form of organisation, *pufQBALM*, in which the *pufC* gene is found elsewhere in the genome. The third type is that of Sphingomonadales, *pufBALM*, which is missing both the *pufQ* and *pufC* genes. The fourth organisation is *pufLMCBA*, found in the NOR5/OM60 strains (Zheng et al., 2011). Interestingly, this *puf* operon does not have the *pufQ* gene and it has also changed the order of genes from the standard *pufBALM* to *pufLMBA* (Fig. 11.7). These differences in gene organisation were suggested to relate to environmental adaptations (Liotenberg et al., 2008; Zheng et al., 2011). One additional possible difference between AAP and their putative PNSB ancestors is that some extant PNSB contain a *pufX* gene, which seems to be absent from the few sequenced AAP (Kortluke et al., 1997). This gene has been found in *Rba. sphaeroides* and *Rba. capsulatus* and is responsible for dimerisation of the RC subunits. Why the PufX protein is absent from AAP and some purple bacteria is a mystery.

The conserved gene organisations *bchFNBHLM–lhaA–puhABC* and *crtF–bchCXYZ–puf*, although present in all AAP tested, are in one of four orientations (Fig. 11.7). These gene organisations do not seem to be separated based on the 16S rRNA phylogeny, implying that the organisation arose independently in each AAP (Zheng et al., 2011).

The *crt* genes are the most variable in both sequence and organisation in the AAP PGC. A variety of crt can be utilised by different strains, and so the gene complements vary accordingly. Some *crt* genes are in the PGC, others are elsewhere in the genome. Many are also interspersed with *bch* genes, perhaps to allow co-regulation of transcription (Zheng et al., 2011).

7. UNSOLVED MYSTERIES AND CONCLUDING REMARKS

Competition for anaerobic habitats must have been fierce when global oxygenation occurred. O_2 forced bacteria to evolve new genes and proteins to allow the use of this gas and to avoid its toxicity. Photosynthesis at the time was anaerobic and anoxygenic. On adapting to aerobic environments, it is likely that AAP ancestors lost their photosynthetic function only to regain it later through lateral gene transfer. Carbon metabolic pathways certainly support this theory, as does the presence of phage DNA closely associated with the PGC in *Cmi. bathyomarinum* strain KT71. In addition, the PGC genes in PNSB and AAP are almost identical except for key elements that make O_2 a requirement for AAP, such as the *acsF* gene in BChl *a* synthesis. Genomic studies on AAP are in their infancy and many questions have yet to be answered. It is still unclear how cells manage to survive the phototoxic effect of light combined with O_2 during BChl *a* synthesis. Identification of AAP-specific genes for use in culture-independent studies of diversity would help to reduce the need for culturing bacteria for phenotypic evaluation. There remains much work to do to fully understand AAP, although science is pointing us in the right direction.

ACKNOWLEDGEMENTS

This work was supported by a Natural Sciences and Engineering Research Council grant held by V. Yurkov.

REFERENCES

Beatty, J. T. (2005). On the natural selection and evolution of the aerobic phototrophic bacteria. In Govindjee, J. T. Beatty, H. Gest & J. F. Allen (Eds.), *Discoveries in photosynthesis* (pp. 1099–1104). Springer.

Biebl, H., Tindall, B. J., Pukall, R., Lünsdorf, H., Allgaier, M., & Wagner-Döbler, I. (2006). *Hoeflea phototrophica* sp. nov., a novel alphaproteobacterium that forms bacteriochlorophyll *a*. *International Journal of Systematic and Evolutionary Microbiology*, *56*, 821–826.

Boldareva, E. N., Tourova, T. P., Kolganova, T. V., Moskalenko, A. A., Makhneva, Z. K., & Gorlenko, V. M. (2009). *Roseococcus suduntuyensis* sp. nov., a new aerobic bacteriochlorophyll *a*-containing bacterium isolated from a low-mineralized soda Lake of Eastern Siberia. *Microbiology*, *78*(1), 92–101.

Cho, J. C., Stapels, M. D., Morris, R. M., Vergin, M. S., Schwalbach, M. S., Givan, S. A., et al. (2007). Polyphyletic photosynthetic reaction centre genes in oligotrophic marine gammaproteobacteria. *Environmental Microbiology*, *9*, 1456–1463.

Cottrell, M. T., Ras, J., & Kirchman, D. (2010). Bacteriochlorophyll and community structure of aerobic anoxygenic phototrophic bacteria in a particle-rich estuary. *International Society of Microbial Ecology Journal: Multidisciplinary Journal of Microbial Ecology*, *4*, 945–954.

Csotonyi, J., Stackebrandt, E., Swiderski, J., Schumann, P., & Yurkov, V. (2011a). *Chromocurvus halotolerans* gen. nov., sp. nov., a gammaproteobacterial obligately aerobic anoxygenic phototroph, isolated from a Canadian hypersaline spring. *Archives of Microbiology, 193*(8), 573–582.

Csotonyi, J., Stackebrandt, E., Swiderski, J., Schumann, P., & Yurkov, V. (2011b). An alphaproteobacterium capable of both aerobic and anaerobic anoxygenic photosynthesis but incapable of photoautotrophy: *Charonomicrobium ambiphototrophicum*, gen. nov., sp. nov. *Photosynthesis Research, 107*, 257–268.

Csotonyi, J. T., Swiderski, J., Stackebrandt, E., & Yurkov, V. (2010). A new environment for aerobic anoxygenic phototrophic bacteria: biological soil crusts. *Environmental Microbiology Reports, 2*(5), 651–656.

Denner, E., Vybiral, D., Koblížek, M., Busse, H. J., & Velimirov, B. (2002). *Erythrobacter citreus* sp. nov., a yellow pigmented bacterium that lacks bacteriochlorophyll *a*, isolated from the western Mediterranean Sea. *International Journal of Systematic and Evolutionary Microbiology, 52*, 1655–1661.

Eiler, A., Beier, S., Säwström, C., Karlsson, J., & Bertilsson, S. (2009). High ratio of bacteriochlorophyll biosynthesis genes to chlorophyll biosynthesis genes in bacteria of humic lakes. *Applied and Environmental Microbiology, 75*(22), 7221–7228.

Feng, F., Jiao, N., Du, H., & Zeng, Y. (2010). Phylogenetic analysis of aerobic anoxygenic phototrophic bacteria and their relatives based on farnesyl pyrophosphate synthase gene. *Acta Oceanologica Sinica, 29*(5), 82–89.

Feng, Y., Lin, X., Mao, T., & Zhu, J. (2011). Diversity of aerobic anoxygenic phototrophic bacteria in paddy soil and their response to elevated atmospheric CO_2. *Microbial Biotechnology, 4*(1), 74–81.

Fürch, T., Preusse, M., Tomasch, J., Zech, H., Wagner-Döbler, I., Rabus, R., et al. (2009). Metabolic fluxes in the central carbon metabolism of *Dinoroseobacter shibae* and *Phaeobacter gallaeciensis*, two members of the marine *Roseobacter* clade. *BioMed Central Microbiology, 9*, 209.

Hügler, M., & Sievert, S. M. (2011). Beyond the Calvin cycle: autotrophic carbon fixation in the ocean. *Annual Review of Marine Science, 3*, 261–289.

Igarashi, N., Harada, J., Nagashima, S., Matsuura, K., Shimada, K., & Nagashima, K. V.P. (2001). Horizontal transfer of the photosynthesis gene cluster and operon rearrangement in purple bacteria. *Journal of Molecular Evolution, 52*, 333–341.

Ivanova, E. P., Bowmanc, J. P., Lysenkod, A. M., Zhukovae, N. V., Gorshkovab, N. M., Kuznetsovab, T. A., et al. (2005). *Erythrobacter vulgaris* sp. nov., a novel organism isolated from the marine invertebrates. *Systematic and Applied Microbiology, 28*, 123–130.

Jaschke, P. R., Hardjasa, A., Digby, E. L., Hunter, C. N., & Beatty, J. T. (2011). A BchD (Mg-chelatase) mutant of *Rhodobacter sphaeroides* synthesizes zinc-bacteriochlorophyll through novel zinc-containing intermediates. *Journal of Biological Chemistry, 286*, 20313–20322.

Jiang, H., Dong, H., Yu, B., Lv, G., Deng, S., Wu, Y., et al. (2009). Abundance and diversity of aerobic anoxygenic phototrophic bacteria in saline lakes on the Tibetan plateau. *FEMS Microbiology Ecology, 67*, 268–278.

Jiao, N., Zhang, R., & Zheng, Q. (2010). Coexistence of two different photosynthetic operons in *Citromicrobium bathyomarinum* JL354 as revealed by whole-genome sequencing. *Journal of Bacteriology, 192*(4), 1169–1170.

Jiao, N. Z., Zhang, Y., Zeng, Y. H., Hong, N., Liu, R. L., Chen, F., Wang, P. X. (2007). Distinct distribution pattern of abundance and diversity of aerobic anoxygenic phototrophic bacteria in the global ocean. *Environmental Microbiology, 9*:3091–3099.

Jung, Y. T., Park, S., Oh, T. K., & Yoon, J. H. (2011). *Erythrobacter marinus* sp. nov., isolated from seawater of the Yellow Sea in Korea. *International Journal of Systematic and Evolutionary Microbiology*. http://dx.doi.org/10.1099/ijs.0.034702-0.

Kim, B. C., Park, J. R., Bae, J. W., Rhee, S. K., Kim, K. H., Oh, J. W., et al. (2006). *Stappia marina* sp. nov., a marine bacterium isolated from the Yellow Sea. *International Journal of Systematic and Evolutionary Microbiology, 56*, 75–79.

Klassen, J. (2009). Pathway evolution by horizontal transfer and positive selection is accommodated by relaxed negative selection upon upstream pathway genes in purple bacterial carotenoid biosynthesis. *Journal of Bacteriology, 191*(24), 7500–7508.

Koblížek, M., Janouškovec, J. J., Oborník, M., Johnson, J. H., Ferriera, S., & Falkowski, P. G. (2011). Genome sequence of the marine photoheterotrophic bacterium *erythrobacter* sp. strain NAP1. *Journal of Bacteriology, 193*(20), 5881–5882.

Kortluke, C., Breese, K., Gad'on, N., Labahn, A., & Drews, G. (1997). Structure of the *puf* operon of the obligately aerobic, bacteriochlorophyll *a*-containing bacterium *Roseobacter denitrificans* OCh114 and its expression in a *Rhodobacter capsulatus puf puc* deletion mutant. *Journal of Bacteriology, 179*, 5247–5258.

Labrenz, M., Lawson, P. A., Tindall, B. J., & Hirsch, P. (2009). *Roseibaca ekhonensis* gen. nov., sp. nov., an alkilitolerant and aerobic bacteriochlorophyll *a*-producing alphaproteobacterium from hypersaline Ekho Lake. *International Journal of Systematic and Evolutionary Microbiology, 59*, 1935–1940.

Lee, Y. S., Lee, D. H., Kahng, H. Y., Kim, E. M., & Jung, J. S. (2010). *Erythrobacter gangjinensis* sp. nov., a marine bacterium isolated from seawater. *International Journal of Systematic and Evolutionary Microbiology, 60*, 1413–1417.

Lehours, A. -C., Cottrell, M. T., Dahan, O., Kirchman, D. L., & Jeanthon, C. (2010). Summer distribution and diversity of aerobic anoxygenic phototrophic bacteria in the Mediterranean Sea in relation to environmental variables. *FEMS Microbiology Ecology, 74*, 397–409.

Liebetanz, R., Hornberger, U., & Drews, G. (1991). Organization of the genes coding for the reaction-centre L and M subunits and B870 antenna polypeptides α and β from the aerobic photosynthetic bacterium *Erythrobacter* species OCh114. *Molecular Microbiology, 5*, 1459–1468.

Liotenberg, S., Steunou, A. S., Picaud, M., Reiss-Husson, F., Astier, C., & Ouchane, S. (2008). Organization and expression of photosynthesis genes and operons in anoxygenic photosynthetic proteobacteria. *Environmental Microbiology, 10*, 2267–2276.

Marrs, B. (1981). Mobilization of the genes for photosynthesis from *Rhodopseudomonas capsulata* by a promiscuous plasmid. *Journal of Bacteriology, 146*, 1003–1012.

Oh, H. -M., Giovannoni, S. J., Ferriera, S., Johnson, J., & Cho, J. -C. (2009). Complete genome sequence of *Erythrobacter litoralis* HTCC2594. *Journal of Bacteriology, 191*(7), 2419–2420.

Ponnampalam, S. N., Elsen, S., & Bauer, C. E. (1998). Aerobic repression of the *Rhodobacter capsulatus bchC* promoter involves cooperative interactions between *crtJ* bound to neighbouring palindromes. *Journal of Biological Chemistry, 273*(46), 30757–30761.

Rathgeber, C., Alric, J., Hughes, E., Vermeglio, A., & Yurkov, V. (2012). The photosynthetic apparatus and photoinduced electron transfer in the aerobic phototrophic bacteria *Roseicyclus mahoneyensis* and *Porphyrobacter meromictius*. *Photosynthesis Research, 110*(3), 193–203.

Rathgeber, C., Beatty, J. T., & Yurkov, V. (2004). Aerobic phototrophic bacteria: new evidence for the diversity, ecological importance and applied potential of this previously overlooked group. *Photosynthesis Research, 81*, 113–128.

Rathgeber, C., Yurkova, N., Stackebrandt, E., Schumann, P., Beatty, J. T., & Yurkov, V. (2005). *Roseicyclus mahoneyensis* gen nov., sp. nov., an aerobic phototrophic bacterium isolated from a meromictic lake. *International Journal of Systematic and Evolutionary Microbiology, 55*, 1597–1603.

Rathgeber, C., Yurkova, N., Stackebrandt, E., Schumann, P., Humphrey, E., Beatty, J. T., et al. (2007). *Porphyrobacter meromictius* sp. nov., an appendaged bacterium, that produces bacteriochlorophyll *a*. *Current Microbiology*, *55*, 356–361.

Raymond, J., & Blankenship, R. E. (2004). Biosynthetic pathways, gene replacement and the antiquity of life. *Geobiology*, *22*, 199–203.

Salka, I., Cuperova, Z., Masin, M., Koblizek, M., & Grossart, H.-P. (2011). *Rhodoferax*-related *pufM* gene cluster dominates the aerobic anoxygenic phototrophic communities in German freshwater lakes. *Environmental Microbiology*, *13*(11), 2865–2875.

Sato-Takabe, Y., Hamasaki, K., & Suzuki, K. (2011). Photosynthetic characteristics of marine aerobic anoxygenic bacteria *Roseobacter* and *Erythrobacter* strains. *Archives of Microbiology*. http://dx.doi.org/10.1007/s00203-011-0761-2.

Shiba, T. (1991). *Roseobacter litoralis* gen. nov., sp. nov., and *Roseobacter denitrificans* sp. nov., aerobic pink-pigmented bacteria which contain bacteriochlorophyll *a*. *Systematic and Applied Microbiology*, *14*(2), 140–145.

Shiba, T., & Simidu, U. (1982). *Erythrobacter longus* gen. nov., sp. nov., an aerobic bacterium which contains bacteriochlorophyll *a*. *International Journal of Systematic Bacteriology*, *32*, 211–217.

Shiba, T., Simidu, U., & Taga, N. (1979). Distribution of aerobic bacteria which contain bacteriochlorophyll *a*. *Applied and Environmental Microbiology*, *38*, 43–45.

Sieracki, M. E., Gilg, I. C., Their, E. C., Poulton, N. J., & Goericke, R. (2006). Distribution of planktonic aerobic photoheterotrophic bacteria in the northwest Atlantic. *Limnology and Oceanography*, *51*, 38–46.

Spring, S., Lunsdorf, H., Fuchs, B. M., & Tindall, B. J. (2009). The photosynthetic apparatus and its regulation in the aerobic gammaproteobacterium *Congregibacter litoralis* gen. nov., sp. nov. *PLoS One*, *4*(3), e4866.

Swingley, W. D., Sadekar, S., Mastrian, S. D., Matthies, H. J., Hao, J., Ramos, H., et al. (2007). The complete genome sequence of *Roseobacter denitrificans* reveals a mixotrophic rather than photosynthetic metabolism. *Journal of Bacteriology*, *189*, 683–690.

Tang, K. H., Feng, X., Tang, Y. J., & Blankenship, R. E. (2009). Carbohydrate metabolism and carbon fixation in *Roseobacter denitrificans* Och114. *PLoS One*, *4*(10), e7233.

Tang, K.-H., Tang, Y., & Blankenship, R. E. (2011). Carbon metabolic pathways in phototrophic bacteria and their broader evolutionary implications. *Frontiers in Microbiology*, *2*(165)http://dx.doi.org/10.3389/fmicb.2011.00165.

Tang, K., Zong, R., Zhang, F., Xiao, N., & Jiao, N. (2010). Characterization of the photosynthetic apparatus and proteome of *Roseobacter denitrificans*. *Current Microbiology*, *60*, 124–133.

Venkata Ramana, V., Sasikala, C., Takaichi, S., & Ramana, C. V. (2010). *Roseomonas aestuarii* sp. nov., a bacteriochlorophyll-*a* containing alphaproteobacterium isolated from an estuarine habitat of India. *Systematic and Applied Microbiology*, *33*, 198–203.

Wakao, N., Yokoi, N., Isoyama, N., Hiraishi, A., Shimada, K., Kobayashi, M., et al. (1996). Discovery of natural photosynthesis using Zn-containing bacteriochlorophyll in an aerobic bacterium *Acidiphilium rubrum*. *Plant Cell Physiology*, *37*, 889–896.

Woese, C. R. (1987). Bacterial evolution. *Microbiology Reviews*, *51*, 221–271.

Wu, H. X., Lai, P. Y., Lee, O. O., Zhou, X. J., Miao, L., Wang, H., et al. (2012). *Erythrobacter pelagi* sp. nov., a member of the family *Erythrobacteraceae* isolated from the Red Sea. *International Journal of Systematic and Evolutionary Microbiology*, *62*, 1348–1353.

Xu, M., Xin, Y., Yu, Y., Zhang, J., Zhou, Y., Liu, H., et al. (2010). *Erythrobacter nanhaisediminis* sp. nov., isolated from marine sediment of the South China Sea. *International Journal of Systematic and Evolutionary Microbiology*, *60*, 2215–2220.

Yoon, J.-H., Kang, K. H., Oh, T. K., & Park, Y. H. (2004). *Erythrobacter aquimaris* sp. nov., isolated from sea water of a tidal flat of the Yellow Sea in Korea. *International Journal of Systematic and Evolutionary Microbiology*, *54*, 1981–1985.

Yoon, J.-H., Kang, K. H., Yeo, S. H., & Oh, T. -K. (2005). *Erythrobacter luteolus* sp. nov., isolated from a tidal flat of the Yellow Sea in Korea. *International Journal of Systematic and Evolutionary Microbiology, 55*, 1167–1170.

Yoon, J.-H., Kim, H., Kim, I. G., Kang, K. H., & Park, Y. H. (2003). *Erythrobacter flavus* sp. nov., a slight halophile from the East Sea in Korea. *International Journal of Systematic and Evolutionary Microbiology, 53*, 1169–1174.

Yoon, B. J., Lee, D. H., & Oh, D. C. (2012). *Erythrobacter jejuensis* 1 sp. nov., isolated from seawater. *International Journal of Systematic and Evolutionary Microbiology*. doi:10.1099/ijs.0.038349-0.

Yoon, J.-H., Oh, T.-K., & Park, Y. H. (2005). *Erythrobacter seohaensis* sp. nov. and *Erythrobacter gaetbuli* sp. nov., isolated from a tidal flat of the Yellow Sea in Korea. *International Journal of Systematic and Evolutionary Microbiology, 55*, 71–75.

Yurkov, V. (1990). Academy of Sciences, Ph.D. thesis. Moscow, Russia.

Yurkov, V., & Beatty, J. T. (1998). Aerobic anoxygenic phototrophic bacteria. *Microbiology and Molecular Biology Reviews, 62*, 695–724.

Yurkov, V., & Csotonyi, J. (2009). New light on aerobic anoxygenic photosynthesis. In C. Neil Hunter, F. Daldal, M. C. Thurnauer & J. T. Beatty (Eds.), *The purple phototrophic bacteria* (pp. 31–55). Springer Science + Business Media B.V.

Yurkov, V. V., & Gorlenko, V. M. (1990). *Erythrobacter sibiricus* sp. nov., a new freshwater aerobic bacterial species containing bacteriochlorophyll *a*. *Microbiology (New York), 59*(1), 120–126.

Yurkov, V. V., & Gorlenko, V. M. (1992a). A new genus of freshwater aerobic bacteriochlorophyll *a*-containing bacteria, *Roseococcus* gen. nov. *Microbiology (New York), 60*(5), 902–907.

Yurkov, V. V., & Gorlenko, V. M. (1992b). New species of aerobic bacteria from the genus *Erythromicrobium* containing bacteriochlorophyll *a*. *Microbiology (New York), 61*(2), 248–255.

Yurkov, V. V., Gorlenko, V. M., & Kompantseva, E. I. (1993). A new type of freshwater aerobic orange colored bacterium *Erythromicrobium* gen. nov., containing bacteriochlorophyll *a*. *Microbiology (New York), 61*(2), 256–260.

Yurkov, V., Krieger, S., Stackebrandt, E., & Beatty, J. T. (1999). *Citromicrobium bathyomarinum*, a novel aerobic bacterium isolated from deep-sea hydrothermal vent plume waters that contains photosynthetic pigment–protein complexes. *Journal of Bacteriology, 181*, 4517–4525.

Yurkov, V. V., Lysenko, A. M., & Gorlenko, V. M. (1991). Hybridization analysis of the classification of bacteriochlorophyll *a*-containing freshwater aerobic bacteria. *Microbiology (New York), 60*(3), 518–523.

Yurkov, V., Stackebrandt, E., Buss, O., Verméglio, A., Gorlenko, V., & Beatty, J. T. (1997). Reorganization of the genus *Erythromicrobium*: description of "*Erythromicrobium sibiricum*" as *Sandaracinobacter sibiricus* gen. nov., sp. nov., and of "*Erythromicrobium ursincola*" as *Erythromonas ursincola* gen. nov., sp. nov. *International Journal of Systematic Bacteriology, 47*, 1172–1178.

Yurkov, V., Stackebrandt, E., Holmes, A., Fuerst, J. A., Hugenholtz, P., Golecki, J., et al. (1994). Phylogenetic positions of novel aerobic, bacteriochlorophyll *a*-containing bacteria and description of *Roseococcus thiosulfatophilus* gen. nov., sp. nov., *Erythromicrobium ramosum* gen nov., sp. nov., and *Erythrobacter litoralis* sp. nov. *International Journal of Systematic Bacteriology, 44*, 427–434.

Yurkov, V., & Van Germeden, H. (1993). Impact of light/dark regimen on growth rate, biomass formation and bacteriochlorophyll synthesis in *Erythromicrobium hydrolyticum*. *Archives of Microbiology, 159*, 84–89.

Yutin, N., Suzuki, M.T., Rosenberg, M., Rotem, D., Madigan, M.T., Suling, J., et al. (2009). BchY-based degenerate primers target all types of anoxygenic photosynthetic bacteria in a single PCR. *Applied and Environmental Microbiology*, *75*(23), 7556–7559.

Yu, Y., Yan, S. L., Li, H. R., & Zhang, X. H. (2011). *Roseicitreum antarcticum* gen. nov., sp. nov., an aerobic bacteriochlorophyll *a*-containing alphaproteobacterium isolated from Antarctic sandy intertidal sediment. *International Journal of Systematic and Evolutionary Microbiology*, *61*, 2173–2179.

Zeng, Y., Shen, W., & Jiao, N. (2009). Genetic diversity of aerobic anoxygenic photosynthetic bacteria in open ocean surface water and upper twilight zones. *Marine Biology*, *156*, 425–437.

Zheng, Q., Zhang, R., Koblížek, M., Boldareva, E. N., Yurkov, V., Yan, S., et al. (2011). Diverse arrangement of photosynthetic gene clusters in aerobic anoxygenic phototrophic bacteria. *PLoS One*, *6*(9), e25050. http://dx.doi.org/10.1371/journal.pone.0025050.

CHAPTER TWELVE

Evolutionary Divergence of Marine Aerobic Anoxygenic Phototrophic Bacteria as Seen from Diverse Organisations of Their Photosynthesis Gene Clusters

Qiang Zheng*, Michal Koblížek, J. Thomas Beatty†, Nianzhi Jiao*,[1]**

*State Key Laboratory of Marine Environmental Science, Xiamen University, Xiamen, PR China
**Institute of Microbiology CAS, Opatovický mlýn, Třeboň, Czech Republic
†Department of Microbiology and Immunology, University of British Columbia, Vancouver, BC, Canada
[1]Corresponding author: E-mail: jiao@xmu.edu.cn

Contents

This chapter summarised our previous works including two papers published in PloS One journal.

Advances in Botanical Research, Volume 66
ISSN 0065-2296, http://dx.doi.org/10.1016/B978-0-12-397923-0.00012-6

Abstract

Aerobic anoxygenic phototrophic bacteria (AAPB) represent an important group of microorganisms inhabiting euphotic zones of the oceans, freshwater or saline lakes. They harvest light using bacterial reaction centres (RCs) containing bacteriochlorophyll (BChl). Their photosynthetic apparatus is encoded by a number of genes organised in a so-called photosynthesis gene cluster (PGC). In this chapter, the organisation of PGCs in 10 AAPB strains belonging to the *Roseobacter*, Eryth–Citro and NOR5/OM60 clades is summarised. The distribution of four types of gene arrangements, based on permutation and combination of the two conserved regions *bchFNBHLM–lhaA–puhABC* and *crtF–bchCXYZ*, does not correspond to the phylogenetic affiliation of individual AAPB. While PGCs of all analysed species contain the same set of genes for BChl synthesis and assembly of photosynthetic centres, they differ greatly in the carotenoid biosynthetic genes. Spheroidenone, spirilloxanthin, and zeaxanthin biosynthetic pathways were found separately in each clade. The AAPB in *Roseobacter* and the NOR5/OM60 clades are closely related to purple non-sulfur bacteria. To date, only the AAPB in the Eryth–Citro clade are obligately aerobic phototrophs with unique carotenoid biosynthetic pathways and pigments, compared to the other clades. Interestingly, members of the Eryth–Citro clade possess the shortest and simplest PGC structure of the known AAPB, and the light-harvesting complex 2 genes are completely absent; perhaps this indicates that the AAPB in this clade arose more recently than their counterparts in other clades. Comparison of two closely related Citromicrobial genomes (98.1% sequence identity of complete 16S rRNA genes), *Citromicrobium* sp. JL354, which contains two copies of RC genes, and Citromicrobial strain JLT1363, which is chemotrophic, revealed evidence for the loss of photosynthesis genes. The incomplete PGC (*pufLMC–puhCBA*) in strain JL354 was located within a putative integrating conjugative element, which indicates a potential mechanism for the horizontal transfer of genes for phototrophy.

1. INTRODUCTION

Anoxygenic phototrophic bacteria were proposed to have emerged approximately 3000 Myr ago and to be the ancestor of all photosynthetic organisms (Blankenship, 1992; Des Marais, 2000; Xiong, Fischer, Inoue, Nakahara, & Bauer, 2000). Aerobic anoxygenic phototrophic bacteria (AAPB), which probably evolved from purple non-sulfur bacteria after the accumulation of oxygen in the earth's biosphere, represent an important group of microorganisms inhabiting the euphotic zone of the world ocean. These phototrophic microorganisms are thought to be important players in oceanic carbon cycling (Jiao et al., 2007; Jiao, Zhang, & Hong, 2010; Koblížek, Mašín, Ras, Poulton, & Prášil, 2007; Kolber et al., 2001). Culture-independent studies have shown that marine AAPB communities are mostly represented by Alpha- and Gammaproteobacteria (Béjà et al., 2002; Yutin et al., 2007). Most cultured marine Alphaproteobacterial AAPB belong to the *Roseobacter* clade or Eryth–Citro clade, the latter of which includes members of the genera *Erythrobacter* and *Citromicrobium* (Koblížek et al., 2003; Shiba, Simidu, & Taga, 1979; Yurkov, Krieger, Stackebrandt, & Beatty, 1999). Isolates of AAPB related to Gammaproteobacteria belong to the clade NOR5/OM60 which contains *Congregibacter litoralis* KT71 (Fuchs et al., 2007; Spring, Lünsdorf, Fuchs, & Tindall, 2009) and strain HTCC2080 (Cho et al., 2007; Thrash et al., 2010).

With the development of high-throughput DNA sequencing technology, a great number of AAPB genomes have been sequenced (Table 12.1, Fig. 12.1). Genome sequencing has revealed that AAPB, like purple photosynthetic bacteria, contain a highly conserved ~40–50 kb photosynthesis gene cluster (PGC) (Zsebo & Hearst, 1984). The heart of anoxygenic phototrophy is the reaction centre (RC), encoded by the *puf* and *puh* operons. So far, there are two main scenarios describing the evolution of Proteobacteria. The first is that phototrophic ancestors lost phototrophy genes and thus became chemotrophic (Swingley, Blankenship, & Raymond, 2009; Woese, 1987). The second is that chemotrophic bacteria acquired phototrophy genes via horizontal gene transfer (HGT) and therefore became phototrophic. For example, the *Rubrivivax gelatinosus* (Betaproteobacteria) phototrophy genes were suggested to have originated in the photosynthetic Alphaproteobacteria (Igarashi et al., 2001; Nagashima, Hiraishi, Shimada, & Matsuura, 1997; Swingley et al., 2009).

The complete PGC contains genes for the photosynthetic RC, light-harvesting (LH) complexes, BChl and carotenoid biosynthesis, as well as some regulatory factors. Despite the fact that the basic set of genes in PGC is conserved, the gene organisation of the operons in PGC largely varies between different AAPB lineages. Two conserved subclusters, *crt–bchCXYZ–puf* (about 10 kb) and *bchFNBHLM–lhaA–puh* (about 12–15 kb) were identified in PGCs of different phototrophic Proteobacteria (Liotenberg et al., 2008; Tuschak, Leung, Beatty, & Overmann, 2005; Waidner & Kirchman, 2005). The orientation of the genes in each subcluster was the same, although the gene order could vary slightly (e.g. *pufBA* and *pufLM*). Interestingly, the regulatory elements such as the transcriptional regulator *ppsR* gene were conserved as well, suggesting that the operons in the PGCs are co-expressed. The organisation of *puf* (approximately 3 kb) operon varies among different AAPB species. The presence/absence of *pufC* and *pufQ*, as well as various gene orders of *puf* genes, were observed (Tuschak et al., 2005; Waidner & Kirchman, 2005; Yutin & Béjà, 2005; Yutin et al., 2007). Further investigation indicated that such gene organisation is crucial for environmental adaptation (Liotenberg et al., 2008).

Table 12.1 Main characteristics of genomes and PGCs of sequenced species

Clade	Organism	Genome size (kb)	PGC size (kb)	Genome GC%	PGC GC%	PGC/ Genome	LH 2
Roseobacter clade	*D. shibae* DFL 12	4417.8	48.1	65	67	1.09%	+
	Rsb. denitrificans OCh 114	4331.2	44.6	58	60	1.03%	+
	Rsb. litoralis Och 149	4678.9	48.3	57	59	1.03%	+
	Roseovarius sp. 217	4762.6	45.1	60	64	0.95%	+
	Jannaschia sp. CCS1	4404.0	45.8	62	62	1.04%	+
	L. vestfoldensis SKA53	3063.7	41.2	59	60	1.34%	+
Eryth–Citro clade	*Erythrobacter* sp. NAP1	3265.3	38.9	61	62	1.19%	−
	Citromicrobium sp. JL354	3273.3	38.7	65	67	1.18%	−
NOR5/ OM60 clade	*Cb. litoralis* KT71	4328.1	44.7	58	59	1.03%	+
	Gamma- HTCC2080	3576.1	43.6	51	53	1.22%	+

GC% = relative percentage of guanine and cytosine nucleotides. PGC/Genome = PGC as % of genome size.
http://dx.doi.org/10.1371/journal.pone.0025050.t001.

A detailed investigation of the gene and operon arrangement of AAPB PGCs has been performed recently (Zheng, Zhang, Koblížek, et al., 2011; Zheng et al., 2012). In this chapter, we summarise the structure and arrangement of PGCs in the AAPB genomes, with the aim of addressing the evolutionary divergence of different clades of AAPB, as well as the differences in carotenoid gene composition and biosynthetic pathways.

2. DIVERSE STRUCTURES OF PGCs

2.1. Composition of PGCs

Ten fully sequenced AAPB's species were analysed for their PGC composition. According to phylogenetic analysis using both 16S rRNA and *pufM* genes, the 10 strains were classified into three main groups: *Roseobacter* clade (order *Rhodobacterales*), *Erythrobacter–Citromicrobium* clade

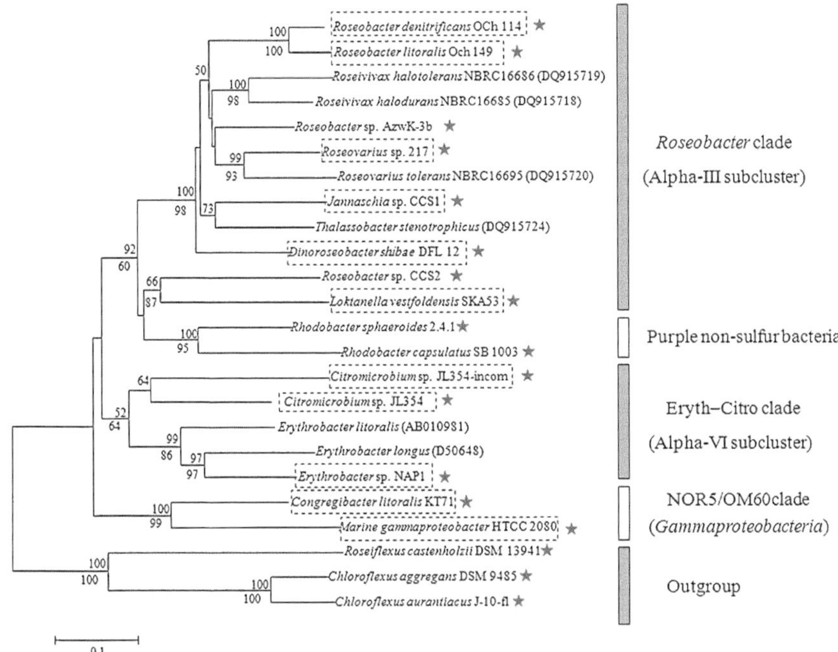

Figure 12.1 ***Phylogenetic analysis of* pufM *gene sequences from GenBank database.*** Symbol '★' indicates that the *pufM* sequence comes from a whole genome sequence. The 10 strains in dashed boxes are used to analyse their PGCs in Fig. 12.2. Bootstrap percentages from both neighbour joining (above nodes) and maximum parsimony (below nodes) methods are shown. Scale bar represents 5% nucleotide substitution percentage. doi:10.1371/journal.pone.0025050.g001.

(order *Sphingomonadales*) and NOR5/OM60 clade (Gammaproteobacteria) *(Fig. 12.1)*. The *Roseobacter* clade contained six strains belonging to five genera: *Roseobacter denitrificans* OCh 114 (Swingley et al., 2007) and *Roseobacter litoralis* Och 149 (Kalhoefer et al., 2011; Pradella et al., 2004), *Loktanella vestfoldensis* SKA53, *Dinoroseobacter shibae* DFL 12 (Wagner-Döbler et al., 2010), *Jannaschia* sp. CCS1, and *Roseovarius* sp. 217 (Baldock, Denger, Smits, & Cook, 2007). Two species belonged to the order *Shingomonadales*: *Erythrobacter* sp. NAP1 (Koblížek et al., 2003, 2011) and *Citromicrobium* sp. JL354 (Jiao, Zhang, & Zheng, 2010). Two species were members of the Gammaproteobacteria: *Cb. litoralis* KT71 (Fuchs et al., 2007; Spring et al., 2009) and marine *Gammaproteobacterium* HTCC2080 (Cho et al., 2007).

The genome size varied from approximately 3064 kb (*L. vestfoldensis*) to 4763 kb (*Roseovarius* sp. 217). The PGCs represented roughly 1% of the genomes (Table 12.1). The Eryth–Citro clade contained comparatively smaller PGCs (~39 kb) whereas the average size of PGC in *Roseobacter* and NOR5/OM60 clades was about 45 kb (Table 12.1). The (guanine–cytosine) GC content in the PGCs varied from 52.9% to 66.7%, which was similar to the total GC contents of the corresponding genomes (Table 12.1). This may indicate that these PGCs have evolved with their genomes long enough to keep homogenous genomic characteristics. The fact that the PGC is a stable part of phototroph genomes is indicated also by the fact that the phylogenetic trees constructed for 16S rRNA, *pufM* gene and concatenated PGC core genes show basically the same topology (Figs. 12.1, 12.3 and 12.6A).

2.1.1. Structure and Arrangement of PGCs

The PGCs have a mosaic structure and consist of five main sets of genes: *bch* genes encoding enzymes of the BChl *a* biosynthetic pathway, *puf* operons encoding proteins forming the RC, *puh* operons encoding an RC protein and also involved in the RC assembly, *crt* genes responsible for biosynthesis of carotenoids, and various regulatory genes (Fig. 12.2). A core set of 27 genes was identified, which were present in all analysed PGCs (Fig. 12.3). Sixteen of them (*bchBCDFGHILMNOPXYZ* and *acsF*) encode enzymes of the BChl *a* biosynthetic pathway. These gene sets, with an exception of 8-vinyl reductase, represent the complete biosynthetic pathway from protoporphyrin XI to BChl *a*. In contrast, there are only two genes (*crtC* and *crtF*) involved in carotenoid synthesis, which are in common to all PGCs (Fig. 12.2). Other shared core genes encode proteins *pufABLM*, *puhA*, and assembly factors *puhBCE* and *lhaA* of the bacterial photosynthetic units (Fig. 12.2).

In general, PGCs of AAPB in the *Roseobacter* clade contain more genes compared to Eryth–Citro or NOR5/OM60 clades (Fig. 12.2). The majority of *Roseobacter*-related species contained all the *puf* genes organised in *pufQBALMC* operon. The only exception was *L. vestfoldensis* SKA53, in which some photosynthesis genes are located outside the PGC and spread throughout the genome. Previously, it was reported that the PGC in *Rsb. litoralis* Och 149 is located on a linear plasmid, with two *RPA* genes between *bchFNBHLM–lhaA–puh* and *crtF–bchCXYZ–puf*, which act as a centromere-like anchor when plasmids replicate (Fig. 12.2) (Petersen, Brinkmann, & Pradella, 2009; Pradella et al., 2004).

The PGC organisation in *Erythrobacter* sp. NAP1 and *Citromicrobium* sp. JL354 (Eryth–Citro clade) is almost identical in terms of gene arrangement and composition (Fig. 12.2). When compared to *Roseobacters*, this group

Figure 12.2 ***PGC structure and arrangement in AAPB species.*** Green, *bch* genes; red, *puf* and regulators genes; pink, *puh* genes; orange, *crt* genes; blue, *hem* and *cyc* gene; yellow, *lhaA* gene; blank, uncertain or unrelated genes; grey, hypothetical protein. The horizontal arrows represent putative transcripts. The gene sets in dotted boxes indicate two conserved regions in all PGCs, and the two genes in the red circle of *Rsb. litoralis* Och 149 are two inserted genes thought to act as a centromere-like anchor when plasmids replicate. http://dx.doi.org/10.1371/journal.pone.0025050.g002. See the color plate.

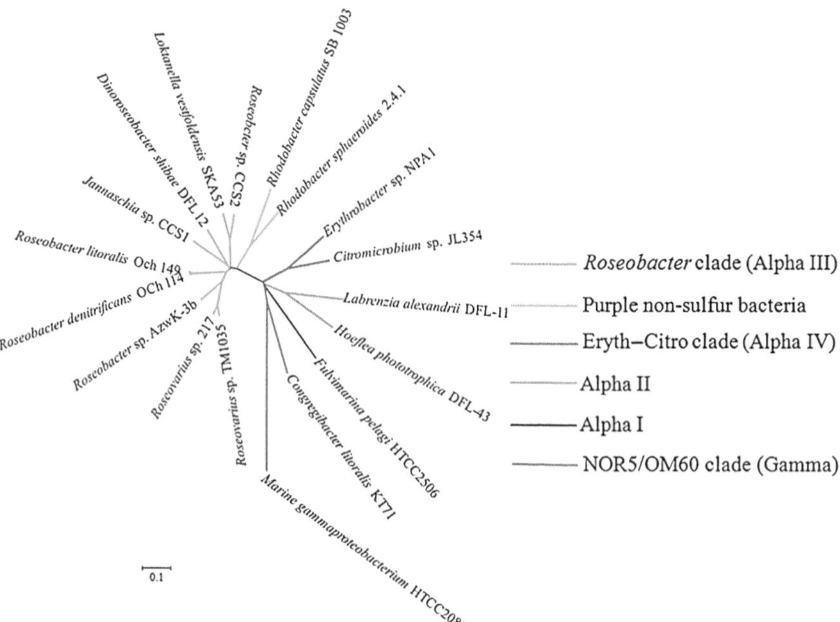

Figure 12.3 ***Neighbour joining phylogenetic analysis of 27 core proteins in PGCs from GenBank database.*** The core proteins are BchBCDFGHILMNOPXYZ–CrtCF–PufABLM–LhaA–PuhABCE–AcsF. Bar, 0.1 substitutions per amino acid position. (For colour version of this figure, the reader is referred to the online version of this book.)

contains fewer carotenoid genes and no LH 2 genes. The presence of fewer photosynthesis genes in Eryth–Citro clade is consistent with the smaller size of their genomes (Table 12.1).

Similarly, PGCs of two NOR5/OM60 strains have very comparable gene composition and organisation (Fig. 12.2). It contains less transcriptional regulators compared to the other groups. Conversely, a BLUF (blue light using flavin adenine dinucleotide sensors) was usually observed in upstream regions of PGCs of NOR5/OM60 clade (Fuchs et al., 2007).

2.2. Four Types of Gene Arrangements in PGCs

Two conserved gene arrangements are found in all analysed PGCs: *bchFNBHLM–lhaA–puhABC* and *crtF–bchCXYZ* (Fig. 12.2). According to the direction of transcription and relative order of these conserved arrangements, 10 PGCs can be divided into three groups: type I (transcribed in the same direction with *bchFNBHLM–lhaA–puh* upstream of *crtF–bchCXYZ–puf*) includes *Rsb. denitrificans* OCh 114 and *Rsb. litoralis* Och 149. *C ongregibacter litoralis* KT71, *Gammaproteobacterium* HTCC2080;

type II includes *Roseovarius* sp. 217 (*bchFNBHLM–lhaA–puh* and *crtF–bchCXYZ–puf* are transcribed in convergent orientation); and type III includes the last five organisms (transcribed in the same direction with *crtF–bchCXYZ–puf* upstream of *bchFNBHLM–lhaA–puh*). The last possible arrangement (type IV, divergent transcription of *bchFNBHLM–lhaA–puh* and *crtF–bchCXYZ–puf*) has not been yet found in an AAPB (or AAPB candidate) genome, however, it is present in the purple non-sulfur anaerobic bacteria *Rhodobacter sphaeroides* and *Rhodobacter capsulatus* (Fig. 12.2). The distribution of PGC types does not correspond to their phylogenetic affiliation. For example, the *Roseobacter* clade shows all three PGC arrangement types observed in AAPB genomes. This suggests that complex operon recombination in PGC occurred after phylogenetic divergence of AAPB genera.

In all the PGCs, there are four conserved regions that contain genes involved in BChl *a* biosynthesis: *bchFNBHLM*, *bchCXYZ*, *bchIDO* and *bchOP*. Gene *bchEJ*, which exist in most *Rhodobacter*, was found in *Cb. litoralis* KT71 (Fig. 12.2). There are carotenoid genes between *bchCXYZ* and *bchIDO*, except in *D. shibae* DLF 12 and *Jannaschia* sp. CCS1. The region between *bchOP* and *bchFNBHLM* contains variable sequences in different AAPB clades. In the *Roseobacter* and Eryth–Citro clades, there are two regulators (*ppsR* and *ppaA*) that are sensitive to light intensity and oxygen concentration (Komiya, Yeates, Rees, Allen, & Feher, 1988). In NOR5/OM60 clade, a *crtJ* gene was found, which controls aerobic repression of BChl and carotenoid production, and LH 2 gene expression (Elsen, Ponnampalam, & Bauer, 1998; Ponnampalam & Bauer, 1997).

Four structural types of *puf* gene organisation were observed in 10 PGCs: *pufQBALMC*, *pufQBALM*, *pufBALM* and *pufLMCBA* (Fig. 12.2). Unlike the purple non-sulfur species *Rba. sphaeroides* and *Rba. capsulatus*, all the AAPB strains studied lack the *pufX* gene in the PGC. The *pufQ* gene is absent in the *puf* operon of NOR5/OM60 and Eryth–Citro clades. In addition, the Eryth–Citro clade and *L. vestfoldensis* SKA53 do not have a *pufC* gene. The gene encoding 1-deoxy-D-xylulose-5-phosphate synthase (DXPS) is always located downstream of *puf* genes in the *Roseobacter* clade. DXPS is a part of the mevalonate-independent pathway for isopentenylpyrophosphate biosynthesis, a precursor for carotenoid and BChl biosynthesis (Rohmer, 1999). Interestingly, a switch of order in the *puf* gene cluster is observed in NOR5/OM60 clade (*pufLMCBA*) compared to the other two AAPB clades (*pufBALMC*).

The structure of *puhABC–hyp–ascF–puhE* is conserved in *Roseobacter* and Eryth–Citro clades. However, in the NOR5/OM60 clade, *puhABC* and *puhE* are located together and *acsF* is at a site near a gene predicted to encode a BLUF domain (Fig. 12.2). The *lhaA* gene, encoding a possible LH 1 assembly protein (Young & Beatty, 1998), is located immediately upstream of the *puhA* gene. Following the *puhE* gene, are *hemN* (NOR5/OM60 clade) or *hemA* (*Roseobacter* and Eryth–Citro clades) genes (Hippler et al., 1997; Wang, Elliott, & Elliott, 1999).

2.3. Variety of *crt* Operons in PGCs

The main difference among PGCs was found in the genes encoding the carotenoid biosynthetic pathway. The set of *crt* genes identified in *Rba. capsulatus* contains *crtAIBKCDEFJ* (Table 12.2). A slightly reduced set of genes (*crtAIBCDEF*) was found in some *Roseobacter* species (Table 12.2). However, the organisation of the *crt* operon in the *Roseobacter* clade is the most variable feature of PGCs (Fig. 12.2). The almost complete structure *crtAIBK–hyp–crtCDEF* is present in the genera *Roseobacter* and *Dinoroseobacter*

Table 12.2 The composition of carotenoid genes in AAPB. 1, *Roseobacter* clade (*Rsb. denitrificans* Och 114, *Rsb. litoralis* OCh 149, *D. shibae* DLF 12, *L. vestfoldensis* SKA53, *Jannaschia* sp. CCS1, *Roseovarius* sp. 217 included). 2, *Rhodobacter* genus (*Rba. sphaeroides* 2.4.1 and *Rba. capsulatus* BEC404). The *crtJ* gene was found only in *Rba. capsulatus* BEC404 and NOR5/OM60 Gammaproteobacteria. 3, *Erythrobacter* sp. NAP1 and *Citromicrobium* sp. JL354. 4, *Erb. litoralis* HTCC2594, *Erythrobacter* sp. SD-21 and Citromicrobial sp. JLT1363. 5, NOR5 clade (*Cb. litoralis* KT71 and *Gammaproteobacterium* HTCC2080 included). The genes located in the PGC are marked by '●'; the genes outside PGC are marked by '○'. The gene *CrtK* is not included in the Table as it does not participate in any known carotenoid biosynthetic pathway

	Alpha				*Gamma*
	Rhodobacterales		*Sphingomonadales*		NOR5
	1	2	3	4	5
crtA	●	●			
crtI	●	●	○	○	●
crtB	●	●	○	○	●
crtC	●	●	●		●
crtD	●	●	●		○
crtE	●	●	○	○	●
crtF	●	●	●		●
crtJ		●			●
crtY			○	○	
crtZ			○	○	
crtW				○	

http://dx.doi.org/10.1371/journal.pone.0025050.t002.

(Fig. 12.2), whereas in *D. shibae*, *crtA* and *crtIBK* they are separated. Homologous recombination appears to have occurred between *crtAIB* and *crtCDEF* in *Jannaschia* sp. CCS1. Comparably, *crtICDEF* and *crtCDF* are missing in the NOR5/OM60 and Eryth–Citro clades, respectively. Rearrangement of *crt* genes may result from events of gene duplication and loss, accounting for the absence of a *crtA* gene in Eryth–Citro and NOR5 clades (Table 12.2), and duplication of some of the *crt* genes, such as the *crtE* and *crtIB* found outside the PGC in *Bradyrhizobium* sp. ORS278 (Swingley et al., 2009).

The key *crt* genes encoding enzymes involved in zeaxanthin (*crtY* and *crtZ*) and nostoxanthin (*crtG*) biosynthesis are not located in the PGCs (Table 12.2). Similarly, the genes *crtBI* involved in lycopene biosynthesis (which are essential for spheroidenone and spirilloxanthin pathways) were found outside the PGC (Table 12.2). Except for the strains *Erythrobacter* sp. NAP1 and *Citromicrobium* sp. JL354, *crtYIB*, *crtWZ* and *crtE* are also found in closely related non-phototrophic strains, *Erythrobacter litoralis* HTCC2594, *Erythrobacter* sp. SD-21 and Citromicrobial strain JLT1363 (Table 12.2).

3. EXPRESSION OF PGC

3.1. Diverse Carotenoid Pigments in AAPB

A typical feature of AAPB is their pigmentation due to abundant carotenoids, which spans from yellow/orange to brown or from pink/red to purple. While some of the carotenoids serve as harvesting pigments, most of them do not participate in the LH likely having a photoprotection function (Yurkov & Beatty, 1998; Yurkov & Csotonyi, 2009). As suggested earlier, spheroidenone is the main LH carotenoid in *Roseobacters* (Koblízek, Falkowski, & Kolber, 2006; Takaichi, 2009; Wagner-Döbler & Biebl, 2006) (Table 12.3). Spheroidenone is also produced by anaerobic purple nonsulfur photoautotrophic organisms such as *Rba. sphaeroides* or *Rhodovulum marinum* when grown under aerobic conditions (Koyama et al., 1982; Lutz, Agalidis, Hervo, Cogdell, & Reiss-Husson, 1978). This is consistent with the closer phylogenetic relationship of these two organisms to *Roseobacter*-related photoheterotrophic species (Figs. 12.1 and 12.3). This indicates the presence of the same carotenoid biosynthetic pathway in all Rhodobacterales. The central biosynthetic pathway for carotenoids in the *Roseobacter* clade is the spheroidene pathway (Fig. 12.4, Table 12.3), and all the necessary genes (*crtAIBCDF*) for it are located in the PGCs (Fig. 12.2 and Table 12.2).

In most studied *Erythrobacter* species, erythroxanthin sulfate was shown to be the main carotenoid (Koblížek et al., 2003; Noguchi, Hayashi, Shimada,

Table 12.3 The major carotenoid composition in AAPB

AAPB	Strain	Main carotenoids	Reference
Roseobacter clade	*Rsb. denitrificans* Och 114	Spheroidenone	Takaichi et al., 1991 and Zheng, Zhang, Koblížek, et al., 2011
	Rsb. litoralis OCh 149	Spheroidenone	Zheng, Zhang, Koblížek, et al., 2011
	D. shibae JL1447	Spheroidenone	Zheng, Zhang, Koblížek, et al., 2011
	Roseobacter sp. COL2P	Spheroidenone	Koblížek, Mlčoušková, Kolber, & Kopecký, 2010
Erythrobacter	*Erythrobacter* sp. NAP1	Erythroxanthin sulfate, Bacteriorubixanthinal, Zeaxanthin and β-carotene	Koblížek et al., 2003
	Erb. longus DSM 6997	Erythroxanthin sulfate, Bacteriorubixanthinal, Zeaxanthin and β-carotene	Noguchi et al., 1992 and Zheng, Zhang, Koblížek, et al., 2011
	Erb. litoralis $T4^T$	Erythroxanthin sulfate, Bacteriorubixanthinal	Yurkov et al., 1994
	Erythrobacter sp. JL475	Erythroxanthin sulfate, Bacteriorubixanthinal, Zeaxanthin and β-carotene	Zheng, Zhang, Koblížek, et al., 2011
Citromicrobium	*Citromicrobium* sp. JL354	Nostoxanthin	Zheng, Zhang, Koblížek, et al., 2011
NOR5/OM60	*Cb. litoralis* KT71	Spirilloxanthin	Spring et al., 2009

Takaichi, & Tasumi, 1992) (Table 12.3), however, it does not participate in the photosynthetic processes (Noguchi et al., 1992). Light is harvested by other pigments such as bacteriorubixanthinal, zeaxanthin and β-carotene (Noguchi et al., 1992). The main carotenoid identified in *Citromicrobium* sp. JL354 was nostoxanthin (Table 12.3). We assume that both species share

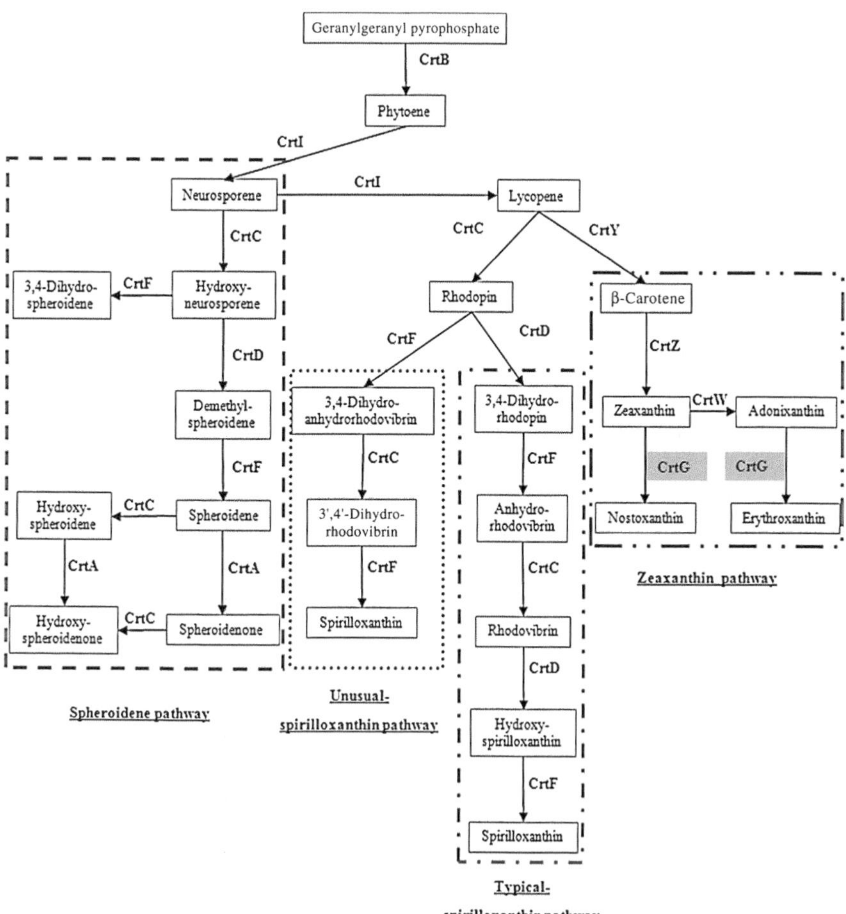

Figure 12.4 ***Tentative carotenoid biosynthesis pathways in AAPB.*** The pathways proposed are based on the identification of indicated genes in genome sequences. doi:10.1371/journal.pone.0025050.g003.

similar carotenoid biosynthetic pathways (Fig. 12.4). First, β-carotene is produced from lycopene by the action of lycopene cyclase (*crtY* gene product). Zeaxanthin is obtained by the two-step hydroxylation of β-carotene catalysed by β-carotene hydroxylase (*crtZ* gene product). Interestingly, the key genes (*crtY* and *crtZ*) for zeaxanthin pathway are not organised in the PGCs, but are spread throughout the chromosome (Table 12.2). Zeaxanthin is then a starting intermediate for the synthesis of both nostoxanthin (in genus *Citromicrobium*) and erythroxanthin (in genus *Erythrobacter*) (Fig. 12.4).

The major carotenoid in *Cb. litoralis* KT71 is spirilloxanthin, the same as in *Rhodospirillum rubrum* DSM 467^{T} (Spring et al., 2009) (Table 12.3).

There are two possible options for spirilloxanthin biosynthesis: a typical spirilloxanthin biosynthetic pathway and an unusual spirilloxanthin pathway (Fig. 12.4). Interestingly, the gene *crtD* was found to be located outside of the PGC in *Cb. litoralis* KT71 (Table 12.2), indicating that *Cb. litoralis* KT71 might use the unusual spirilloxanthin pathway.

3.2. Absorbance Spectra

The membranes of photoheterotrophically grown cultures of *Rsb. denitrificans* OCh 114, *Rsb. litoralis* OCh 149 and *Erb. litoralis* DSM 6997 were isolated and the absorption spectra measured (Fig. 12.5). The photosynthetic apparatus of AAPB in the *Roseobacter* clade is thought to comprise an RC enclosed by a circular LH 1, and LH 2 antenna complexes. Peaks at 806 and 868 nm were found in the absorption spectrum of *Rsb. denitrificans* OCh 114 (Fig. 12.5A). *Roseobacter litoralis* OCh 149 had a major peak around 805 nm, with a very small peak around 865 nm (Fig. 12.5B). Two inserted genes of PGC in strain *Rsb. litoralis* OCh 114 may repress the expression of RC and LH1, which needs further evidence (Daniela et al., 2011; Pradella et al., 2004).

The genome sequences indicated that no complete LH 2 genes exist in the Citro–Eryth clade. The absorption spectrum of *Erb. litoralis* DSM 6997 was different from that of the AAPB in the *Roseobacter* clade and indicated

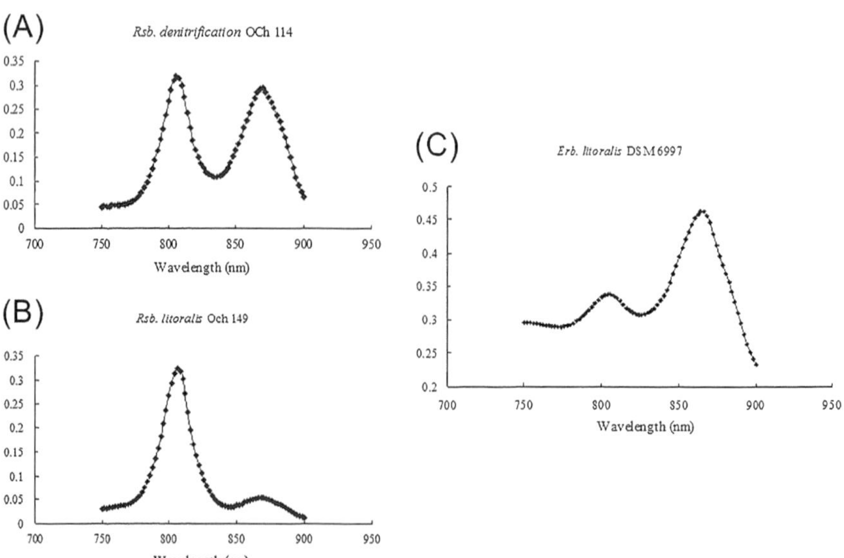

Figure 12.5 ***Absorbance spectra of photosynthetic membranes.*** A, *Rsb. denitrification* OCh 114; B, *Rsb. litoralis* Och 149; C, *Erb. litoralis* DSM 6997.

that the photosynthetic apparatus of strain DSM 6997 comprises only one type of LH antenna complex (LH 1, peak at 876 nm), with the peak at 802 nm attributed to the RC. Compared to AAPB in the *Roseobacter* clade, the absorption spectrum of strain DSM 6997 is a little blue-shifted (Fig. 12.5C). Although the genes encoding LH 2 complex were found in the genome of AAPB in the NOR5/OM60 clade, no significant amount of the LH 2 complex was detected in KT71^{T} cells under laboratory growth conditions (Spring et al., 2009).

4. EVOLUTION OF AAPB

Proteobacteria are thought to have diverged from a phototrophic ancestor, according to the scattered distribution of phototrophy throughout the Proteobacterial clade, and so the occurrence of numerous closely related phototrophic and chemotrophic microorganisms may be the result of the loss of genes for phototrophy (Swingley et al., 2009; Woese, 1987).

4.1. Phylogeny of AAPB

AAPB have been thought to be closely related to the purple non-sulfur bacteria, as exemplified by the presence of AAPB strains in the *Roseobacter* clade (Alpha III), the Alpha II subcluster, the Betaproteobacteria (which includes the photosynthetic *R. gelatinosus*), and the NOR5/OM60 clade (Gammaproteobacteria) (Fig. 12.6A). There are also AAPB that possess atypical characteristics akin to the purple non-sulfur bacteria; e.g. *Rsb. denitrificans* OCh 114 and *D. shibae* DFL 12 are able to grow under anaerobic conditions, and *Cb. litoralis* KT71 induces the expression of its photosynthetic apparatus under semi-aerobic conditions (Spring et al., 2009; Swingley et al., 2007; Wagner-Döbler et al., 2010). In addition, AAPB in the *Roseobacter* and NOR5/OM60 clades utilise LH pigments and carotenoid biosynthesis pathways similar to those used by the purple non-sulfur (photosynthetic) bacteria (Beatty, 2005; Koblížek et al., 2003; Swingley et al., 2009; Zheng, Zhang, Koblížek, et al., 2011).

To date, only AAPB in the order *Sphingomonadales* (Eryth–Citro clade) are true obligate aerobic phototrophs, and contain unusual carotenoid biosynthetic pathways and pigments, compared to the other clades (Koblížek et al., 2003; Zheng, Zhang, Koblížek, et al., 2011). Interestingly, members of the Eryth–Citro clade possess the shortest and simplest PGC structure of the known AAPB (Zheng, Zhang, Koblížek, et al., 2011), and the LH 2

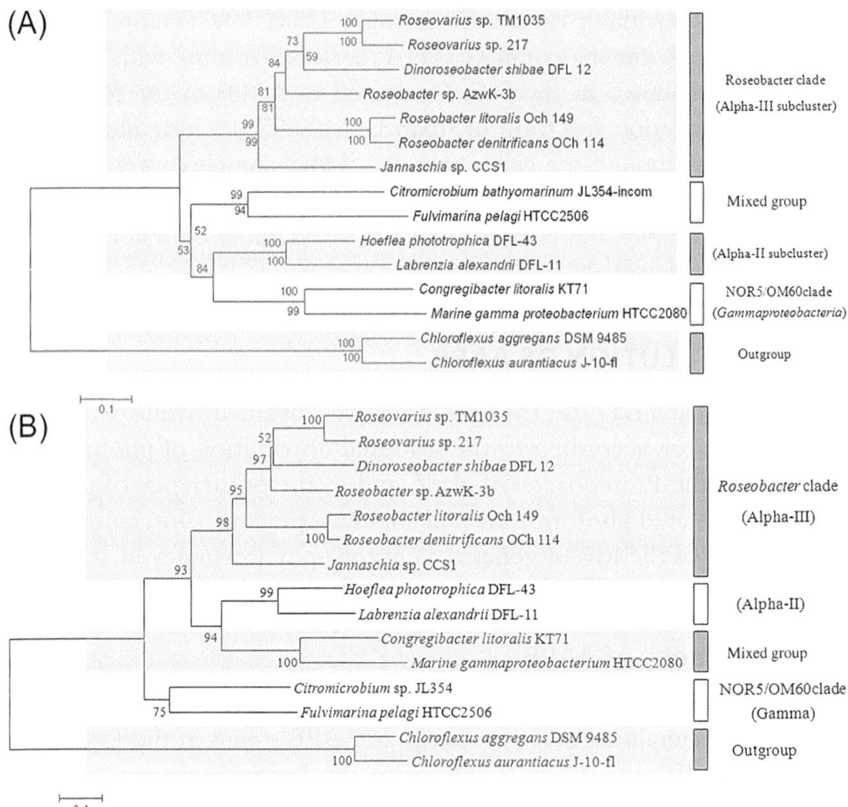

Figure 12.6 ***Phylogenetic comparative analyses of all whole genome-sequenced marine AAPB.*** A, tree of 16S rRNA gene sequences. Scale bar represents 5% nucleotide substitution percentage. B, tree of *pufC* gene sequences. Scale bar represents 10% nucleotide substitution percentage. Bootstrap values ≥50% from both neighbour joining (above) and maximum parsimony (below) are shown. doi:10.1371/journal.pone.0035790.g001.

complex genes are absent, perhaps indicating that the AAPB in this clade are younger than their counterparts in other clades.

4.1.1. Citromicrobium *Genus*

The genus *Citromicrobium* belonging to order Sphinogomonadales (Alpha IV subcluster) contains only one species, *Citromicrobium bathyomarinum* (Fig. 12.6A). The strain JL354, which was isolated from the upper ocean water of the South China Sea (Jiao, Zhang, & Zheng, 2010), shares 99.6% 16S rRNA sequence identity with the strain type *Cmi. bathyomarinum* JF-1 that was isolated from deep sea hydrothermal vent plume waters (Yurkov et al., 1999). A previous study proposed the existence of two PGCs in the *Citromicrobium* sp. JL354 genome sequence – one complete and the other incomplete (Jiao, Zhang, & Zheng, 2010). Subsequently, the Citromicrobial strain JLT1363 was isolated from the South China Sea and was found to have 98.1 and 98.0% 16S rRNA gene sequence identity to *Citromicrobium* sp. JL354 and *Cmi. bathyomarinum* JF-1, respectively (Fig. 12.6A). However, there is no PGC in the genome of the Citromicrobial strain JLT1363 (Zheng, Zhang, & Jiao, 2011). In the strictly taxonomic sense the genus name *Citromicrobium* for strain JLT1363 is incorrect, because species in the *Citromicrobium* genus are defined as producing BChl *a* in RC, LH 1 and LH 2 complexes (Yurkov et al., 1999), and so we here use the name Citromicrobial to emphasise the close evolutionary relatedness of strain JLT1363 to taxonomically genuine members of the genus *Citromicrobium*.

4.2. Gain and Loss of Photosynthesis Genes as Revealed by Comparison of Two Citromicrobial Genomes

Compared to members of the *Roseobacter* clade, which have diverse metabolic capabilities supported by large genomes (Newton et al., 2010; Wagner-Döbler & Biebl, 2006), the oligotrophs of the genus *Citromicrobium* have small, streamlined genomes and no known plasmids, potentially making them less adaptable (Polz, Hunt, Preheim, & Weinreich, 2006). In order to compete and survive in natural environments, *Citromicrobium* species may benefit from alternative strategies to enhance their genomes by HGT. *Citromicrobium* species possess at least four potential mechanisms of HGT, which could result in adaptation to changing conditions: a gene transfer agent; an integrative conjugative element (ICE); a Mu-like prophage; and a type IV secretion system.

By comparing the genomes of the AAPB *Citromicrobium* sp. JL354 and the closely related chemotrophic Citromicrobial strain JLT1363, evidence was obtained to support the idea that JLT1363 evolved from an AAPB by loss of phototrophy genes to become heterotrophic. In addition, it was found that the incomplete PGC of *Citromicrobium* sp. JL354 might have been acquired by HGT.

4.2.1. *Loss of Phototrophy Genes*

The AAPB may be considered as a transitional group between photosynthetic and chemotrophic bacteria, because phototrophy supports no more than 20% of cellular energy requirements under illuminated aerobic conditions (Kolber et al., 2001). The AAPB mainly utilise organic matter for growth. In oligotrophic oceanic regions, AAPB may have some advantages over chemotrophs, but in a eutrophic oceanic area or rich organic medium, some of them (especially AAPB in the Eryth–Citro clade) do not express phototrophy genes (Jiao, Zhang, & Hong, 2010), and so prolonged growth in a eutrophic environment could allow for the loss of phototrophy genes.

The complete PGC in the *Citromicrobium* sp. JL354 genome consists of two conserved subclusters, *crtCDF–bchCXYZ–pufBALM* and *bchFNBHLM–lhaA–puhABC* (Figs. 12.2 and 12.7), flanked by putative genes involved in cell division, transmembrane transport and signal transduction. Interestingly, in the JLT1363 genome, we found exactly the same flanking gene organisation but with a single hypothetical gene in place of the PGC (Figs. 12.2 and 12.7). It appears that either HGT (i.e. JL354 obtained the PGC) or gene loss (i.e. JLT1363 lost the PGC) occurred in *Citromicrobium*. To address these possibilities, phylogenetic analyses of strain JL354 PGC and 16S rRNA genes were carried out, and the trees were evaluated for congruency. The general topology of these phylogenetic trees was similar (Figs. 12.6A and 12.3), indicating coevolution of the PGC with the 16S rRNA gene. These results indicate that *Citromicrobium* sp. JL354 acquired the complete PGC long before 16S rRNA phylogenetic divergence, and that relatively recent HGT does not account for the complete PGC in strain JL354. Therefore, the data suggest that the Citromicrobial strain JLT1363 lost its PGC after divergence from a common ancestor shared with strain JL354.

HGT could not only increase the host genetic information but could also trigger chromosome recombination resulting in gene loss or duplication. For example, Mu-like phage lysogen formation involves transposition into almost random locations within the host genome, and if this were to occur within a gene or an operon it would be likely to cause complete or partial loss of function at the insertion site (Morgan, Hatfull, Casjens, & Hendrix, 2002). Furthermore, transposing phages and other insertion sequences are known to be the key factors in reorganisation and evolution of bacterial genomes (Craig, 1995; Mahillon and Chandler, 1998; Taylor, 1963). The disruption of a single BChl biosynthesis or RC gene in the PGC would lead to loss of phototrophy, and thus there are many potential sites for gene disruption that would lead to the absence of a selective advantage for the PGC as a whole. Subsequently, a non-functional PGC would be an energy drain during genome replication, conferring selective advantage on progeny for the loss of this non-functional PGC in an environment allowing chemotrophic growth. Although further investigation is needed to elucidate the exact mechanism of this proposed PGC loss, our

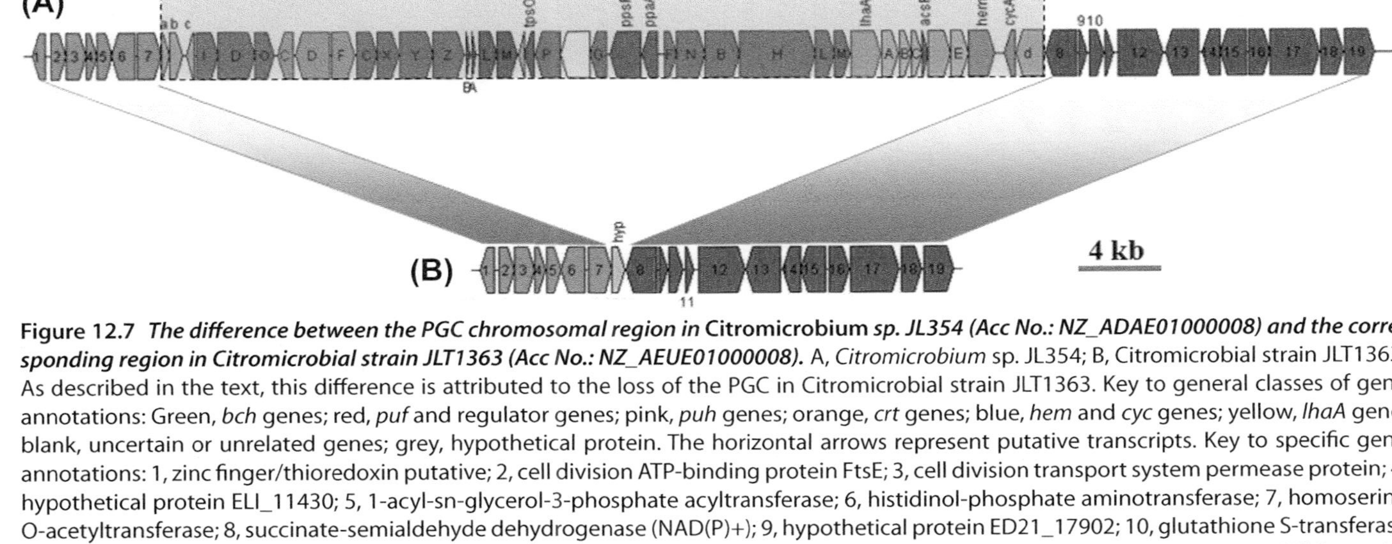

Figure 12.7 ***The difference between the PGC chromosomal region in* Citromicrobium *sp. JL354 (Acc No.: NZ_ADAE01000008) and the corresponding region in Citromicrobial strain JLT1363 (Acc No.: NZ_AEUE01000008).*** A, *Citromicrobium* sp. JL354; B, Citromicrobial strain JLT1363. As described in the text, this difference is attributed to the loss of the PGC in Citromicrobial strain JLT1363. Key to general classes of gene annotations: Green, *bch* genes; red, *puf* and regulator genes; pink, *puh* genes; orange, *crt* genes; blue, *hem* and *cyc* genes; yellow, *lhaA* gene; blank, uncertain or unrelated genes; grey, hypothetical protein. The horizontal arrows represent putative transcripts. Key to specific gene annotations: 1, zinc finger/thioredoxin putative; 2, cell division ATP-binding protein FtsE; 3, cell division transport system permease protein; 4, hypothetical protein ELI_11430; 5, 1-acyl-sn-glycerol-3-phosphate acyltransferase; 6, histidinol-phosphate aminotransferase; 7, homoserine O-acetyltransferase; 8, succinate-semialdehyde dehydrogenase (NAD(P)+); 9, hypothetical protein ED21_17902; 10, glutathione S-transferase family protein; 11, protein-methionine-S-oxide reductase; 12, membrane carboxypeptidase; 13, trypsin-like serine protease; 14, HflC protein; 15, integral membrane proteinase; 16, ATPase; 17, aldehyde oxidase and xanthine dehydrogenase molybdopterin binding; 18, ferrochelatase; 19, cytochrome P450. http://dx.doi.org/10.1371/journal.pone.0035790.g004. See the color plate.

study could provide an explanation for the scattered evolutionary lineage for the presence and absence of phototrophy throughout the *Proteobacteria* clade and may also explain the numerous closely related phototrophic and chemotrophic strains (Swingley et al., 2009; Woese ,1987).

4.2.2. Gain of Phototrophy Genes

Coexistence of two different PGC gene sets in one phototrophic bacterium has been found only in *Citromicrobium* sp. JL354 (Jiao, Zhang, & Zheng, 2010). The complete PGC is typical, whereas the incomplete PGC consists of just the *pufLMC* and *puhABC* genes, which are contiguous (Fig. 12.8). However, in most other PGCs, including the complete PGC elsewhere in the same genome, *pufLM* and *puhABC* are separated (Zheng, Zhang, Koblížek, et al., 2011), and *pufC* is not present in Alpha IV AAPB containing complete PGCs (e.g. *Citromicrobium* sp. JL354 and *Erythrobacter* sp. NAP1). Furthermore, the GC content of the *pufLM* and *puhABC* sequences in the incomplete PGC (62 and 66%, respectively) is lower than that of the complete PGC (63 and 69%, respectively). The nucleotide sequence identity is less than 80% for all shared genes between the two PGCs, and the difference is higher than for a genus level distinction (15%) (Zeng, Chen, & Jiao, 2007). On the basis of the foregoing observations, we speculate that the incomplete PGC in *Citromicrobium* sp. JL354 was obtained via HGT. This hypothesis is strengthened substantially by further genomic analysis which revealed that the

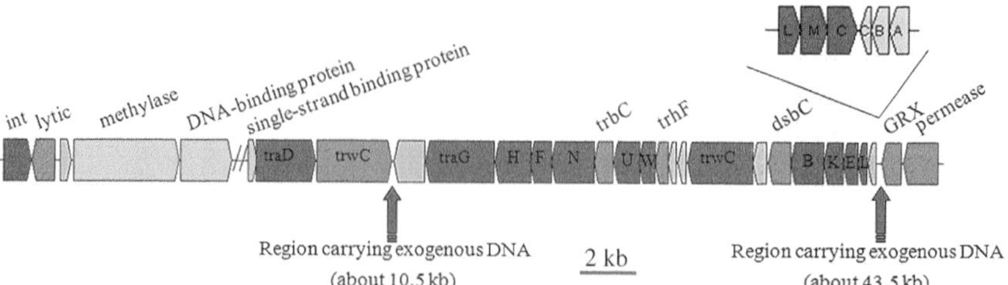

Figure 12.8 ***Genomic organisation of core genes of a putative ICE found in the genome of Citromicrobial strain JLT1363.*** The two regions suggested to carry exogenous DNA are predicted to be involved in phage repression (10.5 kb) and heavy metal transport (43.5 kb) (Acc No.: NZ_AEUE01000004). The ICE in JL354 is in different contigs, but all the core genes after and including 'single-strand binding protein' are present (Acc No.: NZ_ADAE01000017). The incomplete PGC (*pufLMC–puhCBA*) was found only found in an ICE of *Citromicrobium* strain JL354. Red, yellow and pink indicate phage-related genes; green and cyan indicate plasmid-related genes. http://dx.doi.org/10.1371/journal.pone.0035790.g003. See the color plate.

Table 12.4 Highest-scoring BLAST results (PSI-BLAST) using genes from the JL354 incomplete PGC (*pufLMC–puhABC*)

Incomplete PGC gene	Closest match (AA identity)	AA identity with *F. pelagi* HTCC2506
pufL	Gammaproteobacterium NOR51-B 211/275 (77%)	203/271 (75%)
pufM	*F. pelagi* HTCC2506 226/319 (71%)	226/319 (71%)
pufC	*F. pelagi* HTCC2506 156/328 (48%)	156/328 (48%)
puhA	*F. pelagi* HTCC2506 100/250 (40%)	100/250 (40%)
puhB	*H. phototrophica* DFL-43 72/193 (38%)	71/198 (36%)
puhC	*Rps. palustris* DX-1 42/144 (30%)	41/144 (29%)

http://dx.doi.org/10.1371/journal.pone.0035790.t001.

incomplete PGC in *Citromicrobium* sp. JL354 is located within one of the putative ICE intergenic hotspots (Fig. 12.8). Further phylogenetic analysis showed that the *pufC* gene sequence belonging to the incomplete PGC is closest to *Fulvimarina pelagi* HTCC2506 (Fig. 12.6B), which indicates that the incomplete PGC genes may have been acquired from a *Fulvimarina*-related species (Table 12.4) (Kang et al., 2010).

5. SUMMARY AND FUTURE PERSPECTIVES

In summary, this chapter showed that most of the photosynthesis genes in AAPB species are organised in a PGC. Two conserved regions, *bchFNBHLM–lhaA–puhABC* and *crtF–bchCXYZ*, were identified in all the studied PGCs. Based on the orientation of these regions, we divide the studied strains into four different groups (three in the AAPB). The distribution of these three arrangements does not correspond to the phylogenetic affiliation of an individual AAPB. The composition of *bch*, *puf* and *puh* genes in the analysed PGCs was relatively similar, and the main difference was found among *crt* genes. Such variability was mainly connected with different carotenoid biosynthetic pathways present in the AAPB groups. A spheroidenone biosynthetic pathway in *Roseobacters*, a zeaxanthin pathway in Eryth–Citro clade and a spirilloxanthin pathway in the Gammaproteobacterial NOR5/OM60 clade. The genera *Erythrobacter* and *Citromicrobium*, with unique carotenoid biosynthetic pathways and pigments, and containing the shortest and simplest PGC structure, were thought to be the most evolutionarily recent groups among the known AAPB.

Comparison of two Citromicrobial genomes shed light on how an AAPB species may have lost phototrophic competence to become chemotrophic. In addition, we suggest that multiple potential HGT mechanisms provided opportunities for gain as well as loss of phototrophy genes. Therefore, it appears that the influence of HGT on the presence of phototrophy genes has contributed to the genotypes and phenotypes of extant species of *Proteobacteria*. With the increasing scope and depth of genome and metagenome databases, future detailed analysis should further clarify the evolutionary history of phototrophy.

ACKNOWLEDGEMENT

This work was supported by the NSFC project (91028001, 41076063, 40506059), the National Basic Research Program of China (grant No. 2011CB808800), and the SOA project (201105021); and Canadian NSERC and CIHR grants to JTB; and Czech projects GAČR P501/10/0221 and Algatech CZ.1.05/2.1.00/03.0110 to MK.

REFERENCES

Baldock, M. I., Denger, K., Smits, T. H.M., & Cook, A. M. (2007). *Roseovarius* sp. strain 217: aerobic taurine dissimilation via acetate kinase and acetate–CoA ligase. *FEMS Microbiology Letters, 271*, 202–206.

Beatty, J. T. (2005). On the natural selection and evolution of the aerobic phototrophic bacteria. *Discoveries in Photosynthesis, 73*, 1099–1104.

Béjà, O., Suzuki, M. T., Heidelberg, J. F., Nelson, W. C., Preston, C. M., Hamada, T., et al. (2002). Unsuspected diversity among marine aerobic anoxygenic phototrophs. *Nature, 415*, 630–633.

Blankenship, R. E. (1992). Origin and early evolution of photosynthesis. *Photosynthesis Research, 33*, 91–111.

Cho, J. C., Stapels, M. D., Morris, R. M., Vergin, K. L., Schwalbach, M. S., Givan, S. A., et al. (2007). Polyphyletic photosynthetic reaction centre genes in oligotrophic marine Gammaproteobacteria. *Environmental Microbiology, 9*, 1456–1463.

Craig, N. L. (1995). Unity in transposition reactions. *Science, 270*, 253–254.

Des Marais, D. J. (2000). When did photosynthesis emerge on earth? *Science, 289*, 1703–1705.

Elsen, S., Ponnampalam, S. N., & Bauer, C. E. (1998). CrtJ bound to distant binding sites interacts cooperatively to aerobically repress photopigment biosynthesis and light harvesting II gene expression in *Rhodobacter capsulatus*. *Journal of Biological Chemistry, 273*, 30762–30769.

Fuchs, B. M., Spring, S., Teeling, H., Quast, C., Wulf, J., Schattenhofer, M., et al. (2007). Characterization of a marine gammaproteobacterium capable of aerobic anoxygenic photosynthesis. *Proceedings of the National Academy of Sciences, 104*, 2891–2896.

Hippler, B., Homuth, G., Hoffmann, T., Hungerer, C., Schumann, W., & Jahn, D. (1997). Characterization of *Bacillus subtilis* hemN. *Journal of Bacteriology, 179*, 7181–7185.

Igarashi, N., Harada, J., Nagashima, S., Matsuura, K., Shimada, K., & Nagashima, K. V.P. (2001). Horizontal transfer of the photosynthesis gene cluster and operon rearrangement in purple bacteria. *Journal of Molecular Evolution, 52*, 333–341.

Jiao, N., Zhang, F., & Hong, N. (2010). Significant roles of bacteriochlorophyll *a* supplemental to chlorophyll *a* in the ocean. *The ISME Journal, 4*, 595–597.

Jiao, N., Zhang, Y., Zeng, Y., Hong, N., Liu, R., Chen, F., et al. (2007). Distinct distribution pattern of abundance and diversity of aerobic anoxygenic phototrophic bacteria in the global ocean. *Environmental Microbiology, 9*, 3091–3099.

Jiao, N., Zhang, R., & Zheng, Q. (2010). Coexistence of two different photosynthetic operons in *Citromicrobium bathyomarinum* JL354 as revealed by whole-genome sequencing. *Journal of Bacteriology, 192*, 1169–1170.

Kalhoefer, D., Thole, S., Voget, S., Lehmann, R., Liesegang, H., Wollher, A., et al. (2011). Comparative genome analysis and genome-guided physiological analysis of *Roseobacter litoralis*. *BMC Genomics, 12*, 324–339.

Kang, I., Oh, H. M., Lim, S. I., Ferriera, S., Giovannoni, S. J., & Cho, J. C. (2010). Genome sequence of *Fulvimarina pelagi* HTCC2506T, a Mn (II)-oxidizing alphaproteobacterium possessing an aerobic anoxygenic photosynthetic gene cluster and xanthorhodopsin. *Journal of Bacteriology, 192*, 4798–4799.

Koblízek, M., Falkowski, P. G., & Kolber, Z. S. (2006). Diversity and distribution of photosynthetic bacteria in the Black Sea. *Deep Sea Research Part II: Topical Studies in Oceanography, 53*, 1934–1944.

Koblížek, M., Béjà, O., Bidigare, R. R., Christensen, S., Benitez-Nelson, B., Vetriani, C., et al. (2003). Isolation and characterization of *Erythrobacter* sp. strains from the upper ocean. *Archives of Microbiology, 180*, 327–338.

Koblížek, M., Janouškovec, J., Oborník, M., Johnson, J. H., Ferriera, S., & Falkowski, P. G. (2011). Genome sequence of the marine photoheterotrophic bacterium *Erythrobacter* sp. strain NAP1. *Journal of Bacteriology, 193*, 5881–5882.

Koblížek, M., Mašín, M., Ras, J., Poulton, A. J., & Prášil, O. (2007). Rapid growth rates of aerobic anoxygenic phototrophs in the ocean. *Environmental Microbiology, 9*, 2401–2406.

Koblížek, M., Mlčoušková, J., Kolber, Z., & Kopecký, J. (2010). On the photosynthetic properties of marine bacterium COL2P belonging to *Roseobacter* clade. *Archives of Microbiology, 192*, 41–49.

Kolber, Z. S., Gerald, F., Lang, A. S., Beatty, J. T., Blankenship, R. E., VanDover, C. L., et al. (2001). Contribution of aerobic photoheterotrophic bacteria to the carbon cycle in the ocean. *Science, 292*, 2492–2495.

Komiya, H., Yeates, T., Rees, D., Allen, J., & Feher, G. (1988). Structure of the reaction center from *Rhodobacter sphaeroides* R-26 and 2.4. 1: symmetry relations and sequence comparisons between different species. *Proceedings of the National Academy of Sciences, 85*, 9012–9016.

Koyama, Y., Kito, M., Takii, T., Saiki, K., Tsukida, K., & Yamashita, J. (1982). Configuration of the carotenoid in the reaction centers of photosynthetic bacteria. Comparison of the resonance Raman spectrum of the reaction center of *Rhodopseudomonas sphaeroides* G1C with those of cis–trans isomers of [beta]-carotene. *Biochimica et Biophysica Acta (BBA)-Bioenergetics, 680*, 109–118.

Liotenberg, S., Steunou, A. S., Picaud, M., Reiss–Husson, F., Astier, C., & Ouchane, S. (2008). Organization and expression of photosynthesis genes and operons in anoxygenic photosynthetic proteobacteria. *Environmental Microbiology, 10*, 2267–2276.

Lutz, M., Agalidis, I., Hervo, G., Cogdell, R. J., & Reiss-Husson, F. (1978). On the state of carotenoids bound to reaction centers of photosynthetic bacteria: a resonance Raman study. *Biochimica et Biophysica Acta (BBA)-Bioenergetics, 503*, 287–303.

Mahillon, J., & Chandler, M. (1998). Insertion sequences. *Microbiology and Molecular Biology Reviews, 62*, 725–774.

Morgan, G. J., Hatfull, G. F., Casjens, S., & Hendrix, R. W. (2002). Bacteriophage Mu genome sequence: analysis and comparison with Mu-like prophages in *Haemophilus, Neisseria* and Deinococcus. *Journal of Molecular Biology, 317*, 337–359.

Nagashima, K. V.P., Hiraishi, A., Shimada, K., & Matsuura, K. (1997). Horizontal transfer of genes coding for the photosynthetic reaction centers of purple bacteria. *Journal of Molecular Evolution*, *45*, 131–136.

Newton, R. J., Griffin, L. E., Bowles, K. M., Meile, C., Gifford, S., Givens, C. E., et al. (2010). Genome characteristics of a generalist marine bacterial lineage. *The ISME Journal*, *4*, 784–798.

Noguchi, T., Hayashi, H., Shimada, K., Takaichi, S., & Tasumi, M. (1992). *In vivo* states and functions of carotenoids in an aerobic photosynthetic bacterium, *Erythrobacter longus*. *Photosynthesis Research*, *31*, 21–30.

Petersen, J., Brinkmann, H., & Pradella, S. (2009). Diversity and evolution of *repABC* type plasmids in Rhodobacterales. *Environmental Microbiology*, *11*, 2627–2638.

Polz, M. F., Hunt, D. E., Preheim, S. P., & Weinreich, D. M. (2006). Patterns and mechanisms of genetic and phenotypic differentiation in marine microbes. *Philosophical Transactions of the Royal Society B: Biological Sciences*, *361*, 2009–2021.

Ponnampalam, S. N., & Bauer, C. E. (1997). DNA binding characteristics of CrtJ. *Journal of Biological Chemistry*, *272*, 18391–18396.

Pradella, S., Allgaier, M., Hoch, C., Päuker, O., Stackebrandt, E., & Wagner-Döbler, I. (2004). Genome organization and localization of the *pufLM* genes of the photosynthesis reaction center in phylogenetically diverse marine alphaproteobacteria. *Applied and Environmental Microbiology*, *70*, 3360–3369.

Rohmer, M. (1999). The discovery of a mevalonate-independent pathway for isoprenoid biosynthesis in bacteria, algae and higher plants. *Natural Product Reports*, *16*, 565–574.

Shiba, T., Simidu, U., & Taga, N. (1979). Distribution of aerobic bacteria which contain bacteriochlorophyll a. *Applied and Environmental Microbiology*, *38*, 43–45.

Spring, S., Lünsdorf, H., Fuchs, B. M., & Tindall, B. J. (2009). The photosynthetic apparatus and its regulation in the aerobic gammaproteobacterium *Congregibacter litoralis* gen. nov., sp. nov. *PLoS One*, *4*, e4866.

Swingley, W. D., Blankenship, R. E., & Raymond, J. (2009). Evolutionary relationships among purple photosynthetic bacteria and the origin of proteobacterial photosynthetic systems. In C. Neil Hunter, Fevzi Daldal, Marion C. Thurnauer & J. Thomas Beatty (Eds.), *The purple phototrophic bacteria* (pp. 17–29).

Swingley, W. D., Sadekar, S., Mastrian, S. D., Matthies, H. J., Hao, J., Ramos, H., et al. (2007). The complete genome sequence of *Roseobacter denitrificans* reveals a mixotrophic rather than photosynthetic metabolism. *Journal of Bacteriology*, *189*, 683–690.

Takaichi, S. (2009). Distribution and biosynthesis of carotenoids. In C. Neil Hunter, Fevzi Daldal, Marion C. Thurnauer & J. Thomas Beatty (Eds.), *The purple phototrophic bacteria* (pp. 97–117).

Takaichi, S., Furihata, K., Ishidsu, J., & Shimada, K. (1991). Carotenoid sulphates from the aerobic photosynthetic bacterium, *Erythrobacter longus*. *Phytochemistry*, *30*, 3411–3415.

Taylor, A. L. (1963). Bacteriophage-induced mutation in *Escherichia coli*. *Proceedings of the National Academy of Sciences of the United States of America*, *50*, 1043–1051.

Thrash, J. C., Cho, J. C., Ferriera, S., Johnson, J., Vergin, K. L., & Giovannoni, S. J. (2010). Genome sequences of strains HTCC2148 and HTCC2080, belonging to the OM60/NOR5 clade of the gammaproteobacteria. *Journal of Bacteriology*, *192*, 3842–3843.

Tuschak, C., Leung, M. M., Beatty, J. T., & Overmann, J. (2005). The *puf* operon of the purple sulfur bacterium *Amoebobacter purpureus*: structure, transcription and phylogenetic analysis. *Archives of Microbiology*, *183*, 431–443.

Wagner-Döbler, I., Ballhausen, B., Berger, M., Brinkhoff, T., Buchholz, I., Bunk, B., et al. (2010). The complete genome sequence of the algal symbiont *Dinoroseobacter shibae*: a hitchhiker's guide to life in the sea. *The ISME Journal*, *4*, 61–77.

Wagner-Döbler, I., & Biebl, H. (2006). Environmental biology of the marine *Roseobacter* lineage. *Annual Review of Microbiology*, *60*, 255–280.

Waidner, L. A., & Kirchman, D. L. (2005). Aerobic anoxygenic photosynthesis genes and operons in uncultured bacteria in the Delaware River. *Environmental Microbiology*, 7, 1896–1908.

Wang, L., Elliott, M., & Elliott, T. (1999). Conditional stability of the HemA protein (glutamyl-tRNA reductase) regulates heme biosynthesis in *Salmonella typhimurium*. *Journal of Bacteriology*, *181*, 1211–1219.

Woese, C. R. (1987). Bacterial evolution. *Microbiological Reviews*, *51*, 221–271.

Xiong, J., Fischer, W. M., Inoue, K., Nakahara, M., & Bauer, C. E. (2000). Molecular evidence for the early evolution of photosynthesis. *Science*, *289*, 1724–1730.

Young, C., & Beatty, J. (1998). Topological model of the *Rhodobacter capsulatus* light-harvesting complex I assembly protein LhaA (previously known as ORF1696). *Journal of Bacteriology*, *180*, 4742–4745.

Yurkov, V. V., & Beatty, J. T. (1998). Aerobic anoxygenic phototrophic bacteria. *Microbiology and Molecular Biology Reviews*, *62*, 695–724.

Yurkov, V., & Csotonyi, J. T. (2009). New light on aerobic anoxygenic phototrophs. In C. Neil Hunter, Fevzi Daldal, Marion C. Thurnauer & J. Thomas Beatty (Eds.), *The purple phototrophic bacteria* (pp. 31–55).

Yurkov, V. V., Krieger, S., Stackebrandt, E., & Beatty, J. T. (1999). *Citromicrobium bathyomarinum*, a novel aerobic bacterium isolated from deep-sea hydrothermal vent plume waters that contains photosynthetic pigment–protein complexes. *Journal of Bacteriology*, *181*, 4517–4525.

Yurkov, V., Stackebrandt, E., Holmes, A., Fuerst, J. A., Hugenholtz, P., Golecki, J., et al. (1994). Phylogenetic positions of novel aerobic, Bacteriochlorophyll a-containing bacteria and description of *Roseococcus thiosulfatophilus* gen. nov., sp. nov., *Erythromicrobium ramosum* gen. nov., sp. nov., and *Erythrobacter litoralis* sp. nov. *International journal of systematic and evolutionary microbiology*, *44*, 427–443.

Yutin, N., & Béjà, O. (2005). Putative novel photosynthetic reaction centre organizations in marine aerobic anoxygenic photosynthetic bacteria: insights from metagenomics and environmental genomics. *Environmental Microbiology*, 7, 2027–2033.

Yutin, N., Suzuki, M. T., Teeling, H., Weber, M., Venter, J. C., Rusch, D. B., et al. (2007). Assessing diversity and biogeography of aerobic anoxygenic phototrophic bacteria in surface waters of the Atlantic and Pacific Oceans using the global ocean sampling expedition metagenomes. *Environmental Microbiology*, *9*, 1464–1475.

Zeng, Y., Chen, X., & Jiao, N. (2007). Genetic diversity assessment of anoxygenic photosynthetic bacteria by distance-based grouping analysis of *pufM* sequences. *Letters in Applied Microbiology*, *45*, 639–645.

Zheng, Q., Zhang, R., Fogg, P. C.M., Beatty, J. T., Wang, Y., & Jiao, N. (2012). Gain and loss of phototrophic genes revealed by comparison of two *Citromicrobium* bacterial genomes. *PLoS One*, 7, e35790.

Zheng, Q., Zhang, R., & Jiao, N. (2011). Genome sequence of *Citromicrobium* strain JLT1363, isolated from the South China Sea. *Journal of Bacteriology*, *193*, 2074–2075.

Zheng, Q., Zhang, R., Koblížek, M., Boldareva, E. N., Yurkov, V., Yan, S., et al. (2011). Diverse arrangement of photosynthetic gene clusters in aerobic anoxygenic phototrophic bacteria. *PLoS One*, *6*, e25050.

Zsebo, K. M., & Hearst, J. E. (1984). Genetic-physical mapping of a photosynthetic gene cluster from *R. capsulata*. *Cell*, *37*, 937–947.

CHAPTER THIRTEEN

Regressive Evolution of Photosynthesis in the *Roseobacter* Clade

Michal Koblížek*,,[1], Yonghui Zeng*, Aleš Horák†, Miroslav Oborník*,**,†**

*Department of Phototrophic Microorganisms – Algatech, Institute of Microbiology CAS, Třeboň, Czech Republic
**Faculty of Science, University of South Bohemia, České Budějovice, Czech Republic
†Institute of Parasitology CAS, České Budějovice, Czech Republic
[1]Corresponding author: E-mail: koblizek@alga.cz

Contents

Abstract

The *Roseobacter* clade constitutes a significant fraction of marine microbial communities. The clade is functionally heterogeneous as it consists of both photoheterotrophic as well as chemoheterotrophic species. Such functional diversity can be explained by two different evolutionary scenarios: (1) the progenitors of the *Roseobacter* species were originally photoautotrophic, but lost part or all of their photosynthesis genes over time; or (2) they were originally heterotrophs adopting the photosynthesis genes via horizontal gene transfer (HGT). To address these hypotheses, we analysed genomic information from several *Roseobacter* species and compared it with other phototrophic organisms. The analyses suggest that the photosynthesis genes were not acquired via HGT, but rather that *Roseobacter* species present in today's oceans descend from ancient phototrophic bacteria, and radiated after the oxygenation of the oceans during the Neoproterozoic era. Later, several *Roseobacter* lineages lost their photosynthesis genes forming strictly heterotrophic species.

Advances in Botanical Research, Volume 66
ISSN 0065-2296, http://dx.doi.org/10.1016/B978-0-12-397923-0.00013-8

1. INTRODUCTION

The *Roseobacter* clade is one of the largest marine bacterioplankton groups (Brinkhoff, Giebel, & Simon, 2008; Buchan, Gonzáles, & Moran, 2005; Moran et al., 2007; Newton et al., 2010; Wagner-Döbler & Biebl, 2006). The presence of *Roseobacter* species has been reported in diverse marine environments, from eutrophic coastal regions to oligotrophic mid-ocean gyres. These metabolically versatile organisms take part in several important biogeochemical processes including the oxidation of organic matter, dimethylsulfoniopropionate metabolism (González, Kiene, & Moran, 1999; Moran, Gonzáles, & Kiene, 2003) and anoxygenic phototrophy (Oz, Sabehi, Koblížek, Massana, & Béjà, 2005).

The first organism belonging to the *Roseobacter* clade was isolated by Shiba, Simidu, and Taga (1979) from the Bay of Tokyo. This isolate, originally named *Erythrobacter* OCh 114, was later reclassified as *Roseobacter denitrificans* which, together with *Roseobacter litoralis*, formed a new genus *Roseobacter* (Shiba, 1991). *Rsb. denitrificans* is a typical aerobic anoxygenic phototrophic (AAP) bacterium (Yurkov & Csotonyi, 2009; see also chapter by Yurkov and Hughes in this volume) that contains bacteriochlorophyll *a* and lives under aerobic conditions. AAP bacteria constitute an important fraction of the microbial community in various marine environments (Cottrell, Mannino, & Kirchman, 2006; Hojerová et al., 2011; Jiao et al., 2007; Kolber et al., 2001; Kolber, Van Dover, Niederman, & Falkowski, 2000), and contribute significantly to bacterial secondary production in the upper ocean (Koblížek, 2011; Koblížek, Mašín, Ras, Poulton, & Prášil, 2007). Although AAP bacteria represent only a functional group composed of different Proteobacterial taxa, members of the *Roseobacter* clade frequently dominate marine AAP communities (Béjà et al., 2002; Oz et al., 2005; Salka et al., 2008; Yutin et al., 2007).

Over the past decade, the *Roseobacter* clade has attracted the interest of microbiologists culturing novel organisms. This effort resulted in a number of formally described AAP taxa, establishing new genera such as *Dinoroseobacter*, *Roseivivax*, *Roseovarius*, *Roseicyclus*, *Roseisalinus*, *Staleya*, or *Thalassobacter*. Although the genus *Roseobacter* was originally described as containing photoheterotrophic bacteria, isolates of chemoheterotrophic (chemoorganotrophic) strains related to the *Roseobacter* clade revealed that this cluster is functionally heterogeneous (Brinkhoff et al., 2008; Moran et al., 2003). Presently, a number of genera belonging to the *Roseobacter* clade have been described to contain the chemoheterotrophic species: *Marinovum*, *Maritimibacter*, *Oceanicola*, *Octadecabacter*, *Phaeobacter*, *Pelagibaca*, *Silicibacter*, *Sulfitobacter*, and *Ruegeria*. Regarding formal taxonomy, the *Roseobacter* clade belongs to the order Rhodobacterales, family Rhodobacteraceae (Brenner, Krieg, & Staley, 2005). A close inspection of 16S rRNA phylogeny of Rhodobacterales reveals that photoheterotrophic *Roseobacter* species do not form a separate clade, but rather are inlaid with chemoheterotrophic strains (Fig. 13.1). Moreover, the order Rhodobacterales also contains purple non-sulfur species of the genus *Rhodobacter* and *Rhodovulum*, which are capable of photoautotrophic growth under anaerobic conditions, and can also conduct photoheterotrophy under semiaerobic conditions (Brenner et al., 2005).

Such functional diversity raises the question of the evolutionary origin of the clade. The presence of a photosynthetic apparatus has been traditionally considered an important taxonomic marker, however, this criterion completely fails in the case of *Roseobacter* species. The close relationship between photoheterotrophic and chemoheterotrophic strains can be demonstrated in the case of *Jannaschia helgolandensis*, which was described as a chemoheterotrophic strain (Wagner-Döbler & Biebl, 2006), whereas the closely related species, *Jannaschia* sp. CCS1 and *Thalassobacter stenotrophicus* contain photosynthesis genes. Also, the photoheterotroph *Staleya guttiformis* has 98.8% 16S rRNA identity to the chemoheterotrophic strain *Sulfitobacter donghicola*, which led to its reclassification as *Sulfitobacter guttiformis* (Yoon, Kang, Lee, & Oh, 2007). Similarly, *Roseovarius* species *Rva. tolerans* and *Roseovarius* sp. SL25 are photoheterotrophic, whereas *Rva. nubinhibens* is the chemoheterotroph (Fig. 13.1).

There are two possible evolutionary trajectories that could explain such a large functional diversity. The first option is that the progenitors of *Roseobacter* species were originally photoautotrophic organisms that subsequently lost part or all of their photosynthesis genes, and became photoheterotrophic or chemoheterotrophic species (i.e. regressive evolution). The second option is that originally chemohetrotrophic organisms recruited the photosynthesis genes via horizontal gene transfer (HGT) from phototrophic organisms (Fig. 13.2). The idea that heterotrophic members of Proteobacteria evolved from ancient photosynthetic species was proposed by Woese (1987) based on early 16S rRNA sequence data. However, the current literature more often emphasises cases of the horizontal transfer of photosynthesis genes in the evolution of phototrophy (Nagashima, Hiraishi, Shimada, & Matsuura, 1997; Petersen et al., 2012; Raymond, Zhaxybayeva, Gogarten, & Blankenship, 2003).

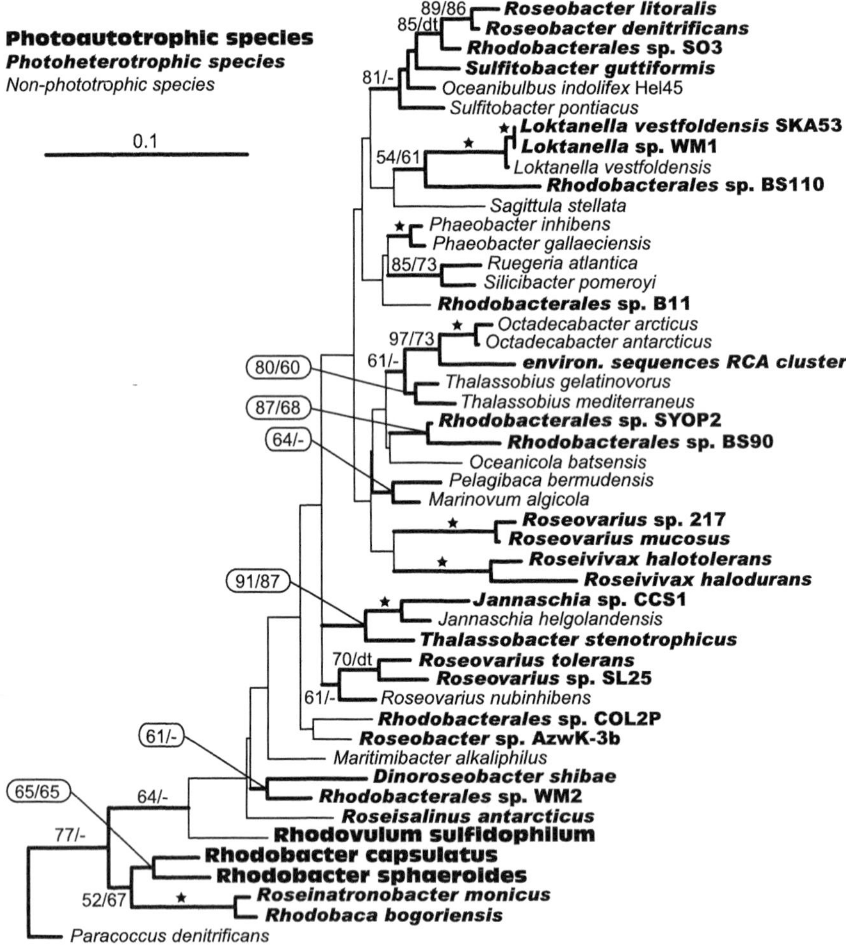

Figure 13.1 ***Maximum-likelihood (ML) (loglk = −10,783.82) phylogenetic tree as inferred from partial 16S rRNA gene sequences.*** Partial 16S rRNA gene sequences (1398–1436 bp in length) were aligned using Kalign (Lassman & Sonhammer, 2005). The tree was constructed using ML (PhyML; Guindon & Gascuel, 2003) and maximum parsimony (MP) (PAUP*; Swofford, 2000). ML trees were computed using GTR model for nucleotide substitutions and discrete gamma distribution in 4 + 1 categories with all parameters estimated from the dataset (gamma shape parameter 0.615 and proportion of invariants 0.657). Numbers above tree branches indicate ML bootstraps (GTR, 300 replicates)/MP (1000 replicates). Nodes with both bootstraps >90% are marked by black stars. Phototrophic strains are in bold.

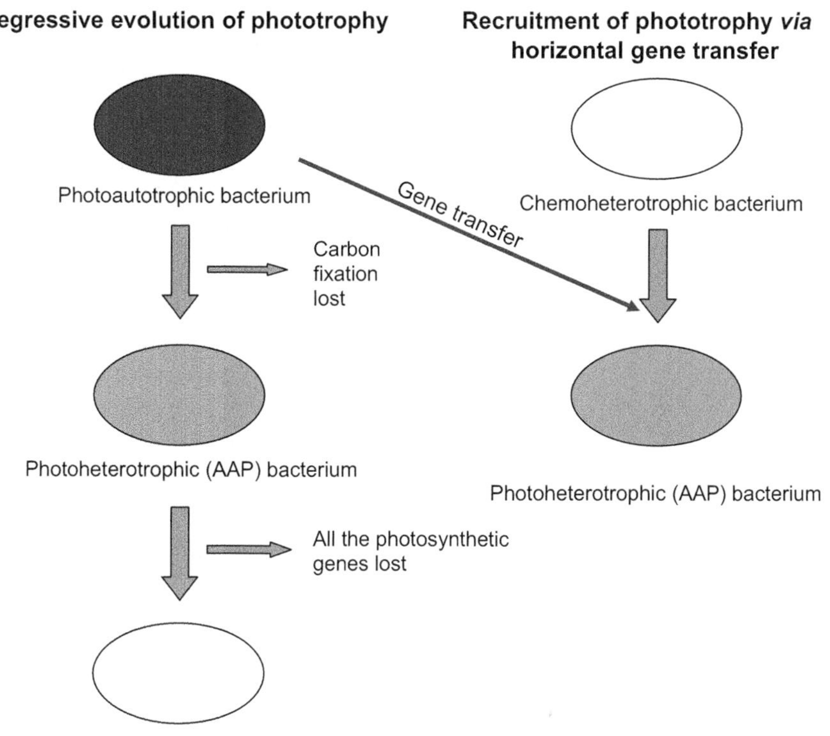

Figure 13.2 ***Two main scenarios of evolutionary origin of photoheterotrophic species.*** Either photoheterotrophs evolved from photoautotrophic species, which lost the carbon fixation capacity (regressive evolution), or they descent from originally heterotrophic species, which recruited the photosynthetic genes via HGT (indicated by the thin arrow). (For colour version of this figure, the reader is referred to the online version of this book.)

2. PHOTOSYNTHESIS GENE CLUSTER

The majority of phototrophic Proteobacteria contain the so-called photosynthesis gene cluster (PGC) (Zsebo & Hearst, 1984). The PGC is a 35–50 kb stretch of DNA that contains genes for bacteriochlorophyll and carotenoid biosynthesis, and genes for photosynthetic reaction centres and light-harvesting complexes as well as some regulatory proteins (Swingley, Blankenship, & Raymond, 2009; Zheng et al., 2011). Although the basic set of genes in PGCs is conserved, the presence of some additional genes in the PGC varies among different Proteobacterial lineages. Other photosynthesis genes, such as some of the carotenoid biosynthesis pathway genes or carbon fixation genes, are encoded outside the PGC. It has been speculated that the PGC organisation might provide an effective framework for the lateral spread of phototrophy (Swingley et al., 2009). Indeed, the HGT of PGC genes was suggested in the case of *Rubrivivax gelatinosus* (Igarashi et al., 2001). The recent identification of PGC-bearing plasmids in *Sulfitobacter* (*Staleya*) *guttiformis* and *Rsb. litoralis* (Petersen et al., 2012) indicates a potential for horizontal spreading of photosynthesis genes and even raises the possibility that originally chemoheterotrophic species may have obtained photosynthesis genes horizontally and become photoheterotrophs. However, there is also an alternative concept of the PGC function. The clustering of the photosynthesis genes might be convenient for efficient regulation and gene expression (Beatty, 1995; Liotenberg et al., 2008). Thus, the organisation of photosynthesis genes in PGCs may be unrelated to HGT.

One way to identify the horizontal transfer of large stretches of DNA is to use nucleotide statistics (Juhas et al., 2009). Therefore, we performed a simple analysis of G + C profiles in 22 species of phototrophic Proteobacteria using a GC profile algorithm (Gao & Zhang, 2006), and calculated the G + C content of the PGCs of all the tested species. In contrast to the hypothesis that PGC might be a subject of frequent horizontal transfer, we did not find any significant difference between G + C content of an entire genome and the corresponding PGC (Fig. 13.3). The average difference

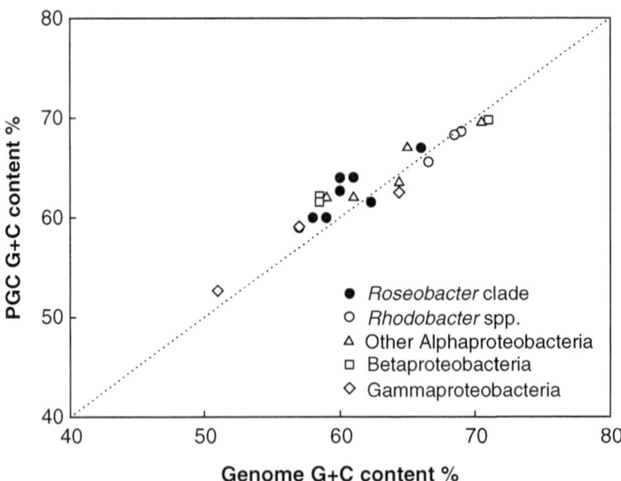

Figure 13.3 ***G + C content in the genomes and PGCs of 22 phototrophic Proteobacteria.*** The plot also contains the G + C content of the plasmid PGC from *Rsb. litoralis* plotted against the G + C content of its genome. The dotted diagonal represents the identity (1:1) line.

between the genome and the PGC G + C content was 1.78% (the maximum difference was 4%). We also did not find any indication of an abrupt change of the G + C content at the boundaries of the PGC clusters. These results indicate that the transfer of the entire PGC between different Proteobacterial groups is at the most a rare event, occurring only on timescales longer than the G + C content shift.

A more detailed inspection of PGCs reveals that all Rhodobacterales species have very similar PGC gene inventories (for details see chapter by Zheng et al. in this volume), which are, with a few exceptions, similar to the PGC composition described for *Rba. sphaeroides* (Choudhary & Kaplan, 2000; Naylor, Addlesee, Gibson, & Hunter, 1999). For example all PGCs of Rhodobacterales species contain the *pufQ* gene, a putative factor involved in light-harvesting complex assembly (Fidai, Kalmar, Richards, & Borgford, 1993), which does not seem to be present in other phototrophic Proteobacteria. The common origin of all Rhodobacterales can also be documented using carotenoid biosynthesis. All the investigated species contain the *crtABCDFI* genes, which encode enzymes of the spheroidene–spheroidenone biosynthetic pathway. This is consistent with the fact that spheroidenone was identified as a main carotenoid in all tested AAP species belonging to the *Roseobacter* clade (Koblížek, Falkowski, & Kolber, 2006; Koblížek, Mlčoušková, Kolber, & Kopecký, 2010; Shiba, 1991; Wagner-Döbler & Biebl, 2006). Spheroidenone is also produced by *Rba. sphaeroides* or *Rhodovulum marinum* when grown under aerobic conditions (Koyama, Takii, Saiki, Tsukida, & Yamashita, 1982; Lutz, Agalides, Hervo, Cogdell, & Reiss-Husson, 1978) which confirms the common origin of carotenoid synthesis in all Rhodobacterales. Similarly, all Rhodobacterales contain a gene for 1-deoxyxylulose-5-phosphate synthase (an initial enzyme of deoxyxylulose-5-phosphate isoprenoid biosynthesis pathway) in the PGC, which is distinct from all other Proteobacteria. All the evidence indicates an in common origin of the PGC in all Rhodobacterales, and makes a distant transfer of the entire PGC from other phototrophic groups unlikely.

3. THE MODE OF EVOLUTION

Although the HGT of the entire PGC appears to be improbable, it is possible that there was horizontal transfer of individual genes or gene operons. To examine this possibility we performed phylogenetic analyses of various genes encoding different parts of the photosynthetic apparatus.

Bacteriochlorophyll biosynthetic pathway genes were suggested as convenient markers for the study of the evolution of photosynthetic organisms (Xiong, Fischer, Inoue, Nakahara, & Bauer, 2000), and so we constructed two phylogenetic trees containing representatives of all major phototrophic Proteobacterial groups. One tree was based on 16S rRNA gene sequences (Fig. 13.4A), whereas the second tree was constructed using concatenated amino acid sequences encoded by the *bchBCDFGHILMNOPXYZ* genes (Fig. 13.4B). The phylogenetic tree contains a monophyletic cluster encompassing all Rhodobacterales species, which is clearly separated from all other sequences (Fig. 13.4B). This implies a common origin for the entire bacteriochlorophyll biosynthesis pathway genes within the Rhodobacterales group and rules out the possibility of any distant HGT of *bch* genes either inside or outside the Rhodobacterales group. Moreover, all *Roseobacter* sequences are part of one single cluster distinct from Rhodobacter species (Fig. 13.4B), which also suggests an early evolutionary divergence of AAP species belonging to the *Roseobacter* clade from photoautotrophic/facultatively photoheterotrophic species belonging to the *Rhodobacter* and *Rhodovulum* genera.

All wild type phototrophic organisms contain carotenoids. Carotenoids are polyisoprenoid molecules which serve as auxiliary light-harvesting pigments that extend the absorption of photosynthetic complexes in the blue-green region of the spectrum. In addition, carotenoids have important structural and protective functions. The basic carotenoid biosynthetic pathway is widely conserved among different groups of phototrophs (Takaichi, 2009). Therefore, we decided to test the origin of photosynthesis in the *Roseobacter* clade also using carotenoid biosynthesis genes. The concatenated tree constructed for sequences encoded by *crtC*, *crtD* and *crtF* genes (which are typically organised inside the PGC) matches the topology of the concatenated tree constructed for the *bch* genes in the Rhodobacterales (Figs. 13.4B and C). A similar topology was also observed for trees constructed for genes *crtB*, *crtE* and *crtI*, encoding the enzymes catalysing earlier steps of the carotenoid biosynthetic pathway (data not shown). The performed analysis confirms the monophyletic origin of phototrophy in Rhodobacterales also based on carotenoid biosynthetic pathway. Moreover, the constructed tree displays a clear separation of *Roseobacter* AAP species and photoautotrophic (mostly anaerobic) *Rhodobacter* species.

Many genetic studies of the purple phototrophic bacteria involve the *pufLM* genes that encode photosynthetic reaction centre proteins L and M. This makes *pufLM* genes a convenient marker for anoxygenic phototrophs containing type-II reaction centres (purple bacteria and green non-sulfur

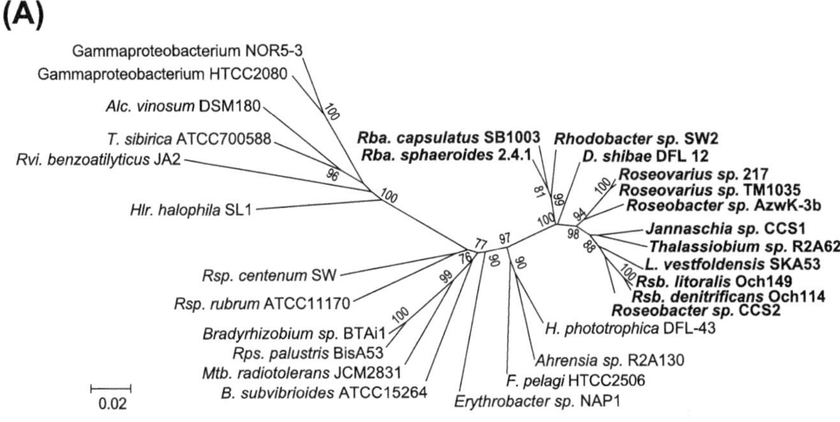

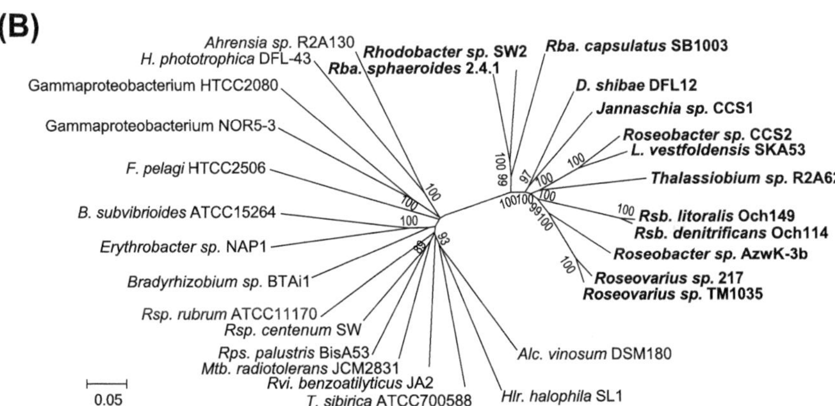

Figure 13.4 ***Neighbour-joining trees constructed for 16S rRNA gene sequences (panel A), and concatenated* bchBCDFGHILMNOPXYZ *(panel B),* crtCDF *(panel C), and* pufLM + puhA *(panel D) protein sequences.*** Full-length sequences for each gene or protein were retrieved from public databases and aligned with ClustalX 2.0 using default settings (Larkin et al., 2007). The concatenated alignments were performed with Geneious 5.5 (Biomatters Ltd, New Zealand). Neighbour-joining trees with 1000 replicates were inferred using MEGA 5.0 (Tamura et al., 2011). The bootstrap values above 70% are shown. Sequences *Rhodobacterales* species are shown in bold. Abbreviations: *Alc., Allochromatium; B., Brevundimonas; D., Dinoroseobacter; F., Fulvimarina; H., Hoeflea; Hlr., Halorhodospira; L., Loktanella; Mtb., Methylobacterium; Rba., Rhodobacter; Rps., Rhodopseudomonas; Rsb., Roseobacter; Rsp., Rhodospirillum; Rvi., Rubrivivax; T., Thiorhodospira.* Note that *Ahrensia* sp. R2A130 does not belong to the Rhodobacterales in any of the constructed trees. Thus, the classification of this genus as a part of *Rhodobacterales* in the *Bergey's manual of systematic bacteriology* (Brenner et al., 2005) is probably incorrect.

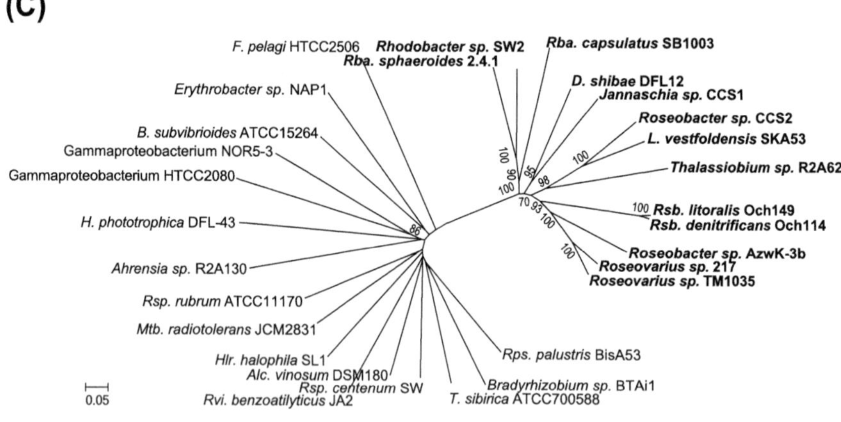

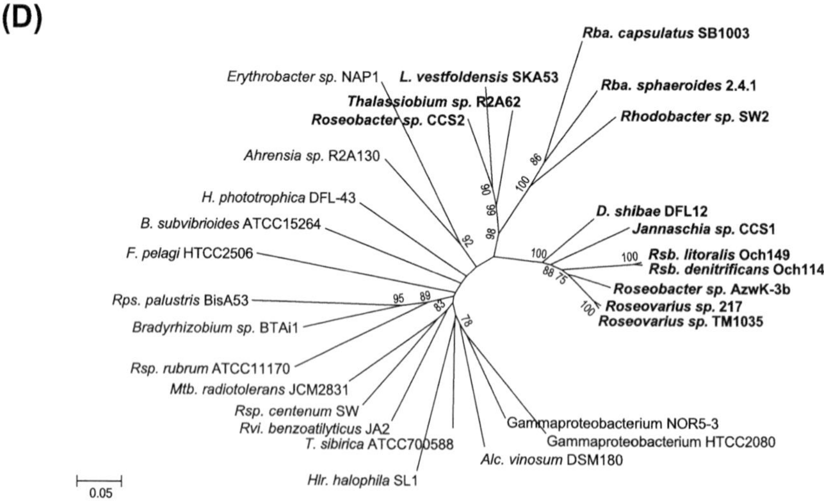

Figure 13.4 Cont'd

bacteria). The development of universal primers made it possible to PCR-amplify and sequence these genes from both cultured species as well as environmental DNA samples (Achenbach, Carey, & Madigan, 2001; Tank, Thiel, & Imhoff, 2009; Yutin, Suzuki, & Béjà, 2005).

In addition to amino acid sequences encoded by the *pufL* and *pufM* genes, we constructed a concatenated tree also using the *puhA* gene encoding subunit H of the photosynthetic reaction centre. Comparison of the concatenated *pufLM/puhA* and 16S rRNA trees shows that while Rhodobacterales form a monophyletic group in both trees, the detailed topology within Rhodobacterales in these two trees differs. In contrast to the 16S rRNA phylogeny, the *pufLM/puhA* sequences clearly split into two main groups (Fig. 13.4D). The first group contains all *Rhodobacter*-related sequences, *Loktanella vestfoldensis* strain SKA53, *Thalassiobium* sp. R2A62 and *Roseobacter* sp. CCS2. This '*Rhodobacter–Loktanella*' group corresponds to the E and F subgroups defined by Yutin et al. (2007) based on the *pufM* sequences. The second '*Roseobacter–Roseovarius*' group corresponds to Yutin's group G and covers all other sequences from AAP species belonging to the *Roseobacter* clade, and also phototrophic species belonging to the genera *Rhodovulum* (not shown). These two groups are clearly different from the 16S-based tree as well as from the *bch*- or *crt*-based phylogenies (Figs. 13.4A–C).

This clustering can be interpreted in two ways. Either as a result of multiple HGT events of the *puf* operon which have occurred within Rhodobacterales or as a result of convergent evolution of these genes driven by their crucial function as reaction centre subunits. Evidence for a horizontal transfer of the *puf* operon may be the presence of the *pufC* gene among members of the *Roseobacter–Roseovarius* group (Yutin's group G) and the absence of this gene in the *Rhodobacter–Loktanella* group.

4. HORIZONTAL GENE TRANSFER

In spite of the fact that members of Rhodobacterales are monophyletic in almost all photosynthesis genes (which implies vertical evolution) there are a few exceptions from this general rule. An example is the *pufL* gene encoding the L subunit of the photosynthetic reaction centre. Here Rhodobacterales species are intercalated with *Erythrobacter* species, which might indicate an ancient gene transfer from Rhodobacterales into *Erythrobacter* genus. The same observation was made previously by Allgaier, Uphoff, Felske, and Wagner-Döbler (2003), who reported a heterogeneous phylogeny of *pufL* genes among various *Roseobacter* isolates.

Another gene with a complicated phylogenetic pattern is the *bchE* gene encoding the oxygen-independent form of Mg-protoporphyrin monomethylester cyclase. This gene is an analogue of the oxygen-dependant form of the cyclase encoded by the *acsF* gene. However, the *bchE* gene does not seem to be essential for phototrophy in *Roseobacter* species. It is present in only about one half of *Roseobacter* AAP species and it is not part of the PGC (Boldareva and Koblížek, *unpublished data*).

Last, a complex phylogeny was also found among genes encoding various subunits of RubisCO. While all marine AAP species lack RubisCO genes (Table 13.1), photoautotrophic *Rhodobacter* and *Rhodovulum* species contain two types of RubisCO, the so-called form I and form II (Tabita et al., 2007). While phylogeny of the type I form is rather complex, the type II form RubisCO is ancestral to Proteobacteria (but also in Dinoflagellates), and presumably represents an ancestral form of the gene (Tabita et al., 2007). Here, *Rhodobacter* species cluster together with other Alphaproteobacteria (Swingley et al., 2009).

5. WHEN DID ROSEOBACTERS EVOLVE?

Assuming that the *Roseobacter* clade evolved from purple non-sulfur bacteria, a question arises: 'When did this evolutionary step occur?' We attempted to date this event based on molecular clock calculations, to place in time a branching point between *Roseobacter* and *Rhodobacter* species.

For the dating analyses, we first selected 27 genomes representing phototrophic members of Proteobacteria, Cyanobacteria, and Chlorobi (see legend of Table 13.2 for complete list). We also searched through the available genomes of Rhodobacterales to select a set of suitable genes. Then, we used the Phylogenie package (Frickey & Lupas, 2004) to automatically identify homologues of these genes. The genes were aligned using MAFFT v.6 (Katoh, Misawa, Kuma, & Miyata, 2002) and ambiguously aligned regions were removed using Gblocks (Castresana, 2000). We constructed maximum-likelihood phylogenies using RAxML 7.2.8 (Stamatakis, 2006) and filtered out topologies where the *Rhodobacter* species, *Roseobacter* clade or Cyanobacteria were not monophyletic. After this step, we selected 13 datasets. Seven genes represented enzymes of the bacteriochlorophyll biosynthetic pathway, three enzymes of carotenoid biosynthetic pathway and three genes were not involved in photosynthesis (Table 13.2).

The molecular-dating analysis was performed using an uncorrelated relaxed log-normal clock algorithm under the WAG + Gamma model of evolution, as implemented in BEAST 1.7 (Drummond, Suchard, Xie, & Rambaut, 2012). For calibration, we used the assumed appearance of Cyanobacteria (2700 Myr ago) and the origin of life (3000–3500 Myr ago). For each dataset, we performed initial Markov chain runs for 10 million generations, then analysed the resulting data using Tracer 1.5, and reanalysed with increased numbers of generations until we obtained sufficient

Table 13.1 Presence of nitrogenase (*nifH*), RubisCO (*cbbL* and *cbbM*), outer light-harvesting complex (*pucAB*), reaction centre (*pufLM*), and carotenoid biosynthesis pathway (*crtCDF*) genes in various species belonging to *Roseobacter* clade

Organism	Genbank no.	*nifH*	*cbbLM*	*pucAB*	*pufLM*	*crtCDF*
Rba. sphaeroides 2.4.1	CP000143	○	○	○	●	●
Rba. capsulatus SB1003	CP001312	○	○	○	●	●
Rhodobacter sp. SW2	ACYY00000000	○	○	○	●	●
Jannaschia sp. CCS1	CP000264	–	–	○	●	●
Rsb. denitrificans Och 114	CP000362	–	–	○	●	●
Rsb. litoralis Och 149*	CP002623 + CP002624	–	–	○	●	●
Dinoroseobacter shibae DFL12	CP000830	–	–	○	●	●
Roseobacter sp. AzwK-3b	ABCR00000000	–	–	–	●	●
Roseovarius sp. TM1035	ABCL00000000	–	–	–	●	●
Roseovarius sp. 217	AAMV00000000	–	–	–	●	●
Roseobacter sp. CCS2	AAYB00000000	–	–	–	●	●
Thalassiobium sp. R2A62	ACOA01000000	–	–	–	●	●
Loktanella vestfoldensis SKA53	AAMS01000000	–	–	–	●	●
Oceanicola batsensis HTCC2597	AAMO00000000	–	–	–	–	–
Silicibacter pomeroi DSS-3**	CP000031	–	–	–	–	–

●, gene present in the PGC; ○, gene present outside of the PGC; –, gene absent.

***pufLM* and *crtA* genes are located on the plasmid, *pucAB* genes are on the chromosome.

**No *puf* genes were found in the following heterotrophic *Rsb.* species: *Maritimibacter alkaliphilus* HTCC2654, *Oceanibulbus indolifex* HEL-45, *Oceanicola granulosus* HTCC2516, *Octadecabacter antarcticus* 238, *Octadecabacter antarcticus* 307, *Pelagibaca bermudensis* HTCC2601, *Phaeobacter gallaeciensis* 2.10, *Phaeobacter gallaeciensis* BS107, *Rhodobacterales* bacterium KLH11, *Rhodobacterales* bacterium HTCC2150, *Roseobacter* sp. GAI-101, *Roseobacter* sp. MED193, *Roseobacter* sp. SK209-2-6, *Roseovarius nubinhibens* ISM, *Silicibacter* sp. TM1040, *Ruegeria* sp. R11, *Sagittula stellata* E37, *Sulfitobacter* sp. EE-36, *Sulfitobacter* sp. NAS-14.1.

Table 13.2 List of genes used in molecular-dating analyses and divergence times of the last common ancestor of *Roseobacter* clade and *Rhodobacter* genus estimated by an uncorrelated relaxed log-normal clock algorithm (see relevant part of text for details)

Enzyme	Gene*	Appearance of *Roseobacter* clade Myr ago
Mg-protoporphyrin IX monomethylester aerobic cyclase	*acsF*	944
Geranylgeranyl bacteriochlorophyll synthase	*bchG*	702
Magnesium chelatase subunit	*bchH*	1220
Light-independent protochlorophyllide reductase iron-sulfur ATP-binding protein	*bchL*	799
S-adenosyl-L-methionine: Mg-protoporphyrin IX methyltransferase	*bchM*	1100
Protochlorophyllide reductase	*bchN*	631
Geranylgeranyl reductase	*bchP*	1152
Carbamoyl phosphate synthase, small subunit	*carA*	980
Cytochrome *c* oxidase subunit I	*coxA*	677
Geranylgeranyl pyrophosphate synthase	*crtE*	1031
Phytoene synthase	*crtB*	1037
Methionine aminopeptidase	*metAP*	992
Ubiquinone–menaquinone biosynthesis methyltransferase	*ubiE*	686
Mean ± st. dev.		919 ± 198

*Datasets were constructed from the following taxa – Alphaproteobacteria: *Rba. sphaeroides* 2.4.1, *Rba. capsulatus*, *Rvu. sulfidophilum*, *Jannaschia* sp. CCS1, *Rsb. denitrificans*, *Rsb. litoralis*, *Dinoroseobacter shibae*, *Roseobacter* sp. AzwK-3b, *Roseovarius* sp. TM1035, *Roseovarius* sp. 217, *Roseobacter* sp. CCS2, *Loktanella vestfoldensis* SKA53, *Oceanicola batsensis* HTCC2597, *Erythrobacter* sp. NAP1, *Methylobacterium radiotolerans*, *Rsp. rubrum*, and *Silicibacter pomeroi* DSS-3; Betaproteobacteria: *Rvi. gelatinosus*; Gammaproteobacteria: *Congregibacter litoralis*; Cyanobacteria: *Thermosynechococcus elongatus*, *Acaryochloris marina* MBIC11017, *Gloeobacter violaceus* PCC 7421, *Arthrospira platensis* str. Paraca, *Trichodesmium erythraeum* IMS101, *Chroococcidiopsis thermalis* and *Prochlorococcus marinus* MIT 9211; Chlorobi: *Chl. tepidum* TLS.

effective sample size for the estimates. Using the molecular clock calculations, the divergence of *Roseobacter* and *Rhodobacter* species was dated 919 ± 198 Myr ago (Table 13.2). This suggests that the photoheterotrophic *Roseobacter* species first appeared during the Neoproteorozoic era.

6. EVOLUTIONARY SCENARIO

Interestingly, in spite of a large body of studies on the evolution of photosynthesis (Raymond et al., 2003; Swingley et al., 2009; Xiong et al., 2000), the regressive loss of photosynthetic properties among prokaryotes has not been studied in detail. The hypothesis that heterotrophic members of Proteobacteria evolved from ancient photoautotrophic species was proposed by Woese (1987) based on 16S rRNA sequence data. In spite of the fact, that the evolution of *Proteobacteria* was much more complex than envisioned originally, the presented evidence suggests that the regressive scenario seems to hold in the case of the *Roseobacter* clade.

The regressive scenario is also consistent with the current biogeochemical models (Fig. 13.5). During the Archaean period the Earth's atmosphere was largely anoxic. At the beginning of the Proterozoic era, oxygen produced by cyanobacteria started to gradually oxygenate the Earth's atmosphere. Based on geological records it was estimated that approximately 2400–2000 Myr ago oxygen levels reached 0.02 atm causing the so-called great oxidation event (Bekker et al., 2004; Holland, 2006). After that time, the concentration of oxygen stayed relatively constant (0.02–0.04 atm) through most of the Proterozoic era until approximately 850 Myr ago. During this period the ocean was probably mildly oxygenated at its surface whereas the deep waters remained anoxic, providing an environment that supported both anoxygenic and oxygenic phototrophs (Johnston, Wolfe-Simon, Pearson, & Knoll, 2009). We assume that the semiaerobic conditions facilitated the radiation of purple non-sulfur species during the Proterozoic era. The strictly anaerobic phototrophs remained, probably, widespread in the deeper waters containing reduced electron donors, such as H_2S or Fe^{2+} (Brocks et al., 2005).

The second rise of oxygen took place during the Neoproterozoic era (1000–542 Myr ago). In spite of the sparse geological data it seems that during this period Earth encountered several dramatic changes (Shields-Zhou & Och, 2011). The supercontinent Rodinia, which was formed at the beginning of the Neoproterozoic, started to fragment. The carbon isotope records indicate an increased organic carbon burial during the Neoproterozoic, which likely facilitated the rise of the oxygen concentration in the atmosphere. The concomitant decline in CO_2 levels probably triggered several large glaciations, which embraced almost the entire planet during the Cryogenian period (Hoffman, Kaufman, Halverson, & Schrag, 1998). These massive environmental changes stimulated a remarkable diversification of various eukaryotic groups – opistokonts, amoebazoa, green and red algae, ciliates and dinoflagellates (Porter, 2004; Knoll, Summons, Waldbauer, & Zumberge, 2007). The appearance of the first metazoan fossils suggests that oxygen levels were high enough to facilitate the development of more complex life forms. Ultimately, the rising oxygen concentration resulted in the complete and irreversible oxidation of the deep ocean by the end of the

Neoproterozoic, which paved the way for the dramatic radiation of all the multicellular life forms during the Cambrian.

Such dramatic environmental changes exerted a strong selection pressure on the originally anaerobic anoxygenic phototrophs, which had to modify their metabolism and photosynthetic apparatus and enzymatic machinery. A question remains whether ancient Rhodobacterales contained a carbon fixation pathway. Either they were photoautotrophic organisms, which later lost in some lineages the carbon fixation capability, or they were semiaerobic species without carbon fixation capacity. In the second case, some of the species adopted the carbon fixation genes and became photoautotrophs, while the modern *Roseobacter* species embarked on a chemoorganotrophic or facultative photoheterotrophic lifestyle. These two options are difficult to reconcile based on genomic data only. However assuming the geological context, it seems that the second scenario is less likely. The increased oxygen levels had a profound impact on the ocean redox chemistry. Based on the available data, Fe^{2+} and H_2S were removed even from the deep ocean by the end of the Neoproterozoic period (Canfield et al., 2008). This means that there was a complete disappearance of all inorganic potential electron donors necessary for anoxygenic photosynthesis. We assume that under this situation the originally photoautotrophic species had either to retract into marginal anoxic environments (sediments, anoxic hypolimnia of freshwater or saline lakes, etc.) or to abandon their autotrophic lifestyle and eliminate the carbon fixation pathway. In addition, oxygenic photosynthesis provided an abundant source of organic carbon, which offered an easy alternative source of metabolic energy via aerobic respiration. This is consistent with our molecular clock calculations suggesting the placement of the split of AAP species from purple nonsulfur species at the beginning of the Neoproterozoic eon (Fig. 13.5).

The loss of carbon fixation capacity gave rise to photoheterotrophic species. Through similar mechanisms the phototrophic species may have lost their nitrogen fixation capacity, which is widespread among photoautotrophic species, but so far has not been reported in AAP bacteria. The next step may have involved the reduction of the light-harvesting capacity, as photophosphorylation became an auxiliary process. Indeed, photoheterotrophic *Roseobacter* species typically contain an order of magnitude less bacteriochlorophyll *a* per cell than photoautotrophs (Koblížek et al., 2010). Many AAP species also lost their peripheral light-harvesting complex encoded by *pucBA* genes (Table 13.1).

Subsequently, some AAP species increasingly relied on pure chemoorganotrophy, which finally led to the complete elimination of photosynthesis

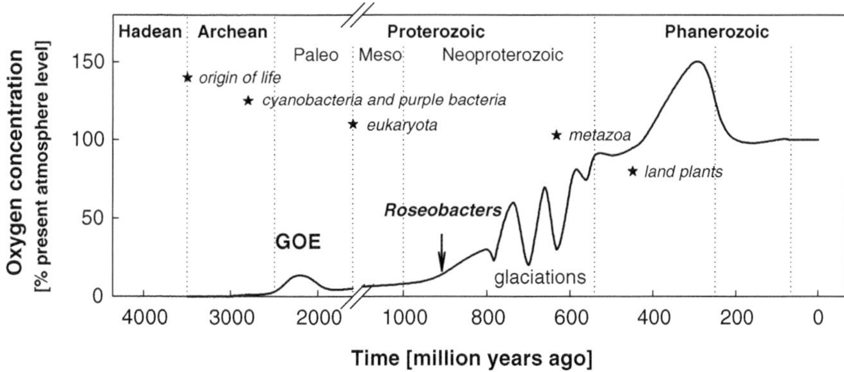

Figure 13.5 ***Schematic presentation of main geological and evolutionary events during Earth history.*** The estimated oxygen level is marked by the thick line. The main evolutionary events are marked by stars. GOE – great oxidation event. Origin of Roseobacters indicated by an arrow. *(Compiled from Holland (2006) and Shields-Zhou and Och (2011)).*

genes and the evolution of strictly chemoheterotrophic species. All the chemoheterotrophic species lack the entire PGC including the complete set of *puf* (encoding the subunits of the bacterial photosynthetic reaction centre), *bch* (coding for enzymes of the bacteriochlorophyll biosynthetic pathway) and *crt* genes (carotenoid biosynthesis pathway) (Table 13.1).

In the long-term perspective, the co-existence of photoheterotrophic and heterotrophic species in the same environment indicates that the negative selection pressure (maintenance cost and genome streamlining processes) and the positive selection pressure (benefits of using light energy) are balanced. In this view, the metabolic investment into assembly and maintenance of a photosynthetic apparatus is compensated by the ability to supplement cellular energy demands using light-derived energy.

7. CONCLUSIONS

Although HGT of photosynthesis genes has been suggested in several phototrophic species, this scenario can be ruled out in the case of the evolution of the *Roseobacter* clade. All the phylogenetic evidence indicates that anoxygenic photosynthesis is endogenous in the Rhodobacterales. We assume that the progenitors of all today's Rhodobacterales inhabiting semiaerobic Proterozoic oceans were phototrophic organisms. The *Roseobacter* clade radiated after the oxygenation of the oceans during Neoproteorozoic era. It contained aerobic photoheterotrophic species with a reduced light-harvesting capacity. Finally, some *Roseobacter* species completely eliminated the photosynthesis genes and retreated to a chemoheterotrophic lifestyle. The phylogenetic patterns suggest that the loss of the photosynthesis genes has occurred independently in several lineages. The evolution of anoxygenic phototrophy after the oxidation of Earth's atmosphere was obviously more complex than just a case of simple gene loss. HGT can be documented in several examples. However, in the perspective of abrupt changes caused by the shift from an anoxic to an oxic world, the main evolutionary line in the *Roseobacter* clade was regressive, whereas the HGT was only of minor importance.

ACKNOWLEDGEMENTS

The authors thank Prof. Paul G. Falkowski for his advice in the initial phases of this project, Dr Vladimíra Moulisová for 16S rRNA sequences of *Roseobacter* species, Dr José M. González and Prof. J. Thomas Beatty for their comments on this text and Jason Dean for the language correction. This research was also supported by GAČR projects P501/10/0221 and P501/12/G055 and project Algatech (CZ.1.05/2.1.00/03.0110).

REFERENCES

Achenbach, L. A., Carey, J., & Madigan, M. T. (2001). Photosynthetic and phylogenetic primers for detection of anoxygenic phototrophs in natural environments. *Applied and Environmental Microbiology*, *67*, 2922–2926.

Allgaier, M., Uphoff, H., Felske, A., & Wagner-Döbler, I. (2003). Aerobic anoxygenic photosynthesis in *Roseobacter* clade bacteria from diverse marine habitats. *Applied and Environmental Microbiology*, *69*, 5051–5059.

Beatty, J. T. (1995). Organization of photosynthetic gene transcripts. In R. E. Blankenship, M. T. Madigan & C. E. Bauer (Eds.), *Anoxygenic photosynthetic bacteria* (pp. 1209–1219). The Netherlands: Kluwer Academic Publishers.

Béjà, O., Suzuki, M. T., Heidelberg, J. F., Nelson, W. C., Preston, C. M., Hamada, T., et al. (2002). Unsuspected diversity among marine aerobic anoxygenic phototrophs. *Nature*, *415*, 630–633.

Bekker, A., Holland, H. D., Wang, P. L., Rumble, D., Stein, H. J., Hannah, J. L., et al. (2004). Dating the rise of atmospheric oxygen. *Nature*, *427*, 117–120.

Brenner, D. J., Krieg, N. R., & Staley, J. T. (2005). (2nd ed.). *Bergey's manual of systematic bacteriology*. (Vol. 2). New York, Berlin, Heidelberg: Springer-Verlag.

Brinkhoff, T., Giebel, H. A., & Simon, M. (2008). Diversity, ecology, and genomics of the *Roseobacter* clade: a short overview. *Archives of Microbiology*, *189*, 531–539.

Brocks, J. J., Love, G. D., Summons, R. E., Knoll, A. H., Logan, G. A., & Bowden, S. A. (2005). Biomarker evidence for green and purple sulphur bacteria in a stratified Palaeoproterozoic sea. *Nature*, *437*, 866–870.

Buchan, A., Gonzáles, J. M., & Moran, M. A. (2005). Overview of the marine *Roseobacter* lineage. *Applied and Environmental Microbiology*, *71*, 5665–5677.

Canfield, D. E., Poulton, S. W., Knoll, A. H., Narbonne, G. M., Ross, G., Goldberg, T., et al. (2008). Ferruginous conditions dominated later Neoproterozoic deep-water chemistry. *Science*, *321*, 949–952.

Castresana, J. (2000). Selection of conserved blocks from multiple alignments for their use in phylogenetic analysis. *Molecular Biology and Evolution*, *17*, 540–552.

Choudhary, M., & Kaplan, S. (2000). DNA sequence analysis of the photosynthesis region of *Rhodobacter sphaeroides* 2.4.1^{T}. *Nucleic Acids Research*, *28*, 862–867.

Cottrell, M. T., Mannino, A., & Kirchman, D. L. (2006). Aerobic anoxygenic phototrophic bacteria in the Mid-Atlantic Bight and the North Pacific Gyre. *Applied and Environmental Microbiology, 72*, 557–564.

Drummond, A. J., Suchard, M. A., Xie, D., & Rambaut, A. (2012). Bayesian phylogenetics with BEAUti and the BEAST 1.7. *Molecular Biology and Evolution, 29*(8), 1969–1973.

Fidai, S., Kalmar, G. B., Richards, W. R., & Borgford, T. J. (1993). Recombinant expression of the pufQ gene of *Rhodobacter capsulatus. Journal of Bacteriology, 175*, 4834–4842.

Frickey, T., & Lupas, A. N. (2004). PhyloGenie: automated phylome generation and analysis. *Nucleic Acids Research, 32*, 5231–5238.

Gao, F., & Zhang, C.T. (2006). GC-Profile: a web-based tool for visualizing and analyzing the variation of GC content in genomic sequences. *Nucleic Acids Research, 34*, 686–691.

González, J. M., Kiene, R. P., & Moran, M. A. (1999). Transformation of sulfur by an abundant lineage of marine bacteria in the α-subclass of the class Proteobacteria. *Applied and Environmental Microbiology, 65*, 3810–3819.

Guindon, S., & Gascuel, O. (2003). A simple, fast, and accurate algorithm to estimate large phylogenies by maximum likelihood. *Systematic Biology, 52*, 696–704.

Hoffman, P. F., Kaufman, A. J., Halverson, G. P., & Schrag, D. P. (1998). A Neoproterozoic snowball Earth. *Science, 281*, 1342–1346.

Hojerová, E., Mašín, M., Brunet, C., Ferrera, I., Gasol, J. M., & Koblížek, M. (2011). Distribution and growth of aerobic anoxygenic phototrophs in the Mediterranean Sea. *Environmental Microbiology, 13*, 2717–2725.

Holland, H. D. (2006). The oxygenation of the atmosphere and oceans. *Philosophical Transactions of the Royal Society B, 361*, 903–915.

Igarashi, N., Harada, J., Nagashima, S., Matsuura, K., Shimada, K., & Nagashima, K. V. P. (2001). Horizontal transfer of the photosynthesis gene cluster and operon rearrangement in purple bacteria. *Journal of Molecular Evolution, 52*, 333–341.

Jiao, N., Zhang, Y., Zeng, Y., Hong, N., Liu, R., Chen, F., et al. (2007). Distinct distribution pattern of abundance and diversity of aerobic anoxygenic phototrophic bacteria in the global ocean. *Environmental Microbiology, 9*, 3091–3099.

Johnston, D. T., Wolfe-Simon, F., Pearson, A., & Knoll, A. H. (2009). Anoxygenic photosynthesis modulated Proterozoic oxygen and sustained Earth's middle age. *Proceedings of the National Academy of Sciences of United States of America, 106*, 16925–16929.

Juhas, M., van der Meer, J. R., Gaillard, M., Harding, R. M., Hood, D. W., & Crook, D. W. (2009). Genomic islands: tools of bacterial horizontal gene transfer and evolution. *FEMS Microbiology Reviews, 33*, 376–393.

Katoh, K., Misawa, K., Kuma, K., & Miyata, T. (2002). MAFFT: a novel method for rapid multiple sequence alignment based on fast Fourier transform. *Nucleic Acids Research, 30*, 3059–3066.

Knoll, A. H., Summons, R. E., Waldbauer, J. R., & Zumberge, J. E. (2007). The geological succession of primary producents in the oceans. In P. G. Falkowski & A. H. Knoll (Eds.), *Evolution of primary producers in the sea* (pp. 133–163). : Elsevier.

Koblížek, M. (2011). Role of photoheterotrophic bacteria in the marine carbon cycle. In N. Jiao, F. Azam & S. Sanders (Eds.), *Microbial carbon pump in the ocean* (pp. 49–51). Washington D.C: Science/AAAS.

Koblížek, M., Falkowski, P. G., & Kolber, Z. S. (2006). Diversity and distribution of photosynthetic bacteria in the Black Sea. *Deep-Sea Research Part II, 53*, 1934–1944.

Koblížek, M., Mašín, M., Ras, J., Poulton, A. J., & Prášil, O. (2007). Rapid growth rates of aerobic anoxygenic prototrophs in the ocean. *Environmental Microbiology, 9*, 2401–2406.

Koblížek, M., Mlčoušková, J., Kolber, Z., & Kopecký, J. (2010). On the photosynthetic properties of marine bacterium COL2P belonging to *Roseobacter* clade. *Archives of Microbiology, 192*, 41–49.

Kolber, Z. S., Plumley, F. G., Lang, A. S., Beatty, J. T., Blankenship, R. E., VanDover, C. L., et al. (2001). Contribution of aerobic photoheterotrophic bacteria to the carbon cycle in the ocean. *Science, 292*, 2492–2495.

Kolber, Z. S., Van Dover, C. L., Niederman, R. A., & Falkowski, P. G. (2000). Bacterial photosynthesis in surface waters of the open ocean. *Nature, 407*, 177–179.

Koyama, Y., Takii, T., Saiki, K., Tsukida, K., & Yamashita, K. J. (1982). Configuration of the carotenoid in the reaction centers of photosynthetic bacteria. Comparison of the resonance Raman spectrum of the reaction centers of *Rhodopseudomonas sphaeroides* G1C with those of cis-trans isomers from β-carotene. *Biochimica et Biophysica Acta, 680*, 109–118.

Larkin, M. A., Blackshields, G., Brown, N. P., Chenna, R., McGettigan, P. A., McWilliam, H., et al. (2007). Clustal W and Clustal X version 2.0. *Bioinformatics, 23*, 2947–2948.

Lassman, T., & Sonhammer, E. L. (2005). Kalign – an accurate and fast multiple sequence alignment algorithm. *BMC Bioinformatics, 6*, 298.

Liotenberg, S., Steunou, A.-S., Picaud, M., Reiss-Husson, F., Astier, C., & Ouchane, S. (2008). Organization and expression of photosynthesis genes and operons in anoxygenic photosynthetic Proteobacteria. *Environmental Microbiology, 10*, 2267–2276.

Lutz, M., Agalides, I., Hervo, G., Cogdell, R. J., & Reiss-Husson, F. (1978). On the state of carotenoids bound to reaction centers of photosynthetic bacteria: a resonance Raman study. *Biochimica et Biophysica Acta, 503*, 287–303.

Moran, M. A., Belas, R., Schell, M. A., González, J. M., Sun, F., Sun, S., et al. (2007). Ecological genomics of marine Roseobacters. *Applied and Environmental Microbiology, 73*, 4559–4569.

Moran, M. A., Gonzáles, J. M., & Kiene, R. P. (2003). Linking a bacterial taxon to sulfur cycling in the sea: studies of the marine *Roseobacter* group. *Geomicrobiology Journal, 20*, 375–388.

Nagashima, K. V. P., Hiraishi, A., Shimada, K., & Matsuura, K. (1997). Horizontal transfer of genes coding for the photosynthetic reaction centers of purple bacteria. *Journal of Molecular Evolution, 45*, 131–136.

Naylor, G. W., Addlesee, H. A., Gibson, L. C. D., & Hunter, C. N. (1999). The photosynthesis gene cluster of *Rhodobacter sphaeroides*. *Photosynthesis Research, 62*, 121–139.

Newton, R. J., Griffin, L. E., Bowles, K. M., Meile, C., Gifford, S., Givens, C. E., et al. (2010). Genome characteristics of a generalist marine bacterial lineage. *ISME Journal, 4*, 784–798.

Oz, A., Sabehi, G., Koblížek, M., Massana, R., & Béjà, O. (2005). *Roseobacter*-like bacteria in Red and Mediterranean Sea aerobic anoxygenic photosynthetic populations. *Applied and Environmental Microbiology, 71*, 344–353.

Petersen, J., Brinkmann, H., Bunk, B., Michael, V., Päuker, O., & Pradella, S. (2012). Think pink: photosynthesis, plasmids and the *Roseobacter* clade. *Environmental Microbiology, 14*, 2661–2672.

Porter, S. M. (2004). The fossil record of early eukaryotic diversification. *Paleontological Society Papers, 10*, 35–50.

Raymond, J., Zhaxybayeva, O., Gogarten, J. P., & Blankenship, R. E. (2003). Evolution of photosynthetic prokaryotes: a maximum-likehood mapping approach. *Philosophical Transactions of the Royal Society B, 358*, 223–230.

Salka, I., Moulisová, V., Koblížek, M., Jost, G., Jürgens, K., & Labrenz, M. (2008). Abundance, depth distribution, and composition of aerobic bacteriochlorophyll *a*-producing bacteria in four deeps of the central Baltic Sea. *Applied and Environmental Microbiology, 74*, 4398–4404.

Shiba, T. (1991). *Roseobacter litoralis* gen. nov., sp. nov. and *Roseobacter denitrificans* sp. nov., aerobic pink-pigmented bacteria which contain bacteriochlorophyll *a*. *Systematic Applied Microbiology, 14*, 140–145.

Shiba, T., Simidu, U., & Taga, N. (1979). Distribution of aerobic bacteria which contain bacteriochlorophyll *a*. *Applied and Environmental Microbiology, 38*, 43–45.

Shields-Zhou, G., & Och, L. (2011). The case for a Neoproterozoic oxygenation event: geochemical evidence and biological consequences. *GSA Today, 21*, 4–11.

Stamatakis, A. (2006). RAxML-VI-HPC: maximum likelihood-based phylogenetic analyses with thousands of taxa and mixed models. *Bioinformatics, 22*, 2688–2690.

Swingley, W. D., Blankenship, R. E., & Raymond, J. (2009). Evolutionary relationship among purple photosynthetic bacteria and the origin of Proteobacterial photosynthetic systems. In C. N. Hunter, F. Daldal, M. C. Thurnauer & J. T. Beaty (Eds.), *The purple phototrophic bacteria* (pp. 17–29). Dordrecht: Springer Verlag.

Swofford, D. L. (2000). *PAUP* phylogenetic analysis using parsimony (*and related methods)*. Sunderland, MA, USA: Sinauer Associates Inc..

Tabita, F. R., Hanson, T. E., Li, H., Satagopan, S., Singh, J., & Chan, S. (2007). Function, structure, and evolution of the RubisCO-like proteins and their RubisCO homologs. *Microbiology and Molecular Biology Reviews, 71*, 576–599.

Takaichi, S. (2009). Distribution and biosynthesis of carotenoids. In C. N. Hunter, F. Daldal, M. C. Thurnauer & J. T. Beaty (Eds.), *The purple phototrophic bacteria* (pp. 97–117). Springer Verlag.

Tamura, K., Peterson, D., Peterson, N., Stecher, G., Nei, M., & Kumar, S. (2011). MEGA5: molecular evolutionary genetics analysis using maximum likelihood, evolutionary distance, and maximum parsimony methods. *Molecular Biology and Evolution, 28*, 2731–2739.

Tank, M., Thiel, V., & Imhoff, J. F. (2009). Phylogenetic relationship of phototrophic purple sulfur bacteria according to *pufL* and *pufM* genes. *International Microbiologzy, 12*, 175–185.

Wagner-Döbler, I., & Biebl, H. (2006). Environmental biology of the marine *Roseobacter* lineage. *Annual Review of Microbiology, 60*, 255–280.

Woese, C. R. (1987). Bacterial evolution. *Microbiology Reviews, 51*, 221–271.

Xiong, J., Fischer, W. M., Inoue, K., Nakahara, M., & Bauer, C. E. (2000). Molecular evidence for the early evolution of photosynthesis. *Science, 289*, 1724–1730.

Yoon, J. H., Kang, S. J., Lee, M. H., & Oh, T. K. (2007). Description of *Sulfitobacter donghicola* sp. nov., isolated from seawater of the East Sea in Korea, transfer of *Staleya guttiformis* Labrenz et al. 2000 to the genus *Sulfitobacter* as *Sulfitobacter guttiformis* comb. nov. and emended description of the genus *Sulfitobacter*. *International Journal of Systematic and Evolutionary Microbiology, 57*, 1788–1792.

Yurkov, V. V., & Csotonyi, J. T. (2009). New light on aerobic anoxygenic phototrophs. In C. N. Hunter, F. Daldal, M. C. Thurnauer & J. T. Beaty (Eds.), *The purple phototrophic bacteria* (pp. 31–55). : Springer Verlag.

Yutin, N., Suzuki, M. T., & Béjà, O. (2005). Novel primers reveal wider diversity among marine aerobic anoxygenic phototrophs. *Applied and Environmental Microbiology, 71*, 8958–8962.

Yutin, N., Suzuki, M. T., Teeling, H., Weber, M., Venter, J. C., Rusch, D. B., et al. (2007). Assessing diversity and biogeography of aerobic anoxygenic phototrophic bacteria in surface waters of the Atlantic and Pacific Oceans using the Global Ocean Sampling expedition metagenomes. *Environmental Microbiology, 9*, 1464–1475.

Zheng, Q., Zhang, R., Koblížek, M., Boldareva, E. N., Yurkov, V., Yan, S., et al. (2011). Diverse arrangement of photosynthetic gene clusters in aerobic anoxygenic phototrophic bacteria. *PLoS ONE, 6*(9), e25050. 10.1371/journal.pone.0025050.

Zsebo, K. M., & Hearst, J. E. (1984). Genetic-physical mapping of a photosynthetic gene cluster from *R. capsulata*. *Cell, 37*, 937–947.

INDEX

Note: Page numbers followed by "f" and "t" indicate figures and tables respectively

A

C

F

P

Q

R

S

T

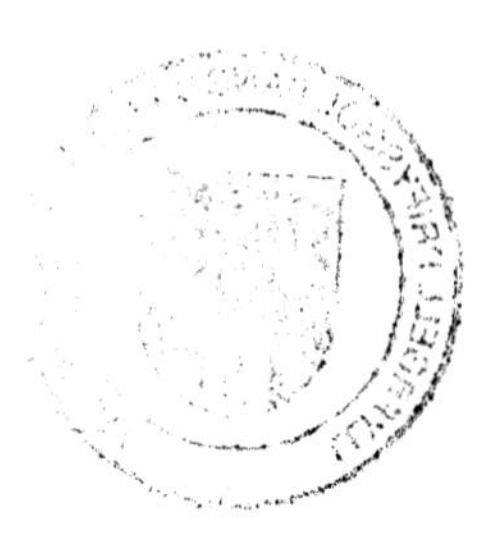

COLOR PLATES

Group	Species	38	④	66	127	⑤	164
NifH homologues (>200/200)	Methanosarcina barkeri	GCDPKADCT	R	LVLGGVAQTTIMDTLRELG	LGDVVCGGFAMPIR	EG	KAQEVYIVASGEMMATYAANNI
	Methanosarcina acetivorans	-----R-S-	-	ILAE-KFIPAVLEEH--QL	---------S----	--	F-E-I-LIC--GF-SI------
	Rhodopseudomonas palustris	-----S-S-	T	ILR--EDLP-VL-S--DS-	----------V---	N-	I-ESAFV-T-SDF--IF----L
	Desulfitobacterium hafniense	-----S-S-	N	TLR--KYIP-VL-----KS	----------I---	--	I-EH-FT-S-SDF-SI--S--L
	Rhodospirillum rubrum	-------S-	-	-I---KP-E-L--V---Q-	--------------	D-	---------------V------
	Chlorobium tepidum	-------S-	-	-L---LQ-K-VL-----E-	--------------	D-	--E-I---C------M------
	Ch. her. II	-------S-	-	-L---LI-K-VL-----E-	--------------	D-	--E-I---V------M------
	Clostridium acetobutylicum	-------S-	-	-L---L--K-VL-----E-	--------------	--	--K-I----------M------
	Azotobacter vinelandii	-------S-	-	-I-HSK--G-V-EMAASA-	--------------	-N	----I---C------M------
	Heliobacterium modesticaldum	-------S-	-	-I-HSK--A-V--LA--K-	--------------	-N	----I---T------M------
	He. chlorum	-------S-	-	-I-HSK--A-V--LA--K-	--------------	-N	----I---T------M------
	Sinorhizobium meliloti	-------S-	-	-I-NAK--D-VLHLAATE-	--------------	-N	----I---M------L------
	Rhodobacter sphaeroides	-------S-	-	-I-NTKL-D-VLHLAA-A-	--------------	-N	----I---M------L------
	Nostoc punctiforme	-------S-	-	-M-HSK----VLHLAA-R-	--------------	--	----I---T------M------
	Trichodesmium erythraeum	-------S-	-	-I-DAK----VLHVAA---	--------------	-N	----I---C------M------
BchX homologues (>70/70)	Heliobacterium modesticaldum	-----H-S-	V	ILFN--NPP-LLEYWA--N	---------GV--S	KS	I-KSIIL--GNDHQSL-V----
	Chloroflexus aurantiacus	-----H-SC	N	ALF--ISLP-LG-VW--FK	----------T-LS	RS	L-E--I-LCGNDRQSL------
	Chloroflexus aggregans	-----H-SC	N	ALF---SLP-LG-VW--FK	----------T-LS	RS	L-E--I-LCGNDRQSL------
	Roseiflexus castenholzii	-----H-SC	N	TIF--HSLP-LG-QW-LFR	----------T-LA	RS	L-EQ-I-LVGHDRQSL------
	Ro. sp. RS-1	-----H-SC	N	TIF--HSLP-LG-QW-LFK	----------T-LA	RS	L-EQ-I-LVGHDRQSL------
	Chlorobium phaeobacteroides	-----H-S-	T	SLF---SLP-VTEVFA-KN	----------T-LA	RS	LSE--ILLTNNDRQSIFT----
	Chlorobium tepidum	-----H-S-	T	SLF--ISLP-VTEVFA-KN	----------T-LA	RS	LSE--LL-T-NDRQSIFTS---
	Prosthecochloris vibrioformis	-----H-S-	T	SLF--MSLP-LT-VFS-KN	----------T-LA	RS	LSE--IL-T-NDRQSIFT----
	Chloroherpeton thalassium	-----H-S-	T	SLF---SLP-VTEVFAKKN	----------T-LS	RS	LCE--IL-V-NDRQSIF-----
	Rhodobacter sphaeroides	-----S-T-	S	-LF--K-CP--IE-SARKK	---------GL--A	RD	M--K-IL-G-NDLQSL-VT--V
	Rhodospirillum rubrum	-----S-T-	S	-LF--R-CP--IE-SSARK	---------GL--A	RD	LC-K-IV-G-NDLQSL-VV--V
	Methylobacterium extorquens	-----S-T-	S	-LF--R-CP--IE-STKKK	---------GL--A	RD	MC-K-IV-G-NDLQSL-V---V
	Rubrivivax gelatinosus	-----S-T-	S	-LF--K-TP--IE-SAKKK	---------GL--A	RD	MC-K-IV-G-NDLQSL-V---V
	Halorhodospira halophila	-----S-T-	S	-LF--R-CP--I--SSRKK	---------GL--A	RD	LC-K-IL-GANDLQSL-VV--V
BchL homologues (0>130)	Synechococcus sp. RC307	-----H-S-		FT-THKMVP-VI-I-E-VD	----------A-LQ		H-NYCL--TANDFDSIF-M-R-
	Prochlorococcus marinus MIT9303	-----H-S-		FT-THRMVP-VI-I-E-VD	----------A-LQ		H-NYCL--TANDFDSIF-M-R-
	Rhodobacter sphaeroides	-----H-S-		FT-T-SLVP-VI-V-KDVD	----------A-LQ		H-DQAVV-TANDFDSI--M-R-
	Methylobac.	-----H-S-		FT-TKRLAP-VI-A-EAVK	----------S-LQ		H-DRAL--TANDFDSIF-M-R-
	Chlorobium phaeobacteroides	-----H-S-		FPIT-KL-K-VIEA-E-VD	---------SA-LN		Y-DYAI-I-TNDFDSIF---RL
	Chloroflexus aurantiacus	-----H-S-		FP-T-HL-P-VI-V-DSVN	---------SA-LN		Y-DYGL-I-CNDFDSIF---RL
	Trichodesmium erythraeum	-----H-S-		FT-T-FLIP--I---Q-KD	----------A-LN		YSDYCM--TDNGFD-LF---R-
	Synechocystis sp. PCC6803	-----H-S-		FT-T-FLIP--I---Q-KD	----------A-LN		Y-DYCL--TDNGFD-LF---R-
	Heliobacterium modesticaldum	-----S-S-		FTIA-RMIP-VVEI-DKFN	----------T-LQ		Y-DLAC--S-NDFD-LF---R-
	Heliobacterium mobilis	-----S-S-		FTIA-KMIP-VVEI-DKFN	----------T-LQ		Y-DLACV-S-NDFD-LF---R-

Figure 2.6 ***Partial sequence alignments of the NifH, BchX and BchL homologues showing two different CSIs in these proteins that are commonly shared by the NifH and BchX homologues, but not found in any of the BchL homologues.*** The numbers below the group names indicate that, of the available homologues from these groups, how many contained or lacked these CSIs. The dashes (-) indicate identity with the amino acid on the top line.

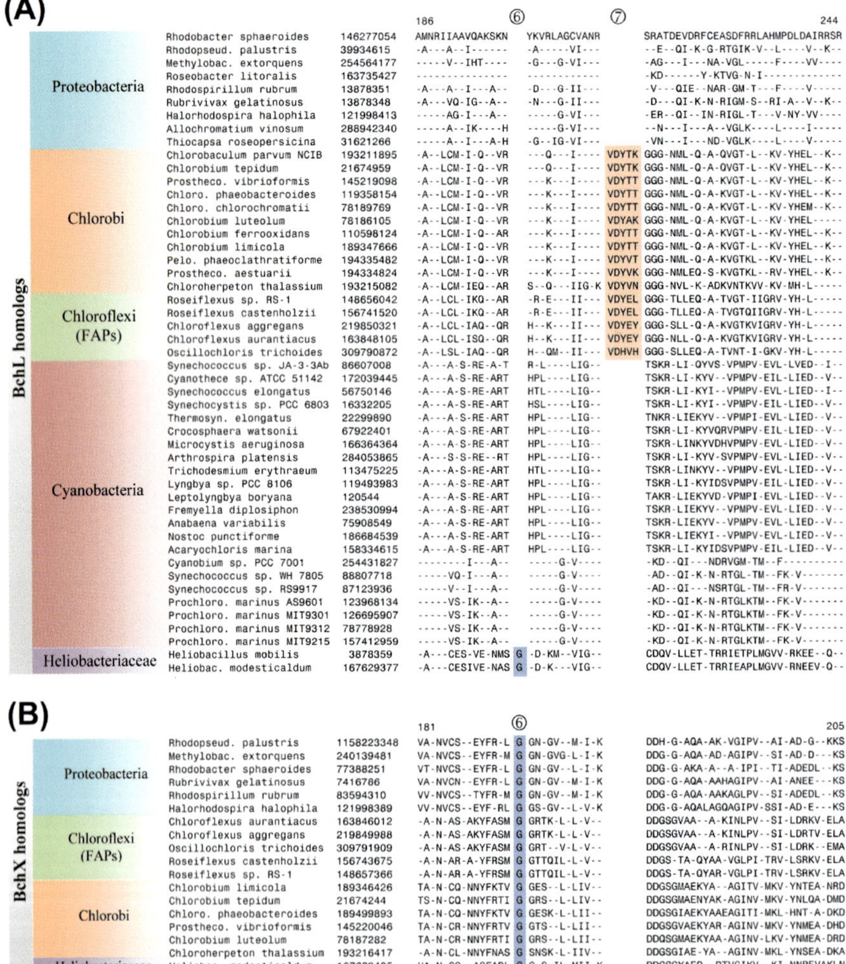

Figure 2.7 ***(A) Excerpts from the sequence alignment of BchL proteins showing two conserved signature indels that are specific for different lineages of phototrophic bacteria.*** The CSI ⑥ is specific for the *Heliobacteriaceae*, whereas CSI ⑦ is commonly shared by different Chlorobi and Chloroflexi homologues. The dashes (-) in these as well as other sequence alignments indicate identity with the amino acid on the top line. (B) A sequence alignment of the BchX homologues from different phototrophic lineages for the same region as shown in part A for the BchL protein sequences.

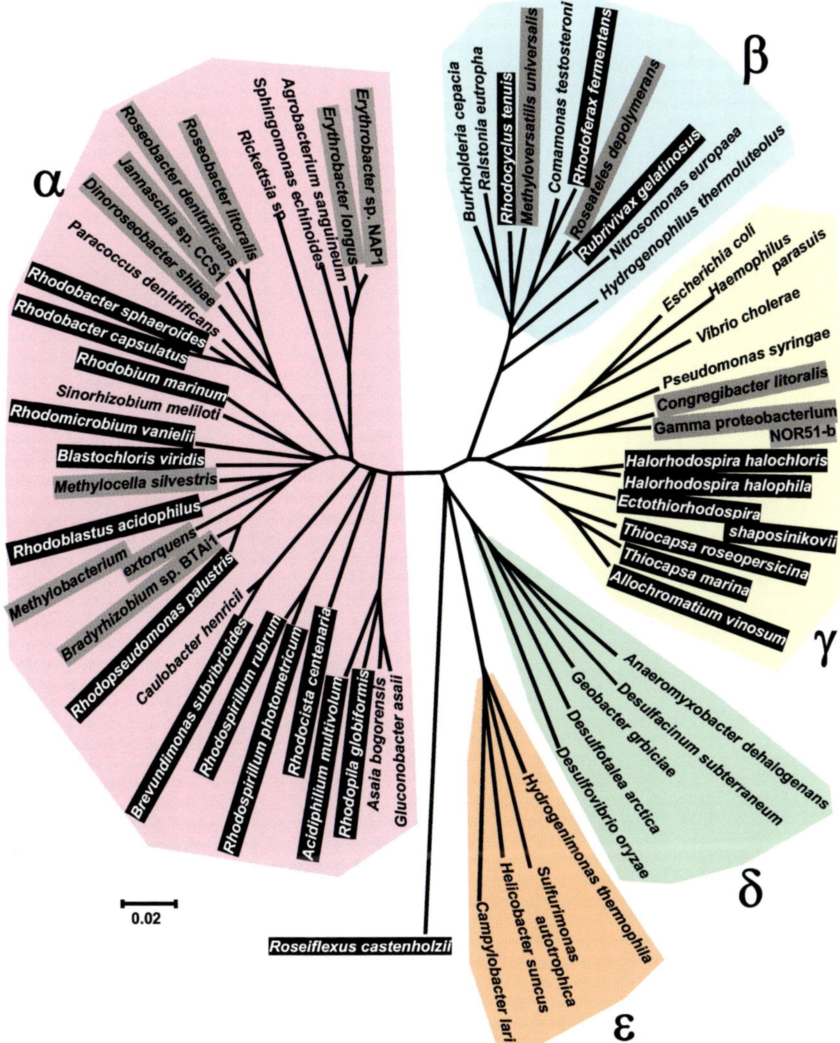

Figure 5.1 ***Phylogenetic tree of purple bacteria (Proteobacteria) based on nucleotide sequences of 16S rRNA.*** Phylogenetic analysis was performed using the programs ClustalX (Thompson, Gibson, Plewniak, Jeanmougin, & Higgins, 1997) and MEGA (Kumar, Tamura, & Nei, 2004). All gaps in the sequence alignment were omitted. The tree was generated by the neighbour-joining method, applying the Kimura 2-parameter distance as a distance estimator. The names of phototrophic species are shown by reverse contrast. Aerobic photosynthetic species are marked with grey background. The sequence of the filamentous anoxygenic phototroph (green filamentous bacteria), *Roseiflexus castenholzii*, was used as an outgroup.

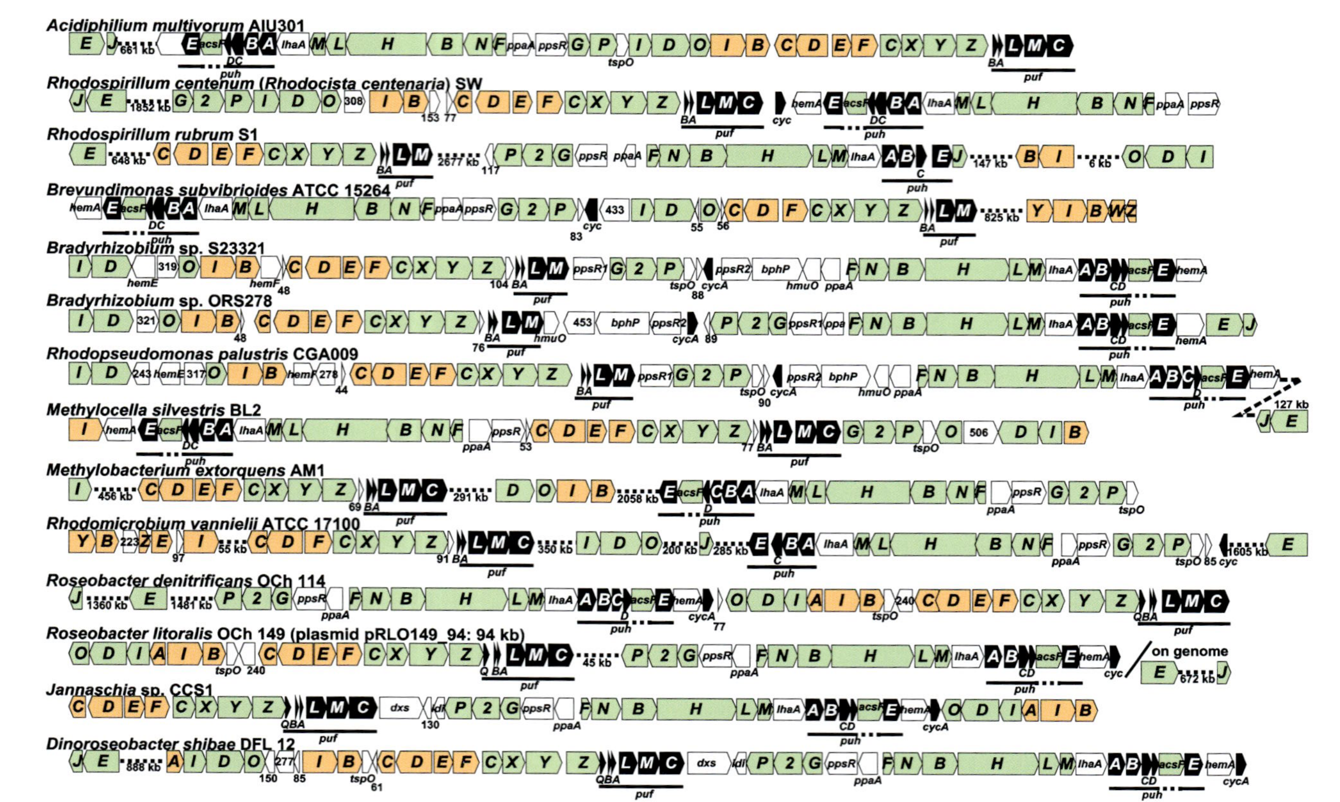

Acidiphilium multivorum AIU301
Rhodospirillum centenum (Rhodocista centenaria) SW
Rhodospirillum rubrum S1
Brevundimonas subvibrioides ATCC 15264
Bradyrhizobium sp. S23321
Bradyrhizobium sp. ORS278
Rhodopseudomonas palustris CGA009
Methylocella silvestris BL2
Methylobacterium extorquens AM1
Rhodomicrobium vannielii ATCC 17100
Roseobacter denitrificans OCh 114
Roseobacter litoralis OCh 149 (plasmid pRLO149_94: 94 kb)
Jannaschia sp. CCS1
Dinoroseobacter shibae DFL 12

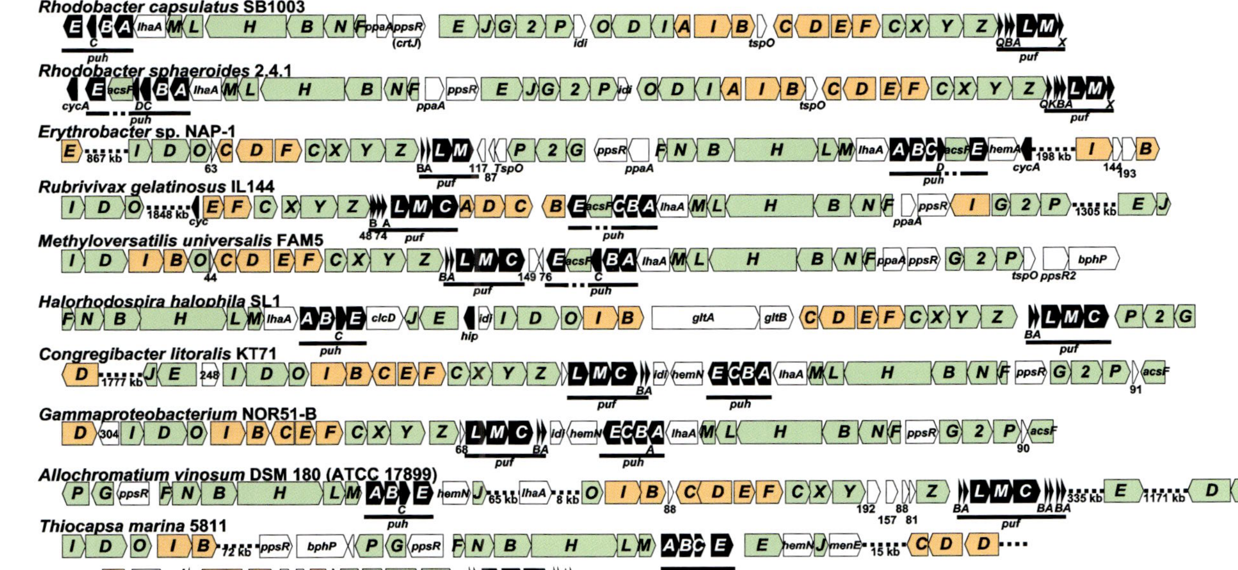

Figure 5.2 ***Arrangements of genes in the PGCs found in the total genomic sequences of 24 species of purple bacteria.*** The genes are presented as arrows pointing in the direction of their transcriptions. Genes coding for the LH and the RC apoproteins and the related products are indicated by solid arrows (*puf* and *puh*). Genes assigned to bacteriochlorophyll (*bch* and *acsF*) and carotenoid (*crt*) biosynthesis genes are shown by green and orange arrows, respectively. ORFs without assigned functions are marked by the lengths of the amino acid sequences of their predicted products. The PGCs are ordered corresponding to the layout of the species in the 16S rRNA tree shown in Fig. 5.3. Each of the PGCs is displayed to direct the *puf* gene transcription at right.

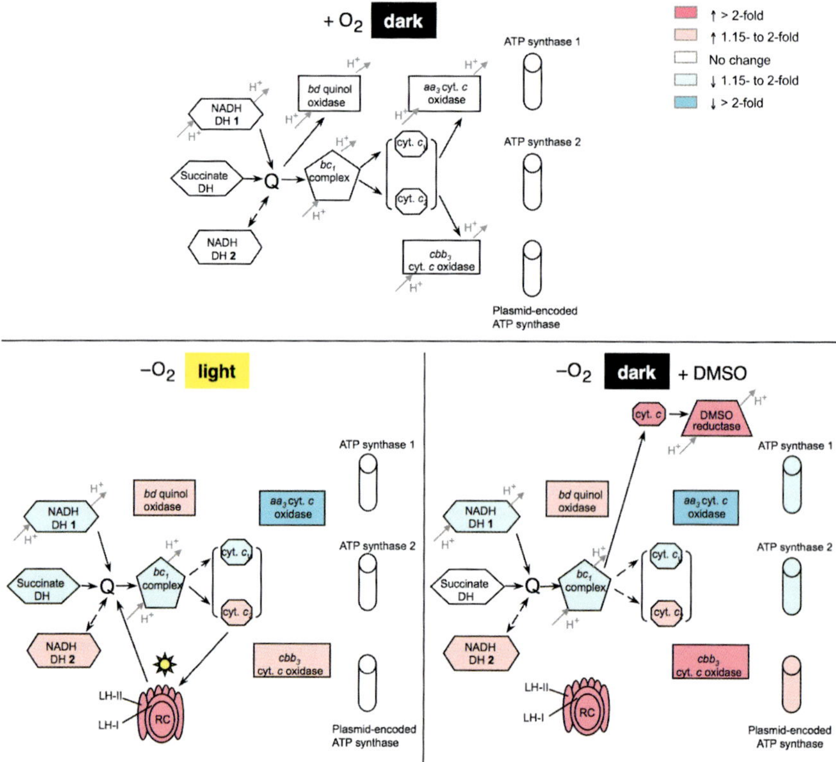

Figure 6.2 R. sphaeroides ***electron transport chain complexes and ATP synthases involved in energy generation under oxic, anoxic-light, and anoxic-dark-DMSO conditions.*** Arrows indicate electron flow; dashed line arrows indicate anticipated direction of electron flow. The known sites for generation of proton motive force are shown. The expression of genes under oxic conditions is set as the background for comparisons (not coloured). Blue corresponds to decreased gene expression of the corresponding proteins under anoxic compared to oxic conditions as follows: light blue, less than 2-fold decrease; dark blue, at least 2-fold decrease. Pink corresponds to increased gene expression compared to the expression under oxic conditions as follows: light pink, less than 2-fold increase; dark pink, at least 2-fold increase. No colour corresponds to anoxic expression that is not significantly different from expression under oxic conditions. Proteins whose genes are expressed below reliable detection are not shown. DH, dehydrogenase; Q, quinone–quinol pool; LH, light-harvesting complex; RC, reaction centre complex. The RC and light-harvesting complexes are not shown under oxic conditions because several *bch* genes involved in bacteriochlorophyll biosynthesis are not expressed under these conditions; hence, no photosynthetic apparatus is made. *(Modified from Pappas et al. (2004)).*

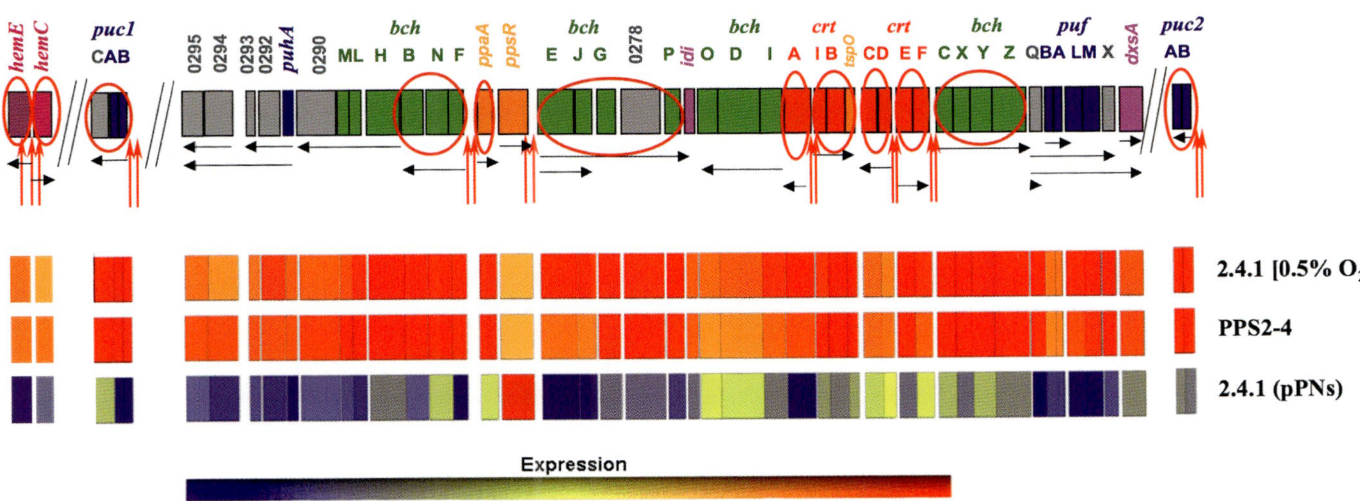

Figure 6.3 ***Composition (upper panel) and expression (lower panel) of the*** **R. sphaeroides** ***PpsR regulon.*** Each gene is represented by a box, coloured according to its function. Green, *bch* genes; red, *crt* genes; blue, genes encoding structural polypeptides of photocomplexes; grey, genes encoding assembly factors or proteins of unknown function; orange, genes encoding regulatory factors; pink, genes encoding enzymes common to bacteriochlorophyll and ubiquinone biosynthesis; magenta, protoporphyrin IX biosynthesis genes. PpsR-binding sites are shown as red vertical arrows. Putative transcripts are shown as black horizontal arrows. Circled genes are repressed by PpsR directly. Relative expression of PpsR-dependant genes, indicated according to the included expression colour scheme, reflect comparisons to the wild type strain grown with 20% O_2. 2.4.1, wild type strain; PPS2-4, *ppsR* point mutant; pPNs, plasmid overexpressing the *ppsR* gene. *Copyright © American Society for Microbiology,* Journal of Bacteriology *(2005),* 187, *2148–2156; doi: 10.1128/JB.187.6.2148-2156.2005.*

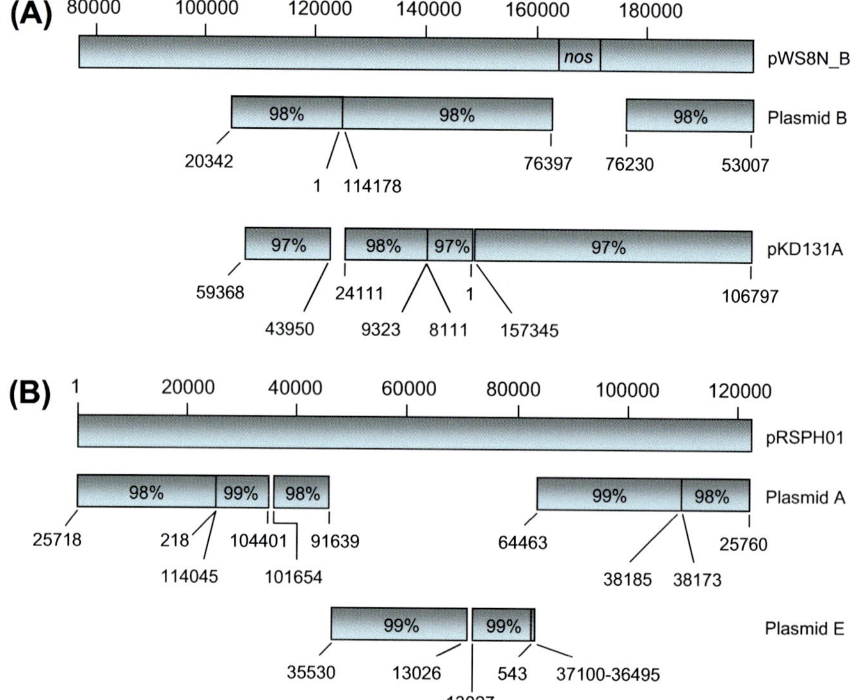

Figure 8.1 A. Alignment of the nucleotide sequences of *R. sphaeroides* plasmids pWS8N-B, plasmid B (strain 2.4.1) and pKD131A, showing an apparent deletion of the region containing the N_2O reductase (*nos*) genes in plasmid B. The percentage sequence identity with pWS8N_B is shown for each segment. For clarity, only part of the alignment is shown. B. Alignment of the nucleotide sequence of plasmid pRSPH01 with plasmids A and E from strain 2.4.1. The gap in the alignment between pRSPH01 and plasmid A is almost completely filled by the sequence of plasmid E.

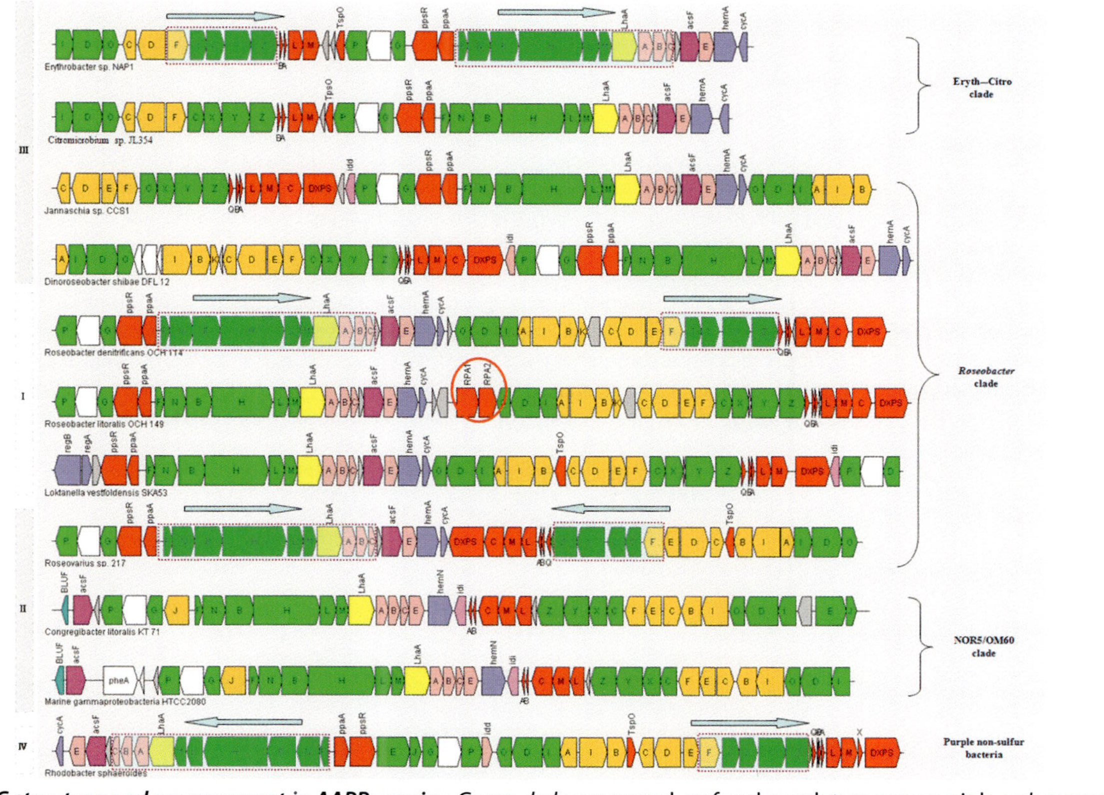

Figure 12.2 ***PGC structure and arrangement in AAPB species.*** Green, *bch* genes; red, *puf* and regulators genes; pink, *puh* genes; orange, *crt* genes; blue, *hem* and *cyc* gene; yellow, *lhaA* gene; blank, uncertain or unrelated genes; grey, hypothetical protein. The horizontal arrows represent putative transcripts. The gene sets in dotted boxes indicate two conserved regions in all PGCs, and the two genes in the red circle of *Rsb. litoralis* Och 149 are two inserted genes thought to act as a centromere-like anchor when plasmids replicate. http://dx.doi.org/10.1371/journal.pone.0025050.g002.

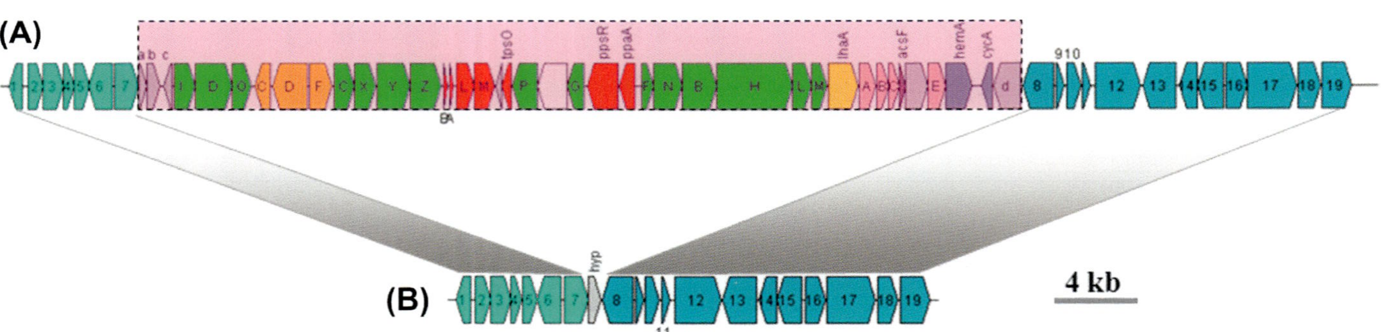

Figure 12.7 ***The difference between the PGC chromosomal region in* Citromicrobium *sp. JL354 (Acc No.: NZ_ADAE01000008) and the corresponding region in Citromicrobial strain JLT1363 (Acc No.: NZ_AEUE01000008).*** A, *Citromicrobium* sp. JL354; B, Citromicrobial strain JLT1363. As described in the text, this difference is attributed to the loss of the PGC in Citromicrobial strain JLT1363. Key to general classes of gene annotations: Green, *bch* genes; red, *puf* and regulator genes; pink, *puh* genes; orange, *crt* genes; blue, *hem* and *cyc* genes; yellow, *lhaA* gene; blank, uncertain or unrelated genes; grey, hypothetical protein. The horizontal arrows represent putative transcripts. Key to specific gene annotations: 1, zinc finger/thioredoxin putative; 2, cell division ATP-binding protein FtsE; 3, cell division transport system permease protein; 4, hypothetical protein ELI_11430; 5, 1-acyl-sn-glycerol-3-phosphate acyltransferase; 6, histidinol-phosphate aminotransferase; 7, homoserine O-acetyltransferase; 8, succinate-semialdehyde dehydrogenase (NAD(P)+); 9, hypothetical protein ED21_17902; 10, glutathione S-transferase family protein; 11, protein-methionine-S-oxide reductase; 12, membrane carboxypeptidase; 13, trypsin-like serine protease; 14, HflC protein; 15, integral membrane proteinase; 16, ATPase; 17, aldehyde oxidase and xanthine dehydrogenase molybdopterin binding; 18, ferrochelatase; 19, cytochrome P450. http://dx.doi.org/10.1371/journal.pone.0035790.g004.

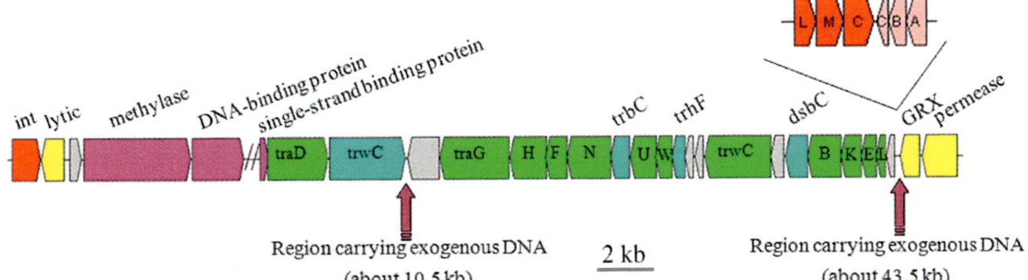

Figure 12.8 ***Genomic organisation of core genes of a putative ICE found in the genome of Citromicrobial strain JLT1363.*** The two regions suggested to carry exogenous DNA are predicted to be involved in phage repression (10.5 kb) and heavy metal transport (43.5 kb) (Acc No.: NZ_AEUE01000004). The ICE in JL354 is in different contigs, but all the core genes after and including 'single-strand binding protein' are present (Acc No.: NZ_ADAE01000017). The incomplete PGC (*pufLMC–puhCBA*) was found only found in an ICE of *Citromicrobium* strain JL354. Red, yellow and pink indicate phage-related genes; green and cyan indicate plasmid-related genes. http://dx.doi.org/10.1371/journal.pone.0035790.g003.